Design, Automation, and Test in Europe

The Most Influential Papers of 10 Years DATE

Rudy Lauwereins • Jan Madsen

Editors

Design, Automation, and Test in Europe

The Most Influential Papers of 10 Years DATE

Springer

Rudy Lauwereins
IMEC vzw and Katholieke Universiteit
  Leuven
Belgium

Jan Madsen
Informatics and Mathematical
  Modelling
Technical University of Denmark
Denmark

ISBN 978-1-4020-6487-6   e-ISBN 978-1-4020-6488-3

Library of Congress Control Number: 2007939398

Printed on acid-free paper

9 8 7 6 5 4 3 2 1

springer.com

Introduction

The Design Automation and Test in Europe, DATE, is Europe's leading international electronic systems design conference for electronic design, automation and test, from system level hardware and software implementation right down to integrated circuit design. It combines the conference with Europe's leading international exhibition for electronic design, automation and test.

To celebrate the tenth anniversary of DATE, we have compiled this book with the aim to highlight some of the most influential technical contributions from ten years of DATE. Selecting 30 papers, only 3 papers from each year, is a challenging endeavor. Although the impact of papers from the first years of DATE can be determined through various citation indexes, the impact from the later years still have to be seen. Together with all 10 Program Chairs, we have made a selection of the most influential papers covering the very broad range of topics which is characteristic for DATE.

Ten Years of DATE

DATE was formed in 1998 on the basis of two previous European conferences: The European Design & Test Conference (ED&TC) and EuroDAC. DATE inherited many of the structures and procedures from the ED&TC conference and from the EDAC exhibition. DATE inherited from ED&TC for example the particular style of the very big Technical Program Committee, which meets every year to put together the final technical program in one intensive day with the size of a small conference on itself. DATE has been alternating between Paris in France and Munich in Germany and since 2007 between Nice, France and Munich.

DATE has grown into a truly international event with paper submissions and participation from all over the globe. DATE has grown from 3 parallel tracks to the currently 7 tracks. From the very beginning, DATE has aimed at being a technical conference of high quality combined with a large exhibition. This is documented not only through the continuing increase of number of submissions and participants, but more importantly through an acceptance rate around 25% and close to 5 reviews per paper.

One of the characteristics of DATE is its continuous innovation, i.e. DATE is constantly exploring new formats, such as Hands-on-Tutorials, Friday workshops, Executive Tracks, and Exhibition Programs. Successful experiments may turn into permanent conference events, such as the Embedded Software Forum. Introduced in 2003 to attract more software people, it was merged into a full Embedded Software track in 2005 with its own set of topics and subTPCs. Many of these innovations have been taken up or copied by other conferences.

Since its beginning, DATE has aimed at not only being the primary event for the researcher, but also for actual designers of chips and systems. Hence, DATE adopted the Designers Forum from ED&TC. The Designers Forum has grown from a parallel event with mainly invited contributions to a full integration within DATE, forming one out of four major areas: **D**esign methods and tools, **A**pplication designs, **T**est and verification and **E**mbedded software.

DATE covers not only chip-level topics, but also system-level design issues. It was an early decision to go system, and as can be seen from the graph below, the number of sessions dealing with system issues has grown in relative terms from 22% to close to 50% without reducing the chip-level design contributions in absolute numbers. The DATE steering group facilitated this move to system level by actively organizing special sessions, keynotes and special focus topic days highlighting various aspects related to systems applications, design and design technology. This has been instrumental in the effort to create a conference with a clear trend toward a holistic view on design, test and embedded software and a comprehensive system design focus.

How the Book was Formed

With more than 200 papers published every year over the last 10 years, selecting the 30 papers with the biggest impact has been a daunting task. Each of the 10 program chairs have been asked to provide a ranked list with around 6 papers from their respective years. For your reference and to acknowledge that they were nominated as candidates for the most influential papers of DATE, they are listed in the appendix at the end of this book. For the oldest years, selection has been mainly based on the number of citations these papers received over the years. For the more recent years, best paper awards and review evaluation figures have been taken into account as well. Needless to say that any selection will miss influential papers. We hence do not claim any completeness of the picture sketched in this book. Nevertheless, we are convinced it presents a fairly balanced view of the historical evolution of the design technology domain.

We then reduced this pre-selection of around 60 papers down to the final list of 30 papers, strictly 3 per year, which appear in this book. Apart from the priority ranking made by the respective program chairs, we used fair coverage of the various domains as a selection criterion. Those papers have been organized in 6 parts:

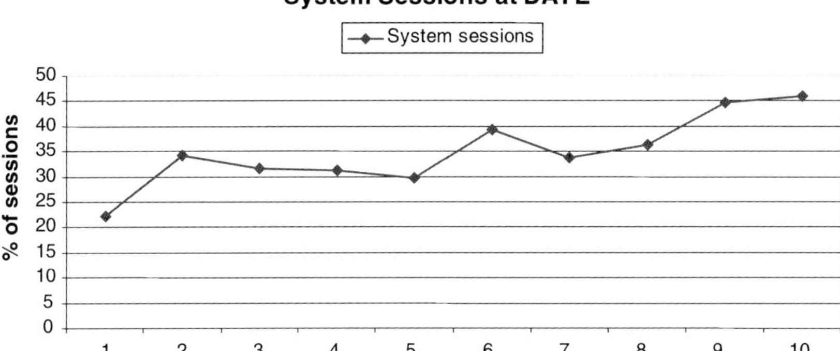

Fig. 1 Relative evolution of the number of sessions allocated to system level design

- System Level Design
- Networks on Chip
- Modeling, Simulation and Run-Time Management
- Digital Technology for Advanced Digital Systems in CMOS and Beyond
- Physical Design and Validation
- Test and Verification

We then invited the winners of the prestigious EDAA Lifetime Achievement Award as well as some other recognized experts in their field to write an introduction to a section, summarizing the history in their domain and indicating how the selected DATE papers contributed to it.

We hope that you will enjoy reading the book as much as we enjoyed organizing DATE and preparing this book. We will like to explicitly thank the program chairs;

Prof. Dr. Franz Ramig, Program Chair 1998
Prof. Dominique Borionne, Program Chair 1999
Prof. Dr. Peter Marwedel, Program Chair 2000
Prof. Dr.-lng. Wolfgang Nebel, Program Chair 2001
Prof. Carlos Delgado Kloos, Program Chair 2002
Prof. Dr.-lng. Norbert Wehn, Program Chair 2003
Prof. Georges Gielen, Program Chair 2004
Prof. Luca Benini, Program Chair 2005
Prof. Donatella Sciuto, Program Chair 2006

We will also like to thank the experts who wrote the introductions, the authors of the selected papers and all attendees of the DATE conferences for their contribution and continued support of our community.

Contents

Introduction v

Part I System Level Design

System Level Design: Past, Present, and Future **3**
D. D. Gajski

**Scheduling of Conditional Process Graphs for the Synthesis
of Embedded Systems** **15**
P. Eles, K. Kuchcinski, Z. Peng, A. Doboli, and P. Pop

**EXPRESSION: A Language for Architecture Exploration
Through Compiler/Simulator Retargetability** **31**
A. Halambi, P. Grun, V. Ganesh, A. Khare, N. Dutt, and A. Nicolau

RTOS Modeling for System Level Design **47**
A. Gerstlauer, H. Yu, and D. D. Gajski

**Context-Aware Performance Analysis for Efficient Embedded
System Design** **59**
M. Jersak, R. Henia, and R. Ernst

**Lock-Free Synchronization for Dynamic Embedded
Real-Time Systems** **73**
H. Cho, B. Ravindran, and E. D. Jensen

What If You Could Design Tomorrow's System Today? **87**
N. Wingen

Part II Networks on Chip

Networks on Chips **105**
G. De Micheli

A Generic Architecture for On-Chip Packet-Switched Interconnections **111**
P. Guerrier and A. Greiner

Trade-offs in the Design of a Router with Both Guaranteed and Best-Effort Services for Networks on Chip **125**
E. Rijpkema, K. G. W. Goossens, A. Rădulescu, J. Dielissen,
J. van Meerbergen, P. Wielage, and E. Waterlander

Exploiting the Routing Flexibility for Energy/Performance-Aware Mapping of Regular NoC Architectures **141**
J. Hu and R. Marculescu

×pipesCompiler: A Tool for Instantiating Application-Specific Networks on Chip **157**
A. Jalabert, S. Murali, L. Benini, and G. De Micheli

A Network Traffic Generator Model for Fast Network-on-Chip Simulation **173**
S. Mahadevan, F. Angiolini, J. Sparsø, M. Storgaard, J. Madsen,
and R. G. Olsen

Part III   Modeling, Simulation and Run-Time Management

Modeling, Simulation and Run-Time Management **187**
E. Macii

Dynamic Power Management for Nonstationary Service Requests **195**
E.-Y. Chung, L. Benini, A. Bogliolo, and G. De Micheli

Quantitative Comparison of Power Management Algorithms **207**
Y.-H. Lu, E.-Y. Chung, T. Šimunić, L. Benini, and G. De Micheli

Energy Efficiency of the IEEE 802.15.4 Standard in Dense Wireless Microsensor Networks: Modeling and Improvement Perspectives **221**
B. Bougard, F. Catthoor, D. C. Daly, A. Chandrakasan, and W. Dehaene

Statistical Blockade: A Novel Method for Very Fast Monte Carlo Simulation of Rare Circuit Events, and its Application **235**
A. Singhee and R. A. Rutenbar

Compositional Specification of Behavioral Semantics **253**
K. Chen, J. Sztipanovits, and S. Neema

Part IV   Design Technology For Advanced Digital Systems in CMOS and Beyond

Design Technology for Advanced Digital Systems in CMOS and Beyond **269**
H. De Man

**Address Bus Encoding Techniques for System-Level
Power Optimization** 275
L. Benini, G. De Micheli, E. Macii, D. Sciuto, and C. Silvano

**MOCSYN: Multiobjective Core-Based Single-Chip
System Synthesis** 291
R. P. Dick and N. K. Jha

**Minimum Energy Fixed-Priority Scheduling for Variable
Voltage Processors** 313
G. Quan and X. S. Hu

**Synthesis and Optimization of Threshold Logic Networks
with Application to Nanotechnologies** 325
R. Zhang, P. Gupta, L. Zhong, and N. K. Jha

Part V Physical Design and Validation

Physical Design and Validation. 347
J. A. G. Jess

Interconnect Tuning Strategies for High-Performance ICs 359
A. B. Kahng, S. Muddu, E. Sarto, and R. Sharma

Efficient Inductance Extraction via Windowing. 377
M. Beattie and L. Pileggi

**Soft-Error Tolerance Analysis and Optimization
of Nanometer Circuits** 389
Y. S. Dhillon, A. U. Diril, and A. Chatterjee

**A Single Photon Avalanche Diode Array Fabricated
in Deep-Submicron CMOS Technology**. 401
C. Niclass, M. Sergio, and E. Charbon

Part VI Test and Verification

**The Test and Verification Influential Papers in the 10 Years
of DATE**. 417
T. W. Williams and R. Kapur

**Cost Reduction and Evaluation of a Temporary Faults-Detecting
Technique** 423
L. Anghel and M. Nicolaidis

An Integrated System-on-Chip Test Framework 439
E. Larsson and Z. Peng

Efficient Spectral Techniques for Sequential ATPG
A. Giani, S. Sheng, M. S. Hsiao, and V. D. Agrawal. 455

BerkMin: A Fast and Robust Sat-Solver **465**
E. Goldberg and Y. Novikov

**Improving Compression Ratio, Area Overhead,
and Test Application Time for System-on-Chip
Test Data Compression/Decompression** **479**
P. T. Gonciari, B. M. Al-Hashimi, and N. Nicolici

**An Effective Technique for Minimizing the Cost of Processor
Software-Based Diagnosis in SoCs** **497**
P. Bernardi, E. Sánchez, M. Schillaci, G. Squillero, and M. S. Reorda

Appendix   Shortlist of Most Influential Papers **511**

Part I
System Level Design

System Level Design: Past, Present, and Future

Daniel D. Gajski

1 Introduction

With complexities of multiprocessor platforms (MPPs) rising almost daily, the design community has been searching for a new methodology that can handle given complexities with increased productivity and decreased time-to-market. The obvious solution that comes to mind is increasing levels of abstraction, or in other words, increasing the size of the basic building blocks. However, it is not clear what these basic blocks should be and how to create a multiprocessor system out of these basic blocks. To make things more difficult, the difference between software and hardware is becoming indistinguishable which, in turn, requires sizable change in the industrial and academic infrastructure.

In order to find a solution for the system level design flow, we must look at the system gap between SW and HW and then try to bridge this gap by developing a design flow that is based on common principles applicable to software and hardware which, in turn, allows us to define any system and its models with clean unambiguous semantics. Such clean semantics allows automatic model generation, simplifies synthesis algorithms and verification techniques, and leads to the ultimate goal of increasing system-design productivity by several orders of magnitude while reducing expertise level needed for system design to the basic principles of design science only.

2 Design Flow Through History

Design flow is defined as a sequence of design steps that are necessary to take the product specification to manufacturing. Design methodology, usually considered to be more detailed definition of design flow, indicates for every design step the set of design decisions, corresponding design model, and the set of tools used on each

---

Center for Embedded Computer Systems University of California at Irvine gajski@uci.edu

R. Lauwereins and J. Madsen (eds.), *Design, Automation, and Test in Europe–The Most Influential Papers of 10 Years DATE*, 3–14
© Springer 2008

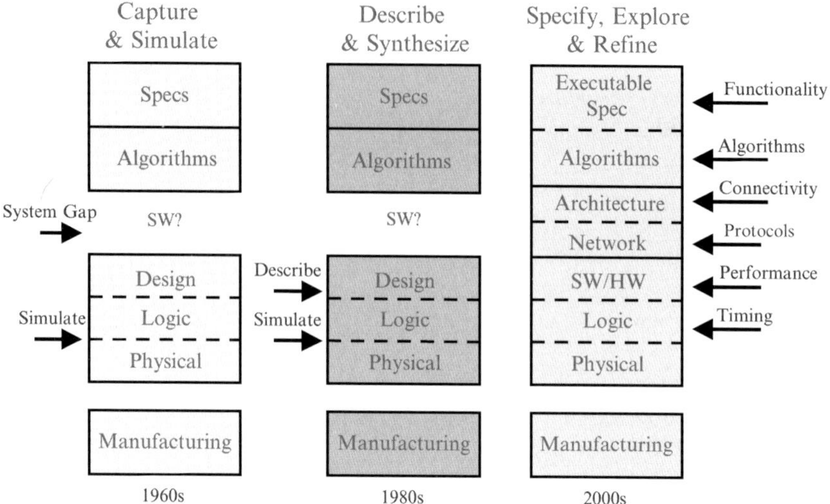

Fig. 1 History of design methodologies

step. Design methodology changes from company to company and from division to division in the same company. It has also changed several times since the start of design automation industry as shown in Fig. 1.

2.1 *Capture-and-Simulate Methodology*

In Capture-and-Simulate methodology (1960–1980s) software and hardware design was separated by a system gap. Software designers tested some algorithms and possibly wrote the requirements document and an initial, incomplete specification.

This specification was given to hardware designers who read it briefly and started immediately system design with a block diagram. They kept refining the block diagram until they reached the gate level. They did not know whether their design will work until the gate level design was captured and simulated. Since simulation was too slow to simulate the whole design they built a prototype and gave it to SW designers to develop the necessary system and application software. SW designers built software around any possible mistakes or bugs in the prototype. Therefore, the need for communication between SW and HW designers appeared twice. However, they did not talk the same language, since software and hardware were two different philosophies, methodologies, education backgrounds, and products.

2.2   *Describe-and-Synthesize Methodology*

Logic synthesis has significantly altered the design philosophy from Capture-and-Simulate into Describe-and-Synthesize methodology (late 1980s to late 1990s). In this methodology designers first specify what they want in terms of Boolean equations or FSM descriptions and the synthesis tools generate the implementation in terms of a gate and flip-flop netlists. Therefore, the design behavior or the function comes first and the structure or implementation second. In other words, we have two logic-level models to simulate: behavior (function) and structure (netlist). The most important new concept introduced here was that specification comes before implementation and they are both simulatable. Furthermore, it is possible to verify their equivalence since they can be both reduced to a canonical form. This new design flow was introduced on gate level first with a hope that it will spread to other levels later. By the late 1990s the logic level has been partially abstracted to cycle-accurate (CA) or register-transfer (RT) level of description and synthesis but not further. Therefore, two abstraction levels (cycle-accurate and logic levels) and two different models on each level (behavioral and structural) were in the design flow by the end of 20th century. However, the system gap persisted since Describe-and-Synthesize methodology was not applied to the entire design flow for hardware and software.

2.3   *Specify-Explore-Refine Methodology*

In order to close the system gap we must increase the level of abstraction to system level and apply the same principles to SW and HW parts of design. On system level we start with executable specification which represents the system behavior (functionality) without identifying SW or HW parts. As we keep refining the design we may use several intermediate models representing partially finished design. Each model is used to prove some system property such as functionality, algorithms, connectivity, and communication. In any case we must deal with several models that mix system behaviors and structure on different level of detail. Each model can be considered to be a specification for the next more accurate model in which more detail in the implementation is added. This kind of Specify-Explore-Refine (SER) methodology (since early 2000 to the present) represents a sequence of models in which each model is a refinement of the previous one. Thus, SER methodology follows the natural design process where designers specify the intent first, explore possibilities and then refine the model according to their design decisions. In general, SER methodology can be viewed as several iterations of the basic Describe-and-Synthesize methodology.

With several iterations of SER the new issue of rewriting or refining the models emerges. Designers would prefer writing one model at the end of design process and prove all its properties at once. However, that is not possible in 21st century with present system design complexities. The other possibility is to automatically

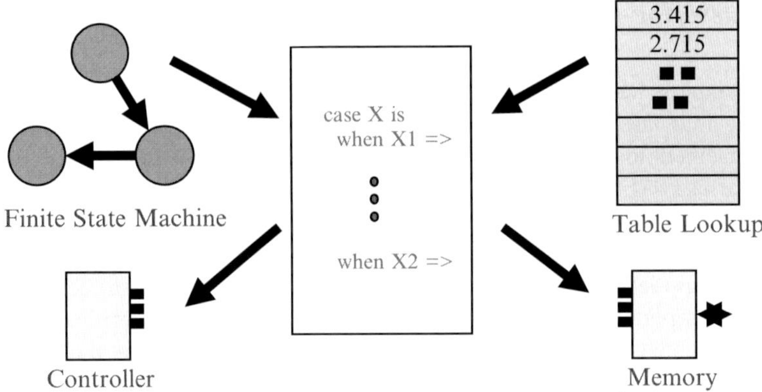

Fig. 2 Ambiguous semantics in hardware-description languages

synthesize each step in the design process, automatically generate or refine models, and automatically verify the refinement.

This designers dream may not be impossible. However, it requires well-defined system and modeling semantics. This is not as simple as it sounds, since design methodologies and EDA industry have been dominated by simulation-based methodologies so far. Unfortunately, hardware-description languages (HDLs), such as Verilog or system level languages, such as SystemC, are simuletable but not synthesizable or verifiable.

As an example of this problem, we can look at a simple case statement in any of the HDLs as shown in Fig. 2. Such case statement can be used to model a *Finite State Machine* (FSM) or a *Look-up Table*, for example. However, FSMs and look-up tables require different implementations: a FSM can be implemented with a *Controller* or logic gates while a look-up table is implemented with a *Memory*. On the other hand, using a memory to implement a FSM or controller to implement a table is not very efficient and not acceptable by any designer. Therefore, the model which uses case statement to model FSMs and tables is good for simulation but ambiguous for implementation since a designer or a synthesis tool can not determine what was really described by the case statement.

Thus, clean and unambiguous model semantics is needed for refinement, synthesis, and verification. Such a clean semantics is missing from most of the simulation-oriented languages. To make things worse, the definition of model semantic was left to each individual designer, design project, or company division. With hundreds of different model semantics, it is impossible to develop uniform set of tools for synthesis and verification.

Therefore, design and model semantics must be defined first. Each model should emphasize different design aspects or design detail. With this in mind, an acceptable system level methodology would naturally start with a well-defined executable specification model that serves as the golden reference. The specification is gradually refined to a detailed model that can be fed to traditional simulation,

Fig. 3 Refinement methodology

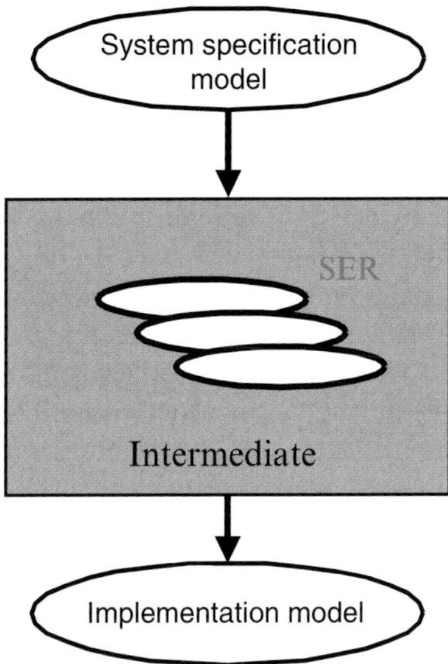

synthesis and verification tools. The gradual refinement as shown in Fig. 3 produces some intermediate models depending on the choice of methodology. The details that are added to models during refinement depend on the designer decisions. Each decision corresponds to one or more model transformations. If all the transformations are well defined, the refinement process can be automated thus, enabling the possibility for automatic synthesis and verification.

3   Modeling Algebra

We can define design models on principles similar to those used in ordinary arithmetic algebras required in high-school programs. An arithmetic algebra is a set of numbers and a set of operations,

$$\text{Algebra} = \langle \text{numbers, operations} \rangle$$

For example, numbers can be a set of integers and operations can be addition (+) and multiplication (*). Then such algebra defines the set of all possible expressions connected by + and *. In addition, such algebra has a set of laws, such as laws of associativity, commutativity, or distributivity of operations. These laws define equivalence of two expressions and the possibility of replacing one expression with another. Similar to arithmetic algebra we can define model algebra as a set of objects and a set of composition rules that allow combining those objects into larger objects. In other words,

Model algebra = <objects, compostions>

In case of well-known finite-state-machine (FSM) model, objects are states while composition rules allow combining two states, the present state and the next state with a transition arc that is labeled with the condition under which the transition will take place. For example, FSM algebra represents all possible FSM models. One of the axioms of FSM algebra defines two states to be equivalent if and only if both states produce the same output for every input and if both states transition to equivalent states for every possible input. This axiom is used to prove the equivalence of two FSM states. Furthermore, if two states are equivalent then we can combine them into one state and reduce the number of states in the FSM by one. If we combine all the equivalent states in the similar manner, then we can replace the larger FSM with the smaller equivalent FSM while reducing the cost of FSM implementation.

3.1   *Model Definition*

From the above definition, a model is a set of objects that are combined together according to the set of composition rules. For example, a specification level model would have objects like behaviors for computation and channels for communication, while an system platform model would have components such as processors, memories, and IPs for computation and buses for communication. Composition rules allow combining these objects hierarchically in a sequential, parallel, pipelined, or state fashion into larger objects in order to create a complete system model.

Such model algebra represents the set of all possible models that use the same set of objects and the same set of composition rules. Obviously, such a set of all possible models is infinite. A particular model is defined by specifying for each composition rule the objects and their composition in the given model. For example, the transition arcs between states in the FSM model can be defined by a subset of the cross product $S \times S$.

3.2   *Model Transformations and Refinement*

A model transformation can be viewed as a rearrangement of objects or a replacement of an object with another object. For instance, in order to distribute the behaviors in a specification model across components in the given platform, we need to partition the behaviors into groups and assign each group to one component in the platform. Similarly, in order to refine communication over abstract channels to communication over system buses, we need to partition channels into groups, assign each group to one bus and replace abstract channels with protocol channels. Another example of a transformation is a replacement of sequential execution with concurrent execution that requires adding synchronization to insure that the proper data values are used

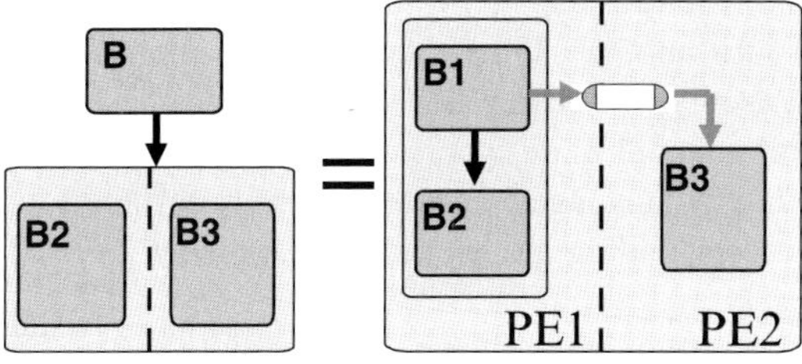

(b) distributivity of behaviors in modeling

Fig. 4 Distributive law in arithmetic and modeling

every time. Yet another example is decomposition of data into a set of messages for transmission over a limited capacity bus on the sender side and data assembly on the receiver side.

Intuitively, we can draw an analogy between distributive law of numbers in arithmetic and distribution of behaviors overprocessing elements (PEs) in architecture modeling. In Fig. 4a we see that the a* (b + c) is equal to a*b + a*c by law of distributivity. Similarly, the model on the left-hand side in Fig. 4b is equal to the model on the right-hand side by law of distributed execution. The specification model on the left-hand side is a sequential composition of behavior B1 with the behavior consisting of concurrent composition of behaviors B2 and B3. We can distribute the concurrent execution over PE1 and PE2. In this case P1 will execute B1 followed by B2 while PE2 will execute B1 followed by B3. The behavior B1 in PE2 can be replaced by a data passing channel that will transfer necessary data from B1 to B3 after B1 is executed. The equivalence of two models is established by the same results for the same inputs after the model execution. The same results are guaranteed by enforcing the same data dependencies in both models.

A model refinement can be defined as a specific sequence of transformations t1, t2 …,tn. We say that model B is a refinement of A if and only if B can be derived from A by applying transformations t1 through tn,

$$B = tn (...ti(...t2(t1(A))$$

Model transformations correspond usually to design's decisions leading to desired system structure. Therefore, the new refined model has more implementation details. On the other hand, not every sequence of transformations is a useful refinement. However, if each transformation is an equivalence preserving rearrangement or replacement, then model refinement generates an equivalent model. Furthermore, if all transformations can be automated then refinement will also automatically produce a new system model that is equivalent to the original model.

3.3 Refinement-based Environments

Any design methodology can be described by a set of designs and their models, a set of design decisions and corresponding refinements (sequences of transformations) from one model into another. If new models are automatically generated then designers do not need to learn how to write different design models except the original specification model. If designers do not write models then there is no need for different modeling languages. In other words, designers only make design decisions, while corresponding model transformation can be used to automatically generate new models.

Such a generic environment is shown graphically in Fig. 5. Given the *Model A*, designers profile the model, estimate key design metrics, and define how to refine the design. One type of design refinement means replacing each component in the design with set of subcomponents. One example of this kind of refinement would be replacing an adder with a set of Boolean equations or a netlist of gates. The other type of design refinement would be a rearrangement of existing components. One example of this type of refinement would be changing connections between several components or changing the protocol of a bus. Different components, their attributes and their models are in the *Component Library*. Designers make decisions which components to take and how to connect them. The selection of components and their configuration must be reflected in the new model. The *Refinement Tool* performs the model transformations one at the time according to *Designer*

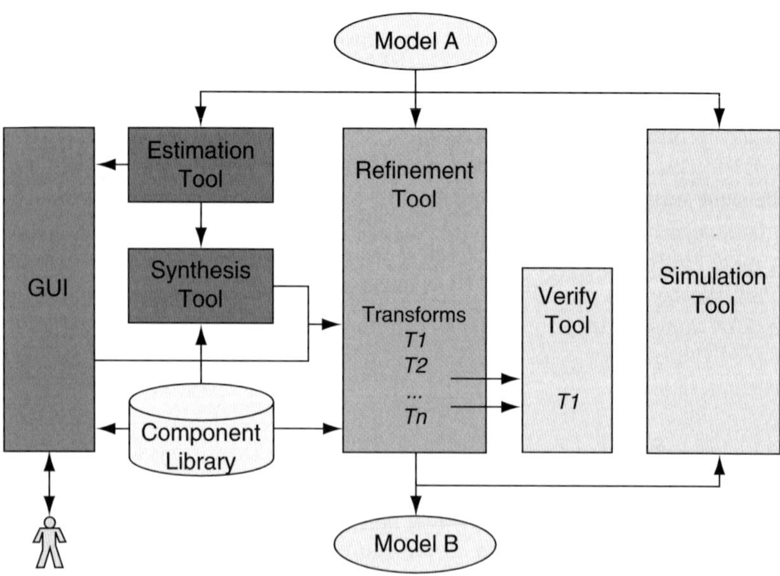

Fig. 5 Generic design methodology

Decisions. If design decisions are only rearrangement of existing components or replacement of one component by a netlist of subcomponents from the library, the transformations are easily automated. Thus *Refinement Tool* will generate *Model B* automatically. Thus, such model automation frees designers from rewriting the models.

3.4 System Level Modeling

This generic methodology can be applied to any two models. The main challenge is to define necessary and sufficient set of models and refinements for any application specific methodology. Obviously, the number of models should be small in order to simplify design flow, but large enough to simplify algorithms for refinement, synthesis and verification. Remember, closer the models, fewer decisions and transformations are needed for refinement of one model into another. Our past experiments have shown that three models may be sufficient for system-level modeling. They are shown in Fig. 6 with respect to OSI levels: system specification (untimed), transaction-level (timed) and pin accurate (cycle-accurate) models.

These three models represent the given system design on three different levels of abstraction. The general system architecture is given in Fig. 7. It consists of any number of processing elements (PEs) and memories connected with any number of busses. Each PE can be a standard processor, an application-specific processor or any

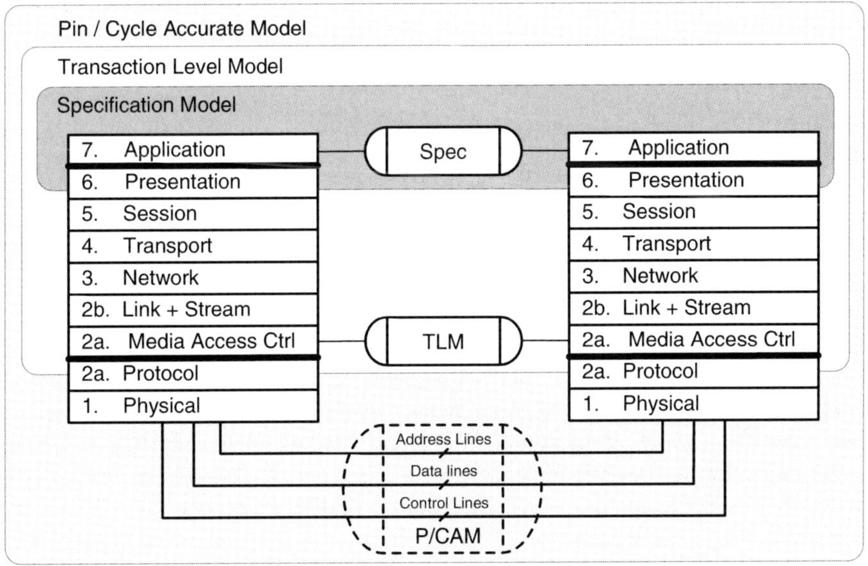

Fig. 6 Three basic system level models

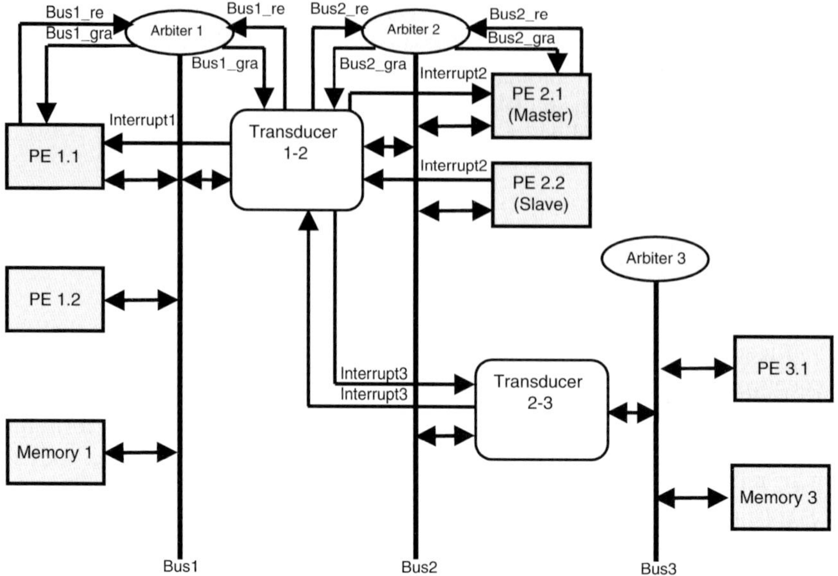

Fig. 7 Generic platform architecture

IP. Each bus has an arbiter if there is more then one PE on the bus. If there are two busses with different protocols a transducer or a bridge is inserted between them.

Therefore, this general system architecture consists of four components (processors, memories, transducers, and arbiters) and buses. Obviously, busses, arbiters, and transducers can be replaced by any other network-on-chip (NoC).

3.5   System Synthesis

Synthesis is the process of converting the behavioral description of a black box into a structure of the black box. The behavioral description defines the black box functionality by specifying the outputs as a function of inputs and time while structural description consists of connected components from a given library. During this process designers must make several different design decisions such as selecting components and their connectivity, mapping computation and communication descriptions to components, ordering or scheduling computations and communication of different messages, adding new objects, inserting synchronization to preserve dependences to name a few. The design process is a sequence of such design decisions. Since for every design decision there is a model transformation, a sequence of design decisions will result in a sequence of transformations or model refinement as explained earlier. Therefore, every synthesis task can be defined as a set of design decisions plus corresponding model refinement.

This concept is demonstrated in Fig. 5 where designer's estimations of design metrics and designers decisions are replaced by *Estimation* and *Synthesis Tools*. The *Estimation Tool* runs through Model A and generates different design-quality metrics needed for making corresponding design decisions. The design algorithms in the *Synthesis Tool* select proper components from *Component Library* and define their connectivity while *Refinement Tool* generates the refined model as explained earlier.

3.6   *System Verification*

In a refinement-based system level methodology, each model is produced through a sequence of transformations. As shown in Fig. 5 a set of designer decisions is used to add details to a model in order to refine it to the next lower level of abstraction. If each transformation in the refinement preserves the model equivalence then the refined model is equal to the original one. If transformations are performed automatically by the *Refinement Tool* the source code in it may not be bug free and may require additional verification tool. Therefore, we may use the *Verify Tool* which verifies that models before and after specific transformation are equivalent.

The notion of model equivalence comes from the simulation semantics of the model. Two models are equivalent is they have the same input-to-output results. This translates to existence of the same or equivalent objects in both models, the same partial order of execution of tasks, same input and output data for each task and same partial order of data transactions. In other words, all data dependences must be preserved.

Note that verifying each refinement, does not necessarily mean that the final model is completely correct. We also need to use traditional verification techniques to verify that the starting specification model is correct and we also must demonstrate correctness of the models in the component library. Furthermore, if we replace one object in the model with the other, we must also demonstrate that this replacement will not change the final result.

4   Summary

Present efforts by the academia and industry to move to the next level of abstraction have been partially successful at best. Without well-defined models and clean semantics only very little progress can be expected. In order to establish adequate set of models we can use the algebra principles which in turn simplify simulation, synthesis, and verification. Although the basic principles look attractive, the main challenge in the future is the definition of application specific methodologies. In other words, for each specific application we need to define the set of appropriate models, design decisions, transformations, and refinements before the field of system level design automation becomes a profitable success.

Some early results in this direction have been collected in this section which offers some of the most influential results of the last 10 years at DATE Conference. In the first paper by Eles et al. authors presented a new scheduling algorithm for synthesis of custom RTOS for arbitrary multiprocessor platforms based on new data structure process graphs. In a paradigm shift toward language-based design methodology, the second paper by Halambi et al. describes the EXPRESSION language for capturing multiprocessor information used for generation of retargetable compilers and simulators for application specific processors. The third paper by Gerstlauer et al. deals with model refinement for multiprocessor platform after RTOS type has been decided by designers. The fourth paper by Jersak et al. deals with performance estimation for multiprocessor platforms which guarantees better corner-case coverage and faster execution then standard simulation. The paper by Cho et al. presents a new synchronization synthesis technique for real-time systems under resource overloads. Finally, the paper by Neal Wingen gives series of practical strategies for integration of platform-based designs.

Reference

[1] D. Gajski et al.: Embedded System Design: Concepts and Tools ASP-DAC Tutorial, Yokohama, Japan, January 23, 2007 (http://www.cecs.uci.edu/pub_slides.htm).

Scheduling of Conditional Process Graphs for the Synthesis of Embedded Systems

Petru Eles[1,2], Krzysztof Kuchcinski[1], Zebo Peng[1], Alexa Doboli[2], and Paul Pop[2]

Abstract We present an approach to process scheduling based on an abstract graph representation which captures both data-flow and the flow of control. Target architectures consist of several processors, ASICs and shared busses. We have developed a heuristic which generates a schedule table so that the worst case delay is minimized. Several experiments demonstrate the efficiency of the approach.

1 Introduction

In this paper we concentrate on process scheduling for systems consisting of communicating processes implemented on multiple processors and dedicated hardware components. In such a system in which several processes communicate with each other and share resources, scheduling is a factor with a decisive influence on the performance of the system and on the way it meets its timing constraints. Thus, process scheduling has not only to be performed for the synthesis of the final system, but also as part of the performance estimation task.

Optimal scheduling, in even simpler contexts than that presented above, has been proven to be an NP complete problem [13]. In our approach, we assume that some processes can be activated if certain conditions, computed by previously executed processes, are fulfilled. Thus, process scheduling is further complicated since at a given activation of the system, only a certain subset of the total amount of processes is executed and this subset differs from one activation to the other. This is an important contribution of our approach because we capture both the flow of data and that of control at the process level, which allows an accurate and direct modeling of a wide range of applications.

Performance estimation at the process level has been well studied in the past years [10, 12]. Starting from estimated execution times of single processes, performance estimation and scheduling of a system containing several processes can

---

[1] Department of Computer and Information Science Linköping University Sweden

[2] Computer Science and Engineering Department Technical University of Timisoara Romania

be performed. In [14] performance estimation is based on a preemptive scheduling strategy with static priorities using rate-monotonic-analysis. In [11] scheduling and partitioning of processes, and allocation of system components are formulated as a mixed integer linear programing problem while the solution proposed in [8] is based on constraint logic programing. Several research groups consider hardware/software architectures consisting of a single programmable processor and an ASIC. Under these circumstances deriving a static schedule for the software component practically means the linearization of a dataflow graph [2, 6].

Static scheduling of a set of data-dependent software processes on a multi-processor architecture has been intensively researched [3, 7, 9]. An essential assumption in these approaches is that a (fixed or unlimited) number of identical processors are available to which processes are progressively assigned as the static schedule is elaborated. Such an assumption is not acceptable for distributed embedded systems which are typically heterogeneous.

In our approach we consider embedded systems specified as interacting processes which have been mapped on an architecture consisting of several processors and dedicated hardware components connected by shared busses. Process interaction in our model is not only in terms of dataflow but also captures the flow of control under the form of conditional selection. Considering a nonpreemptive execution environment we statically generate a schedule table for processes and derive a worst case delay which is guaranteed under any conditions.

The paper is divided into 7 sections. In Section 2 we formulate our basic assumptions and introduce the graph-based model which is used for system representation. The schedule table and the general scheduling strategy are presented in Sections 3 and 4. The algorithm for generation of the schedule table is presented in Section 5. Section 6 describes the experimental evaluation and Section 7 presents our conclusions.

2 Problem Formulation and the Conditional Process Graph

We consider a generic architecture consisting of *programmable processors* and application specific *hardware processors* (ASICs) connected through several *busses*. These busses can be shared by several communication channels connecting processes assigned to different processors. Only one process can be executed at a time by a programmable processor while a hardware processor can execute processes in parallel. Processes on different processors can be executed in parallel. Only one data transfer can be performed by a bus at a given moment. Computation and data transfer on several busses can overlap.

In [4] we presented algorithms for automatic hardware/software partitioning based on iterative improvement heuristics. The problem we are discussing in this paper concerns performance estimation *of a given design alternative* and scheduling of processes and communications. Thus, we assume that each process is assigned to a (programmable or hardware) processor and each communication channel which connects processes assigned to different processors is assigned to a bus. Our goal is to derive a worst case delay by which the system completes execution, so that this delay is as small as possible, and to generate the schedule which guarantees this delay.

As an abstract model for system representation we use a directed, acyclic, polar graph $\Gamma(V, E_S, E_C)$. Each node $P_i \in V$ represents one process. E_S and E_C are the sets of simple and conditional edges respectively. $E_S \cap E_C = \varnothing$ and $E_S \cup E_C = E$, where E is the set of all edges. An edge $e_{ij} \in E$ from P_i to P_j indicates that the output of P_i is the input of P_j. The graph is polar, which means that there are two nodes, called *source* and *sink*, that conventionally represent the first and last process. These nodes are introduced as dummy processes so that all other nodes in the graph are successors of the source and predecessors of the sink respectively.

The mapping of processes to processors and busses is given by a function M: $V \rightarrow PE$, where $PE = \{pe_1, pe_2, .., pe_{Npe}\}$ is the set of processing elements. For any process $P_i, M(P_i)$ is the processing element to which P_i is assigned for execution.

Each process P_i, assigned to processor or bus $M(P_i)$, is characterized by an execution time t_{Pi}. In the process graph depicted in Fig. 1, P_0 and P_{32} are the source and sink nodes respectively. Nodes denoted $P_1, P_2, .., P_{17}$, are "ordinary" processes specified by the designer. They are assigned to one of the two programmable processors pe_1 and pe_2 or to the hardware component pe_3. The rest are so called *communication processes* $(P_{18}, P_{19}, .., P_{31})$. They are represented in Fig. 1 as black dots and are introduced for each connection which links processes mapped to different processors. These processes model inter-processor communication and their execution time is equal to the corresponding communication time.

An edge $e_{ij} \in E_C$ is a *conditional edge* (thick lines in Fig. 1) and it has an associated condition. Transmission on such an edge takes place only if the associated condition is satisfied. We call a node with conditional edges at its output a *disjunction node* (and the corresponding process a *disjunction process*). Alternative paths starting from a disjunction node, which correspond to a certain condition, are disjoint and they meet in a so called *conjunction node* (with the corresponding process called *conjunction process*). In Fig. 1 circles representing conjunction and disjunction nodes are depicted with thick borders. We assume that conditions are independent.

A boolean expression X_{Pi}, called guard, can be associated to each node P_i in the graph. It represents the necessary condition for the respective process to be activated. In Fig. 1, for example, $X_{P3}=true$, $X_{P14}=D \wedge K$, $X_{P17}=true$, $X_{P5}=\overline{C}$. Two nodes P_i and P_j, where P_j is not a conjunction node, can be connected by an edge e_{ij} only if $X_{Pj} \Rightarrow X_{Pi}$ (which means that X_{Pi} is true whenever X_{Pj} is true). This restriction avoids specifications in which a process is blocked because it waits for a message from a process which will not be activated. If P_j is a conjunction node, predecessor nodes P_i can be situated on alternative input paths.

According to our model, we assume that a process, which is not a conjunction process, can be activated only after all its inputs have arrived. A conjunction process can be activated after messages coming on one of the alternative paths have arrived. All processes issue their outputs when they terminate. If we consider the activation time of the source process as a reference, the activation time of the sink process is the delay of the system at a certain execution.

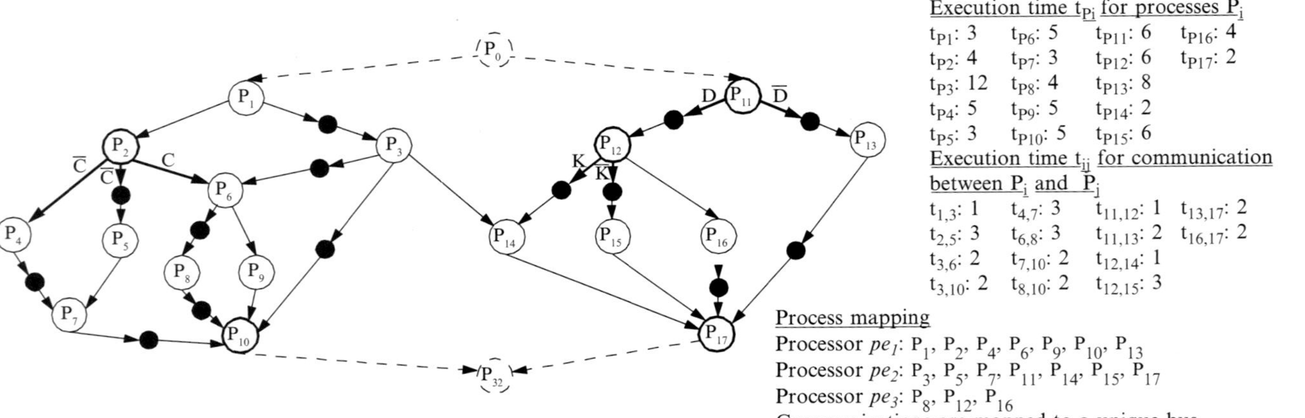

Fig. 1 Conditional Process Graph with execution times and mapping

3   The Schedule Table

For a given execution of the system, a subset of the processes is activated which corresponds to the actual path through the process graph. This path depends on certain conditions. For each individual path there is an optimal schedule of the processes which produces a minimal delay. Let us consider the process graph in Fig. 1. If all three conditions, C, D, and K are true, the optimal schedule requires P_1 to be activated at time $t = 0$ on processor pe_1, and processor pe_2 to be kept idle until $t=4$, in order to activate P_3 as soon as possible (see Fig. 4a). However, if C and D are true but K is false, the optimal schedule requires to start both P_1 on pe_1 and P_{11} on pe_2 at $t=0$; P_3 will be activated in this case at $t=6$, after P_{11} has terminated and, thus, pe_2 becomes free (see Fig. 4b). This example reveals one of the difficulties when generating a schedule for a system like that in Fig. 1. As the values of the conditions are unpredictable, the decision on which process to activate on pe_2 and at which time, has to be taken without knowing which values the conditions will later get. On the other side, at a certain moment during execution, when the values of some conditions are already known, they have to be used in order to take the best possible decisions on when and which process to activate. An algorithm has to be developed which produces a schedule of the processes so that the worst case delay is as small as possible. The output of this algorithm is a so called *schedule table*. In this table there is one row for each "ordinary" or communication process, which contains activation times for that process corresponding to different values of the conditions. Each column in the table is headed by a logical expression constructed as a conjunction of condition values. Activation times in a given column represent starting times of the processes when the respective expression is true.

Table 1 shows part of the schedule table corresponding to the system depicted in Fig. 1. According to this schedule processes P_1, P_2, P_{11} as well as the communication process P_{18} are activated unconditionally at the times given in the first column of the

Table 1 Part of schedule table for the graph in Fig. 1

	true	D	$D \wedge C$	$D \wedge C \wedge \bar{K}$	$D \wedge C \wedge K$	$D \wedge \bar{C}$	$D \wedge \bar{C} \wedge \bar{K}$	$D \wedge \bar{C} \wedge K$	$\bar{D}$	$\bar{D} \wedge C$	$\bar{D} \wedge \bar{C}$
P_1	0										
P_2	3										
P_{10}				34	34		26	26		34	26
P_{11}	0										
P_{14}					35			24			
P_{17}				29	37		30	26		22	24
P_{18} $1{\to}3$	3										
P_{19} $2{\to}5$						9					10
P_{20} $3{\to}10$				28	20		21	21		22	18
D	6										
C		7							7		
K			15			15					

table. No condition has yet been determined to select between alternative schedules. Process P_{14}, on the other hand, has to be activated at $t=24$ if $D \wedge \overline{C} \wedge K = true$ and $t=35$ if $D \wedge C \wedge K = true$. To determine the worst case delay, δ_{max}, we have to observe the rows corresponding to processes P_{10} and P_{17}: $\delta_{max} = \max\{34 + t_{10}, 37 + t_{17}\} = 39$.

The schedule table contains all information needed by a distributed run time scheduler to take decisions on activation of processes. We consider that during execution a very simple non-preemptive scheduler located on each programmable/communication processor decides on process and communication activation depending on actual values of conditions. Once activated, a process executes until it completes. To produce a deterministic behavior which is correct for any combination of conditions, the table has to fulfill several requirements:

1. If for a certain process Pi, with guard X^{Pi}, there exists an activation time in the column headed by expression Ek, then $Ek \Rightarrow XP_i$, this means that no process will be activated if the conditions required for its execution are not fulfilled.
2. Activation times have to be uniquely determined by the conditions. Thus, if for a certain process P_i there are several alternative activation times then, for each pair of such times (τ_{Pi}^{Ej}, τ_{Pi}^{Ek}) placed in columns headed by expressions E_j and E_k, $E_j \wedge E_k = false$.
3. If for a certain execution of the system the guard X_{Pi} becomes true then P_i has to be activated during that execution. Thus, considering all expressions E_j corresponding to columns which contain an activation time for P_i, $\vee E_j = X_{Pi}$.
4. Activation of a process P_i at a certain time t has to depend only on condition values which are determined at the respective moment t and are known to the processing element $M(P_i)$ which executes P_i.

The value of a condition is determined at the moment τ at which the corresponding disjunction process terminates. Thus, at any moment t, $t \geq \tau$, the condition is available for scheduling decisions on the processor which has executed the disjunction process. However, in order to be available on any other processor, the value has to arrive at that processor. The scheduling algorithm has to consider both the time and the resource needed for this communication.

The following strategy has been adopted for scheduling the communication of conditions: after termination of a disjunction process the value of the condition is broadcasted from the corresponding processor to all other processors; this broadcast is scheduled as soon as possible on the first bus which becomes available after termination of the disjunction process. For this task only busses are considered to which all processors are connected and we assume that at least one such bus exists.[1] The time τ_0 needed for this communication is the same for all conditions and depends on the features of the employed buses. Given the minimal amount of transferred information, the time τ_0 is smaller than (at most equal to) any other communication

---

[1] This assumption is made for simplification of the further discussion. If no bus is connected to all processors, communication tasks have to be scheduled on several busses according to the actual interconnection topology.

time. The transmitted condition is available for scheduling decisions on all other processors τ_0 time units after initiation of the broadcast. For the example given in Table 1 communication time for conditions has been considered $\tau_0=1$. The last three rows in Table 1 indicate the schedule for communication of conditions C, D, and K.

4   The Scheduling Strategy

Our goal is to derive a minimal worst-case delay and to generate the corresponding schedule table for a process graph $\Gamma(V, E_S, E_C)$, a mapping function $M: V{\rightarrow}PE$, and execution times t_{Pi} for each process $P_i{\in} V$. At a certain execution of the system, one of the N_{alt} alternative paths through the process graph will be executed. Each alternative path corresponds to one subgraph $G_k{\in}\Gamma$, $k=1,2,...,N_{alt}$. For each subgraph there is an associated logical expression L_k (the label of the path) which represents the necessary conditions for that subgraph to be executed.

If at activation of the system all the conditions *would be* known, the processes *could be* executed according to the (near)optimal schedule of the corresponding subgraph G_k. Under these circumstances the worst case delay δ_{max} would be $\delta_{max} = \delta_M$, with $\delta_M = \max\{\delta_k, k = 1,2,...,N_{alt}\}$, where δ_k is the delay corresponding to subgraph G_k.

However, this is not the case as we do not assume any prediction of the conditions at the start of the system. Thus, what we can say is only that[2]: $\delta_{max} \geq \delta_M$.

A scheduling heuristic has to produce a schedule table for which the difference $\delta_{max}-\delta_M$ is minimized. This means that the perturbation of the individual schedules, introduced by the fact that the actual path is not known in advance, should be as small as possible. We have developed a heuristic which, starting from the schedules corresponding to the alternative paths, produces the global schedule table, as result of a, so called, *schedule merging operation*. Hence, we perform scheduling of a process graph as a succession of the following two steps:

1. Scheduling of each individual alternative path
2. Merging of the individual schedules and generation of the schedule table

We present algorithms for scheduling of the individual paths in [5]. In this paper we concentrate on the generation mechanism of the global schedule table.

5   The Table Generation Algorithm

The input for the generation of the schedule table is a set of N_{alt} schedules, each corresponding to an alternative path, labeled L_k, through the process graph τ. Each such schedule consists of a set of pairs (P_i, τ_{Pi}^{Lk}), where P_i is a process activated on

---

[2] This formula to be rigorously correct, δ_M has to be the maximum of the *optimal* delays for each subgraph.

path L_k and τ_{Pi}^{Lk} is the start time of process P_i according to the respective schedule. The schedule table generated as output fulfills the requirements presented in Section 3.

The schedule merging algorithm is guided by the length of the schedules produced for each alternative path. While progressively constructing the schedule table, at each moment, priority is given to the requirements of the schedule corresponding to that path, among those which are still reachable, that produces the largest delay. Thus, we induce perturbations into the short delay paths and let the long ones proceed as similar as possible to their (near)optimal schedule.

5.1 Schedule Merging

The generation algorithm of the schedule table proceeds along a binary decision tree corresponding to all alternative paths, which is explored in a depth first order. Figure 2 shows the decision tree explored during generation of Table 1. The nodes of the decision tree correspond to the states reached when, according to the actual schedule, a disjunction process has been terminated and the value of a new condition has been computed. The algorithm is guided by the following basic rules:

1. Start times of processes are fixed in the table according, with priority, to the schedule of that path which is reachable from the current state and produces the longest delay.
2. The start time τ_{Pi}^{Lk} of a process P_i is placed in a column headed by the conjunction of all condition values known at τ_{Pi}^{Lk} on the processing element $M(P_i)$, according to the current schedule. If such a column does not yet exist in the table, it will be generated.
3. After a new path has been selected, its schedule will be adjusted by enforcing the start times of certain processes according to their previously fixed values. This can be the case of a process P_i which is part of the current path labelled L_k ($L_k \Rightarrow X_{Pi}$), and of a previously handled path labelled L_q ($L_q \Rightarrow X_{Pi}$). When handling path L_q an activation time for process P_i has been fixed in a column headed by expression E. If E depends exclusively on conditions corresponding to tree

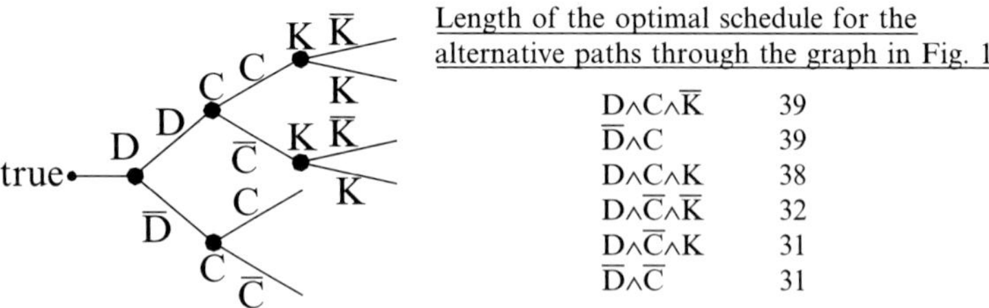

Length of the optimal schedule for the alternative paths through the graph in Fig. 1

$D \wedge C \wedge \overline{K}$	39
$\overline{D} \wedge C$	39
$D \wedge C \wedge K$	38
$D \wedge \overline{C} \wedge \overline{K}$	32
$D \wedge \overline{C} \wedge K$	31
$\overline{D} \wedge \overline{C}$	31

Fig. 2 Decision tree explored for the graph in Fig. 1

nodes which are predecessors of the branching node between the two paths, then the schedule of the current path, L_k, has to be adjusted by taking into consideration the previously fixed activation time of P_i.

4. Further readjustments of the current schedule are performed in order to avoid violation of requirement 2 in Section 3. This aspect will be discussed in subsection 5.2.

At the beginning, start times of processes are placed into Table 1 according to the schedule which corresponds to the path labeled $D{\wedge}C{\wedge}\bar{K}$. After the first back-step, to node K (Fig. 2), the schedule corresponding to path $D{\wedge}C{\wedge}K$ becomes the actual one. New start times will be fixed into the schedule table according to an adjusted version of this schedule. The next back-step is to node C. Two schedules are now reachable taking the branch $\bar{C}$, which are labelled $D{\wedge}\bar{C}{\wedge}\bar{K}$ and $D{\wedge}\bar{C}{\wedge}K$ respectively. $D{\wedge}\bar{C}{\wedge}\bar{K}$, which produces a larger delay, will be selected first as the actual schedule. It will be followed until the next back-step has been performed.

The algorithm for generation of the schedule table is briefly described, as a recursive procedure, in Fig. 3. An essential aspect of this algorithm is that, after each back-step, a new schedule has to be selected as the current one. The selection rule gives priority to the path with the largest delay, among those which are reachable from the current node in the decision tree. Further start times of processes will be placed into the schedule table according to an *adjusted* version of the new current schedule. Processes which satisfy the condition of rule three presented above have to be moved to their previously fixed start time. They are considered as locked in this new position. As result of a simple rescheduling procedure the start times of the other, unlocked, processes are changed to the earliest moment which is allowed, taking in consideration data dependencies. Relative priorities of unlocked processes assigned to the same non-hardware processor are kept as in the original schedule.

In Fig. 4 we show the adjustment of the schedule labeled $D{\wedge}C{\wedge}K$ performed after the back-step to node K. At this moment start times of processes P_1, P_2, P_{11}, P_3, P_{12}, P_{18}, P_{27}, and of the communication processes for conditions D, C, and K have already

```
BuildScheduleTable(current_schedule, back_step)
    if back_step then
        Select new current_schedule
        Adjust current_schedule
        Check for conflicts and readjust current_schedule
    end if
    while not(EndOfSchedule or
            arrived at moment so that a disjunction process is terminated) do
        Take following process in current_schedule and
                            place start time into ScheduleTable
    end while
    if EndOfSchedule then return end if
    BuildSchedule Table(current_schedule, false)
    Buildschedule Table(current_schedule, true)
    end BuildScheduleTable
```

Fig. 3 Algorithm for generation of a schedule table

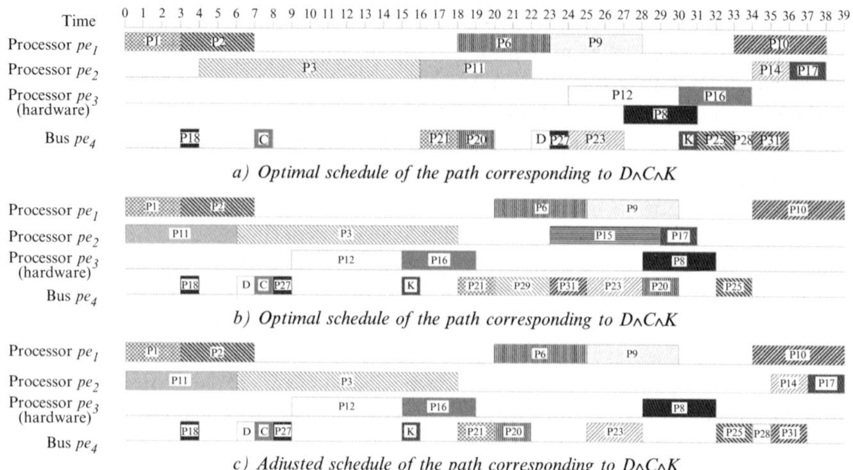

a) *Optimal schedule of the path corresponding to* $D \wedge C \wedge K$

b) *Optimal schedule of the path corresponding to* $D \wedge C \wedge K$

c) *Adjusted schedule of the path corresponding to* $D \wedge C \wedge K$

Fig. 4 Optimal and adjusted schedules for paths extracted from the process graph in Fig. 1

been placed in the table according to the schedule of path $D \wedge C \wedge \bar{K}$ which is shown in Fig. 4b. The activation time of these processes has been placed in columns headed by expressions *true*, C or $D \wedge C$, and consequently they are mandatory also for path $D \wedge C \wedge K$ (both node C and D are predecessors of node K which is the branching node between the two paths). Under these circumstances some of the other processes have to be moved from their original position in this schedule, shown in Fig. 4a, to their position in the adjusted schedule of Fig 4c. This adjusted version is used in order to fix start times of further processes until the next back-step.

5.2 *Conflict Handling at Table Generation*

Suppose we are currently handling a path labeled L_k. According to the adjusted schedule of this path we place an activation time τ_{Pi}^{Lk} of process P_i into the table, so that the respective column is headed by an expression E. The problem is how to preserve the coherence of the table in the sense introduced by requirement 2 defined in Section 3. If there is no activation time previously introduced in the row corresponding to P_i no conflicts can be generated. If, however, the respective row contains activation times, there exists a potential of conflict between the column headed by E and columns which already include activation times of P_i. Let us consider that such a column is headed by an expression F. According to requirement 2, we have a conflict between columns E and F if there exists no condition C so that $E = q \wedge C$ and $F = q' \wedge \bar{C}$. Intuitively, such a conflict means that for two or more paths the same process P_i is scheduled at different times but the conditions known on the processing element $M(P_i)$ do not allow to the scheduler to identify the current path and to take a deterministic decision on activation of P_i.

If placement of an activation time for process P_i in a column headed by expression E produces a conflict, the current schedule has to be *readjusted* so that an expression E' will head the column that hosts the new activation time of P_i and no conflict is induced in the schedule table. As shown in the algorithm presented in Fig. 3, after adjustment of a schedule, unlocked processes are checked if their placement in the table produces any conflicts. If this is the case, the process will be moved to a new activation time and the schedule is readjusted by changing the start time of some unlocked processes (similar to the operation performed at the initial adjustment). The main problem which has to be solved is to find the new activation time for P_i so that conflicts are avoided. In [5] we demonstrated the following two theorems in the context of our table generation algorithm:

Theorem 1

Consider a process P_i which is part of two paths, L_k and L_q, with activation times $\tau_{P_i}^{L_k}$ and $\tau_{P_i}^{L_q}$ respectively. If the set of predecessors of P_i is different in the two paths, then no conflict is possible between the columns corresponding to the two activation times.

As a consequence of this theorem readjustments for conflict handling can not impose an activation time of a process which is not feasible for the respective path.

Theorem 2

Consider a process P_i so that placement of its activation time $\tau_{P_i}^{L}$, corresponding to the current path L, into the schedule table produces a conflict. There exists an activation time τ' of P_i, corresponding to one of the previously handled paths with which the current one is in conflict, so that τ' has the following property: if P_i is moved to activation time τ' in the current schedule, all conflicts are avoided.

Consider W the set of columns with which there exists a conflict at placement of the activation time for P_i. Based on Theorem 2 we know that one of the times $\tau_{P_i}^{F}$ placed in a column $F \in W$, represents a correct solution for conflict elimination. Thus, the following loop over the set W can produce the new activation time of a process P_i so that all conflicts are avoided:

for all columns $F \in W$ **do**
    if moving P_i to $\tau_{P_i}^{F}$ all conflicts are avoided **then**
        return $\tau_{P_i}^{F}$
    end if
end for

6 Experimental Evaluation

The strategy we have presented for generation of the schedule table guarantees that the path corresponding to the largest delay, δ_M, will be executed in exactly δ_M time. This, however, does not mean that the worst case delay δ_{max}, corresponding to the generated global schedule, is always guaranteed to be δ_M. Such a delay can not be guaranteed in theory. According to our scheduling strategy δ_{max} will be worse than δ_M if the schedule corresponding to an initially faster path is disturbed at adjustment or conflict handling so that its delay becomes larger than δ_M.

For evaluation of the schedule merging algorithm we used 1080 conditional process graphs generated for experimental purpose. 360 graphs have been generated for each dimension of 60, 80, and 120 nodes. The number of alternative paths through the graphs is 10, 12, 18, 24, or 32. Execution times were assigned randomly using both uniform and exponential distribution. We considered architectures consisting of one ASIC and one to eleven processors and one to eight busses [5]. Experiments were run on a SPARCstation 20.

Figure 5 presents the percentage increase of the worst-case delay δ_{max} over the delay δ_M of the longest path. The average increase is between 0.1% and 7.63% and, practically, it does not depend on the dimension of the graph but only on the number of merged schedules. It is worth to be mentioned that a zero increase ($\delta_{max} = \delta_M$) was produced for 90% of the graphs with 10 alternative paths, 82% with 12 paths, 57% with 18 paths, 46% with 24 paths, and 33% with 32 paths. In Fig. 6 we show the average execution time for the schedule merging algorithm, as a function of the number of merged schedules. The time needed for scheduling of the individual paths depends on the employed algorithm. As demonstrated in [5], good-quality results can

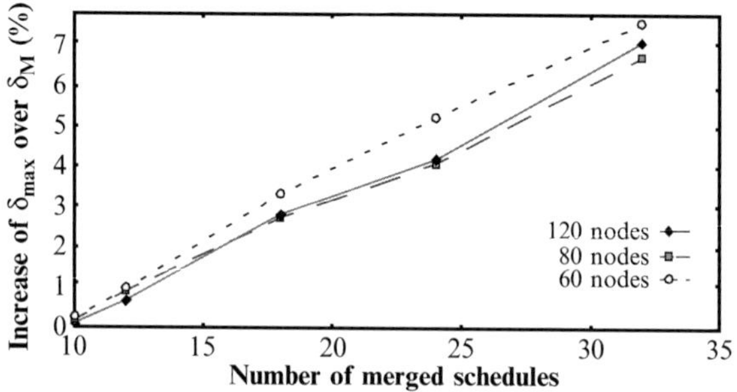

Fig. 5 Increase of the worst-case delay

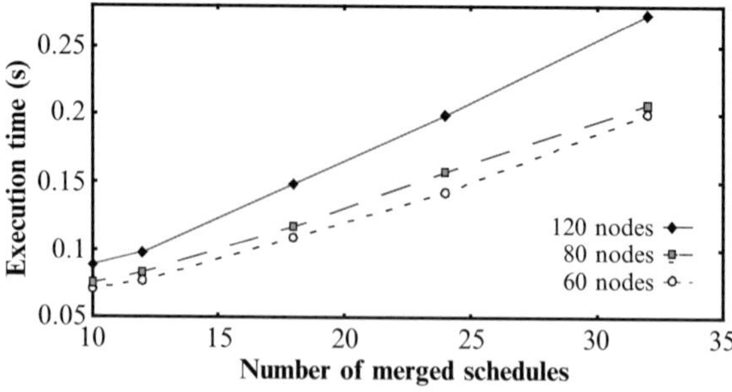

Fig. 6 Execution time for schedule merging

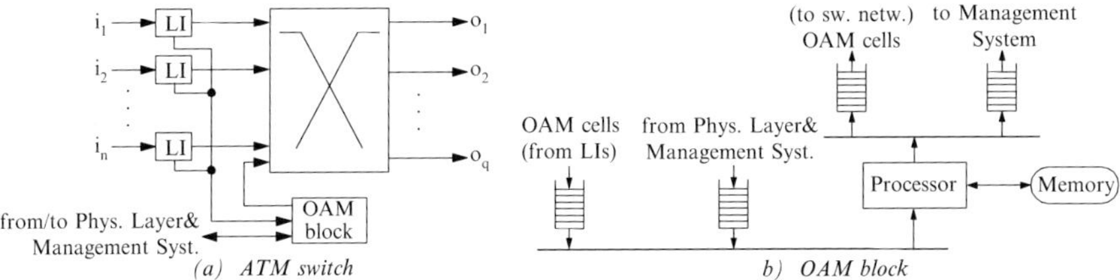

Fig. 7 ATM switch with OAM block

be obtained with a list scheduling based algorithm which needs less than 0.003 s for graphs having 120 nodes.

Finally, we present a real-life example which implements the operation and maintenance (OAM) functions corresponding to the F4 level of the ATM protocol layer [1]. Figure 7a shows an abstract model of the ATM switch. Through the switching network cells are routed between the n input and q output lines. In addition, the ATM switch also performs several OAM related tasks.

In [4] we discussed hardware/software partitioning of the OAM functions corresponding to the F4 level. We concluded that filtering of the input cells and redirecting of the OAM cells towards the OAM block have to be performed in hardware as part of the line interfaces (LI). The other functions are performed by the *OAM block* and can be implemented in software.

We have identified three independent modes in the functionality of the OAM block. Depending on the content of the input buffers (Fig. 7b), the OAM block switches between these three modes. Execution in each mode is controlled by a statically generated schedule table for the respective mode. We specified the functionality corresponding to each mode as a set of interacting VHDL processes. Table 2 shows the characteristics of the resulting process graphs. The main objective of this experiment was to estimate the worst-case delays in each mode for different alternative architectures of the OAM block. Based on these estimations as well as on the particular features of the environment in which the switch will be used, an appropriate architecture can be selected and the dimensions of the buffers can be determined.

Figure 7b shows a possible implementation architecture of the OAM block, using one processor and one memory module (1P/1M). Our experiments included also architecture models with two processors and one memory module (2P/1M), as well as structures consisting of one respectively two processors and two memory modules (1P/2M, 2P/2M). The target architectures are based on two types of processors: 486DX2/80 MHz and Pentium/120 MHz. For each architecture, processes have been assigned to processors taking into consideration the potential parallelism of the process graphs and the amount of communication between processes. The worst case delays resulting after generation of the schedule table for each of the three modes, arc given in Table 2. As expected, using a faster processor reduces the delay in each of the three modes. Introducing an additional processor, however, has no

Table 2 Worst-case delays for the OAM block

| | Model | | 1P/1M | | 1P/2M | | 2P/1M | | | 2P/2M | | |
| | | | | | | | | | | | | |
mode	nr. proc.	nr. paths	486	Pent.	486	Pent.	2× 486	2× Pent.	486+ Pent.	2× 486	2× Pent.	486+ Pent.
1	32	6	4471	2701	4471	2701	2932	2131	2532	2932	1932	2532
2	23	3	1732	1167	1732	1167	1732	1167	1167	1732	1167	1167
3	42	8	5852	3548	5852	3548	5033	3548	3548	5033	3548	3548

The header spans: "Worst-case delay (ns)" over the last nine data columns, with "2P/1M" over the 2× 486 / 2× Pent. / 486+ Pent. group and "2P/2M" over the last 2× 486 / 2× Pent. / 486+ Pent. group.

effect on the execution delay in *mode 2* which does not present any potential parallelism. In *mode 3* the delay is reduced by using two 486 processors instead of one. For the Pentium processor, however, the worst case delay can not be improved by introducing an additional processor. Using two processors will always improve the worst case delay in *mode 1*. As for the additional memory module, only in *mode 1* the model contains memory accesses which are potentially executed in parallel. Table 2 shows that only for the architecture consisting of two Pentium processors providing an additional memory module pays back by a reduction of the worst case delay in *mode 1*.

7 Conclusions

We have presented an approach to process scheduling for the synthesis of embedded systems. The approach is based on an abstract graph representation which captures, at process level, both dataflow and the flow of control. A schedule table is generated by a merging operation performed on the schedules of the alternative paths. The main problems which have been solved in this context are the minimization of the worst-case delay and the generation of a logically and temporally deterministic table, taking into consideration communication times and the sharing of the busses. The algorithms have been evaluated based on experiments using a large number of graphs generated for experimental purpose as well as real-life examples.

References

[1] T.M. Chen, S.S. Liu, *ATM Switching Systems*, Artech House Books, 1995.
[2] P. Chou, G. Boriello, Interval Scheduling: Fine-Grained Code Scheduling for Embedded Systems, *Proc. ACM/IEEE DAC*, 1995, 462–467.
[3] E.G. Coffman Jr., R.L. Graham, Optimal Scheduling for two Processor Systems, *Acta Informatica*, V1, 1972, 200–213.
[4] P. Eles, Z. Peng, K. Kuchcinski, A. Doboli, System Level Hardware/Software Partitioning Based on Simulated Annealing and Tabu Search, *Des. Aut. for Emb. Syst.*, V2, N1, 1997, 5–32.

[5] P. Eles, K. Kuchcinski, Z. Peng, A. Doboli, P. Pop, Process Scheduling for Performance Estimation and Synthesis of Embedded Systems, Research Report, Department of Computer and Information Science, Linköping University, 1997.

[6] R.K. Gupta, G. De Micheli, A Co-Synthesis Approach to Embedded System Design Automation, *Des. Aut. for Emb. Syst.*, V1, N1/2, 1996, 69–120.

[7] H. Kasahara, S. Narita, Practical Multiprocessor Scheduling Algorithms for Efficient Parallel Processing, *IEEE Trans. Comp.*, V33, N11, 1984, 1023–1029.

[8] K. Kuchcinski, Embedded System Synthesis by Timing Constraint Solving, *Proc. Int. Symp. Syst. Synth.*, 1997.

[9] Y.K. Kwok, I. Ahmad, Dynamic Critical-Path Scheduling: An Effective Technique for Allocating Task Graphs to Multiprocessors, *IEEE Trans. Par. Distr. Syst.*, V7, N5, 1996, 506–521.

[10] Y.S. Li, S. Malik, Performance Analysis of Embedded Software Using Implicit Path Enumeration, *Proc. ACM/IEEE Design Automation Conference*, 1995, 456–461.

[11] S. Prakash, A. Parker, "SOS: Synthesis of Application-Specific Heterogeneous Multiprocessor Systems", *J. Parallel Distrib. Comp.*, V16, 1992, 338–351.

[12] K. Suzuki, A. Sangiovanni-Vincentelli, Efficient Software Performance Estimation Methods for Hardware/Software Codesign, *Proc. ACM/IEEE DAC*, '96, 605–610.

[13] J.D. Ullman, NP-Complete Scheduling Problems, *J. Comput. Syst. Sci.*, V10, 384–393, 1975.

[14] T.Y. Yen, W. Wolf, *Hardware-Software Co-Synthesis of Distributed Embedded Systems*, Kluwer Academic, 1997.

EXPRESSION: A Language for Architecture Exploration Through Compiler/Simulator Retargetability*

Ashok Halambi[1], Peter Grun[2], Vijay Ganesh[3], Asheesh Khare[4], Nikil Dutt[5], and Alex Nicolau[6]

Abstract We describe **EXPRESSION**, a language supporting architectural design space exploration for embedded Systems-on-Chip (SOC) and automatic generation of a retargetable compiler/simulator toolkit. Key features of our language-driven design methodology include: a mixed behavioral/structural representation supporting a natural specification of the architecture; explicit specification of the memory subsystem allowing novel memory organizations and hierarchies; clean syntax and ease of modification supporting architectural exploration; a single specification supporting consistency and completeness checking of the architecture; and efficient specification of architectural resource constraints allowing extraction of detailed reservation tables for compiler scheduling. We illustrate key features of EXPRESSION through simple examples and demonstrate its efficacy in supporting exploration and automatic software toolkit generation for an embedded SOC codesign flow.

1 Introduction

The advent of System-on-Chip (SOC) technology has resulted in a paradigm shift for the design process of embedded systems employing programmable processors with custom hardware. Modern system-level design libraries frequently consist of Intellectual Property (IP) blocks such as processor cores that span a spectrum of architectural styles, ranging from traditional DSPs and superscalar RISC, to VLIWs and hybrid ASIPs. Furthermore, SOC technologies permit the incorporation of novel on-chip memory organizations (including the use of on-chip DRAM, frame buffers, streaming buffers, and partitioned register files), allowing a wide range of

---

*This work was partially supported by grants from NSF (MIP-9708067) and ONR (N00014-93-1-1348).

Department of Information and Computer Science, University of California, Irvine, CA 92697-3425, USA

[1] ahalambi@ics.uci.edu; [2] pgrun@ics.uci.edu; [3] ganesh@ics.uci.edu; [4] akhare@ics.uci.edu; [5] dutt@ics.uci.edu; [6] nicolau@ics.uci.edu

memory organizations and hierarchies to be explored and customized for the specific embedded application.

The embedded SOC designer is thus faced with the dual tasks of (1) rapidly exploring and evaluating different architectural and memory configurations, and (2) using a cycle-accurate simulator and retargetable optimizing compiler to adapt the application and architecture with the goal of meeting system-level performance, power, and cost objectives. Furthermore, shrinking time-to-market cycles create an urgent need to perform the traditionally sequential tasks of hardware and software design in parallel. An effective embedded SOC codesign flow must therefore support automatic software toolkit generation, without loss of optimizing efficiency. This has resulted in a paradigm shift towards a *language-based design methodology* for embedded SOC optimization and exploration. Consequently there is tremendous interest in using Architectural Description Languages (ADLs) to drive design space exploration and automatic compiler/simulator toolkit generation.

As with an HDL-based ASIC design flow, several benefits accrue from a language-based design methodology for embedded SOC exploration, including the ability to perform (formal) verification and consistency checking, to modify easily the target architecture and memory organization for design space exploration, and to drive automatically the backend toolkit generation from a single specification. In this paper we describe **EXPRESSION**, an ADL that effectively supports these dual goals of SOC exploration, as well as automatic generation of a high-quality software toolkit for embedded SOC. Section 2 describes our goals and approach. In Section 3, we describe related work on ADLs and compare them with EXPRESSION. Sections 4 and 5 present a brief overview of EXPRESSION using an example architecture. Section 6 describes EXPRESSION's support for detailed scheduling and resource constraint specification. Section 7 illustrates the ease of modification in EXPRESSION to support design space exploration, while Section 8 concludes this paper.

2   Goals and Approach

SOC designers spend a lot of time and effort exploring candidate processor architectures. The availability of a variety of processor core IP libraries (including DSP, VLIW, SS/RISC, and ASIP) presents the system designer with a large exploration space for the choice of a base processor architecture. Thus, toolkits which allow the designer to perform rapid exploration of various processor alternatives are necessary. These toolkits must provide the designer with quantitative performance measurements in order for him to perform intelligent trade-offs. Furthermore, the stringent performance, power, code density, and cost constraints mandated by modern embedded systems necessitate the development of a high-quality software toolkit, including, at a minimum, a cycle-accurate simulator, and

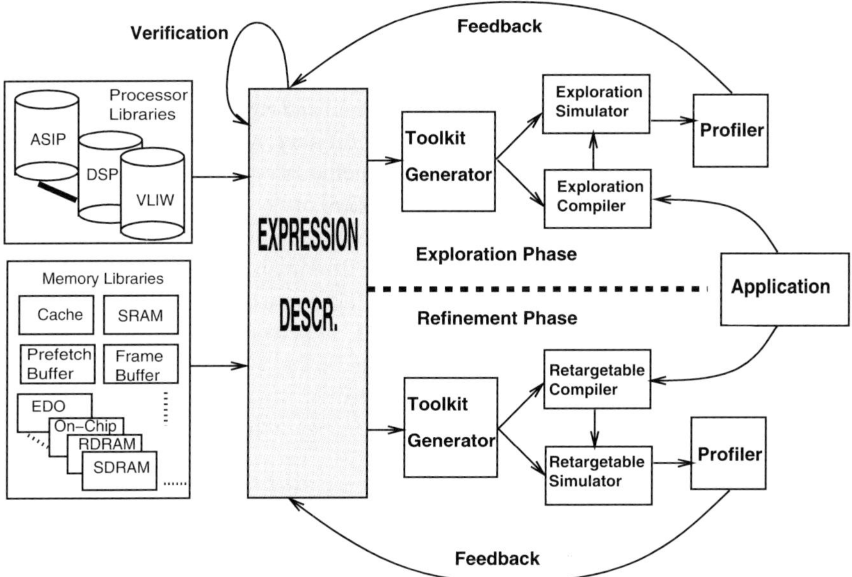

Fig. 1 EXPRESSION design flow

an optimizing Instruction-Level Parallelism (ILP) compiler that can exploit novel memory organizations.

The system designer also requires the ability to customize the base processor by changing parameters of the processor core (e.g. number of functional units, operation latencies). The memory-intensive nature of many embedded applications (e.g. multimedia and network) further exacerbates the traditionally critical memory bottleneck. This requires the ability to explore (and optimize for) novel on-chip and off-chip memory organizations and hierarchies to improve memory bandwidth (examples include the use of on-chip DRAM, frame-buffers, queues, novel cache hierarchies, etc.). An important aspect of such an exploration (not taken into account by most other approaches) is the ability to also customize the compiler concurrently with the processor such that a "best-fit" is obtained.

Figure 1 shows our language-based design methodology using EXPRESSION. An EXPRESSION description of an embedded SOC architecture can be used in two modes. In the *Exploration Phase*, the system designer explores and evaluates different base processor candidates (selected from the Processor Libraries), and different memory organizations and hierarchies (with components selected from the Memory Libraries). In the exploration phase, the toolkit generator is used to produce an *Exploration Simulator* and a *Exploration Compiler*. The goal here is to support rapid Design Space Exploration (DSE) with fast (possibly functional) simulation, and using the compiler in an estimation mode for comparative evaluation of candidate base processors and memory organizations. In the *Refinement Phase*, the EXPRESSION description is used to generate a cycle-accurate simulator and an optimizing ILP compiler that allows the system designer to tune the base processor characteristics, as well as to tune the memory subsystem hierarchy.

EXPRESSION was designed to provide a natural and easy-to-specify mechanism for capturing the information needed to support this ADL-based design space exploration and software toolkit generation methodology. As shown in Fig. 1, EXPRESSION facilitates the automatic generation of an optimizing compiler and simulator. The retargetable compiler exploits the parallelism and pipelining available, while the simulator provides accurate timing and utilization information. Furthermore, since the description of complex processors is cumber-some and error-prone, EXPRESSION provides the ability to perform consistency checking and verification of the input specification.

3   Related Work

Traditionally, ADLs have been classified into two categories depending on whether they primarily capture the Instruction-Set (IS) or the structure of the processor.

nML [8] and **ISDL** [2] are examples of IS ADLs. In nML, the processor's IS is described as an attributed grammar with the derivations reflecting the set of legal instructions. nML has been used by the retargetable code generation environment CHESS [1] to describe DSP and ASIP processors. However, nML does not directly support multi-cycle or multi-word instructions. ISDL also describes the processor in terms of its IS, with the goal of deriving a code-generator [12], assembler, and simula-tor. In ISDL, constraints on parallelism are explicitly specified through illegal operation groupings. This could be tedious for complex architectures like DSPs which permit operation parallelism (e.g. Motorola 56K) and VLIW machines with distributed regis-ter files (e.g. TI C6X). The retargetable compiler system by Yasuura et al. [3] produces code for RISC architectures starting from an instruction set processor description, and an application described in **Valen-C**. Valen-C is a C language extension supporting explicit and exact bit width for integer type declarations, targeting embedded software. The processor description represents the instruction set, but does not appear to capture resource conflicts, and timing information for pipelining.

MIMOLA [13] is an example of an ADL which primarily captures the structure of the processor wherein the net-list of the target processor is described in a HDL like language. One advantage of this approach is that the same description is used for both processor synthesis, and code generation. The target processor has a micro-code architecture. The net-list description is used to extract the instruction set [6], and produce the code generator. Extracting the instruction set from structure may be difficult for complicated instructions, and may lead to poor-quality code. MIMOLA descriptions are generally very low-level, and laborious to write. It is not clear how they generate a cycle-accurate simulator using the MIMOLA description.

Recently, languages which capture both the structure and the behavior of the proc-essor, as well as detailed pipeline information (typically specified using Reservation

Tables) have been proposed. **LISA** [7] is one such ADL whose main characteristic is the operation-level description of the pipeline. LISA seems to have been designed primarily for retargeting simulators. Also, it does not support specification of detailed constraint information needed for compiler instruction scheduling. **RADL** [16] is an extension of the LISA approach that focuses on explicit support of detailed pipeline behavior to enable generation of production quality cycle- and phase-accurate simulators. **FLEXWARE** [5] and **MDes** [18] have a mixed-level structural/behavioral representation. FLEXWARE contains the CODESYN code-generator and the Insulin simulator for ASIPs. The simulator uses a VHDL model of a generic parameterizable machine. The application is translated from the user-defined target instruction set to the instruction set of this generic machine. However, it is not clear how the resource conflicts between parallel/pipelined instructions, needed in scheduling, are captured. Furthermore, explicit specification of the memory subsystem does not appear to be possible.

The **MDes** [11] language used in the Trimaran [18] system is closest in its goals to our approach. It is a mixed-level ADL, intended for DSE. Information is broken down into sections (such as format, resource-usage, latency, operation, register, etc.), based on a high-level classification of the information being represented. MDes has good constructs for preprocessing which enable concise descriptions. However, MDes allows only a restricted retargetablility of the simulator to the HPL-PD processor family. MDes permits the description of the memory system, but is limited to the traditional hierarchy (register files, caches, etc.). In our approach we target more general memory organizations and hierarchies, including on-chip DRAM, frame buffers, partitioned memory address spaces, etc. MDes captures constraints between operations with explicit Reservation Tables [4], using a hierarchical description for compactness. Because this hierarchical specification allows instruction set and structure information to be specified at any level of the hierarchy, (e.g. usage, latency, operation), local changes to the architecture (for exploration) could propagate through the hierarchy, and require global changes to the specification making it cumbersome and error-prone.

In EXPRESSION we also follow a mixed-level approach (behavioral and structural) to facilitate DSE. Furthermore, we provide support for specification of novel memory subsystems. We avoid explicit representation of the reservation tables by extracting them from the structural description (using the algorithm outlined in Section 6) Our approach also eliminates redundancy by using the same net-list information to drive both the compiler and the simulator.

4   EXPRESSION Overview

EXPRESSION captures enough information to retarget a cycle-accurate structural simulator and an optimizing ILP compiler. The system is described both in terms of its instruction-set (Behavior Specification) and its structure as shown in Fig. 2. Section 5 illustrates these with the aid of an example.

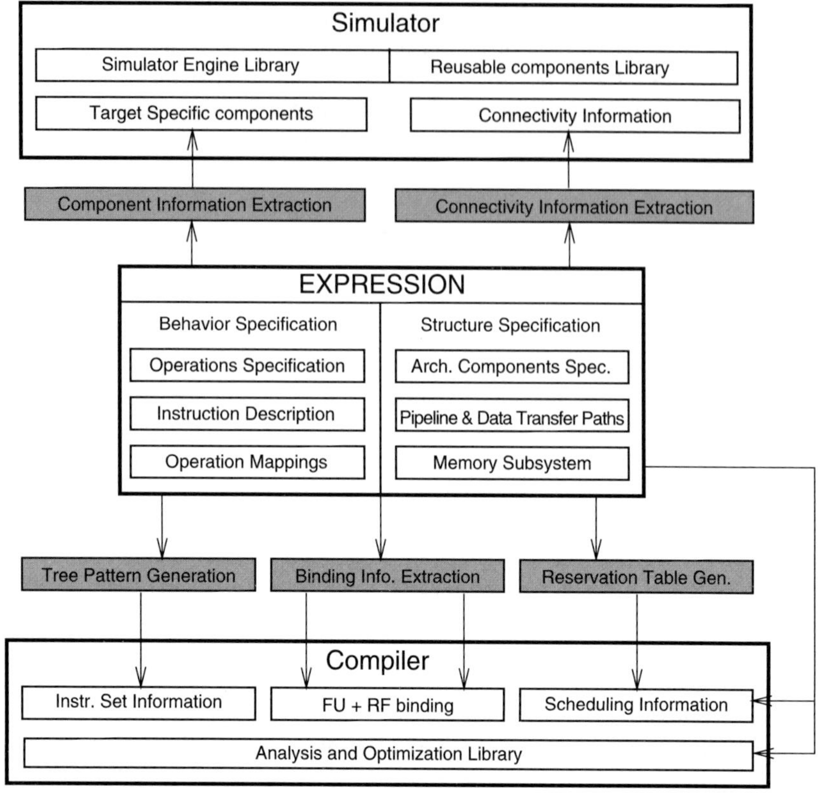

Fig. 2 Interaction between EXPRESSION and the retargetable compiler/simulator toolkit

Figure 2 shows the interaction between EXPRESSION specification and the retargetable compiler/simulator. The dark shaded boxes represent generators that read in the appropriate sections of the EXPRESSION specification and generate the information required by the compiler/simulator components. The structural simulator is retargeted by changing the datapath netlist and the execution semantics of the architectural components. The ILP compiler is retargeted by changing the machine dependent parameters of Instruction Selection, Resource Allocation, ILP Scheduling, and Memory Optimization techniques.

A key feature required by an ADL supporting aggressive and accurate compiler scheduling is the ability to capture detailed resource constraint information, typically in the form of reservation tables. Previous approaches (e.g. MDes) required the user to specify reservation tables on a per-operation basis. This makes specification of reservation tables cumbersome for architectures containing a lot of instruction types (e.g. instructions with varied accessing modes). Furthermore, for VLIW architectures with DSP-style features, the IS may not be well structured; thus, one cannot use hierarchical composition rules to simplify

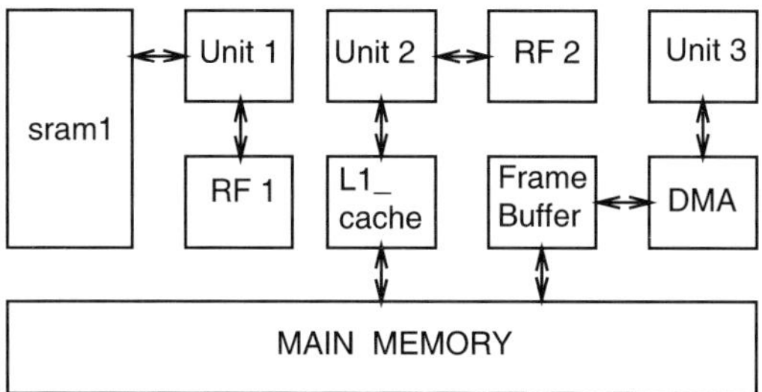

Fig. 3 An example Memory Subsystem

the specification of reservation tables. In EXPRESSION we solve this problem by permiting concise specification of resource constraints at an abstract level, from which detailed reservation tables can be automatically generated on a per-operation basis.

EXPRESSION allows the user to specify the RT–level datapath netlist[1] (i.e. netlist without the control signals) of the processor at an abstract level. First, each RT-level architectural component is specified. Then, pipeline paths and all valid data-transfer paths are specified. Using this information, both the netlist for simulator and reservation tables for compiler are generated. Our resource constraint specification scheme not only reduces the complexity of specification, but also allows for consistency and completeness checking on the specification.

Another key feature of EXPRESSION is the ability to specify novel memory subsystems. SOC technology permits customization of processor cores with different memory architectures and thus requires the exploration of novel memory organizations and hierarchies. Figure 3 shows an example memory system with the memory address space divided between a SRAM and a main memory (DRAM). The main memory is accessed through the data cache and the frame buffer. EXPRESSION's Storage Subsytem specification is used to retarget compiler optimizations that exploit memory organization.

5   EXPRESSION Organization

EXPRESSION employs a simple LISP-like syntax to ease specification and enhance readability. A detailed description of EXPRESSION can be found in [10]. We illustrate salient features of EXPRESSION using the example in Fig. 4.

---

[1] subsequently, netlist will refer to RT–level datapath netlist.

5.1   Example Architecture

Figure 4 shows a simplified version of the Motorola DSP 56000 processor to which an additional multiplier unit has been added. The DSP 56000 is a bus-based architecture and can execute one ALU operation and upto two additional parallel moves in one cycle. Two Address Generation Units AGU1 and AGU2 (not shown in the figure) generate the addresses required for the parallel moves.[2]

5.2   Sections in EXPRESSION

As shown in Fig. 2, an EXPRESSION description is composed of two main sections: Behavior (or IS), and Structure, which are further subdivided into three subsections each: Operations, Instruction, Operation mappings, and Components, Pipeline/Data–transfer paths, Memory subsystem respectively. Each subsection is illustrated with examples below. (refer Fig. 4)

5.2.1   Operations Specification

This subsection describes the IS of the processor. Each operation of the processor is described in terms of its opcode and operands. The types and possible destinations of each operand are also specified.

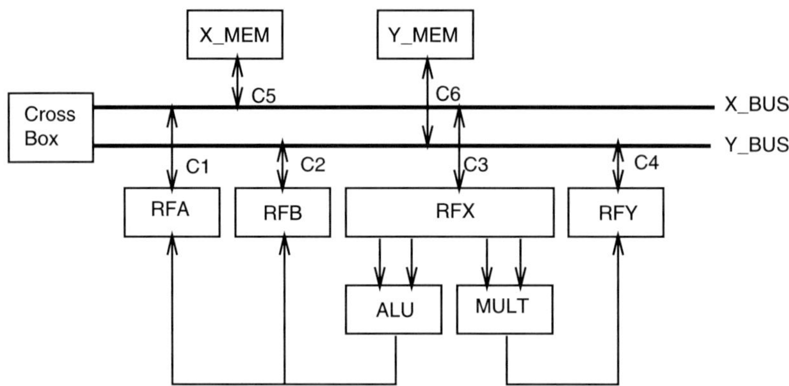

Fig. 4   An example architecture

---

[2]While the 56K does not contain explicit opcodes to model the parallel moves, for purposes of illustration, we denote moves by opcodes 'pmove1' and 'pmove2'.

//Specifying the parallel move operations
(OP_GROUP pmove1_ops
(OPCODE pmove1 (OPERANDS src1 dst1)))
//The var groups (src1, dst1) used above are defined here.
(VAR_GROUPS
(src1 (OR X_mem Y_mem)) (dst1 (OR RFA RFB RFX RFY)))

5.2.2   Instruction Description

This subsection captures the parallelism available in the architecture. An Instruction
is viewed as containing operations that can be executed in parallel. Each Instruction contains a list of slots (to be filled with operations), with each slot corresponding to a Functional Unit.

//The instr has 4 slots (alu, mult and 2 parallel moves)
(INSTR (SLOTS (UNIT ALU) (UNIT MULT)
(UNIT AGU1) (UNIT AGU2)))

5.2.3   Operation Mappings

In this subsection the user specifies information needed by Instruction Selection
and architecture-specific optimizations of the compiler.

//mapping a compiler operation to a target specific op.
(OP_MAPPING ((GENERIC (iadd src1 src2 dst))
(TARGET (add src1 src2 dst)))
//Multiply x by 2 can be replaced with ADD xx
((GENERIC(mult src1 #2 dst))(TARGET(add src1 src1 dst)))

5.2.4   Components Specification

This subsection describes each RT-level component in the architecture. The
components can be any of Pipeline units, Functional units, Storage elements, Ports,
Connections, and Busses. For multi-cyle or pipelined units, the timing behavior is
also specified.

(ExUnit ALU() (OPCODES alu_ops))//Instantiate ALU.
//alu_ops is defined in the Operations section.

5.2.5   Pipeline and Data-Transfer Paths Description

This subsection describes the netlist of the processor. The *pipeline description*
provides a mechanism to specify the units which comprise the pipeline stages,
while the *data-transfer paths description* provides a mechanism for specifying the

valid data-transfers. This information is used to both retarget the simulator, and to generate reservation tables needed by the scheduler, as shown in Section 6.

> *(PIPELINE FETCH DECODE EX)// 3 stage pipeline*
> *(EX (PARALLEL ALU Mult AGU1 AGU2)*
> *//Describe datapath transfers. Type: uni/bi-directional.*
> *//(Source, sink, components activated during transfer)*
> *(DTPATHS (TYPE B1)*
> *(RFA X_bus C1) (RFB Y_bus C2) (RFX X_bus C3)*
> *(RFY Y_bus C4) (X_mem X_bus C5) (Y_mem Y_bus C6)*
> *(X_bus Y_bus CrossBox))*

5.2.6   Memory Subsystem

This subsection describes the types and attributes of various storage components (like Register Files, SRAMs, DRAMs, Caches, etc). Note that the netlist also contains connectivity information for the memory subsystem.

> *(STORAGE_PARAMETERS*
> *(L1_cache (TYPE cache) (SIZE 1024) (LINE 32)*
> *(ASSOCIATIVITY 2) (ADDRESS_RANGE 0511)*
> *(ACCESS_TIMES 1)))*

6   Reservation Table Generation

As mentioned before, the compiler's scheduler needs reservation tables to test for resource conflicts between operations whose execution cycles overlap. Rather than requiring the user to laboriously specify the reservation tables on a per-operation basis, we use the structural information in EXPRESSION to automatically generate them. The key idea behind the reservation table generation approach in EXPRESSION is that every operation proceeds through a pipeline path and accesses storage units through some data-transfer paths. In effect, it is possible to trace the execution of the operation through the pipeline and data-transfer segments and thus generate accurate reservation tables. This frees the user from the burden of having to specify reservation tables for each operation and additionally, makes it possible to check for consistency and completeness in the specification.

In Fig. 5 we outline the reservation table generation scheme, given an operation and the structural description. We also illustrate the generation of the reservation table for a parallel move operation (opcode 'pmove1' and functional unit 'AGU1').

As shown in Fig. 1, EXPRESSION is designed to support both rapid DSE (exploration phase requiring fast turnaround time), as well as high-quality code generation with cycle-accurate simulation (refinement phase). Thus, in different phases of Fig. 1's design flow, the computation speed requirement shifts from total exploration time to compilation and simulation time. To account for these conflicting

Steps Needed for Reservation Table Generation	Example to illustrate Reservation Table Generation
Input: Operation (OP) ⋯⋯⋯⋯⋯⋯⋯⋯⋯⋯⋯	Input: *Pmove1(AGUI) X_mem[100] RFB*
Output: Reservation Table (RT) for operation OP.	Output: /*Reservation Table for the operaation*/
	/* Cyc stands for Cycle*/
1. Find pipeline path which contains OP's F.U.⋯⋯⋯	⋯⋯ Pipeline path - Fetch: Decode; AGU1;
2. Generate partial RT from the pipeline path. ⋯⋯⋯	⋯⋯ Partial RT - Cyc1: Fetch; Cyc2: Decode;
3. For each data-transfer initiated by OP	Cyc3: AGU1;
Deteermine the source and sink components.⋯⋯⋯⋯	⋯Source: X_mem, Sink: RFB
the data transer	⋯⋯⋯⋯
Find transfer segments which implement⋯⋯	⋯⋯⋯(C5 X_bus) (Cross Box) (Y_bus C2)
Add objects in the segment to RT.	RT - Cyc1 ... Cyc3:AGU1, X_mem, C5, X_bus,
	Cross Box, Y_bus C2 RFB
/* Output the completed reservation table*/	

Fig. 5 The steps in Reservation Table generation illustrated with an example (also refer Fig. 4)

goals, Reservation Tables can either be generated on-the-fly during compilation, or they can be generated beforehand and stored in a database. During the exploration phase computing reservation tables on-the-fly is beneficial. Once the architecture has been fixed, and a quality toolkit is needed, the reservation tables are computed apriori to reduce compilation time.

7 Support for Design Space Exploration

DSE allows the system designer to perform tradeoffs between various competing goals like cost, performance and power. The objective during DSE is to evaluate numerous processor and memory configurations w.r.t. the targeted applications. Thus, an ADL for DSE should capture both structural and behavioral aspects of the system. Hence, EXPRESSION follows a mixcd-lcvel approach (to enable changes to structure or IS or the memory subsystem). Furthermore, the ADL should be able to easily reflect the changes made to the system. For example, the designer may vary the processor parameters like number of functional units, register files, busses etc. Another example could be varying the IS by adding/deleting operations, changing the operand types, etc. Changes to the behavior are fairly easy in EXPRESSION as well as in other ADLs that are behavior-centric (e.g. ISDL) or which employ a mixed-level approach (e.g. MDes). On the other hand, structural changes in EXPRESSION are quite simple while structural changes in other ADLs (e.g. ISDL, MDes) become complicated, and in some cases (e.g. for novel memory organization and hierarchies) they are not feasible.

We evaluated EXPRESSION to see how easy it is to make local changes to the architecture and to add/delete operations and compared this with equivalent changes in MDes. We illustrate this on the example architecture shown in Fig. 4.

For this example, the initial MDes specification required approximately 100 lines while EXPRESSION required approximately 75 lines.[3]

Although the MDes specification appears to be fairly concise, this comes at a large cost: modifications to the architecture in MDes may necessitate a change in many sections. This increases the possibility of errors creeping into the specification. In EXPRESSION, changes to the architecture usually involve making changes only to certain sections. Furthermore, we can automatically generate detailed reservation table information from EXPRESSION's structural description, whereas MDes requires a per-operation change in the affected reservation tables. Additionally, the consistency and completeness checking mechanism alerts the designer to errors (if any) in the EXPRESSION specification.

We evaluate the effect of splitting the register file RFX in Fig. 4 into two register files RFX1 and RFX2. (This could be useful, for instance, to reduce the cost since it is expensive to build register files with many ports). The modified architecture is shown in Fig. 6.

This change in the architecture impacts the resource constraints specification of both MDes and EXPRESSION. Shown below are the additions[4] made to the specifications of both MDes and EXPRESSION in order to incorporate the changes to the example architecture (Fig. 6).

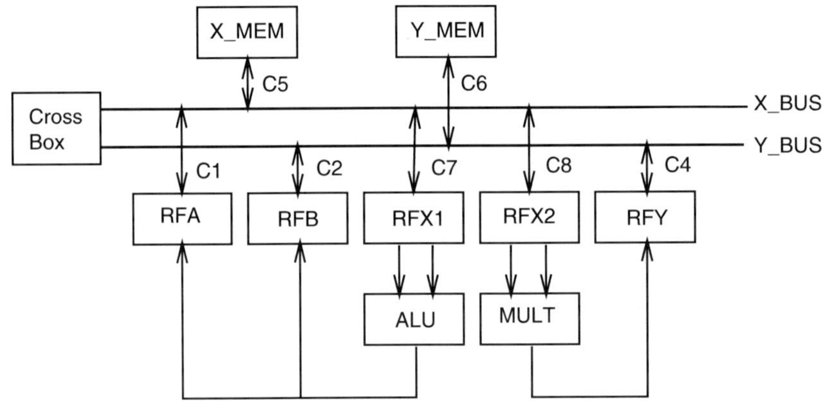

Fig. 6 The modified example with RFX split into RFX1 & RFX2, connection C3 deleted, C7 & C8 added

---

[3] Due to space limitations, the complete specfications for MDes and EXPRESSION were omitted. These can be viewed at http://www.ics.uci.edu/~ahalambi/EXPRESS/ADL/main.html

[4] The deletions made to remove the original register file were slightly more extensive in MDes than in EXPRESSION.

7.1 *Changes to MDes:*

```
SECTION Resource{RFX1(); RFX2(); C7(); C8();}
SECTION Resource_Usage{
$for (C7 C8 RFX1 RFX2){
RU_${CASE} (use(${CASE}) time(0));}}
SECTION Resource_Units{
RFX1_unit (use(RFX1 C7 X_bus));
RFX2_unit (use(RFX2 C8 X_bus));}
SECTION Table_Option{
any_X_reg_transfer (one_of(RFX1_unit RFX2_unit));}
SECTION Reservation_Table{
RT_PMOVE2_X2 (use(RU_AGU2 RFA_unit RU_RFX2));
RT_PMOVE2_X1_X2 (use(RU_AGU2 RFX1_unit RU_RFX2));}
SECTION Scheduling_Alternative{
$for (CASE in PMOVE2_X2 PMOVE2_X1_X2){
ALT_${CASE} (resv(RT_${CASE}));}}
SECTION Operation{
$for (CASE in PMOVE2_X2 PMOVE2_X1_X2){
OP_${CASE} (alt(ALT_${CASE}));}}
```

7.2 *Changes to EXPRESSION*

```
(RegFile RFX2()) (Connection C7() C8())
(DTPATHS (TYPE BI
(RFX1 X_bus C7) (RFX2 X_bus C8)))
```

For this local architectural change, MDes required modifications across 7 sections in the hierarchy (18 lines added), whereas in EXPRESSION the changes were more localized, requiring modifications to 2 sections (3 lines added). More complex changes to the architecture (such as adding busses, connection, units, and memories) would require drastic changes in MDes, but would be rather simple in EXPRESSION. Thus, as can be seen from the above example, EXPRESSION seems to be better suited as a language for rapid DSE.

8 Summary

In this paper we present EXPRESSION, a new ADL used for rapid DSE, and retargeting a high-quality compiler/simulator toolkit. To support fast iteration through the design cycles of complex Systems-on-Chip, fast retargeting of the compiler and simulator is necessary. We use the EXPRESSION language to retarget the optimizing compiler and cycle-accurate simulator, through a set of toolkit generation algorithms. The compiler exploits the parallelism/pipelining available in the architecture, while the simulator provides accurate profiling feedback to the designer, to allow good system-level trade-offs.

We also propose a new mechanism to concisely specify resource information required to generate reservation tables used in scheduling parallel/pipeline/multi-cycle operations from a structural description of the architecture. This eliminates redundancy, reducing the amount of description needed from the user, and makes the process less error-prone.

The memory-intensive nature of many multimedia applications requires exploration of different memory organizations in order to meet the cost/performance/power goals. EXPRESSION is able to capture these memory organizations for memory-aware compiler optimizations.

Table 1 summarizes the comparison between EXPRESSION and some other ADLs. The rows represent characteristics of the ADLs. In many respects the MDes language is similar to our approach. However, for cycle-accurate simulation, MDes provides a limited retargeting for the HPL-PD processor family, while we employ an approach similar to LISA, providing more control over the modeled architecture. We differ from LISA in the extensive support for optimizing compiler algorithms. Unlike MDes, we extract reservation tables from the netlist. Also, unlike other ADLs, EXPRESSION is able to capture novel memory subsystems. In EXPRESSION we employ a mixed-level specification combining both behavioral and structural information to efficiently support DSE and retargetability through automatic compiler/simulator toolkit generation.

EXPRESSION descriptions for a range of architectures including VLIW (TI C6X [17]) and DSP (Motorola 56K [14]) have been implemented. EXPRESSION is currently used to drive several projects at U.C. Irvine, including MEMOREX (a memory estimation and exploration environment) [9,15] and EXPRESS (an optimizing, retargetable compiler and exploration environment for Embedded Systems-on-Chip). Our ongoing work targets strengthening the coupling between the compiler and the architecture, by inserting more architecture-dependent optimizations and providing an architecture-based self-adapting compiler flow.

Table 1 Comparison between different ADLs

Attributes	nML	LISA	ISDL	MIMOLA	MDes	EXPRESSION
Compiler supp	✓	-	✓	✓	✓	✓
Cycle-acc sim	-	✓	?	?	-	✓
Pipeline supp	?	✓	-	✓	✓	✓
Mult-cyc supp	-	✓	✓	✓	✓	✓
Net-list info	-	-	-	✓	✓	✓
Instr-set info	✓	✓	✓	-	✓	✓
Reserv tables	-	✓	-	✓	✓	✓
Memory hier	-	-	-	-	?	✓

References

[1] G. Goosens et al. CHESS: retargetable code generation for embedded DSP processors. In *Code Generation for Embedded Processors*. Kluwer, 1997.

[2] G. Hadjiyiannis et al. ISDL: an instruction set description language for retargetability. In *Proc. DAC*, 1997.

[3] H. Yasuura et al. A programming language for processor based embedded systems. In *Proc. APCHDL*, 1998.

[4] J. Gyllenhaal et al. Optimization of machine descriptions for efficient use. In *Proc. 29th Annual International Symposium on Microarchitecture*, 1996.

[5] P. Paulin et al. Flex Ware: a flexible firmware development environment for embedded systems. In *Proc. Dagstuhl Code Generation Workshop*, 1994.

[6] R. Leupers et al. Retargetable generation of code selectors from HDL processor models. In *Proc. EDTC*, 1997.

[7] V. Zivojnovic et al. LISA – machine description language and generic machine model for HW/SW co-design. In *IEEE Workshop on VLSI Signal Processing*, 1996.

[8] M. Freericks. The nML machine description formalism. Technical Report TR SM-IMP/DIST/08, TU Berlin CS Dept., 1993.

[9] P. Grun, F. Balasa, and N. Dutt. Memory size estimation for multimedia applications. In *Proc. CODES/CACHE*, 1998.

[10] P. Grun, A. Halambi, A. Khare, V. Ganesh, N. Dutt, and A. Nicolau. EXPRESSION: An ADL for system level design exploration. Technical Report TR 98–29, University Of California, Irvine, 1998.

[11] J. Gyllenhaal. A machine description language for compilation. Master's thesis, Dept. of EE, UIUC, IL., 1994.

[12] S. Hanono and S. Devadas. Instruction selection, resource allocation, and scheduling in the AVIV retargetable code generator. In *Proc. DAC*, 1998.

[13] R. Leupers and P. Marwedel. Retargetable code generation based on structural processor descriptions. *Design Automation for Embedded Systems*, 3(1), 1998.

[14] MOTOROLA Inc. *DSP56000 24-Bit Digital Signal Processor Family Manual*, 1995.

[15] P. Panda, N. Dutt, and A. Nicolau. Architectural exploration and optimization of local memory in embedded systems. In *Proc. ISSS*, 1997.

[16] C. Siska. A processor description language supporting retargetable multi-pipeline dsp program development tools. In *Proc. ISSS*, December 1998.

[17] TEXAS INSTRUMENTS. *TMS320C62x/C67x CPU and Instruction Set Reference Guide*, 1998.

[18] Trimaran Release: http://www.trimaran.org. *The MDES User Manual*, 1998.

RTOS Modeling for System Level Design

Andreas Gerstlauer, Haobo Yu, and Daniel D. Gajski

Abstract System level synthesis is widely seen as the solution for closing the productivity gap in system design. High-level system models are used in system level design for early design exploration. While real-time operating systems (RTOS) are an increasingly important component in system design, specific RTOS implementations can not be used directly in high-level models. On the other hand, existing system level design languages (SLDL) lack support for RTOS modeling. In this paper we propose a RTOS model built on top of existing SLDLs which, by providing the key features typically available in any RTOS, allows the designer to model the dynamic behavior of multitasking systems at higher abstraction levels to be incorporated into existing design flows. Experimental result shows that our RTOS model is easy to use and efficient while being able to provide accurate results.

1   Introduction

In order to handle the ever increasing complexity and time-to-market pressures in the design of systems-on-chip (SOCs), raising the level of abstraction is generally seen as a solution to increase productivity. Various system level design languages (SLDL) [1, 2] and methodologies have been proposed in the past to address the issues involved in system level design. However, most SLDLs offer little or no support for modeling the dynamic real-time behavior often found in embedded software. In the implementation, this behavior is typically provided by a real-time operating system (RTOS) [3, 4]. At an early design phase, however, using a detailed, real RTOS implementation would negate the purpose of an abstract system model. Furthermore, at higher levels, not enough information might be available to target a specific RTOS. Therefore, we need techniques to capture the abstracted RTOS behavior in system level models.

Center for Embedded Computer Systems, University of California, Irvine,
Irvine, CA 92697, USA E-mail: {gerstl, haoboy, gajski}@cecs.uci.edu

R. Lauwereins and J. Madsen (eds.), *Design, Automation, and Test in Europe–The Most Influential Papers of 10 Years DATE*, 47–58
© Springer 2008

In this paper, we address this design challenge by introducing a high level RTOS model for system design. It is written on top of existing SLDLs and does not require any specific language extensions. It supports all the key concepts found in modern RTOS like task management, real-time scheduling, preemption, task synchronization, and interrupt handling [5]. On the other hand, it requires only a minimal modeling effort in terms of refinement and simulation overhead. Our model can be integrated into existing system level design flows to accurately evaluate a potential system design (e.g. in respect to timing constraints) for early and rapid design space exploration.

The rest of this paper is organized as follows: Section 2 gives an insight into the related work on software modeling and synthesis in system level design. Section 3 describes how the RTOS model is integrated with the system level design flow. Details of the RTOS model, including its interface and usage as well as the implementation are covered in Section 4. Experimental results are shown in Section 5 and Section 6 concludes this paper with a brief summary and an outlook on future work.

2   Related Work

A lot of work recently has been focusing on automatic RTOS and code generation for embedded software. In [8], a method for automatic generation of application-specific operating systems and corresponding application software for a target processor is given. In [6], a way of combining static task scheduling and dynamic scheduling in software synthesis is proposed. While both approaches mainly focus on software synthesis issues, their papers do not provide any information regarding high level modeling of the operating systems integrated into the whole system.

In [10], a technique for modeling fixed-priority preemptive multitasking systems based on concurrency and exception handling mechanisms provided by SpecC is shown. However, their model is limited in its support for different scheduling algorithms and inter-task communication, and its complex structure makes it very hard to use.

Our method is similar to Desmet et al. where they present a high-level model of a RTOS called SoCOS. The main difference is that our RTOS model is written on top of existing SLDLs whereas SoCOS requires its own proprietary simulation engine. By taking advantage of the SLDL's existing modeling capabilities, our model is simple to implement yet powerful and flexible, and it can be directly integrated into any system model and design flow supported by the chosen SLDL.

3   Design Flow

System level design is a process with multiple stages where the system specification is gradually refined from an abstract idea down to an actual heterogeneous multi-processor implementation. This refinement is achieved in a stepwise manner through several levels of abstraction. With each step, more implementation detail is introduced through a refined system model. The purpose of high-level, abstract models is the early validation of system properties before their detailed implementation, enabling rapid exploration.

Figure 1 shows a typical system level design flow. The system design process starts with the specification model. It is written by the designer to specifiy and validate the desired system behavior in a purely functional, abstract manner, i.e. free of any unnecessary implementation details. During system design, the specification functionality is then partitioned onto multiple processing elements (PEs), some or all of the concurrent processes mapped to a PE are statically scheduled, and a

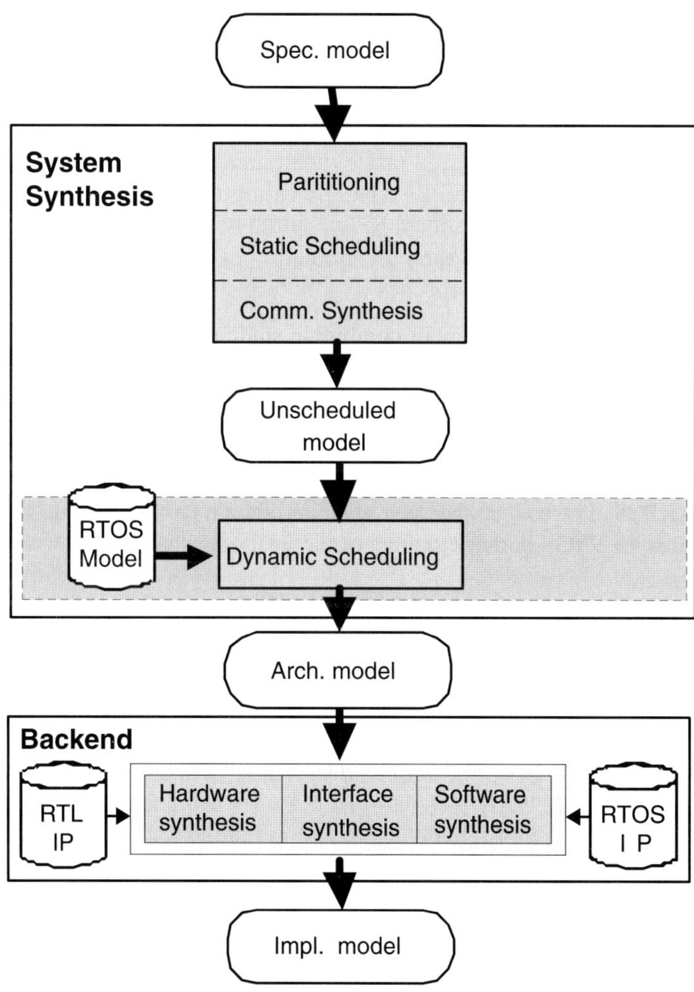

Fig. 1 Design flow

communication architecture consisting of busses and bus interfaces is synthesized to implement communication between PEs. Note that during communication synthesis, interrupt handlers will be generated inside the PEs as part of the bus drivers.

Due to the inherently sequential nature of PEs, processes mapped to the same PE need to be serialized. Depending on the nature of the PE and the data interdependencies, processes are scheduled statically or dynamically. In case of dynamic scheduling, in order to validate the system model at this point a representation of the dynamic scheduling implementation, which is usually handled by a RTOS in the real system, is required. Therefore, a high-level model of the underlying RTOS is needed for inclusion into the system model during system synthesis. The RTOS model provides an abstraction of the key features that define a dynamic scheduling behavior independent of any specific RTOS implementation.

The dynamic scheduling step in Fig. 1 refines the unscheduled system model into the final architecture model. In general, for each PE in the system a RTOS model corresponding to the selected scheduling strategy is imported from the library and instantiated in the PE. Processes inside the PEs are converted into tasks with assigned priorities. Synchronization as part of communication between processes is refined into OS-based task synchronization. The resulting architecture model consists of multiple PEs communicating via a set of busses. Each PE runs multiple tasks on top of its local RTOS model instance. Therefore, the architecture model can be validated through simulation or verification to evaluate different dynamic scheduling approaches (e.g. in terms of timing) as part of system design space exploration.

In the backend, each PE in the architecture model is then implemented separately. Custom hardware PEs are synthesized into a RTL description. Bus interface implementations are synthesized in hardware and software. Finally, software synthesis generates code from the PE description of the processor in the architecture model. In the process, services of the RTOS model are mapped onto the API of a specific standard or custom RTOS. The code is then compiled into the processor's instruction set and linked against the RTOS libraries to produce the final executable.

4   The RTOS Model

As mentioned previously, the RTOS model is implemented on top of an existing SLDL kernel. Figure 2 shows the modeling layers at different steps of the design flow. In the specification model (Fig. 2a), the application is a serial-parallel composition of SLDL processes. Processes communicate and synchronize through variables and channels. Channels are implemented using primitives provided by the SLDL core and are usually part of the communication library provided with the SLDL.

In the architecture model (Fig. 2b), the RTOS model is inserted as a layer between the application and the SLDL core. The SLDL primitives for timing and synchronization used by the application are replaced with corresponding calls to the RTOS layer. In addition, calls of RTOS task-management services are inserted.

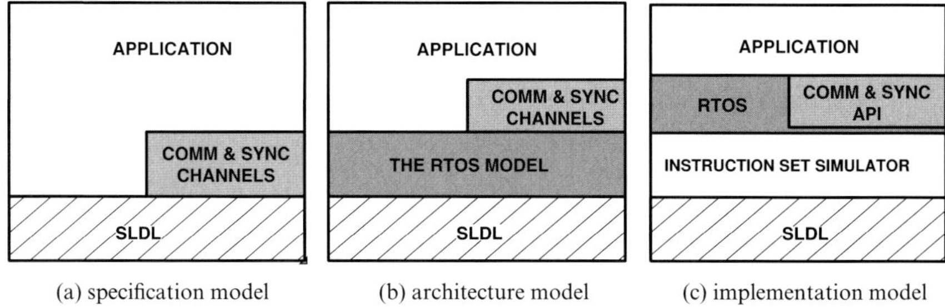

(a) specification model   (b) architecture model   (c) implementation model

Fig. 2 Modeling layers

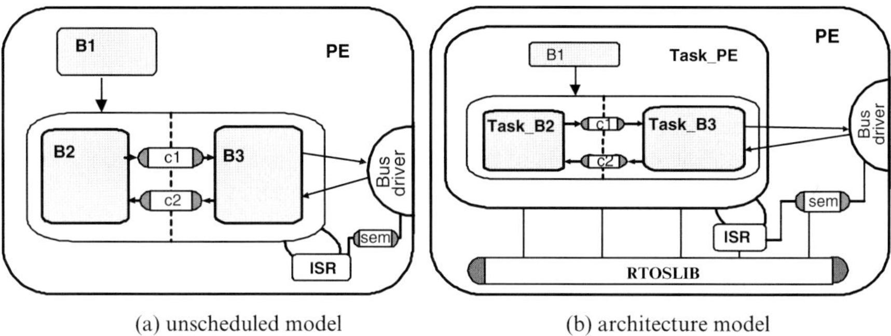

(a) unscheduled model   (b) architecture model

Fig. 3 Model refinement example

The RTOS model implements the original semantics of SLDL primitives plus additional details of the RTOS behavior on top of the SLDL core. Existing SLDL channels (e.g. semaphores) from the specification are reused by refining their internal synchronization primitives to map to corresponding RTOS calls. Using existing SLDL capabilities for modeling of extended RTOS services, the RTOS library can be kept small and efficient. Later, as part of software synthesis in the backend, channels are implemented by mapping them to an equivalent service of the actual RTOS or by generating channel code on top of RTOS primitives if the service is not provided natively.

Finally, in the implementation model (Fig. 2c), the compiled application linked against the real RTOS libraries is running in an instruction set simulator (ISS) as part of the system co-simulation in the SLDL.

We implemented the RTOS model on top of the SpecC SLDL [1]. In the following sections we will discuss the interface between application and the RTOS model, the refinement of specification into architecture using the RTOS interface, and the implementation of the RTOS model. Due to space restrictions, implementation details are limited. For more information, please refer to [11].

4.1 RTOS Interface

Figure 4 shows the interface of the RTOS model. The RTOS model provides four categories of services: operating system management, task management, event handling, and time modeling.

Operating system management mainly deals with initialization of the RTOS during system start where *init* initializes the relevant kernel data structures while *start* starts the multi-task scheduling. In addition, *interrupt_return* is provided to notify the RTOS kernel at the end of an interrupt service routine.

Task management is the most important function in the RTOS model. It includes various standard routines such as task creation (*task_create*), task termination (*task_terminate, task_kill*), and task suspension and activation (*task_sleep, task_activate*). Two special routines are introduced to model dynamic task forking and joining: *par_start* suspends the calling task and waits for the child tasks to finish after which *par_end* resumes the calling task's execution. Our RTOS model supports both periodic hard real time tasks with a critical deadline and non-periodic real time tasks with a fixed priority. In modeling of periodic tasks, *task_endcycle* notifies the kernel that a periodic task has finished its execution in the current cycle.

Event handling reimplements the semantics of SLDL synchronization events in the RTOS model. SpecC events are replaced with RTOS events (allocated and deleted

```
1   interface RTOS {
2       /* OS management */
3       void init(void);
4       void start (int sched_alg);
5       void interrupt_return(void);
6       /* Task management */
7       proc task_create(char *name, int type,
8              sim_time period, sim_time wcet);
9       void task_terminate (void);
10      void task_sleep (void);
11      void task_activate (proc tid);
12      void task_endcycle (void);
13      void task_kill (proc tid);
14      proc par_start (void);
15      void par_end(proc p);
16      /* Event handling */
17      evt event_new(void);
18      void event_del(evt e);
19      void event_wait (evt e);
20      void event_notify (evt e);
21      /* Time modeling */
22      void time_wait (sem_time nsec);
23   };
```

Fig. 4 Interface of the RTOS model

through *event_new* and *event_del*). Two system calls *event_notify* and *event_wait* are used to replace the SpecC primitives for event notify and event wait.

During simulation of high-level system models, the logical time advances in discrete steps. SLDL primitives (such as waitfor in SpecC) are used to model delays. For the RTOS model, those delay primitives are replaced by *time_wait* calls which model task delays in the RTOS while enabling support for modeling of task preemption.

4.2   Model Refinement

In this section, we will illustrate application model refinement based on the RTOS interface presented in the previous section through a simple yet typical example of a single *PE* (Fig. 3). In general, the same refinement steps are applied to all the PEs in a multiprocessor system. The unscheduled model (Fig. 3a) executes behavior *B1* followed by the parallel composition of behaviors *B2* and *B3*. Behaviors *B2* and *B3* communicate via two channels *c1* and *c2* while *B3* communicates with other PEs through a bus driver. As part of the bus interface implementation, the interrupt handler *ISR* for external events signals the main bus driver through a semaphore channel *sem*.

The output of the dynamic scheduling refinement process is shown in Fig. 3b. The RTOS model implementing the RTOS interface is instantiated inside the PE in the form of a SpecC channel. Behaviors, interrupt handlers and communication channels use RTOS services by calling the RTOS channel's methods. Behaviors are refined into three tasks. *Task_PE* is the main task which executes as soon as the system starts. When *Task_PE* finishes executing *B1*, it spawns two concurrent child tasks, *Task_B2* and *Task_B3*, and waits for their completion.

4.2.1   Task Refinement

Task refinement converts parallel processes/behaviors in the specification into RTOS-based tasks in a two-step process. In the first step (Fig. 5), behaviors are converted into tasks, e.g. behavior *B2* (Fig. 5a) is converted into *Task_B2* (Fig. 5b). A method *init* is added for construction of the task. All waitfor statements are replaced with RTOS *time_wait* calls to model task execution delays. Finally, the main body of the task is enclosed in a pair of *task_activate / task_terminate* calls so that the RTOS kernel can control the task activation and termination.

The second step (Fig. 6) involves dynamic creation of child tasks in a parent task. Every par statement in the code (Fig. 6a) is refined to dynamically fork and join child tasks as part of the parent's execution (Fig. 6b). The *init* methods of the children are called to create the child tasks. Then, *par_start* suspends the calling parent task in the RTOS layer before the children are actually executed in the par statement. After the two child tasks finish execution and the par exits, *par_end* resumes the execution of the parent task in the RTOS layer.

```
1   behavior B2 ( ) {
2       void main (void)  {
3       . . .
4       waitfor (500);
5       . . .
6       }
7   };
```

(a) specificataion model

```
1    behavior task_B2 (RTOS os) implements Init {
2       proc me;
3       void init (void) {
4          me = os.task_create("B2", APERIODIC, 0, 500);
5       }
6       void main(void)  {
7          os.task_activate(me);
8          . . .
9          os.time_wait(500);
10         . . .
11         os.task_terminate( );
12      }
13   };
```

(b) architecture model

Fig. 5 Task modeling

```
1                                 1    . . .
2    . . .                        2    task_b2.init( );
3                                 3    task_b3.init( );
4    par                          4    os.par_start( );
5    {                            5    par {
6        b2.main( );              6        task_b2.main( );
7        b3.main( );              7        task_b3.main( );
8    }                            8    }
9                                 9    os.par_end( );
10   . . .                        10   . . .
```

(a) before   (b) after

Fig. 6 Task creation

4.2.2 Synchronization Refinement

In the specification model, all synchronization in the application or inside communication channels is implemented using SLDL events. Synchronization refinement replaces all events and event-related primitives with corresponding event-handling routines of the RTOS model (Fig. 7). All event instances are replaced with instances of RTOS events *evt* and wait/notify statements are replaced with RTOS *event_wait / event_notify* calls.

<table>
<tr><td>

```
1   channel c_queue( ) {
2     event eRdy, eAck
3     void send(. . .)
4     { . . .
5       notify eRdy;
6       wait(eAck);
7       . . . }
8   };
```

</td><td>

```
1     channel c_queue (RTOS os) {
2       evt eRdy, eAck;
3       void send (. . .)
4       { . . .
5           os.event_notify(eRdy);
6           os.event_wait(eAck);
7           . . . }
8     };
```

</td></tr>
<tr><td align="center">(a) before</td><td align="center">(b) after</td></tr>
</table>

Fig. 7 Synchronization refinement

After model refinement, both task management and synchronization are implemented using the system calls of the RTOS model. Thus, the dynamic system behavior is completely controlled by the the RTOS model layer.

4.3 Implementation

The RTOS model library is implemented in 2000 lines of SpecC channel code [11]. Task management in the RTOS model is implemented in a customary manner [5] where tasks transition between different states and a task queue is associated with each state. Task creation (*task_create*) allocates the RTOS task data structure and *task_activate* inserts the task into the ready queue. The *par_start* method suspends the task and calls the scheduler to dispatch another task while *par_end* resumes the calling task's execution by moving the task back into the ready queue.

Event management is implemented by associating additional queues with each event. Event creation (*event_new*) and deletion (*event del*) allocate and deallocate the corresponding data structures in the RTOS layer. Blocking on an event (*event_wait*) suspends the task and inserts it into the event queue whereas *event_notify* moves all tasks in the event queue back into the ready queue.

In order to model the time-sharing nature of dynamic task scheduling in the RTOS, the execution of tasks needs to serialized according to the chosen scheduling algorithm. The RTOS model ensures that at any given time only one task is running on the underlying SLDL simulation kernel. This is achieved by blocking all but the current task on SLDL events. Whenever task states change inside a RTOS call, the scheduler is invoked and, based on the scheduling algorithm and task priorities, a task from the ready queue is selected and dispatched by releasing its SLDL event. Note that replacing SLDL synchronization primitives with RTOS calls is necessary to keep the internal task state of the RTOS model updated.

In high level system models, simulation time advances in discrete steps based on the granularity of waitfor statements used to model delays (e.g. at behavior or basic block level). The time-sharing implementation in the RTOS model makes sure that delays of concurrent task are accumulative as required by any model of serialized task execution. However, additionally replacing waitfor statements with corresponding RTOS time modeling calls is necessary to accurately model preemption. The *time_wait* method is a wrapper around the waitfor statement that allows the RTOS kernel to reschedule and switch tasks whenever time increases, i.e. in between regular RTOS system calls.

Normally, this would not be an issue since task state changes cannot happen outside of RTOS system calls. However, external interrupts can asynchronously trigger task changes in between system calls of the current task in which case proper modeling of preemption is important for the accuracy of the model (e.g. response time results). For example, an interrupt handler can release a semaphore on which a high-priority task for processing of the external event is blocked. Note that, given the nature of high-level models, the accuracy of preemption results is limited by the granularity of task delay models.

Figure 8 illustrates the behavior of the RTOS model based on simulation results obtained for the example from Fig. 3. Figure 8a shows the simulation trace of the unscheduled model. Behaviors *B2* and *B3* are executing truly in parallel, i.e. their simulated delays overlap. After executing for time d_1, *B3* waits until it receives a message from *B2* through the channel *c1*. Then it continues executing for time d_2

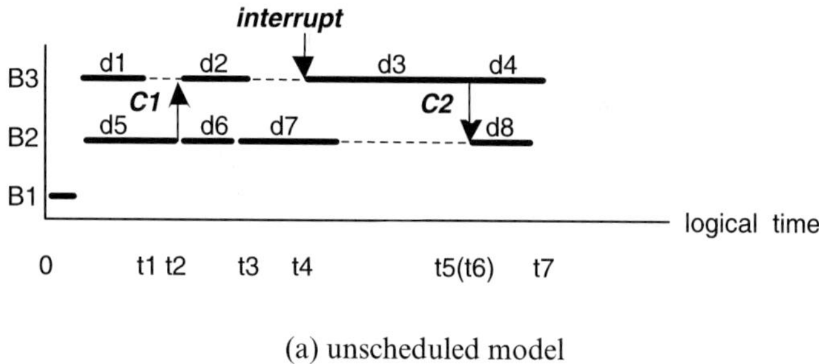

(a) unscheduled model

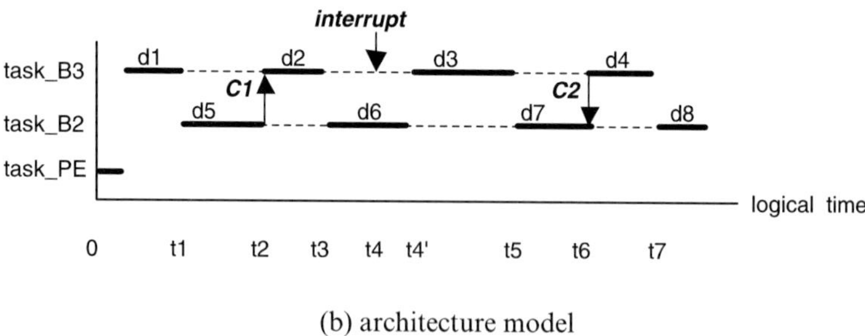

(b) architecture model

Fig. 8 Simulation trace for model example

and waits for data from another PE. *B2* continues for time $(d_6 + d_7)$ and then waits for data from *B3*. At time t_4, an interrupt happens and *B3* receives its data through the bus driver. *B3* executes until it finishes. At time t_5, *B3* sends a message to *B2* through the channel *c2* which wakes up *B2* and both behaviors continue until they finish execution.

Figure 8b shows the simulation result of the architecture model for a priority-based scheduling. It demonstrates that in the refined model *task_B2* and *task_B3* execute in an interleaved way. Since *task_B3* has the higher priority, it executes unless it is blocked on receiving or sending a message from/to *task_B2* (t_1 through t_2 and t_5 through t_6), waiting for an interrupt(t_3 through t_4), or it finishes (t_7) at which points execution switches to *task_B2*. Note that at time t_4, the interrupt wakes up *task_B3* and *task_B2* is preempted by *task_B3*. However, the actual task switch is delayed until the end of the discrete time step d_6 in *task_B2* based on the granularity of the task's delay model. In summary, as required by priority-based dynamic scheduling, at any time only one task, the ready task with the highest priority, is executing.

5   Experimental Results

We applied the RTOS model to the design of a voice codec for mobile phone applications [9]. Table 1 shows the results for this vocoder consisting of two tasks for encoding and decoding running in software. For the implementation model, the model was compiled into assembly code for the Motorola DSP56600 processor and the RTOS model was replaced by a small custom RTOS kernel, described in more detail in [9]. The transcoding delay is the latency when running encoder and decoder in back-to-back mode and it is related to response time in switching between encoding and decoding tasks.

The results show that refinement based on the RTOS model requires only a minimal effort. Refinement into the architecture model was done by converting relevant SpecC statements into RTOS interface calls following the steps described in Section 4.2. For this example, manual refinement took less than 1 h and required changing or adding 104 lines or less than 1% of code. Moreover, we have developed a tool that performs the refinement of unscheduled specification models into RTOS-based architecture models automatically.

Table 1 Vocoder experimental results

	Unscheduled	Archetecturel	Implementation
Lines of code	13,475	15,552	79,096
Execution time	24.0 s	24.4 s	5 h
Context switches	0	326	326
Transcoding delay	9.7 ms	12.5 ms	11.7 ms

The simulation overhead introduced by the RTOS model is negligible while providing accurate results. Compared to the huge complexity required for the implementation model, the RTOS model enables early and efficient evaluation of dynamic scheduling implementations.

6 Summary and Conclusions

In this paper, we proposed a RTOS model for system level design. To our knowledge, this is the first attempt to model RTOS features at such high abstraction levels integrated into existing languages and methodologies. The model allows the designer to quickly validate the dynamic real time behavior of multitask systems in the early stage of system design by providing accurate results with minimal overhead. Using a minimal number of system calls, the model provides all key features found in any standard RTOS but not available in current SLDLs. Based on this RTOS model, refinement of system models to introduce dynamic scheduling is easy and can be done automatically. Currently, the RTOS model is written in SpecC because of its simplicity. However, the concepts can be applied to any SLDL (SystemC, Superlog) with support for event handling and modeling of time.

Future work includes implementing the RTOS interface for a range of custom and commercial RTOS targets, including the development of tools for software synthesis from the architecture model down to target-specific application code linked against the target RTOS libraries.

References

[1] QNX[online]. Available: http://www.qnx.com/.
[2] SpecC[online]. Available: http://www.specc.org/.
[3] SystemC[online]. Available: http://www.systemc.org/.
[4] VxWorks[online]. Available: http://www.vxworks.com/.
[5] G. C. Buttazzo. *Hard Real-Time Computing Systems*. Kluwer Academic, 1999.
[6] J. Cortadella. Task generation and compile time scheduling for mixed data-control embedded software. In *DAC*, Jun 2000.
[7] D. Desmet et al. Operating system based software generation for system-on-chip. In *DAC*, Jun 2000.
[8] L. Gauthier et al. Automatic generation and targeting of application-specific operating systems and embedded systems software. *IEEE Trans. on CAD*, Nov 2001.
[9] A. Gerstlauer et al. Design of a GSM Vocoder using SpecC Methodology. Technical Report ICS-TR-99-11, UCI, Feb 1999.
[10] H. Tomiyama et al. Modeling fixed-priority preemptive multi-task systems in SpecC. In *SASIMI*, Oct 2001.
[11] H. Yu et al. RTOS Modeling in System Level Synthesis. Technical Report CECS-TR-02-25, UCI, Aug 2002.

Context-Aware Performance Analysis for Efficient Embedded System Design

Marek Jersak, Rafik Henia, and Rolf Ernst

Abstract Performance analysis has many advantages in theory compared to simulation for the validation of complex embedded systems, but is rarely used in practice. To make analysis more attractive, it is critical to calculate tight analysis bounds. This paper shows that advanced performance analysis techniques taking correlations between successive computation or communication requests as well a correlated load distribution into account can yield much tighter analysis bounds. Cases where such correlations have a large impact on system timing are especially difficult to simulate and, hence, are an ideal target for formal performance analysis.

1 Introduction

Performance validation is key during architecture design and function implementation of state-of-the-art embedded systems. It has to consider the host of nonfunctional dependencies introduced through processor and bus scheduling. Most existing performance validation approaches rely on simulation and hence suffer from high running times, incomplete coverage and failure to identify corner cases. The problems are aggravated for applications with many scenarios of operation, since combinatorially more cases have to be simulated.

A promising alternative to simulation is formal performance analysis. It can calculate conservative lower and upper bounds for performance values over a range of scenarios and thus guarantees corner-case coverage. Additionally, performance analysis often runs considerably faster than simulation, making it well suited for exploration in early design stages.

If performance analysis seems so attractive in theory compared to simulation, why is it rarely used in practice? A major reason is the fact that performance analysis can be very pessimistic, because it ignores certain correlations between consecutive task activations. Such pessimism sheds a negative light on performance analysis in

Technische Universität Braunschweig Institut für Datentechnik und Kommunikationsnetze (IDA)
D-38106 Braunschweig, Germany {jersak, henia, ernst} @ida.ing.tu-bs.de

R. Lauwereins and J. Madsen (eds.), *Design, Automation, and Test in Europe–The Most Influential Papers of 10 Years DATE*, 59–72
© Springer 2008

the embedded systems community, which does not accept overdimensioned solutions. Therefore, to make performance analysis more attractive, it is critical to calculate tight analysis bounds which are very close to the true performance corner-cases.

A typical performance analysis technique assumes that every activation of a task leads to the worst-case execution time of the task. That is, it ignores that there may be paths through the control-flow graph of the task leading to shorter execution times. Performance analysis also typically assumes a very pessimistic worst-case load distribution over time. That is, it ignores that certain task activations often cannot happen at the same time, which evens out the load distribution to a certain extent. We call such correlations *system contexts*.

This paper shows that advanced performance analysis techniques taking system contexts into account can yield much tighter analysis bounds, making analysis an attractive validation option during embedded system design. The designer then automatically profits from the fundamental advantages that analysis has over performance simulation. Performance analysis can also reveal system-level effects due to small local changes, e.g. a slight change in the worst-case execution time of a task.

After a brief overview of the state-of-the-art in performance analysis, two different types of contexts and the analysis improvements that can be obtained are discussed in Sections 4 and 5. In Section 6 it is shown that additional improvement is possible if contexts can be combined. The design of a hypothetical set-top box (Section 3) is used as an example. The paper concludes with a summary and outlook.

2 State-of-the-Art in Performance Analysis

The goal of performance analysis is to verify that an embedded system implementation meets all response-time and throughput constraints, e.g. end-to-end deadlines. Additionally, it can be used to obtain reliable values for worst-case processor and bus loads, required memories etc.

The performance of tasks sharing a single component (e.g. a CPU or a bus) can be analyzed using so-called scheduling analysis techniques. Consider the following equation which gives the worst-case response time r_i of a lower priority task i due to a worst-case number of interrupts by all higher-priority task $j \in hp(i)$.

$$r_i = C_i + \sum_{\forall j \in hp(i)} n_j(\tau_i) \times C_j \tag{1}$$

$n_j(r_i)$ is the maximum number of activating events for task j arriving during r_i. C_i, C_j are the worst-case core execution times (WCET), i.e. assuming no interrupts, of tasks i and j. The equation holds only if r_i is smaller than the minimum distance between two activations of task i. For more general cases see e.g. Tindell [9].

Since r_i appears on both sides of the equation, it is usually calculated iteratively, as shown in the following algorithm.

```
1      WorstCaseResponseTime(LowPriorTask){
2        NewResponseTime = LowPriorTask.WCET;
3        Do{
4          OldResponseTime = NewResponseTime;
5          for(int i = 0; i < HighPriorTasksList.size; i++){
6            HighPriorTask = HighPriorTasksList.get(i);
7            MaxActivations = max number of activations
             of HighPriorTask during OldResponseTime;
8            InterruptTime =
             MaxActivations * HighPriorTask.WCET;
9            NewResponseTime =
             NewResponseTime + InterruptTime;
10           }
11       }while (NewResponseTime > OldResponseTime)
12       return NewResponseTime;
13     }
```

Scheduling analysis algorithms are not directly applicable to a heterogeneous multicomponent architecture with different scheduling and resource-sharing strategies. Eles et al. [5] have extended existing techniques to allow performance analysis of special heterogeneous architectures, e.g. fixed-priority-scheduled CPUs connected via a TDMA-scheduled bus. Richter [6, 7] takes a more general approach that allows to couple different existing performance analysis techniques for an arbitrarily complex architecture. Chakraborty et al. [2] take a somewhat similar approach using a real-time calculus. Jersak [3] has extended [6, 7] to allow performance analysis for applications with complex task dependencies, including multiple activating inputs, and multi-rate data dependencies with intervals.

3   Set-Top Box Design Example

The SoC implementation of a hypothetical set-top box shown in Fig. 1 is used as an example throughout this paper. The set-top box can simultaneously process two MPEG-2 video streams, which can either come from the RF-module or from the hard disk, and can either be shown on a TV screen or saved to the hard disk. A decryption unit allows to decrypt encrypted video streams. We assume that MPEG frames arrive periodically from the RF-module. When two MPEG streams are received simultaneously, the exact order of interleaved frame types is unknown. The set-top box can additionally process IP traffic and download web content either to the screen or to the hard disk.

We will focus on load analysis and response-time analysis for the system bus. We will assume *priority-based bus scheduling* of communication tasks sharing the system bus. We are going to demonstrate the improvement in analysis tightness that can be obtained when different types of system contexts are considered.

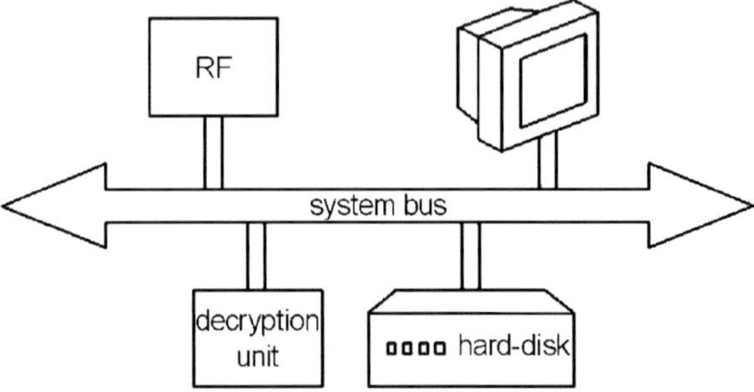

Fig. 1 Set-top box example

4   Intra Event Stream Contexts

Context-blind analysis assumes that in the worst case, every scheduled task executes with its worst-case execution time for each activation. In reality, different events often activate different behaviors of a computation task with different WCET, or different bus loads for a communication task. Therefore, a lower maximum load (and a higher minimum load) can be determined for a sequence of successive activations of a higher-priority task if the types of the activating events are considered. This in turn leads to a shorter calculated worst-case response time (and a longer best-case response time) of lower-priority tasks. We call a sequence of different activating events an *intra event stream* context.

Mok and Chen [4] introduced this idea in [1] and showed promising results for MPEG-streams where the average load for a sequence of I-, P-, and B-frames is much smaller than in a stream that consists only of large I-frames, which is assumed by a context-blind worst-case response time analysis [1]. However, the periodic sequence of types of activating events was supposed to be completely known.

In reality, intra event stream contexts can be more complicated. If no complete information is available about the types of the activating events, it is no longer possible to apply Mok's and Chen's [4] approach. However, partial information may be available. Specifically, it may be possible to specify minimum and maximum conditions for the occurrence of each event type. In this case, a single worst-case and a single best-case sequence of events can be determined from the given min- and max-conditions that can be used to calculate the worst- and best-case load due to n consecutive activations of the task. n is an arbitrary integer value.

We have extended response-time calculation to exploit this idea. In the following, we show the worst-case calculation. Since min-conditions represent the lowest bound for the occurrence of an event-type in the sequence, our algorithm fulfills them first.

```
1    FulfillMinConditions () {
2    for (i = 0; i < MinConditionsList.size; i++) {
3    MinCondition = MinConditionsList (i);
4    while (MinCondition not fulfilled in Sequence)
5    Sequence.add(MinCondition.type);
6    }
7    return Sequence
8    }
```

After this step, types have to be assigned to the remaining events to complete the sequence. This is reflected by the following method.

```
1    CompleteSequence(){
2    next = 0;
3    while (EventsToAdd >= 0){
4    type = WeightSortedEventTypeList.get(next);
5    Do as long as ((EventsToAdd >= 0) and
     (MaxConditions are not violated in Sequence)){
6    Sequence.add(type);
7    EventsToAdd = EventsToAdd - 1;
8    }
9    next = next + 1;
10   }
11   return Sequence;
12   }
```

Finally the resulting sequence is sorted by weight.

Intra-event stream contexts can now be exploited for scheduling analysis. In case of a static priority preemptive scheduler, the following extension of Eq. (1) gives the worst-case response time for task i. $C_{j,k}$ is the WCET of the kth activation of task j in a worst-case sequence of activations due to an intra event stream context. If j is a task without intra event stream context information, $C_{j,k}$ equals C_j.

$$r_i = C_i + \sum_{\forall j \in hp(i)} \sum_{k=1}^{n_j(r_i)} C_{j,k} \qquad (2)$$

The code in Section 2 has to be modified accordingly. Line 9 is replaced by the following call which calculates the 2nd sum in Eq. (2):

$$\text{InterruptTime} = \text{MaxLoad(HighPriorTask,MaxActivations)};$$

This method iterates through the weight-sorted sequence starting from the first event, adding up loads until the worst-case load for n activations of the task has been calculated. If n is bigger than l, the sequence length, the method goes only through n mod l events and adds the resulting load to the load of the whole sequence multiplied by n div l.

```
1    MaxLoad(Task, n) {
2    MaxLoad = 0;
3    for (i = 0;i <= n mod Task.Sequence.Length;i++) {
4    MaxLoad =
     MaxLoad + Task.Sequence.get(i).getLoad;
5    }
6    MaxLoad = MaxLoad + Task.Sequence.Load *
     (n div Task.Sequence.Length);
7    return MaxLoad;
8    }
```

Let us apply this approach to our set-top box example. Suppose that the RF sends two multiplexed MPEG-2 video streams to the bus: the first video stream is displayed on the TV while the second is saved on the hard disk. In each video stream, one of the following common MPEG-2 frame patterns is repeated: IBBPBB, IBPBPB, or IBPB. In addition, the exact order of interleaved frame types in the multiplexed video stream is unknown. It is thus impossible to provide a fixed sequence of successive frame types in the multiplexed video stream. However, we can give minimum- and maximum-conditions for the occurrence of each frame type.

We determine min- and max-conditions for a multiplexed video stream with a length that is the smallest common multiple of the length of all patterns, 12 in this example. The following table shows the determined conditions.

frame type	min # of frames	max # of frames
I	2	4
P	2	4
B	6	8

We skip the details how to obtain these numbers. We assume that the size of each frame-type is fixed independent of the pattern that it appears in. Otherwise, we would simply specify more conditions.

The resulting partial sequence after fulfilling the minconditions is IIPPBBBBBB. Worst-case frame types now have to be assigned to the remaining two frames to complete the sequence. I-frames generally have the largest size in MPEG-streams, B-frames have the smallest size. Since the upper bound on the occurrence of I-frames is 4 in our example, the remaining two frames have to be assumed to be I-frames. Therefore, the resulting sequence after this step is IIPPBBBBBBII. Finally, the sequence is sorted by weight (in this case frame size), resulting in IIIIPPBBBBBB.

Let us now observe the response-time analysis improvements through intra event stream contexts in our set-top box example. Suppose that in addition to the multiplexed video stream, IP traffic is sent via the bus to the hard disk. We assume that the multiplexed video stream S_{mux} has a higher priority on the bus than the IP traffic S_{ip}. As an example, Fig. 2 shows for both context-blind (a) and intra-event stream context (b) cases the calculated worst-case response time of S_{ip} due to interrupts by S_{mux} for an IP traffic of 127, I-frame size of 106, P-frame size of 85 and B-frame size of 27. As can

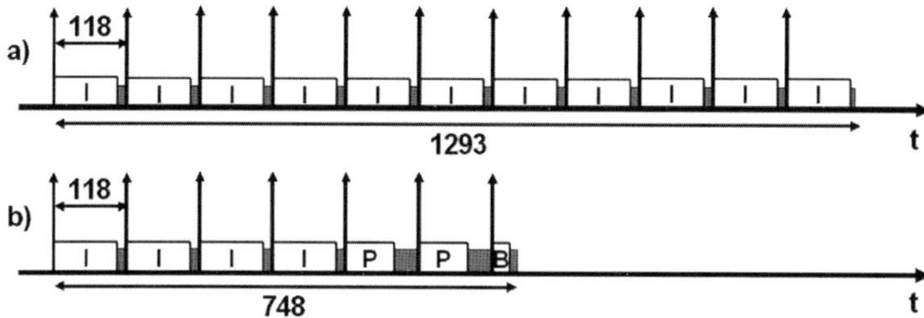

Fig. 2 Worst-case response time calculation for IP traffic (gray boxes) (a) without and (b) with *intra*-event stream context information

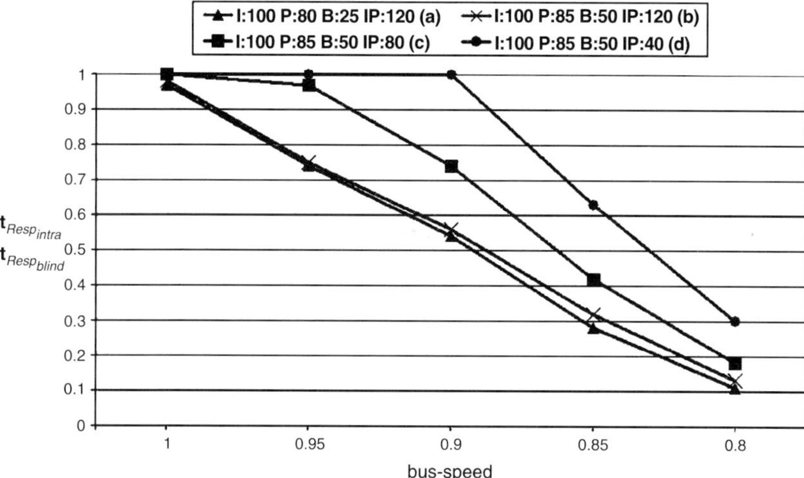

Fig. 3 Improved worst-case response-time analysis for IP-traffic due to *intra*-event stream contexts

bc sccn, when considering the available intra event stream context information for the multiplexed video stream, gaps between successive executions of S_{mux} are larger than in the context-blind case. Since IP traffic is transmitted during these gaps (gray boxes in Fig. 2), a smaller number of interrupts of S_{ip} by S_{mux} is calculated considering intra contexts in comparison to the context-blind case. This leads to a reduction of the calculated worst-case response time of S_{ip}.

In Fig. 3, worst-case response time analysis improvements using intra event stream context information compared to the context-blind case are shown for S_{ip} as a function of the bus-speed. Typical frame size ratios are assumed for the multiplexed video streams. The I-frame size is normalized to 100 and the bus-speed is normalized to 1.

Curves **a** and **b** show the reduction of the calculated worst-case response time of S_{ip} for an IP traffic size of 120. Reducing the bus-speed increases the time needed to transmit one MPEG frame. This in turn leads to smaller gaps between successive transmissions of frames, leading to a larger number of interrupts of S_{ip}. The number of required gaps increases considerably faster in the context-blind case than in the intra-event stream context case. Therefore, a greater reduction in the calculated worst-case response time (up to 90% in our example) is obtained in the intra-event stream context case for slower bus speeds. This observation is important since the designer usually strives to reduce bus bandwidth as much as possible to save cost and power consumption.

Curves **c** and **d** show the reduction of the calculated worst-case response time of S_{ip} for smaller IP traffic sizes (80 and 40). When reducing the bus-speed, we observe generally a lower reduction of the calculated worst-case response time of S_{ip} for smaller IP traffic sizes. This is due to the fact that S_{ip} is interrupted less often by S_{mux} for smaller IP traffic sizes.

Another form of partial intra-event stream context information are subsequences. Suppose that in our example, in addition to the existing min- and max-conditions we know that the subsequence [IIB] is always available in each sequence of 12 successive frames in the multiplexed video stream. Since the subsequence already assigns types to three frames, when evaluating the min- and max-conditions we only have to assign types to the remaining nine frames keeping in mind the types already used in the subsequence. Therefore, some min-conditions are totally (min 2 I-frames out of 12 frames) or partially (min 6 B-frames out of 12 frames) fulfilled by the subsequence. The resulting sequence after fulfilling the remaining min conditions is [IIB]PPBBBBB.

Types still have to be assigned to two frames. Since the upper bound for the occurrence of I-frames is four (max 4 I-frames out of 12 frames), and two I-frames already exist in the sequence, the remaining two frames have to be assumed to be I-frames. Therefore, the resulting sequence after this step is [IIB]PPBBBBB.

The 9 frames outside the subsequence are sorted by weight, however the complete sequence cannot be sorted. When executing MaxLoad, all insertion positions of the sub-sequence have to be tried to obtain the the worst-case sequence of length n.

5   Inter-Event Stream Contexts

Context-blind analysis assumes that in the worst case all scheduled tasks are activated simultaneously. In reality, activating events are often time-correlated, which rules out simultaneous activation of all tasks. This in turn may lead to a lower maximum number (and higher minimum number) of interrupts of a lower-priority task through higher-priority tasks, resulting in a shorter worst-case response time (and longer best-case response time) of the lower-priority task.

Inter-event stream contexts capture information about time-correlated events in different event streams in a way that can be exploited by performance analysis. Tindell [8] introduced this idea for tasks scheduled by a static priority preemptive scheduler [8]. Each set of time-correlated tasks is grouped into a so-called transaction. Each task is activated when an offset elapses after the activation of the transaction to which it belongs. Transactions are activated by periodic external events.

Tindell [8] showed that the worst-case contribution of a transaction to the response time of a lower-priority task occurs when the activation time of the lower-priority task happens as soon as possible after the *critical instant* of the transaction. The critical instant is the activation time of one of the transaction's higher-priority tasks. Subsequent activations of higher-priority tasks belonging to the transaction also have to happen as soon as possible after the critical instant.

Since all activation times of all higher-priority tasks belonging to a transaction are candidates for the critical instant of the transaction, the worst-case response time of a lower-priority task has to be calculated for all possible combinations of all critical instants of all transactions that contain higher-priority tasks, to find the absolute worst case.

The following WorstCaseResponseTime algorithm considers inter-event stream contexts to compute the worst-case response time of a lower-priority task that does not belong to any transaction. If the lower-priority task itself belongs to a transaction, the interrupt-time by higher-priority tasks belonging to the same transaction has to be calculated separately.

```
1     WorstCaseResponseTime (LowPriorTask){
2     MaxResponseTime = LowPriorTask.WCET;
3     for each combination of critical instants
      determined by higher-priority tasks
      belonging to different transactions{
4     NewResponseTime = LowPriorTask.WCET;
5     Do{
6     OldResponseTime = NewResponseTime;
7     for each transaction{
8     Contribution = 0;
9     for each HighPriorTask of transaction{
10    MaxActivations = max activations of
      HighPriorTask during OldResponseTime
      considering HighPriorTask.Offset;
11    InterruptTime =
      MaxActivations * HighPriorTask.WCET;
12    Contribution =
      Contribution + InterruptTime;
13    }
14    NewResponseTime =
      NewResponseTime + Contribution;
15    }
21    }while (NewResponseTime > OldResponseTime)
22    if (NewResponseTime > MaxResponseTime)
23    MaxResponseTime = NewResponseTime;
24    }
25    return MaxResponseTime;
26    }
```

In comparison to the context-blind algorithm presented in Section 2, the response time has to be calculated for all combinations of critical instants to get the worst-case response time (line 3). For each task in a transaction, the maximum number of activations is calculated taking its offset into account (line 10). For tasks not belonging to a transaction, instead of executing lines 7–15, lines 5–10 of the context-blind algorithm in section 2 are executed.

Let us apply Tindell's [8] approach to our set-top box example. Suppose that the RF sends an encrypted MPEG-2 video stream to the bus. The decryption unit decrypts the stream, which we assume takes a fixed amount of time, and forwards it with the resulting time-offset via the bus to the TV. In addition, IP traffic is sent via the bus to the hard disk. We assume that the encrypted video stream S_{enc} has the highest priority, the decrypted video stream S_{dec} has the middle priority and that the IP traffic S_{ip} has the lowest priority.

In order to show in isolation the analysis improvement due to inter-event stream contexts, we will assume for now that all video-frames are I-frames. Since S_{enc} and S_{dec} are time-correlated, they are grouped into a transaction T.

Let us observe the worst-case contribution of the transaction to the worst-case response time of S_{dec} and S_{ip}. As an example, Fig. 4 shows for both context-blind (a) and inter-event stream context (b) cases the calculated worst-case response time of S_{ip} and its interrupts through S_{enc} and S_{dec} for an IP traffic (gray boxes) of 50. The transaction period is normalized to 100, the I-frame size is set to 30. As can be seen, when considering the available inter-event stream context information for the video streams, one interrupt less of S_{ip} by S_{dec} is calculated. In general, when analyzing S_{ip}, we have to try both critical points of transaction T, which are the activation times of S_{enc} and S_{dec}. In Fig. 5, analysis improvements with inter-event stream context information in relation to the context-blind case are shown as a function of the

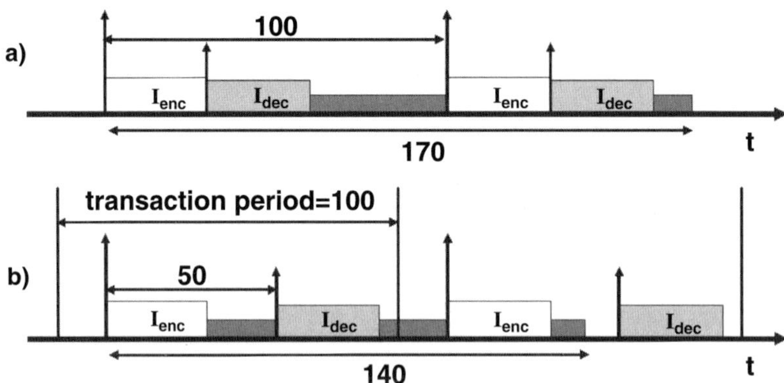

Fig. 4 Worst-case response-time calculation for IP traffic (gray boxes) (a) without and (b) with *inter*-event stream context information

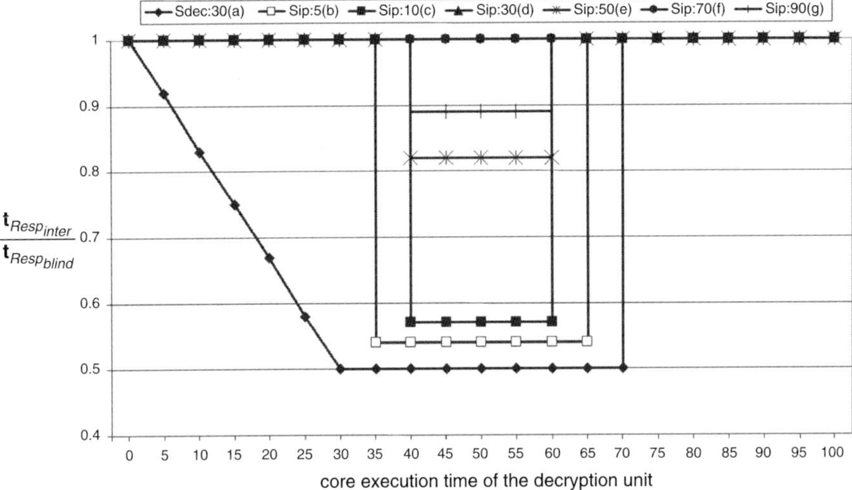

Fig. 5 Improved worst-case response-time analysis for IP-traffic and packet decryption due to *inter*-event stream contexts

offset between S_{enc} and S_{dec}, which is equal to the execution time of the decryption unit. The bus load due to the video streams is 60%. Curve **a** shows the reduction of the calculated worst-case response time of S_{dec}. Depending on the offset, S_{dec} is either partially (offset value less than 30), completely (offset value more than 70) or not interrupted at all by S_{enc} (offset value between 30 and 70). The latter case yields a maximum reduction of 50%.

Curves **b–g** show the reduction in the calculated worst-case response time of S_{ip} for different IP traffic sizes. The reduction is visible in the curves as dips. If no gaps exists between two successive executions of S_{enc} and S_{dec}, no worst-case response time reduction of S_{ip} can be obtained (offset value less than 30 or more than 70). If a gap exists, then sometimes one interrupt less of S_{ip} can be calculated (either through S_{enc} or S_{dec}), or there is no gain at all (curves **d** and **f**). Since the absolute gain that can be obtained equals the smaller worst-case execution time of S_{enc} and S_{dec}, the relative worst-case response time reduction is bigger for shorter IP-traffic.

An important observation is that inter-event stream context analysis reveals the dramatic influence that a small local change, in our example the speed of the decryption unit reading data from the bus and writing the results back to the bus, can have on system performance, in our example the worst-case transmission time of lower-priority IP traffic. Systematically identifying such system-level influences of local changes is especially difficult using simulation due to the large number of implementations that would have to be synthesized and executed. On the other hand, formal performance analysis can systematically and quickly identify such corner cases.

6   Combination of Contexts

Inter-event stream contexts allow to calculate a tighter number of interrupts of a lower-priority task through higher-priority tasks. *Intra*-event stream contexts allow to calculate a tighter load for a number of successive activations of a higher-priority task. The two types of contexts are orthogonal: the worst-case response time of a lower-priority task is reduced both because fewer high-priority task activations can interrupt its execution during a certain time interval, and because the time required to process a sequence of activations of each higher-priority task is reduced. Therefore, performance analysis can be further improved if it is possible to consider both types of contexts in combination.

Let us apply the combination of event stream contexts to the scenario presented in the previous section, by additionally considering intra-event stream contexts with the frame pattern IBBPBB. Therefore, both S_{enc} and S_{dec} have an intra and an inter context. In Fig. 6, we show analysis improvements considering both inter and intra event stream contexts in relation to the context-blind case as a function of the offset between S_{enc} and S_{dec}. Curve **a** shows the reduction of the calculated worst-case response time of S_{dec}. Since S_{dec} is interrupted at most once by S_{enc}, and the worst-case load produced due to one activation of S_{enc} is the transmission time of one I-frame, no improvement is obtained through the context combination in comparison to curve **a** in Fig. 5.

Curves **b**–**g** show the reduction of the calculated worst-case response time of S_{ip} for different IP traffic sizes. When comparing curves **b** and **c** (IP traffic sizes of 5

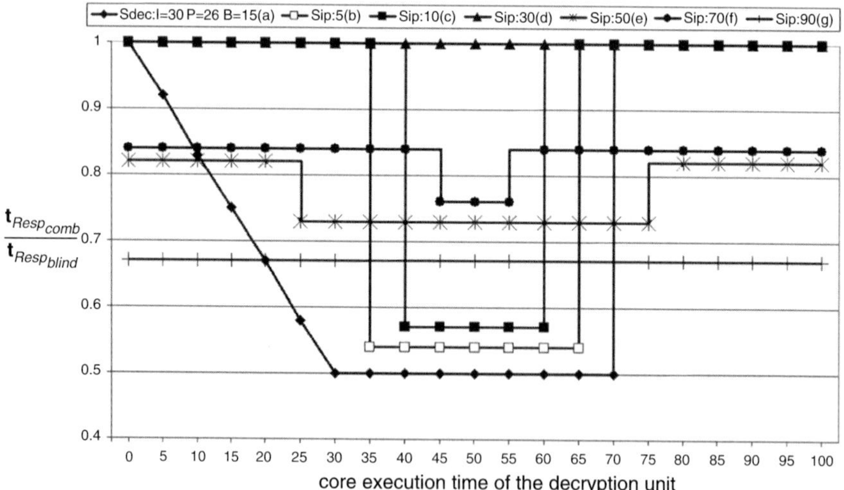

Fig. 6 Analysis improvement due to the *combination of intra-and inter*-event stream contexts

and 10) to curves **b** and **c** in Fig. 5), it can be seen that no improvement is obtained through the context combination. This is due to the fact that S_{ip} is interrupted at most once by S_{enc} and at most once by S_{dec}. Therefore, as in case **a**, the calculated worst-case load produced by the video streams is the same no matter whether the available intra-event stream context information is considered or not.

Curve **d** shows that for an IP traffic size of 30, no improvements are obtained through the context combination in comparison to the *context-blind* case. This is due to the fact that for all offset-values, S_{ip} is interrupted exactly once by S_{enc} and exactly once by S_{dec}, and that the calculated worst-case load produced by the video streams due to one activation is the same no matter if intra event stream contexts are considered or not.

Curve **e** and **f** show that for IP traffic sizes of 50 and 70 improvements are obtained as a result of the context combination in comparison to both the intra and inter event stream context analysis. Let us focus on curve **e**. Since intra- and inter-event stream contexts are orthogonal, the reduction of the calculated worst-case response time of S_{ip} due to the intra event stream context is constant for all offset values. Since no reduction due to inter event stream context can be obtained for an offset value of 0 (equivalent to the inter-event stream context-blind case), we are sure that the reduction shown in the curve for this offset value is only a result of the intra-event stream context. On the other hand, the additional reduction between the offset values 25 and 75 is obtained due to the inter-event stream context.

Curve **g** shows that for an IP traffic size of 90, even though the inter-event stream context leads to an improvement (see curve **g** in Fig. 5), the improvement due to the intra-event stream context dominates, since no dip exists in the curve. I.e. no additional improvements are obtained due to the context combination in comparison to the intra-event stream context case.

7 Conclusion

In this paper we demonstrated that considering intra-event stream contexts and inter-event stream contexts can yield considerably tighter performance analysis bounds compared to a context-blind analysis. We used a priority-based system bus of a hypothetical set-top box as an example. We focused on realistic streams of bus load, where intra event stream contexts can be described only by minimum conditions, maximum conditions and possibly subsequences. We explored the analysis improvements for various bus speeds and various load distributions, by varying the speed of a decryption unit. The results show that the improvement due to intra-event stream contexts is largest for high bus loads (up to 90% in our example), which is a design point a designer is striving for. Considering inter-event stream contexts improves analysis up to 50% in our example.

Equally important, performance analysis reveals the dramatic influence that a small local change can have on system-performance. Systematically identifying such system-level effects due to local changes is especially difficult using simulation. On the other hand, the most complicated curves we presented took a couple of hundred milliseconds to compute on a standard desktop computer. These numbers show that formal performance analysis is ideal for design space exploration and corner-case identification.

References

[1] S. K. Baruah, D. Chen, and A. K. Mok. Static-priority scheduling of multiframe tasks. In *Euromicro Conference on Real-Time Systems*, June 1999.

[2] S. Chakraborty, S. Künzli, and L. Thiele. A general framework for analysing system properties in platform-based embedded system designs. In *Proc. DATE'03*, Munich, Germany, Mar. 2003.

[3] M. Jersak and R. Ernst. Enabling scheduling analysis of heterogeneous systems with multi-rate data dependencies and rate intervals. In *Proceeding 40th Design Automation Conference*, Annaheim, CA, June 2003.

[4] A. Mok and D. Chen. A multiframe model for real-time tasks. *IEEE Transactions on Software Engineering*, 23(10):635–645, 1997.

[5] T. Pop, P. Eles, and Z. Peng. Holistic scheduling and analysis of mixed time/event-triggered distributed embedded systems. In *Tenth International Symposium on Hardware/Software Codesign (CODES'02)*, Estes Park, CO, May 2002.

[6] K. Richter, M. Jersak, and R. Ernst. A formal approach to mpsoc performance verification. *IEEE Computer*, 36(4), Apr. 2003.

[7] K. Richter, R. Racu, and R. Ernst. Scheduling analysis integration for heterogeneous multi-processor SoC. In *Proceedings 24th International Real-Time Systems Symposium (RTSS'03)*, Cancun, Mexico, Dec. 2003.

[8] K. W. Tindell. Adding time-offsets to schedulability analysis. Technical Report YCS 221, University of York, 1994.

[9] K. W. Tindell. An extendible approach for analysing fixed priority hard real-time systems. *Journal of Real-Time Systems*, 6(2):133–152, Mar 1994.

Lock-Free Synchronization for Dynamic Embedded Real-Time Systems*

Hyeonjoong Cho[1], Binoy Ravindran[2], and E. Douglas Jensen[3]

Abstract We consider lock-free synchronization for dynamic embedded real-time systems that are subject to resource overloads and arbitrary activity arrivals. We model activity arrival behaviors using the unimodal arbitrary arrival model (or UAM). UAM embodies a stronger "adversary" than most traditional arrival models. We derive the upper bound on lock-free retries under the UAM with utility accrual scheduling—the first such result. We establish the trade-offs between lock-free and lock-based sharing under UAM. These include conditions under which activities' accrued timeliness utility is greater under lock-free than lock-based, and the consequent upper bound on the increase in accrued utility that is possible with lock-free. We confirm our analytical results with a POSIX RTOS implementation.

1 Introduction

Embedded real-time systems that are emerging in many domains such as robotic systems in the space domain (e.g. NASA/JPL's Mars Rover [7]) and control systems in the defense domain (e.g. airborne trackers [6]) are fundamentally distinguished by the fact that they operate in environments with dynamically uncertain properties. These uncertainties include transient and sustained resource overloads due to context-dependent activity execution times and arbitrary activity arrival patterns. Nevertheless, such systems' desire the strongest possible assurances on activity timeliness behavior. Another important distinguishing feature of these systems is their relatively long execution time magnitudes—e.g., in the order of milliseconds to minutes.

---

*This work was supported by the Office of Naval Research under Grant N00014-05-1-0179 and The MITRE Corporation under Grant 52917.

[1] ECE Dept., Virginia Tech, Blacksburg, VA 24061, USA. hjcho@vt.edu

[2] ECE Dept., Virginia Tech, Blacksburg, VA 24061, USA. binoy@vt.edu

[3] The MITRE Corporation, Bedford, MA 01730, USA. jensen@mitre.org

R. Lauwereins and J. Madsen (eds.), *Design, Automation, and Test in Europe–The Most Influential Papers of 10 Years DATE*, 73–85
© Springer 2008

When resource overloads occur, meeting deadlines of all activities is impossible as the demand exceeds the supply. The urgency of an activity is typically orthogonal to the relative importance of the activity—e.g. the most urgent activity can be the least important, and vice versa; the most urgent can be the most important, and vice versa. Hence when overloads occur, completing the most important activities irrespective of activity urgency is often desirable. Thus, a clear distinction has to be made between urgency and importance, during overloads. (During under-loads, such a distinction need not be made, because deadline-based scheduling algorithms such as EDF are optimal.)

Deadlines by themselves cannot express both urgency and importance. Thus, we consider the abstraction of time/utility functions (or TUFs) [10] that express the utility of completing an activity as a function of that activity's completion time. We specify deadline as a binary-valued, downward "step" shaped TUF; Fig. 1a shows examples. Note that a TUF decouples importance and urgency—i.e. urgency is measured on the *x*-axis, and importance is denoted (by utility) on the *y*-axis.

Many embedded real-time systems also have activities that are subject to *non-deadline* time constraints, such as those where the utility attained for activity completion *varies* (e.g. decreases, increases) with completion time. This is in contrast to deadlines, where a positive utility is accrued for completing the activity anytime before the deadline, after which zero, or infinitively negative utility is accrued. Figures 1 show example such time constraints from two real applications [6]. When activity time constraints are specified using TUFs, the scheduling criteria are based on accrued utility, such as maximizing the total activity attained utility. We call such criteria, *utility accrual* (or UA) criteria, and scheduling algorithms that optimize them, as UA scheduling algorithms.

UA algorithms that maximize total utility under step TUFs (see algorithms in [15]) default to EDF during under-loads, since EDF satisfies all deadlines during under-loads. Consequently, they obtain the maximum total utility during under-loads. During overloads, they favor more important activities (since more utility can be attained from them), irrespective of urgency. Thus, deadline scheduling's optimal timeliness behavior is a special-case of UA scheduling.

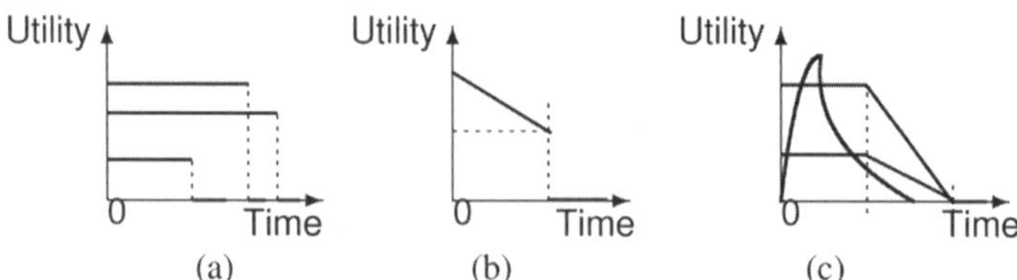

Fig. 1 Example TUF time constraints

1.1 Shared Data and Synchronization

Most embedded real-time systems involve mutually exclusive, concurrent access to shared data objects, resulting in contention for those objects. Resolution of the contention directly affects the system's timeliness behavior. Mechanisms that resolve such contention can be broadly classified into: (1) lock-based schemes— e.g. [16], see algorithms in [15]; and (2) non-blocking schemes including wait-free protocols (e.g. [11, 4, 9]) and lock-free protocols (e.g. [3]).

Lock-based protocols have several disadvantages such as serialized access to shared objects, resulting in reduced concurrency [3]. Further, many lock-based protocols typically incur additional run-time overhead due to scheduler activations that occur when activities request locked objects [16, 15]. Also, deadlocks can occur when lock holders crash, causing indefinite starvation to blockers. Many (real-time) lock-based protocols also require a priori knowledge of the ceilings of locks [16], which may be difficult to obtain for dynamic applications, resulting in reduced flexibility [3]. These drawbacks have motivated research on non-blocking objects in embedded real-time systems.

Especially, lock-free objects guarantee that some object operations will complete in a finite number of steps. It implies other operation may have to retry a potentially infinite number of time. Instead of acquiring locks, lock-free operation continuously accesses the object, checks, and retries until it becomes successful. Inevitably, lock-free protocols incur additional time costs due to their retries, which is antagonistic to timeliness optimization. Prior research has shown how to mitigate these time or space costs. On the other hand, wait-free objects guarantee any operation on the objects will complete in a bounded number of steps, regardless of interferences. Wait-free protocols may incur additional time or space costs for their intrinsic mechanisms.

An excellent survey of the prior research can be found in [9]. We summarize some important efforts in the context of our work here: Anderson et al. [3] show how to bound the retry loops of lock-free protocols through judicious scheduling. Lock-free objects on single- and multiprocessors under quantum-based scheduling is presented in [1]. This work assumes that each task can be preempted at most once during a single quantum; thus each object access needs to be retried at most once. Anderson et al. [2] present wait-free schemes for single- and multiprocessors, where a task which announces its intention to share objects also helps other tasks to complete their access.

Efficient lock-free objects, such as queues and stacks, have been presented in [14, 13]. Valois [18] presents lock-free linked lists. Treiber [17] presents lock-free stack algorithms. Kopetz et al. [11] present an analysis on non-blocking synchronization in real-time systems. This work was later improved by Chen et al. [4], and subsequently by Huang et al. [9] and by Cho et al. [5].

1.2 Contributions

In this paper, we focus on *dynamic* embedded real-time systems on a single processor. By dynamic systems, we mean those subject to resource overloads due to context-dependent activity execution times and arbitrary activity arrivals. To account for the variability in activity arrivals, we describe arrival behaviors using the unimodal arbitrary arrival model (UAM) [8]. UAM specifies the maximum number of activity arrivals that can occur during any time interval. Consequently, the model subsumes most traditional arrival models (e.g. periodic, sporadic) as special cases.

We consider lock-free sharing under the UAM. The past work on lock-free sharing upper bounds the retries under restrictive arrival models like periodic [3]—retry bounds under the UAM are not known. Moreover, we consider the UA criteria of maximizing the total utility, while allowing most TUF shapes including step and non-step shapes, and mutually exclusive concurrent object sharing. We focus on the Resource-constrained Utility Accrual (RUA) scheduling algorithm [19], as it is the only algorithm for that model. RUA allows arbitrarily shaped TUFs and concurrent object sharing using locks. For the special case of step TUFs, no object sharing, and under-loads, RUA defaults to EDF.

We derive the upper bound on lock-free RUA's retries under the UAM—the first ever retry bound under a non-periodic arrival model. Since lock-free sharing incurs additional time overhead due to the retries (as compared to lock-based), we establish the conditions under which activity sojourn times are shorter under lock-free RUA than under lock-based, for the UAM. From this result, we establish the maximum increase in activity utility under lock-free RUA over lock-based. Further, we implement lock-free and lock-based RUA on a POSIX RTOS. Our implementation measurements strongly validate our analytical results.

The rest of the paper is organized as follows: in Section 2, we derive the upper bound on retries under lock-free RUA. We compare lock-free and lock-based RUA, and establish the tradeoffs between the two in Section 3. Section 4 discusses our implementation experience. We conclude the paper in Section 5. In describing our work, we skip some proofs for brevity; they can be found in [20].

2 Bounding Retries Under UAM

Figure 2 shows the three dimensions of the task model that we consider in the paper. The first dimension is the arrival model. We consider the UAM, which is more relaxed than the periodic model, but more regular than the aperiodic model. Hence, it falls in between these two ends of the (regularity) spectrum of the arrival dimension. For a task T_i, its arrival using UAM is defined as a tuple $\langle l_i, a_i, W_i \rangle$, meaning that the maximal number of job arrivals of the task during any sliding time window W_i is a_i and the minimal number is l_i [8]. Jobs may arrive simultaneously. The periodic model is a special case of UAM with $\langle 1, 1, W_i \rangle$.

We refer to the jth job (or invocation) of task T_i as $J_{i,j}$. The basic scheduling entity that we consider is the job abstraction. Thus, we use J to denote a job without being task specific, as seen by the scheduler at any scheduling event. A job's time constraint, which forms the second dimension, is specified using a TUF. A TUF subsumes deadline as a special case (i.e. binary-valued, downward step TUF). Jobs

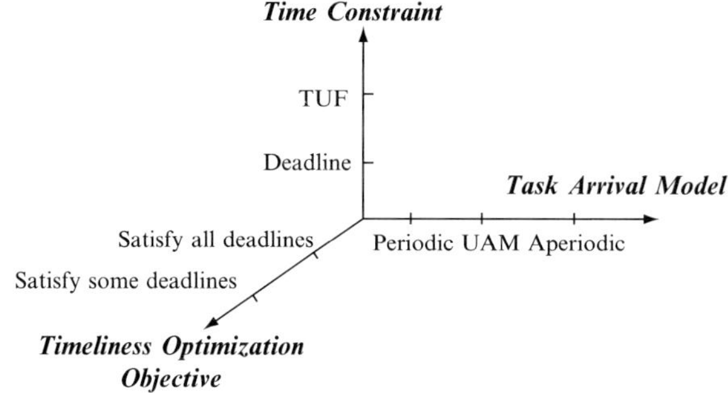

Fig. 2 Three dimensions of task model

of a task have the same TUF. $U_i(\cdot)$ denotes task T_i's TUF; thus, completion of T_i at time t will yield $U_i(t)$ utility.

TUFs can take arbitrary shapes, but must have a (single) "critical time." Critical time is the time at which the TUF has zero utility. For the special case of deadlines, critical time corresponds to the deadline time. We denote the critical time of task i's $U_i(\cdot)$ as C_i, and assume that $C_i \le W_i$.

The third dimension is feasibility. Feasibility includes under-load situations, during which all tasks can be completed before their critical times, and overloads, where only a subset of the tasks (including possibly none) can be done so. Our model includes overloads, the UAM, and TUFs.

Finally, the resource model we consider here does not allow the nested critical section which may cause deadlock in the lock-based synchronization.

2.1 Overview of Lock-based RUA

RUA [19] targets dynamic applications, where activities are subject to arbitrary arrivals and resource overloads. Further, activities may access (logical and physical) resources arbitrarily—e.g. the resources that will be needed, the length of time for which they will be needed, and the order of accessing them are all statically unknown. Thus, the set of activities to be scheduled and their resource dependencies may change over time. Consequently, RUA performs scheduling entirely online. RUA considers activities subject to arbitrarily shaped TUF time constraints, concurrent object sharing under mutual exclusion constraints, and the scheduling objective of maximizing the total utility.

With n jobs, RUA's asymptotic cost is $O(n^2 logn)$. This is higher than that of many traditional real-time scheduling algorithms. However, this high cost is justified for application with longer execution time magnitudes such as those that we focus here. (Of course, this high cost cannot be justified for every application.)

Nevertheless, it is desirable to reduce the cost so that the resources utilized by the scheduling algorithm can yield greater benefit in terms of improved scheduling from the application's point of view.

2.2 Preemptions Under UA Schedulers

Under fixed priority schedulers such as rate-monotonic (RM), a lower-priority job can be preempted at most once by each higher-priority job if no resource dependency (that arises due to concurrent object sharing) exists. This is because a job does not change its execution eligibility until its completion time. However, under UA schedulers such as RUA, execution eligibility of a job dynamically changes.

In Fig 3, assume that two jobs $J_{1,1}$ and $J_{2,1}$ arrive at time t_0, and scheduling events occur at t_1, t_2, and t_3. $J_{2,1}$ occupies CPU at t_0 and is preempted by $J_{2,1}$ at t_1. Subsequently, $J_{1,1}$ is preempted by $J_{2,1}$ at t_2, which does not happen under RM scheduling. Under RUA, a simple condition causing this *mutual preemption*, where a job which has preempted another job can be subsequently preempted by the other job, is when TUFs of the two jobs are increasing.

This potential mutual preemption distinguishes RUA from traditional schedulers such as RM, where a job can be preempted by another job at most once. Hence, the maximum number of preemptions under RM that a job may suffer can be bounded by the number of releases of other jobs during a given time interval (see [2]). However, this is not true with RUA, as one job can be preempted by another job more than once. Thus, the maximum number of preemptions that a job may experience under RUA is bounded by the number of events that invoke the scheduler.

Lemma 1 (Preemptions Under UA scheduler). *During a given time interval, a job scheduled by a UA scheduler can experience preemptions by other jobs at most the number of the scheduling events that invoke the scheduler.*

Lemma 1 helps compute the upper bound on the number of retries on lock-free objects for our model. This is also true with other UA schedulers [15], because it is impossible for more preemptions to occur than scheduling events.

2.3 Bounded Retry Under UAM

Under our model, jobs with TUF constraints arrive under the UAM, and there may not always be enough CPU time available to complete all jobs before their critical times. We now bound lock-free RUA's retries under this model.

Theorem 2 (Lock-free Retry Bound Under UAM). *Let jobs arrive under the UAM $< 1, a_i, W_i >$ and are scheduled by RUA. When a job J_i accesses more than one lock-free object, J_i's total number of retries, f_i, is bounded as:*

$$f_i \leq 3a_i + \sum_{j=1, j \neq i}^{N} 2a_j \left(\left\lceil \frac{C_i}{W_j} \right\rceil + 1 \right)$$

Fig. 3 Mutual preemption under RUA

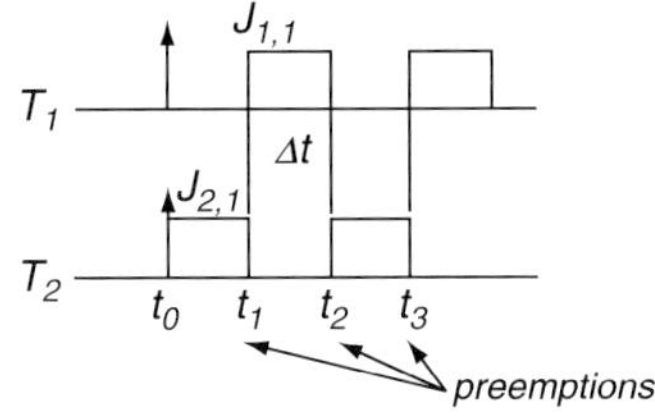

Fig. 4 Interferences under UAM

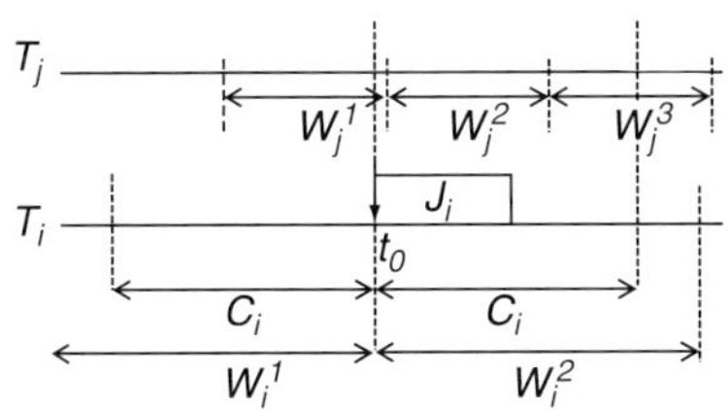

where N is the number of tasks.

Proof. In Fig. 4, J_i is released at time t_0 and has the absolute critical time $(t_0 + C_i)$. After the critical time, J_i will not exist in the ready queue, because it will be either completed by that time or aborted by RUA. Thus, by Lemma 1, the number of retries of J_i is bounded by the maximum number of all scheduling events that occur within the time interval $[t_0, t_0 + C_i]$. The scheduling events that J_i suffers can be categorized into those occurring from other tasks, $T_j, j \neq i$ and those occurring from T_i. We consider these two cases:

Case 1 (Scheduling events from other tasks): to account for the worst-case event occurrence, we assume that all instances of T_j in the window W_j^1 are released right after time t_0, and all instances of T_j in the window W_j^3 are released before time $(t_0 + C_i)$. Thus, the maximum number of releases of T_j within $[t_0, t_0 + C_i]$ is $\lceil C_i/W_j \rceil + 1$. It also holds when $W_j > C_i$ as $\lceil C_i/W_j \rceil + 1 = 2$. All jobs of T_j before the first window W_j must depart either by completion or by abortion before t_0. All released jobs must be completed or aborted, so that the number of scheduling events that a job can create is at most 2. Hence, $a_j (\lceil C_i/W_j \rceil + 1)$ is multiplied by 2.

Case 2 (Scheduling events from the same task): in the worst-case, all jobs of T_i are released and completed within $[t_0, t_0 + C_i]$, which results in at most $2a_i$ events. T_i's jobs, which are released during $[t_0 - C_i, t_0]$ also cause events within $[t_0, t_0 + C_i]$ by completion. Thus, the total number of events that are possible is $3a_i$.

The sum of case 1 and case 2 is the upper bound on the number of events. It is also the maximum number of total retries of J_i's objects.□

Note that f has nothing to do with the number of lock-free objects in J_i in Theorem 2, even when J_i accesses more than one lock-free object. This is because no matter how many objects J_i accesses, f cannot exceed the number of events. Further, even if J_i accesses a single object, the retry can occur only as many times as the number of events.

Theorem 2 also implies that the sojourn time of J_i is bounded. The sojourn time of J_i is computed as the sum of J_i's execution time, the interference time by other jobs, the lock-free object accessing time, and f retry time.

3 Lock-based Versus Lock-free

We now formally compare lock-based and lock-free sharing by comparing job sojourn times. We do so, as sojourn times directly determine critical time-misses and accrued utility. The comparison will establish the tradeoffs between lock-based and lock-free sharing: lock-free is free from blocking times on shared object accesses and scheduler activations for resolving those blockings, while lock-based suffers from these. However, lock-free suffers from retries, while lock-based does not.

We introduce some notations, which are the same as those in [2]. We assume that all accesses to lock-based objects require r time units, and to lock-free objects require s time units. The computation time c_i of a job J_i can be written as $c_i = u_i + m_i \times t_{acc}$, where u_i is the computation time not involving accesses to shared objects; m_i is the number of shared object accesses by J_i; and t_{acc} is the maximum computation time for any object access—i.e. r for lock-based objects and s for lock-free objects.

The worst-case sojourn time of a job with lock-based is $u_i + I + r \cdot m_i + B$, where B is the worst-case blocking time and I is the worst-case interference time. In [19], it is shown that a job J_i under RUA can be blocked for at most $min(m, n)$ times, where n is the number of jobs that could block J_i and m is the number of objects that can be used to block J_i. Thus, B can be computed as $r \cdot min(m, n)$. On the other hand, the worst-case sojourn time of a job with lock-free is $u_i + I + s \cdot m_i + R$, where R is the worst-case retry time. R can be computed as $s \cdot f$ by Theorem 2. Thus, the difference between $r \cdot m_i + B$ and $s \cdot m_i + R$ is the sojourn time difference between lock-based and lock-free.

Theorem 3 (Lock-based versus Lock-free Sojourn). *Let jobs arrive under the UAM and be scheduled by RUA. If*

$$\begin{cases} \left[\left(\frac{s}{r} < \frac{2}{3} \right) \wedge \left(\frac{1}{\frac{2r}{x} - 1}(3a_i + 2x) < m_i < 2a_i + x \right), & m_i \le n \\[2em] \left[\left(\frac{s}{r} < 1 \right) \wedge \left(\frac{s}{r}(3a_i + 2x) < (1 - \frac{s}{r})m_i + n \right), & m_i > n, \end{cases}$$

then J_i's sojourn time with lock-free is shorter than that with lock-based, where

$$x = \sum_{j=1, j \ne i}^{N} a_j \left(\left\lceil \frac{C_i}{W_j} \right\rceil + 1 \right).$$

Theorem 3 shows that at least $\frac{s}{r} < \frac{2}{3}$ for jobs to obtain a shorter sojourn time under lock-free. Anderson et al. [2] show that s is often much smaller than r in comparison with the vast majority of lock-free objects in most systems, such as buffers, queues, and stacks.

Corollary 4 (Special Case Sojourn). *Let jobs arrive under the UAM and be scheduled by RUA. Let r be much larger than s enough for $\frac{s}{r}$ to converge to zero. Now, if:*

$$((0 < m_i < 2a_i + x) \wedge (m_i \leq n)) \vee (m_i > n),$$

then J_i's sojourn time with lock-free is shorter than that with lock-based, where

$$x = \sum_{j=1, j \neq i}^{N} a_i \left(\left\lceil \frac{C_i}{W_j} \right\rceil + 1 \right).$$

Shorter sojourn times always yields higher utility with nonincreasing TUFs, but not always with increasing TUFs. However, it is likely to improve performance at system level because each job can save more CPU time for other jobs.

When r is larger than s, lock-free sharing is more likely to perform better than lock-based sharing, according to Corollary 4. Further, when m_i is larger than n, the sojourn time of a job with lock-free object always outperforms its lock-based counterpart. Thus, lock-free sharing is very attractive for UA scheduling as most UA schedulers' object sharing mechanisms have higher time complexity.

Shorter sojourn times for a job under lock-free sharing will only increase the job's accrued utility under nonincreasing TUFs. However, this does not directly imply that lock-free sharing will yield higher *total* accrued utility than lock-based sharing. This is because, Theorem 3 does not reveal anything regarding aggregate system-level performance, but only job-level performance. Since RUA's objective is to maximize total utility, we now compare lock-based and lock-free sharing in terms of accrued utility ratio (or AUR). AUR is the ratio of the actual accrued total utility to the maximum possible total utility.

Lemma 5 (Lock-based versus Lock-free AUR). *Let jobs arrive under the UAM and be scheduled by RUA. If all jobs are feasible and their TUFs are nonincreasing, then the difference in AUR between lock-free and lock-based sharing, $\Delta AUR = AUR_{lf} - AUR_{lb}$ is:*

$$\sum_{i=1}^{N} \frac{U_i(s_{0,lf} + R) - U_i(s_{0,lb})}{U_i(0)} \leq \Delta AUR$$

$$\leq \sum_{i=1}^{N} \frac{U_i(s_{0,lf}) - U_i(s_{0,lb} + B)}{U_i(0)}$$

where $U_i(t)$ is task i's utility at time t, $s_{0,lf} = u_i + I + s \cdot m_i$, $s_{0,lb} = u_i + I + r \cdot m_i$, and N is the number of tasks.

4 Implementation Experience

We implemented lock-based and lock-free objects with RUA in the *meta-scheduler* scheduling framework [12], which allows middleware-level real-time scheduling atop POSIX RTOSes. Our motivation for implementation is to verify our analytical results. We used QNX Neutrino 6.3 RTOS running on a 500 MHz, Pentium-III processor in our implementation, which provides an atomic memory-modification operation, the CAS (Compare-And-Swap) instruction. In our study, we used queues, one of the common shared objects, to validate our theorems. We used the lock-free queues introduced in [14] in our implementation.

4.1 Object Access Times, Critical Load

As Theorem 3 shows, the tradeoff between lock-based and lock-free sharing under RUA depends upon the time needed for object access—i.e. lock-based object access time r, and lock-free object access time s. We measure r and s with a 10 task set. Each measurement is an average of approximately 2000 samples.

Figure 5a shows r and s under increasing number of shared objects accessed by jobs, not allowing any nested critical section. From the figure, we observe that r is significantly larger than s. As the number of shared objects accessed by each job, r is increasing. Note that r includes the time for lock-based RUA's resource sharing mechanism.

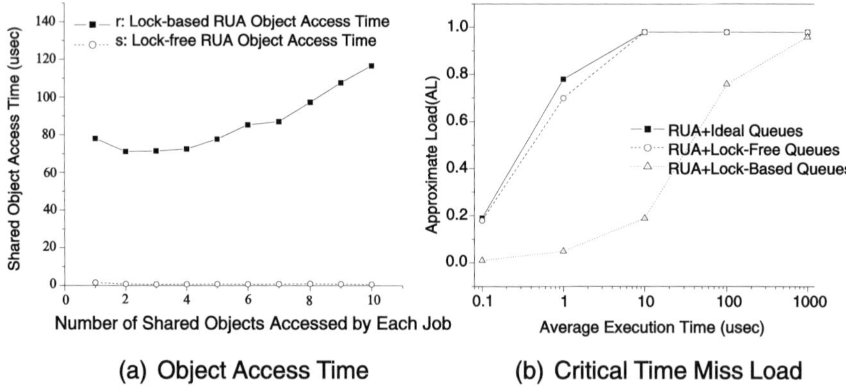

(a) Object Access Time (b) Critical Time Miss Load

Fig. 5 Object access times and critical time miss load

When $r >> s$, Theorem 3 implies that lock-free is likely to perform better than lock-based. Furthermore, when the number of objects m_i increases, it increases the likelihood of satisfying Corollary 4 and lock-free becomes increasingly advantageous over lock-based.

The impact of r and s on lock-based RUA's and lock-free RUA's performance, respectively, can be measured by evaluating the load at which the schedulers miss task critical times. We measure it using a metric called *Critical time-Miss Load* (CML). The CML of a scheduler is defined as the approximate load *after which* the scheduler begins to miss task critical times. We define approximate load as $AL = \Sigma_{i=1}^{n} u_i/C_i$, where u_i is the task computation time excluding object access time, and C_i is the task critical time.

We exclude object access time in CML, because an ideal implementation of objects for synchronization must have negligibly small—almost zero—object access time. If so, the implementation is ideal in the sense that the scheduler's performance with the (ideal) implementation is the same as that under no object sharing. We call it, the ideal RUA.

We consider a task set of 10 tasks, accessing 10 shared queues, and measure the CML of lock-free, lock-based, and ideal RUA under increasing average job execution times. Figure 5b shows the results. We observe that lock-free RUA yields almost the same CML as that of ideal RUA, as it exploits the eliminated blocking times and achieves almost the same performance of RUA without object sharing. Note that the ideal queue and RUA achieve the CML of 1, only at ≈ 10 us of average execution time. This is because of the algorithm's overhead for scheduling. (RUA's CML of 1 is valid at zero job execution times when assuming no system overheads, which is not true in practice.)

On the other hand, lock-based RUA's CML converges to 1, only at ≈ 1 ms. This is precisely because of lock-based RUA's complex operations for resolving jobs' contention for object locks and consequent higher overhead, as manifested by its higher asymptotic complexity and higher object access times in Figure 5a.

4.2 Accrued Utility, Critical Time-Meets

We now measure the AUR and the CMR of lock-free RUA and lock-based RUA for average job execution times in the range of 30 us–1000 us. CMR is the ratio of the number of tasks that meet their critical times to the total number of task releases. We consider a task set of 10 tasks, accessing 10 shared queues. Each experiment is repeated to obtain AUR and CMR averages from more than 5000 task arrivals. We consider a heterogenous class including step, parabolic, and linearly decreasing shapes.

Figures 6 and 7 show lock-based and lock-free RUA's performance, respectively, under heterogenous TUFs, $AL = \approx 1.1$, and increasing number of shared objects. As expected, the figures show that lock-based RUA's AUR and CMR sharply decreases, eventually reaching 0% during overloads, as the number of objects increases. This is because, as the number of objects increases, greater number of task blockings'

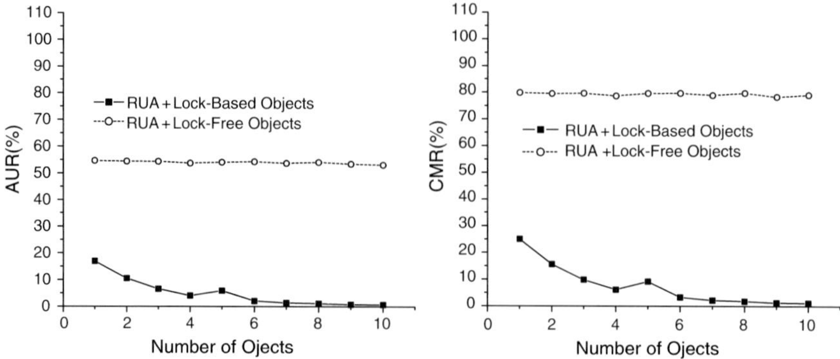

Fig. 6 AUR/CMR during overload, heterogeneous TUFs

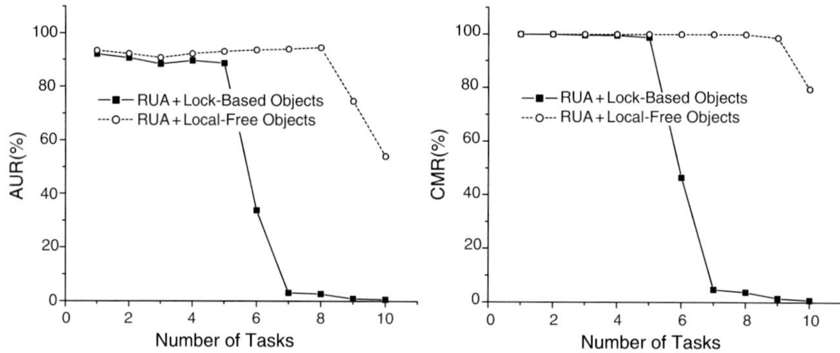

Fig. 7 AUR/CMR during increasing readers, heterogeneous TUFs

occurs, due to the large r, resulting in increased sojourn times, critical time-misses, and consequent abortions. The performance of lock-free RUA, on the contrary, does not degrade as the number of objects increases. This is directly due to the short s of lock-free objects, which results in few retries and thus reduced interferences.

We repeated similar experiments for increasing number of reader tasks (instead of increasing shared objects) and observed exact similar trends and consistent results in Fig. 7 (Heterogeneous TUFs, AL=0.1−1.1), further illustrating lock-free RUA's superiority over lock-based. We omit more results as they show the same trend and consistency.

5 Conclusions

In this paper, we consider non-blocking synchronization for embedded real-time systems that are subject to resource overloads and arbitrary activity arrivals. We consider lock-free synchronization for the multi-writer/multi-reader problem that occurs in such systems. We establish the tradeoffs between lock-free and lock-based object sharing under the UAM, including the conditions under which activity timeliness utility is greater under lock-free than under lock-based, and the upper bound on this utility increase—the first such result. Our implementation experience on a POSIX RTOS strongly validates our analytical results.

Future work includes extending the results to algorithms that provide activity-level timing assurances, the snapshot abstraction, and multiprocessor and distributed systems.

References

[1] J. Anderson, R. Jain, and K. Jeffay. Efficient object sharing in quantum-based real-time systems. In *RTSS*, 1998.

[2] J. H. Anderson, R. Jain, and S. Ramamurthy. Wait-free object-sharing schemes for real-time uniprocessors and multiprocessors. In *IEEE RTSS*, pp. 111–122, Dec. 1997.

[3] J. H. Anderson, S. Ramamurthy, and K. Jeffay. Real-time computing with lock-free shared objects. *ACM TOCS*, 15(2): 134–165, 1997.

[4] J. Chen and A. Burns. Asynchronous data sharing in multiprocessor real-time systems using process consensus. In *10th Euromicro Workshop on Real-Time System*, 1998.

[5] H. Cho et al. A space-optimal, wait-free real-time synchronization protocol. In *IEEE ECRTS*, 2005.

[6] R. Clark, E. D. Jensen, et al. An adaptive, distributed airborne tracking system. In *IEEE WPDRTS*, April 1999.

[7] R. K. Clark, E. D. Jensen, and N. F. Rouquette. Software organization to facilitate dynamic processor scheduling. In *IEEE WPDRTS*, April 2004.

[8] J.-F. Hermant and G. L. Lann. A protocol and correctness proofs for real-time high-performance broadcast networks. In *IEEE ICDCS*, p. 360–369, 1998.

[9] H. Huang, P. Pillai, and K. G. Shin. Improving wait-free algorithms for interprocess communication in embedded real-time systems. In *USENIX Annual*, p. 303–316, 2002.

[10] E. D. Jensen, C. D. Locke, and H. Tokuda. A time-driven scheduling model for real-time systems. In *IEEE RTSS*, p. 112–122, December 1985.

[11] H. Kopetz and J. Reisinger. The non-blocking write protocol nbw: A solution to a real-time synchronisation problem. In *IEEE RTSS*, p. 131–137, 1993.

[12] P. Li, B. Ravindran, et al. A formally verified application-level framework for real-time scheduling on posix real-time operating systems. *IEEE TSE*, 30(9):613–629, Sept. 2004.

[13] M. Herlihy. A methodology for implementing highly concurrent data objects. *ACM TOPLAS*, 15(5):745–770, 1993.

[14] M. M. Michael and M. L. Scott. Non-blocking algorithms and preemption-safe locking on multiprogrammed shared memory multiprocessors. *JPDC*, 51(1): 1–26, May 1998.

[15] B. Ravindran, E. D. Jensen, and P. Li. On recent advances in time/utility function real-time scheduling and resource management. In *IEEE ISORC*, p. 55–60, May 2005.

[16] L. Sha, R. Rajkumar, and J. P. Lehoczky. Priority inheritance protocols: An approach to real-time synchronization. *IEEE Trans. Computers*, 39(9): 1175–1185, 1990.

[17] R. K. Treiber. System programming: Copying with parallelism. Technical report, IBM Almaden Research Center, April 1986. RJ 5118.

[18] J. Valois. Lock-free linked list using compare-and-swap. In *ACM PODC*, pp. 214–222, 1995.

[19] H. Wu, B. Ravindran, et al. Utility accrual scheduling under arbitrary time/utility functions and multiunit resource constraints. In *IEEE RTCSA*, August 2004.

[20] H. Cho. Utility Accrual Scheduling with Non-Blocking Synchronization on Uniprocessors and Multiprocessors. In *PhD Dissertation Proposal*, ECE Dept., Virginia Tech, 2005, http://www.ee.vt.edu/~realtime/cho_proposal05.pdf.

What If You Could Design Tomorrow's System Today?

Neal Wingen

Abstract This paper highlights a series of proven concepts aimed at facilitating the design of next generation systems. Practical system design examples are examined and provide insight on how to cope with today's complex design challenges.

1 Introduction

In today's economy, consumers are always searching for the "next experience". Consumers have grown to expect a steady stream of new and constantly improving products like cell phones, televisions, automobiles, computers, and other electronic gadgets. To meet this demand, the semiconductor design community must be able to release next generation systems with a regular "heart beat".

There are several techniques and methodologies that have been employed to aid in this effort. A hierarchical design methodology known as "Islands of Synchronicity" (IoS) in tandem with advanced System-on-Chip (SoC) infrastructures can be used to break down large, complex systems into independent, locally synchronous islands. The Islands of Power methodology takes this one step further and accounts for the advanced power management required in today's systems. Platform based design becomes a reality when these methodologies and the infrastructures are fully leveraged and the right chip-to-chip interconnect is available. Finally, the latest design automation tooling can speed the integration and verification of these systems.

NXP Semiconductors, USA

R. Lauwereins and J. Madsen (eds.), *Design, Automation, and Test in Europe–The Most Influential Papers of 10 Years DATE*, 87–101
© Springer 2008

2 Islands of Synchronicity

2.1 *Synchronous Design Limitations*

As SoC designs grow bigger, the globally synchronous design grows slower. In a globally synchronous environment, the maximum operating frequency of on-chip interconnect is limited by the absolute delay between the source and destination. Interconnect delays are increasing with each new geometry node and wire delay now dominates logic delay for long traces. Buffer insertion to balance clock trees, to resolve edge rates and to avoid cross-talk problems can further limit the operating speed of the interconnect.

Another ominous trend for synchronous design regards clock distribution. As SoC designs grow bigger and contain more clock domains, clock tree balancing becomes very difficult to achieve at any operating point, and is further complicated by variations in process, voltage, and temperature across the die and intentional variations in voltage for power management. Globally synchronous designs with deep clock buffer trees can suffer from large instantaneous power surges and Electro Magnetic Interference (EMI) coincident with the clock switching activity.

This combination of interconnect performance limitations and global clocking problems requires new solutions that scale with technology and alleviate timing closure issues.

2.2 *Hierarchical Design Trend*

The IoS methodology provides unique solutions for the hierarchical design process. The move to a hierarchical approach is inevitable for most large SoC designs. Designs are just too large and complex to be completed flat. Rather than push EDA tools and compute resources beyond their capabilities, it makes sense to break designs into smaller, manageable pieces. Different design teams in different locations can then implement the design, each with its own area of expertise, in a concurrent engineering environment.

2.3 *SoC Infrastructure "Seams"*

The key to hierarchical design is efficient partitioning and reassembly. The SoC infrastructure must have "seams" to guide the partitioning of the design and identify possible clock and voltage domain boundaries. In Fig. 1, the seams embedded in the infrastructure can be split to allow the design to be partitioned into "islands". Each seam straddles two islands and provides asynchronous or source synchronous communication between them.

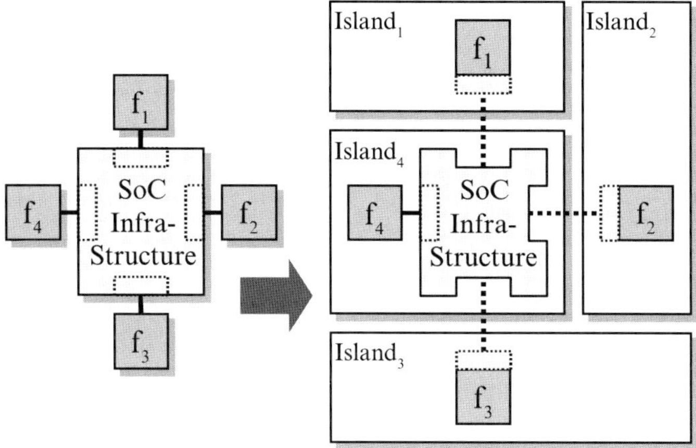

(a) "Seams" in SoC Infrastructure (b) "Seams" split to make islands

Fig. 1 Communication centric system-on-chip with "Seams"

The nonsynchronous interisland communication keeps the timing of the islands independent of the top-level design. It also removes the need for global synchronous clocking and allows clock trees to be implemented locally within islands, making them easier to balance with fewer levels of buffering. In fact, clock balancing between islands is not needed. The overall result is a SoC that is globally asynchronous, but locally synchronous within the islands.

2.4   Seam Classification

There are several varieties of seams based on the possible relationships two islands or domains can have with one another. A "phase seam" is needed between two islands that operate at the same frequency and voltage, but different clock phase. When two islands operate at different frequencies, a "frequency seam" is needed to enable the communication at different rates. A "voltage seam" provides isolation logic necessary when two islands communicate but one or both sides are capable of turning off. A combination "voltage-frequency" seam is needed for communication between islands with varying voltages and frequencies, which is common in designs that utilize Dynamic Voltage Frequency Scaling (DVFS).

2.5   Islands of Power Methodology

New power management techniques like DVFS require further advancements beyond IoS. Islands of Power (IoP) is an evolving methodology to enable multiple variable supply voltage (MVSV) designs. IoP is a comprehensive collection and

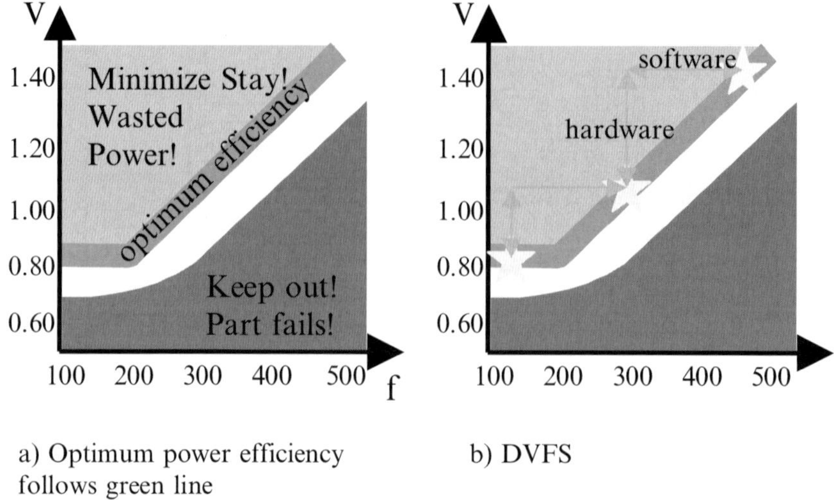

a) Optimum power efficiency
follows green line

b) DVFS

Fig. 2 Dynamic voltage frequency scaling applied to a hypothetical major power consumer

alignment of many disciplines, including power aware: library components, memories, power infrastructure IP, SoC interconnect, compute engines, EDA tooling and flows, board components and software.

In DVFS, the major power consumers in the SoC operate as close to their optimum efficiency curve with respect to voltage and frequency (Fig. 2a). By operating as close to the failure point as possible, the power consumption is as small as possible. In a closed loop mode of operation, hardware automatically finds the optimum voltage given any desired frequency (Fig. 2b).

To raise performance, one must wait for the hardware to find the new voltage before we increase the frequency via software. Conversely, to lower performance, one can lower the frequency immediately while the hardware finds the lowest operating voltage automatically. The optimum-efficiency curve can change with process and temperature and closed loop DVFS hardware can account for this fluctuation and can still find the optimum point for any die at any operating condition.

An open loop mode of operation utilizes a look up table of frequencies and voltage stored in software. In the open loop model, the performance must be chosen for worst-case parts and does not adjust for processing and temperature fluctuations.

Since DVFS controls the frequency and voltage components of the power equation, both dynamic and static power can be reduced.

2.6 Voltage Seams

Voltage seams account for when one side of the seam turns off voltage. When this happens, the undriven input signals need to be clamped and held to their reset

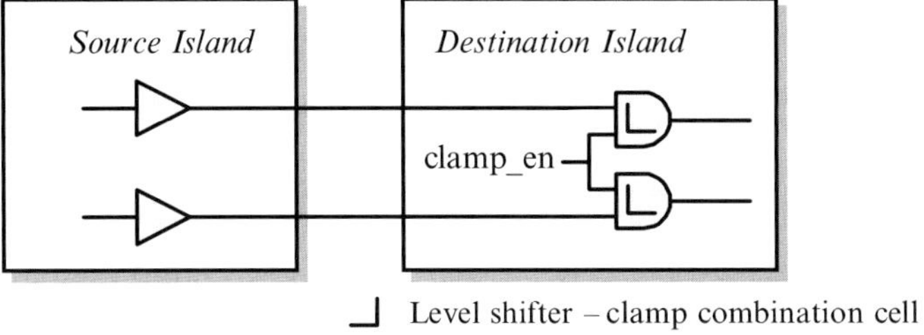

Fig. 3 Voltage only seam

states. Signals traversing from low to high voltage domain also need level shifters when the voltage difference is great enough. Isolation logic is inserted in receiving domain. In Fig. 3, the source island can turn off so the destination island utilizes a combination level shift-clamp cells to provide the necessary isolation.

2.7 Asynchronous-based Voltage Frequency Seams

An asynchronous link uses a two-phase or four-phase handshake mechanism to pass data safely between islands. Figure 5 shows a typical implementation with a request (REQ) and acknowledge (ACK) signal pair performing the handshake. The handshake signals are synchronized to the clock domain they travel to.

The 4-phase handshake seams are naturally voltage tolerant, as the handshake protocol always returns to a known state. A 2-phase handshake seam is toggle based and more difficult to predict. When voltage domains turn off and on, a false handshake trigger could occur. To guard against this, both sides of a 2-phase handshake need to be held in reset when either side powers down.

2.8 Source Synchronous-based Voltage Frequency Seams

A source synchronous link (Fig. 4a) transfers the clock with the data payload from the source island to the destination island. To limit power consumption, the clock only needs to be sent when data is present. Tokens or sideband signals can handle flow control. Typically, storage or data buffering is best kept on the destination side of the seam. Data storage size can be tuned to match cache line size, for instance, and can minimize impact of the latency associated with the link.

Source synchronous links can operate at very high speeds across long distances. The speed of the link is limited by clock and the data payload skew alignment (Fig. 4b), not propagation delays like a synchronous connection. Source synchronous links can be easy to partition if the source and destination sides are delivered in separate modules.

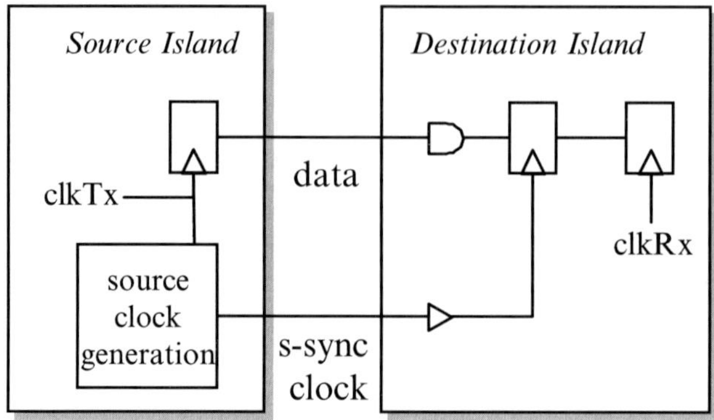

(a) Source Synchronous Seam Structure

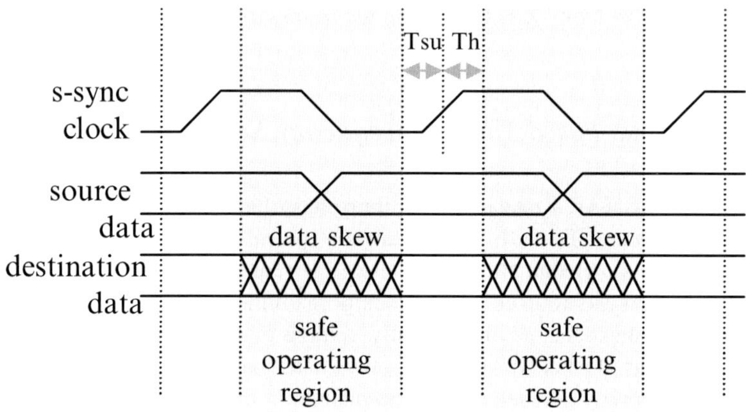

Fig. 4 Source synchronous seam

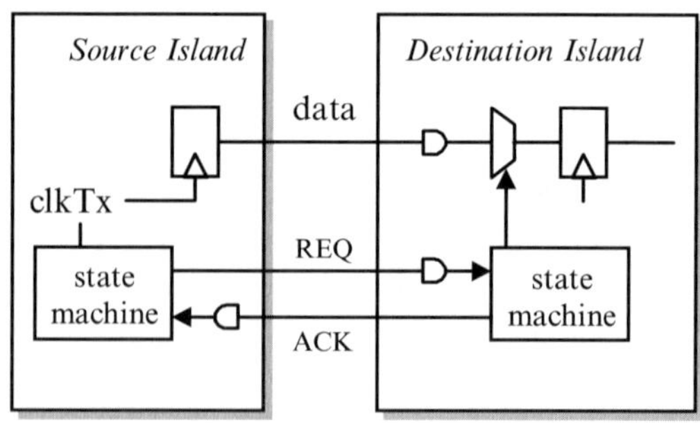

Fig. 5 Asynchronous seam

Source synchronous links may also need provisions to reset both sides of the link whenever one side powers down to maintain correct pointer and token values.

2.9 IoS–IoP Design Considerations

2.9.1 Partitioning Criteria

The chip architect must understand the performance requirements, floor plan, connectivity, and use case information to create effective partitions. There is currently no automated way to determine the optimum partitioning for a large SoC.

Island sizes will vary based on the maximum operating frequency. A smaller island will typically run faster than a larger island. The island size should also be suitable for EDA tools to produce short run times.

Performance or latency requirements will also determine partition placement. Partition seams have latency associated with them. Communication that is latency critical should be self-contained in an island and not travel across a seam to another island.

Voltage domains also influence the partitioning strategy. Major power consumers may be controlled using DVFS techniques and will require hierarchy levels to contain the specific voltage domains. Use cases may require voltage domains to turn off in certain situations to meet power budget requirements. Power supply unit capabilities and internal or external voltage switches will further determine hierarchy and may force voltages to be combined or separated.

Partitioning also needs to consider the physical floor plan. Pin intensive blocks will naturally prefer to be close to the pads and not combined with a large embedded block. The design should be partitioned in a way to limit the top-level wires or the connections on the island boundaries. Major power consumers should have dedicated power supplies and should be placed as close as possible to the voltage source. IR-drop also needs to be tightly limited otherwise the voltage margin needed for effective DVFS operation is reduced.

2.9.2 Performance

The source synchronous and asynchronous links exhibit more latency cycles than the synchronous equivalents. But caches can do an excellent job of hiding these latencies. The latency impact is only exposed when a cache miss occurs. In a real-life example that shows the benefit of caching, a VLIW DSP running an H.264 video decode benchmark with a source synchronous link towards the main memory controller had virtually identical performance when compared to a direct synchronous connection. Of course, the source synchronous implementation did not require clock balancing and was easier to timing close. It is very difficult to theorize cache hits and misses without running an application, but reasonable cache hit rates would suggest only a modest impact if an embedded application processor had source synchronous links to the memory network. Another consideration is that asynchronous design can run faster than synchronous design, especially for large devices as simplified timing closure yields higher frequencies. Asynchronous design can actually be higher overall performance than synchronous equivalent in some cases.

The added delay needed for the isolation logic to separate voltage domains is absorbed well in a voltage seam since the added propagation delay is of no consequence. This would not be the case for a synchronous design.

2.9.3   Area Impact

It is difficult to say categorically that an IoS methodology increases chip area. In some cases, asynchronous versions of networks can be smaller in area and net count when compared to a synchronous counterpart. Source synchronous links do have an area overhead associated with them; however, for large SoC designs, the added area is typically negligible.

3   Platform Based Design

In most cases, the next generation product is not a revolutionary new architecture, but merely an extension or improvement of the current product. Therefore, to make the next generation product, new functions need to be added to the existing product. In platform-based design, a platform is built of the generation N product and this platform is used to create as many derivatives (generation N+1) as possible.

3.1   *History*

Making derivatives from a platform is not a new idea. The automotive and personal computer industries are good examples of successful platform based design. An automotive platform can be applied to several different derivative car styles and models. Automotive designers can concentrate on the differentiating features of the automobile that return the maximum value and not the underlying infrastructures. The automotive industry has proven that platform-based design reduces engineering costs, speeds up development time and makes it possible to offer a greater variety of new models. Platform-based design has never flourished within the semiconductor industry; however, new industry chip-to-chip interconnect standards can drive the concept into the mainstream.

3.2   Hardware Platforms

Nexperia is the NXP brand for a unique group of platforms and products that streamline development of next-generation, connected multimedia appliances. From highly integrated, programmable system-on-chip (SoC) and companion ICs to reference designs, system software, and development tools, flexible Nexperia solutions help manufacturers meet demands for new products. Figure 6 shows one example of how the Nexperia platform concept has been applied to digital television sets. Manufacturers use the base platform to build their basic sets. The platform is extended to add features necessary for their higher end television sets. The extension, in this case, is a companion IC; however, an FGPA is also a common choice. A high bandwidth, low pin count chip-to-chip interconnect enables the two IC's to communicate with each other.

3.3   Chip-to-chip Interconnect

The chip-to-chip interconnect used in Fig. 6 is from a family of NXP proprietary links called Transaction Transport Tunnels. These tunnel mechanisms serve as a bridge to make the extended logic appear as if it is actually on the original base platform or reference design.

In Fig. 7, IC1 represents the generation N product and includes a tunnel to extend functionality in the future. The tunnel contains several layers of abstraction, including transaction, transport and data link layers and just appears as an extra port

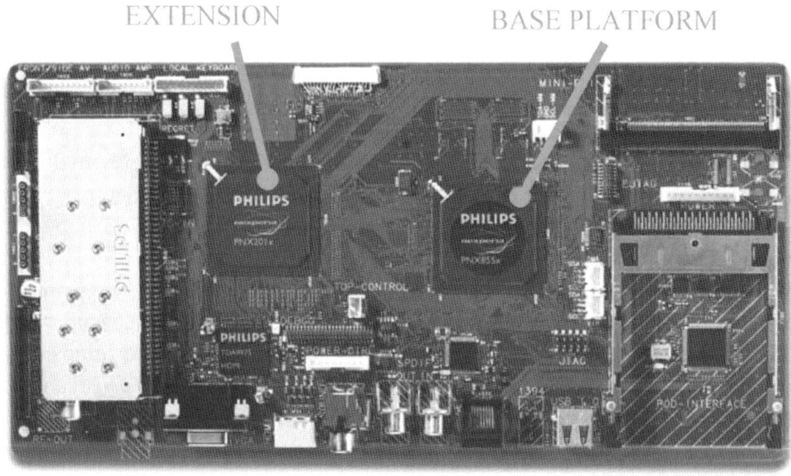

Fig. 6 Nexperia TV810 digital television platform

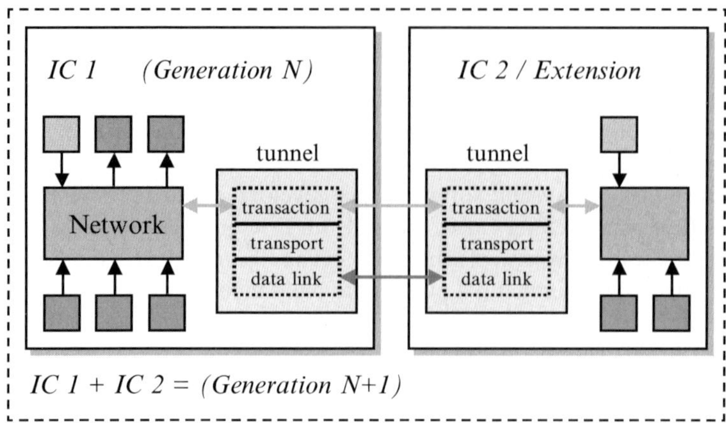

Fig. 7 Platform extension with chip-to-chip interconnect

on the internal bus network. By connecting IC2 to IC1 via the tunnel, the next generation product, generation N+1, is created.

In the transaction layer, native bus transactions, such as AXI, APB, and AHB, are passed implicitly between the IC's. From the IC perspective, only these native bus transaction passing between the IC's are visible.

The transport layer converts the bus transactions into packet formats needed to pass data, address, general-purpose signaling, credits, and error correction information. The general purpose signaling is useful for passing power state information, interrupts or any other relatively low-frequency signaling back and forth. These general-purpose signals can be transported in a dedicated packet, which can be inserted anywhere in the packet stream in order to keep the latency reasonable.

The physical communication link between the two IC's occurs in the data link layer. The link information can be specially encoded and decoded for DC balancing, to increase noise immunity and minimize power. Low swing I/O are also a popular choice to improve speed and further reduce power. In some cases, a PHY device can be used to boost performance and further minimize pin count. By making this layer source-synchronous, both sides of the link are free of synchronous frequency and phase relationship constraints and the link must only balance skew between the clock and data payload.

The tunnel has no software view and is transparent from the application perspective. The IC's only see the native transaction protocol, not the special transport protocol or data link layer. Therefore, the tunnel logic can simply be removed from the final product without any impact. Of course, the tunnel mechanisms can be left in, as well, to serve as a means for possible product differentiation in the future.

The tunnel data width can be configured or tuned to meet the performance and bandwidth requirements for the application. Since tunnels are packet based, fewer pins are necessary, especially when compared to a traditional, on-chip bus being brought off-chip. By using source synchronous techniques, the tunnels can maintain or exceed the full speed of the internal bus network, despite traveling off-chip. The internal, synchronous bus networks are not suited for use off-chip; resulting in a slower performance and higher pin cost.

Since the base platform and the extension logic can operate at full speed, real-time prototyping can be utilized. By operating at the target system speed, real-time system problems can be exposed and many orders of magnitude more patterns can be exercised than traditional simulation. In addition, full speed, real-time operation enables system capabilities to be fully demonstrated. Engineers are more efficient at debugging video applications when they can view video streams and hear audio streams real time. Also, the actual silicon running at speed can be more accurate than a model.

3.4 Tunnels

The AXI Tunnel (also called TAXI, for Tunnel-AXI) was created by NXP as a way to extend an on-chip AXI bus or interconnect to a second, off-chip AXI interconnect. Because information from all five channels of AXI must be sent through a single connection, bandwidth through the link is limited. For example, with an 8-bit interface operating at 150 MHz, the peak bandwidth is about 200 MB/s (100 MB/s in each direction). With a 48-bit interface, the bandwidth can be six times higher, but at a cost of nearly six times more pins. The AXI tunnel gate count is 16–20K gates, depending on the configuration.

The CTL12 Tunnel utilizes an NXP proprietary transaction layer. The transport layer is comprised of 12-bit symbols while the data link layer utilizes 6b8b decoding. This tunnel can achieve a relatively high bandwidth (1066 MB/s) with a modest amount of pins (39).

In the future, several industry standard serial interconnects may become excellent tunnel candidates. The Mobile Industry Processor Interface (MIPI) Alliance has defined the UniPro 1.0 interface as a mobile terminal device standard and an 8-bit interface is capable of 250 MB/s. Inter-Chip USB extends the scope of USB 2.0 to include the use of USB within embedded systems. A full speed IC-USB link can accommodate 1.5 MB/s given a few pins, while the high-speed version will handle 60 MB/s. PCI Express can deliver an impressive 500 MB/s through a 4-bit interface.

In general, the industry standards are very robust and have a larger footprint or gate count than the proprietary interfaces. Nevertheless, as these interfaces become commodities and silicon integration capabilities continue their expansion, new opportunities such as Platform-based design, become more feasible.

3.5 Extension Examples

The extension logic provides additional functionality; either not considered at the time of the original IC design or that is optional and well suited for a separate IC.

For prototyping purposes, a new IC could be prototyped using an existing IC together with an FGPA. Also, the companion IC could be in a different technology than the base platform.

3.6 *Architecture Considerations*

When designing the platform, it is important to make sure the available bandwidth of the tunnel can accommodate all the possible extensions. In a processor-centric platform, the CPU would typically require a majority of the available bandwidth through the tunnel device. A product centric platform, where a majority of the functionality already exists in the platform, may only require a small additional bandwidth through the tunnel. If the extension will process a data stream, the tunnel must be able to accommodate the data bandwidth with the appropriate margin above and beyond.

4   SoC Integration and Verification

4.1 *Nx-Builder*

Today's SoC designs are a myriad of reusable IP components including peripherals, infrastructures, compute engines, subsystems, and even entire ICs. As a result, SoC integration and verification have become monumental tasks and the next frontier for the EDA industry. To further facilitate this effort, the Spirit Consortium [1] has established a set of IP and tool integration standards with the IP-XACT specification [2] as the first embodiment. Automation and standards have the potential to reduce the many months of effort required for SoC integration and verification to weeks, directly impacting time to market. Nx-Builder is an NXP proprietary SoC design environment built on top of Platform Express from Mentor Graphics and uses a simple "enter, build and verify" approach to system architecture development.

4.1.1   Design Entry

Design entry is performed with a high-level schematic editor or a table-based entry mechanism. IP is added to the design with a drag-and-drop approach. Generally, an IP block is selected from the library window and dropped onto the bus it is to connect to. If the IP is configurable, the user will be prompted to provide any necessary parameters. Each IP block has an "electronic data-sheet" in eXtensible Markup Language (XML) [3] form that allows the IP to evolve with the design.

There are a number of automation mechanisms in place to simplify and speed entry. A large design will contain a daunting number of configuration parameters; however, as many parameters as possible are automatically derived from the system context leaving minimal user input. The tool is also capable of automatically adding interface IP (bus bridges, adapters) and infrastructure IP (clocks, resets, interrupts, interconnect networks) during the course of design entry. In addition, the tool maintains an automated interaction with the IP repository to track the abstraction of the IP and flag any possible errors.

4.1.2   Design Build

Once the design has been entered and defined, the underlying design files can be generated and compiled. The front-end views (RTL, models, documentation, etc.) are extracted from the IP repository given the configuration parameters defined in the design entry phase. A standardized file structure helps assemble the individual IP directories in to a cohesive database. In addition, batch mode flow scripts are created for synthesis, simulation and design-for-test insertion. Simple "make" files control the generation and compilation activities, including provisions to choose a specific hardware model type or source.

4.1.3   Design Verification

Software is automatically created to exercise the system hardware. Each piece of IP includes a software module that tests that particular IP when instantiated in a system. Special software module "stubs" exercise external interfaces via software, too. All of the software modules are combined and compiled for the target system to make a self-checking executable ready for hardware–software co-simulation or real-time verification in an FGPA environment. Different software packages corresponding to various use cases can also be created. Software to stress performance, quality of service properties, and power management use cases will become available in the future.

The generated software can verify the integrity of the IP interconnect to the system and give confidence that the design has been connected correctly. Also, the SoC infrastructure (busses, bridges, register space, memories) can be stressed and verified. Modules can be exercised simultaneously, to verify interoperability of the system.

Other design flow scripts for synthesis, static timing analysis, formal equivalence checking and Design-for-Test insertion can be executed in this stage as well.

4.2   Methodology Reuse

Nx-Builder is created based upon designers' experience and standards. It essentially encapsulates the best engineering knowledge and makes this available to the entire user community, providing a generic framework that enables different SoC designers to implement their systems in a consistent, flexible and easy-to-use way.

4.3   Architecture Reuse

Nx-Builder also raises the level of design reuse, beyond the traditional IP level. Integration reuse determines how the IP interconnections are made; including the connections to the SoC infrastructures and how to constrain these interconnect nets. The delivery of simulation and synthesis scripts specific to the architecture specified enforces methodology reuse. The SoC infrastructure is automatically generated based on the system context providing a means for infrastructure reuse. As architectures become IP, architecture reuse and software reuse mature.

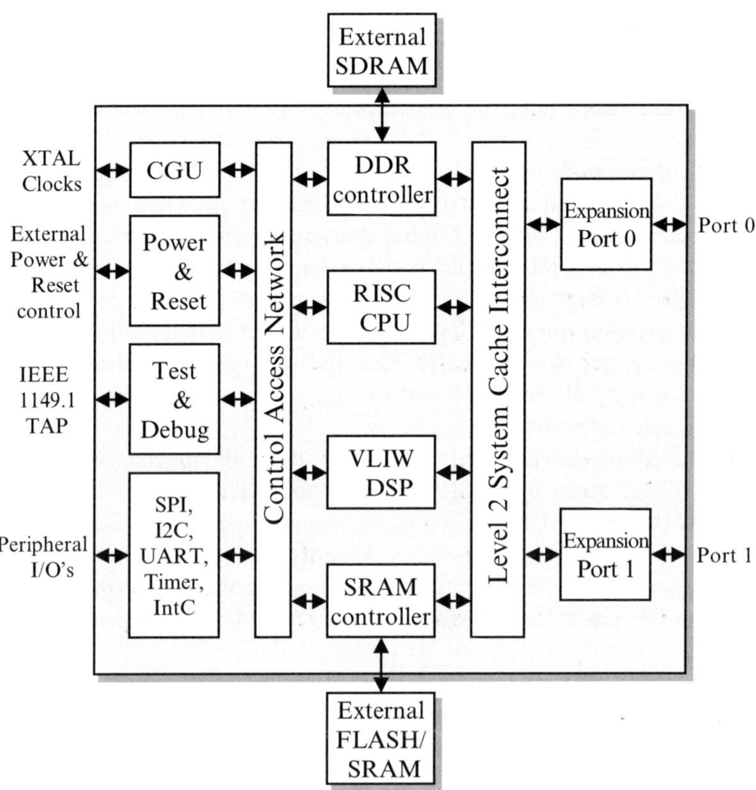

Fig. 8 Energizer II platform chip

4.4   Virtual platforms

Nx-Builder provides a flexible way to define systems or "virtual" platforms and makes it easy to build and evaluate platform and derivative architectures, including software, without the manufacturing commitment. Virtual platforms are cheaper than hardware platforms and can be put on the workstation of every software developer. Once the initial design is available in Nx-Builder, SoC designs, or derivatives can be developed and verified in a minimum amount of time.

5   Conclusion

A platform device has been developed that utilizes all of the concepts included herein (Fig. 8).

The SoC consists of 11 islands. The 3 major power consumers (RISC CPU, VLIW DSP, and L2 System Cache) are controlled using DVFS. High bandwidth expansion ports enable the platform to be extended, with graphics or cellular modem subsystems for example. Nx-Builder also played a role in the integration and verification effort. This platform chip is $42\,mm^2$ in a $65\,nm$ process.

Acknowledgment   Tim Pontius, Greg Ehmann, George Spatz, Marino Strik, Ramon Baas, and Marc Heijligers.

References

[1] SPIRIT Consortium, http://www.spiritconsortium.com
[2] SPIRIT Schema Working Group, "IP-XACT User Guide", Version 1.2, July 2006
[3] eXtensible Markup Language [Online]. Available at http://www.w3schools.com/xml/xml-syntax.asp

Part II
Networks on Chip

Networks on Chips

Giovanni De Micheli

We are witnessing a growing interest in *Networks on Chips* (NoC) that is related to the evolution of integrated circuit technology and to the growing requirements in performance and portability of electronic systems. Current integrated circuits contain several processing cores, and even relatively simple systems, such as cellular telephones, behave as multiprocessors. Moreover, many electronic systems consist of heterogeneous components and they require efficient on-chip communication. In the last few years, multiprocessing platforms have been developed to address high performance computation, such as image rendering. Examples are Sony's emotion engine [OKA] and IBM's cell chip [PHAM] where on-chip communication efficiency is key to the overall system performance.

The introduction of NoCs in *Systems on Chips* (SoCs) has been a gradual process, which can be seen as an evolution of bus interconnect technology [DEMI]. For example, there is not a strict distinction between multilayer busses and crossbar NoCs. Overall, NoCs have become a broad topic of research and development in the new millennium. This is witnessed by a large number of technical contributions that address the system-level solution to cope with technological limitations, such as wiring complexity and propagation delay spread. It is indeed the need for achieving *design closure* with an unpredictable interconnect that motivates the use of NoCs. Compared to traditional interconnects the structure of the NoC and the wiring complexity can be well controlled. When the interconnect is structured, the number of timing violations that occur during the physical design phase is minimum. Such design predictability is critical for today's SoCs for achieving timing closure. It leads to faster design cycle, reduction in the number of design respins, and faster time-to-market.

When considering network technology in general, it is possible to tame the network complexity and to provide reliable service in presence of local delays and/or malfunctions. Thus, networking technology has been able to provide us with *Quality of Service* (QoS), despite the heterogeneity and variability of the interconnect nodes and links. It is then obvious that networking technology can be

EPF Lausanne

R. Lauwereins and J. Madsen (eds.), *Design, Automation, and Test
in Europe–The Most Influential Papers of 10 Years DATE*, 105–110
© Springer 2008

instrumental for the improvement of *Very Large Scale Integration* (VLSI) circuit/ system design technology.

On the other hand, the challenges in marrying network and VLSI technologies are in leveraging the essential features of networking that are crucial to obtaining fast and reliable on-chip communication. On-chip communication must be fast, and thus networking techniques must be simple and effective. Bandwidth, latency and energy consumption for communication must be traded off in the search for the best solution.

Fortunately VLSI chips have wide availability of wires on many layers that can be used to carry data and control information. Having on-chip links with wide data-width realizes the parallel transport of information. Moreover, data and control do not need to be transported by the same means, as in networked computers. Local proximity of computational and storage units on chip makes transport extremely fast. Overall, the wire-oriented nature of VLSI chips makes on-chip networking both an opportunity and a challenge.

Another main reason for using on-chip networking is to achieve performance using a system perspective of communication. This motivation is supported by the fact that simple on-chip communication solutions do not scale up when the number of processing and storage arrays on chip increases. For example, on-chip busses can serve a limited number of units, and beyond that, performance degrades due to the bus parasitic capacitance and the complexity of arbitration. But NoCs' potential benefits go beyond outperforming buses. Indeed they support modular and regular interconnection design as well as a means to increase SoC reliability by layered design.

There are several ways of classifying SoCS according to the required functionality and market: in general, they can be classified in terms of their versatility (i.e. support for programing) and application domains [DEMI]. A simple regrouping is described next.

- *General-purpose on-chip multiprocessors* are high-performance chips that benefit from spatial locality to achieve high performance. They are designed to support various applications, and thus the processor core usage and traffic patterns may vary widely. They are the evolution of onboard multiprocessors, and they are characterized by having a homogeneous set of processing and storage arrays. For this reasons, on-chip network design can benefit from the experience on many architectures and techniques developed for onboard multiprocessors, with the appropriate adjustments to operate on a silicon substrate. A recent example is a 80-tile NoC-connected SoC manufactured by Intel [VANG]

- *Application-specific SoCs* are hardware chips dedicated to an application. In some cases, as for all mobile applications, energy consumption is a major concern. Most application-specific SoCs are programmable, but their application domain is limited and the software characteristics are known a priori. Thus, some knowledge of the traffic pattern is available when NoCs are designed. In many cases, these systems contain fairly heterogeneous computing elements, such as processors, controllers, DSPs, and a number of domain-specific HW accelerators.

This heterogeneity may lead to specific traffic patterns and requirements, thus requiring NoCs with specialized architectures and protocols.

- *SoC platforms* are application-specific SoCs dedicated to a family of applications in a specific domain. Examples are SoCs for UMTS telephony support and platforms for automotive control. A platform is more versatile in nature, as it can be used in different systems (e.g. embedded systems) by different manufacturers. Thus, versatility and programmability are preferred to customization, yielding SoCs that can be produced in high volumes and thus offset the *nonrecurrent engineering* (NRE) costs. Whereas the processing and storage unit may differ in nature and performance, the traffic patterns are harder to guess a priori as the application software may vary widely.
- *Field programmable gate arrays* (FPGAs) are hardware systems where the functionality is determined after manufacturing by connecting and configuring components. Components vary in size and in functionality and are connected by reprogrammable networks. These networks are simple and provide bit-level connectivity with little or no control. Nevertheless we expect FPGAs to grow substantially over the coming years and require effective NoC communication.

NoCs differ from other wide-area networks because of local proximity and because their behavior is still much more predictable as compared to distributed systems. Indeed, despite the undesirable variability features of deep submicron CMOS technologies, it is still possible to predict many physical and electrical parameters with reasonable accuracy.

On the other hand, on-chip networks have a few distinctive characteristics, namely, *low communication latency, energy consumption constraints*, and *design-time specialization*. Latency of communication on chip needs to be small, i.e. in the order of few clock periods. The shortest latency implementations can be achieved by fully hard-wired implementations, which defeat the flexibility required by on-chip networks. Clearly, complex protocols for communication may add to the latency of the signals. Thus, to be competitive in performance, NoCs require streamlined protocols.

Energy consumption in NoCs is often a major concern, because whereas computation and storage energy greatly benefits from device scaling (smaller gates, smaller memory cells), the energy for global communication does not scale down. On the contrary, projections based on current delay optimization techniques for global wires [THEI] show that global communication on chip will require increasingly higher energy consumption. Hence, communication-energy minimization will be a growing concern in future technologies. Furthermore, network traffic control and monitoring can help in better managing the power consumed by networked computational resources. For instance, clock speed and voltage of end nodes can be varied according to available network bandwidth.

Design-time specialization is another facet of NoC design and it is relevant to application-specific and platform SoCs. Whereas macroscopic networks emphasize general-purpose communication and modularity, in NoCs these constraints are less restrictive, because most on-chip solutions are proprietary. Thus, NoC implementation may separate data from control, use arbitrary bus width and control flow schemes. Such a flexibility needs to be mitigated at the NoC boundary, i.e. where the communication infrastructure connects to end nodes (e.g. processors). Existing standards like the *Open Core Protocol* (OCP) are extremely useful in defining the interface between processor/storage arrays and NoCs [PULLI]. Interestingly enough, the flexibility in tailoring the NoC to the specific application can be used effectively to design low-energy communication schemes.

The history of NoCs, though rooted in some early work of Dally's at Caltech for onboard multiprocessor [DALL1, DALL2], came to light in the year 2000, with a paper by Guerrier and Greiner [GUER] at the *Design and Test Europe* (DATE) conference. This paper, presented the concept and first implementation of a NoC: the *scalable, programmable integrated network*, or SPIN. Whereas SPIN was not yet implemented on silicon, the contribution was significant because it brought out the pros and cons of NoCs and outlined an implementation that encompassed all basic issues: the network topology concept (a FAT tree), the packet-based on-chip communication and split transactions as well as the notion of on-chip switching and routing.

Since the year 2000, the concept and use of NoCs developed rapidly. Several industrial and academic research centers started developing NoC solutions. Problems of all kinds surfaced, and researchers started addressing them systematically. The Æ therreal project at Philips research focused on creating a complete design methodology for NoCs, especially for multi-media applications. Within this project, Rijpkema et al. [RIJP] addressed the important issue on quality of service in NoCs. They proposed a solution based on the design of a router, which combines guaranteed throughput and best effort services. An efficient NoC design is achieved by timing sharing the network resources across the different traffic streams. This paper is significant because it addresses the QoS problem, which is of extreme importance, and solves it elegantly within the NoC design paradigm.

Several researchers addressed the on-chip routing problem, which even though inspired by network routing, has to account for the physical limitations of silicon. Hu and Marculescu [Hu] explored the routing problem in conjunction with the mapping of an abstract communication task graph onto a tile-based architecture. The design process is driven by the goal of finding solutions in the power/ performance spectrum. When comparing Hu's mapping to traditional floorplanning, it is clear that the communication function of NoCs has a profound influence on the best position of each mapped tile. The routing function, which supports the communication on the mapped substrate, is then key in finding an optimal solution and is therefore defined concurrently with mapping.

Modularity and flexibility of NoCs allow designers to come up with creative and complex solutions, at the cost of long design time. Such a design process can be facilitated by the design of library component for NoCs. One of the early works in designing a parameterizable library for NoC is xPipes [DALL], designed at Universita' di Bologna, where all necessary elements to construct a NoC are modeled in SystemC. Jalabert et al. [JAL] designed a synthesis system for NoCs, which takes a high-level description of the NoC and maps it into an interconnection of xPipes library elements that is ready for cycle-accurate simulation and logic synthesis. xPipesCompiler performed additional optimization steps, such as sizing the switches, to adapt the library elements to the overall functional requirements. This project can be seen as one of the early works in creating synthesis tools for NoCs and the progenitor of design flows that help integrating NoCs into complex SoCs [MURA].

Performance and power evaluation of NoCs is extremely important to asses their effectiveness. Within design and development, NoCs need to be exercised with data whose properties match the data that will run on the complete and final product, and which of course is context dependent. Traffic generators are modules that abstract components on chip, and, for example, emulate the core behavior. Mahadevan et al. [MAHA] developed traffic generators that can produce realistic workloads, and that can be used as productivity tools to understand quickly the implication of specific architectural choices. Traffic generators have been used ubiquitously in most NoC design systems.

Beyond 2005 the field of NoCs has further evolved, with a few companies in this space and a specialized meeting: the NoC Symposium. For this evolution, much is owed to the seminal papers published in this book that were a strong stimulus to the research community.

Indeed NoCs were conceived with the goal of boosting system performance by taking a system view of on chip communication and by rationalizing communication resources. At the same time, NoCs promise to solve some of the problems of deep submicron technology, by providing a means to deal with signal delay variability and with unreliability of the physical interconnect. The SoC design community has been slowly responding to the challenge of *design closure*. NoCs are also well poised to be instrumental in solving this problem, by providing a clean and functional separation between computation and communication, thus, realizing the much stressed orthogonolization of concerns.

A few advanced chips already use NoCs as a communication substrate [VANG], and it is conjectured that NoCs will be the backbone of all SoCs of significant complexity designed with the 65 nm technology node and below. NoCs will be an integral part of SoC platforms dedicated to specific application domains, and programing platforms with NoCs will be simpler due to regularity and predictability. Moreover, FPGAs are becoming larger and ever-increasing sector of the semiconductor market. FPGAs represent another embodiment of NoCs. Within advanced FPGA architectures, both hardware computing elements and their interconnect are programmable, thus achieving an unprecedented flexibility.

Some design and implementation issues are still in search of efficient solutions. Problems get harder as we move up from the physical layer, because most hardware design problems are well understood while system software issues, from handling massive parallelism to generating predictable code are still being investigated. We expect major seminal contributions in NoCs to come from the system and software levels and in the design of methods to integrate the software with the management of communication in a seamless way. We expect this area of R&D to remain an excellent playground for new ideas and for successful start up companies.

References

[DEMI] G. De Micheli and L. Benini, Networks on Chips, *Morgan Kaufmann*, 2006.

[DALL1]W. Dally and C. Seitz, The Torus Routing Chip, *Distributed Processing*, Vol 1, pp. 187–196, 1996.

[DALL2] W. Dally and B. Towles, Route Packets, Not Wires: On-Chip Interconnection Networks, *Proceedings of the 38th Design Automation Conference*, pp. 684–689, 2001

[DALL] M. Dall'Osso, G. Biccari, L. Giovannini, D. Bertozzi and L. Benini, Xpipes: A Latency. Insensitive Parameterized Network-on-Chip Architecture for Multi-Processor SoCs, *International Conference on Computer Design*, pp. 536–539, 2003.

[GUER] P. Guerrier and A. Greiner, A Generic Architecture for On-Chip Packet-Switched Interconnections, *DATE- Proceedings of the Design Automation and Test in Europe Conference*, pp. 250–256, 2000.

[HU] J. Hu and R. Marculescu, Exploiting the Routing Flexibility for Energy/Performance Aware Mapping of Regular NoC Architectures, *DATE- Proceedings of the Design and Test Europe Conference*, pp. 10688–10693, 2003.

[JALA] A. Jalabert, S. Murali, L. Benini and G. De Micheli, xpipesCompiler: a tool for instantiating application specific Networks on Chips, *DATE- Proceedings of the Design and Test Europe Conference*, pp. 884–889, 2004.

[MAHA] S. Mahadevan, F. Angiolini, M. Storgaard, R.G. Olsen, J. Sparsø and J. Madsen, A Network Traffic Generator Model for fast Network on-Chip Simulation, *DATE- Proceedings of the Design and Test Europe Conference*, pp. 780–785, 2005.

[MURA] S. Murali, P. Meloni, F. Angiolini, D. Atienza, S. Carta, L. Benini, G. De Micheli and L. Raffo, Designing Application-specific Networks on Chips with Floorplan Information, *International Conference on Computer Aided Design*, pp. 355–362, 2006.

[OKA] M. Oka and M. Suzuoki, Designing and Programming the Emotion Engine, *IEEE Micro*, Vol. 19, No. 6, 20–28, Nov.–Dec, 1999.

[PULLI] A. Pullini, F. Angiolini, P. Meloni, D. Atienza, S. Murali, L. Raffo, G. De Micheli and L. Benini, NoC Design and Implementation in 65nm Technology, *International Symposium on Networks on Chips*, pp. 273–282, 2007.

[RIJP] E. Rijpkema, K. Goossens, A. R dulescu, J. Dielissen, J. van Meerbergen, P. Wielage and E Waterlander, Trade Offs in the Design of a Router with Both Guaranteed and Best Effort Services for Networks on Chips, *DATE- Proceedings of the Design and Test Europe Conference*, pp. 10350–10355, 2003

[THEI] T. Theis, The future of Interconnection Technology, *IBM Journal of Research and Development*, Vol. 44, No. 3, 379–390, May 2000.

[VANG] S. Vangal et al., An 80-Tile 1.28TFLOPS Network-on-Chip in 65nm CMOS. *International Solid-State Circuits Conference*, pp. 98–99, 2007.

A Generic Architecture for On-Chip Packet-Switched Interconnections

Pierre Guerrier and Alain Greiner

Abstract This paper presents an architectural study of a scalable system-level interconnection template. We explain why the shared bus, which is today's dominant template, will not meet the performance requirements of tomorrow's systems. We present an alternative interconnection in the form of switching networks. This technology originates in parallel computing, but is also well suited for heterogeneous communication between embedded processors and addresses many of the deep submicron integration issues. We discuss the necessity and the ways to provide high-level services on top of the bare network packet protocol, such as dataflow and address-space communication services. Eventually we present our first results on the cost/performance assessment of an integrated switching network.

1 Introduction

In the year 1999, the first $0.18\,\mu$ fabs have moved into volume production. With such process technology, chip designers can create systems-on-a-chip (SoC) by incorporating several dozens of IP blocks, each in the 50–100 kGate range, together with large amounts of embedded DRAM. Those IPs can be CPU or DSP cores, video stream processors, and high-bandwidth I/O (like IEEE 1394, Gigabit Ethernet or DACs).

SoC designs pose a number of daunting methodology problems: how to specify the system? How to map the specification onto a collection of available IPs ? How to evaluate design options ? These are mostly CAD issues, and some commercial solutions are appearing. However, they build on the implicit assumption that SoC will stick to the *Central Bus* architecture template, or slightly enhanced multi-bus variants.

Yet virtually every IP in the portfolio evoked above, has I/O requirements in the Gbit/s range: fast CPUs, fast network controllers, and multimedia processors. Figure 1 is a synopsis of the latency and throughput required by a

Université Pierre et Marie Curie 4, place Jussieu, F-75252 PARIS CEDEX 05 name.surname@lip6.fr

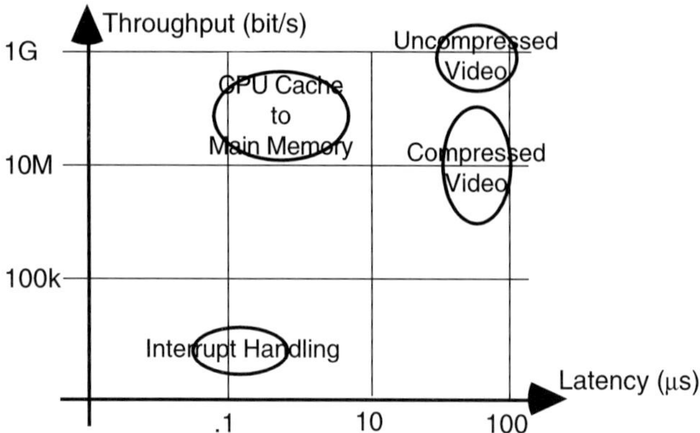

Fig. 1 Requirements for some traffics

few kinds of traffic. They may all be featured in a single consumer equipment, for instance an advanced multimedia PDA. If we want to exploit task-level parallelism between processing IPs, with concurrent reconfigurable communications, then aggregated interconnection throughputs like 50 Gbit/s are needed. This requirement will continue to grow as the number of IPs incorporated on chip will increase, as well as the individual performance of each IP (with faster processors, better video quality, etc.).

Bus-based architectures will not meet this requirement because a bus is *inherently non-scalable*. The bandwidth of a bus is shared by all attached devices, and it is simply not sufficient, firstly because the bus width cannot reasonably exceed a hundred bits, and secondly because the clocking frequency of global wiring becomes tightly constrained by the electrical properties of deep submicron processes [1]. New bus proposals are still being made, e.g. in the framework of [2], but in our opinion their only clear advantage is to standardize the IP interfaces. These architectures advocate multiple on-chip busses, requiring case-specific grouping of IPs and the design of transversal bridges, which does not make for a truly scalable and reusable interconnection.

This paper proposes another generic interconnection template that addresses the performance and scalability requirements of systems-on-chip, using integrated switching networks. We aim to prove that present manufacturing technology already enables the realization of a switching network with the wrappers to carry on both address-space and dataflow communications.

The paper is organized as follows: Section 2 provides the reader with a practical understanding of switching concepts. Then Section 3 explains how they should be tuned for on-chip environment. Section 4 is a cost analysis of a tentative realization of such a network. Section 5 analyzes its performance through simulation benchmarks. Section 6 shows how hardwired protocols can make the network compatible with the synchronous dataflow I/O semantics as it is used in high-performance video processors, or the address-space semantics of the VCI standard.

Eventually Section 7 will discuss the present shortcomings of the proposed architecture and possible solutions.

2 Switching Network Basics

We call switching network, a set of switching elements connected together by full duplex point-to-point links. The reader is referred to [3] for a milestone presentation. The switching elements can operate in space (S), by establishing physical connections between their terminals. The typical S-switch is the crossbar. Other switches operate in time (T), using buffers to swap the order of timeslices on time-division multiplexed links. Historically, combinations of S and T switches have been used to build telephony networks.

Such networks implement a *circuit-switching technique*, where connections are established between two terminals by assigning them a set of time-slices on the network links. This set has to be determined by clever computations when the connection is requested, and subsequently remains constant during the entire connection. The main advantage is the *formal guarantee of bandwidth* resulting from the static establishment of the circuit. The PROPHID architecture template [4] has demonstrated a remarkable application of circuit-switching to on-chip communication. PROPHID uses a T-S-T switch (Fig. 2) to cascade dataflow video processings. It is also the only example to our knowledge of a non-bus SoC interconnection. (We do not consider prototyping platforms such as [5], which featured a single S-switch connecting dataflow computing resources, because those systems were not intended for production and could only be reconfigured at start-up time.)

However, the drawback of circuit-switching is the lack of reactivity against rapidly changing communications. For instance, the PROPHID interconnection cannot dynamically increase a bandwidth allocation to match bursts in an MPEG

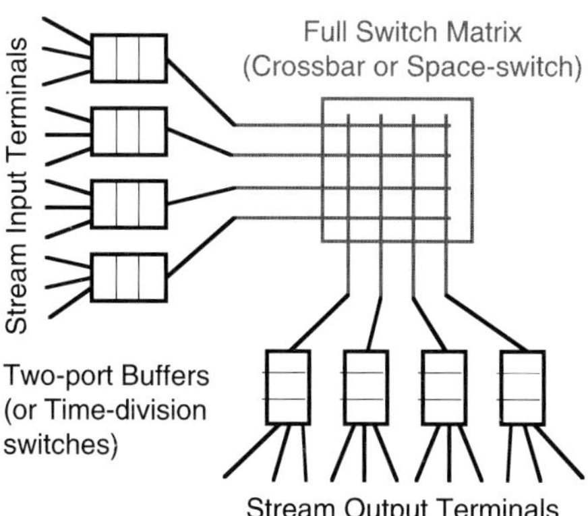

Fig. 2 The PROPHID interconnection

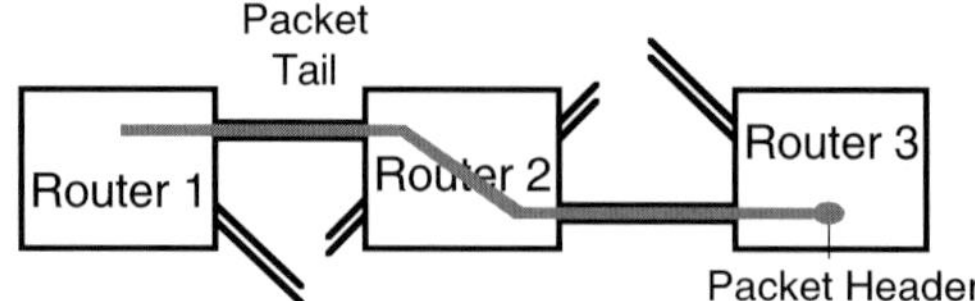

Fig. 3 A wormhole packet switch

bitstream. It is even less suited to the random traffic between a CPU master and several slaves on a bus. For this reason, networks in parallel computers or LANs have used a different technique known as *packet-switching*.

A packet-switched network moves data from one of its terminals to another, in small formatted chunks called packets. These consist in a header identifying the destination of data, followed by a payload (the data itself), unambiguously terminated by a trailer. The switching elements of the network are called *routers* and operate in space. When it receives a packet, a router forwards it to one of his neighbors (chosen according to the header information). Packets repeatedly undergo this process until they reach their final destination. Since routing decisions are distributed over the routers, the network can remain very reactive even for very large sizes.

Despite the many routing hops, low latency can be maintained if routers forward the header of packets ASAP, without waiting for the trailer (Fig. 3). This technique is called *wormhole routing* and has been used extensively in high-performance parallel computer networks [6]. In a low-cost wormhole switch, buffers are small and packets typically span a handful of routers. If a packet requests a link which is already busy, and this packet cannot be entirely buffered in the router, the macropipeline formed by the other links and routers spanned by the packet will be stalled. This may result in many links being inefficiently kept busy. Such *cascaded contentions* practically prevents any packet-switched network from delivering its full theoretical bandwidth.

3   An On-Chip Switched Network

This section is an overview of the global design options for a *Scalable, Programmable, Integrated Network* (SPIN) for packet-switched system-on-chip interconnections. The design decisions concern the nature of elementary links, the topology of the network, the packet structure, and the network access protocols.

- The point-to-point link should be able to stand the throughput of a bus devoted to a single master. Therefore we decided on a parallel link built of two one-way 32-bit data paths. In contrast to a bus, there are no bi-directional wires. We use a credit-based flow control on the paths: buffer overflows at the target end of a path are checked at the source, using a counter to track the amount of free buffer space. The receiver notifies the sender of every datum consumed, with a dedicated

feedback wire. This inexpensive mechanism provides latency-independence: the links can be pipelined transparently to achieve the desired clock speed in a submicron process.

- The network topology impacts the complexity of the distributed routing decisions. Regular topologies like meshes or hypercubes make the decision functions simple and identical for every router, so the same macrocell can be reinstanciated to build the network. Among those, we preferred the Fat-Tree (Fig. 4) because Leiserson has proved formally that it is the most cost-efficient for VLSI realizations [7].

A fat-tree is a tree structure with routers on the nodes and terminals on the leaves, except that every node has replicated fathers. If there are as many fathers as children on all nodes, then there are as many roots as leaves, and the bisection throughput is preserved: the network is non-blocking. We chose a four-child tree because the 8×8 router seemed most convenient for VLSI implementation. The size of this network grows like $(n.\log n)/8$ with the number of terminals. Table 1 lists the costs of some network instances.

- Packets consist of sequences of words of 32 bits. This width allows the header to fit in a single word. A byte in this word identifies the destination (this allows 256 terminals) and other bits are used for packet tagging and routing options.

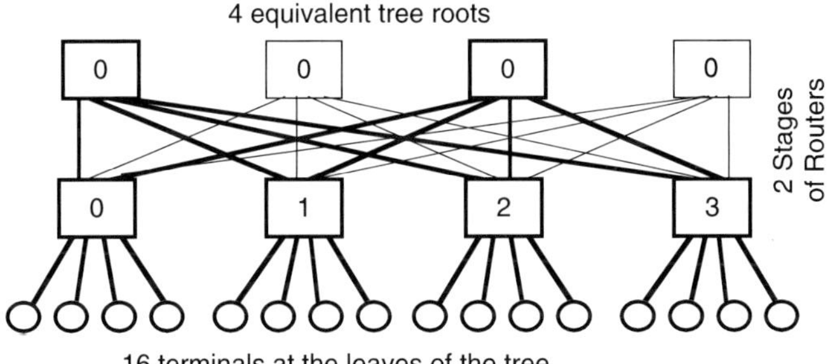

Fig. 4 A fat-tree network

Table 1 Scaling the SPIN fat-tree

Attached units	Resources	
	Routers	Links
8	2	12
16	8	32
32	16	96
64	48	192
128	96	448

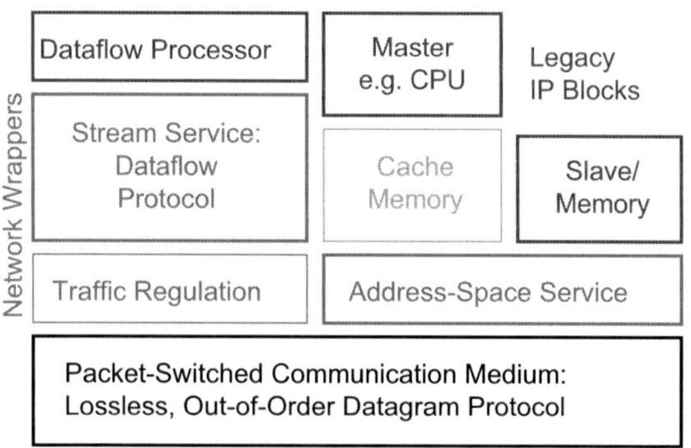

Fig. 5 Hardwired protocol stack

The routers are free to use any of the redundant paths in the fat-tree to route a packet. This feature is called adaptivity and reduces contention hot-spots. The packet payload may be of any size, and possibly infinite. Finally the trailer is a special word marked by a dedicated control line. It contains a checksum of the payload data.

- Packets natively implement the message passing communication model. Messages can be used to build protocols emulating other models like dataflow streams and address-space. We believe that the future system-on-chip will be heterogeneous, featuring at least these two communication mechanisms. They could be used together on some units, since they serve different purposes: for instance, application software configures and monitors dataflow processors through addressable registers. Hence, we specified a modular network access terminal, that can be shared by several wrappers, each providing a different service to the attached unit. The packets are tagged to associate them with services. Each service uses a private payload format for its packets. Figure 5 illustrates the parallel between this modular, *fully hardwired* approach, and a layered stack of APIs.

4 Router Design and Cost

The critical component of the network described in Section 3 is the router macrocell. On one hand, it must be carefully optimized for area, because it is reinstanciated many times in a single system. On the other hand, the packet buffering strategy of individual routers strongly impacts the global performance of the network: the most conservative model where packets are queued in FIFOs at the router inputs is known to generate the highest contention [8]. But the smarter buffering schemes used in general purpose discrete routers require expensive hardware to clear the inputs by queuing packets on additional crossbar channels called *output buffers*.

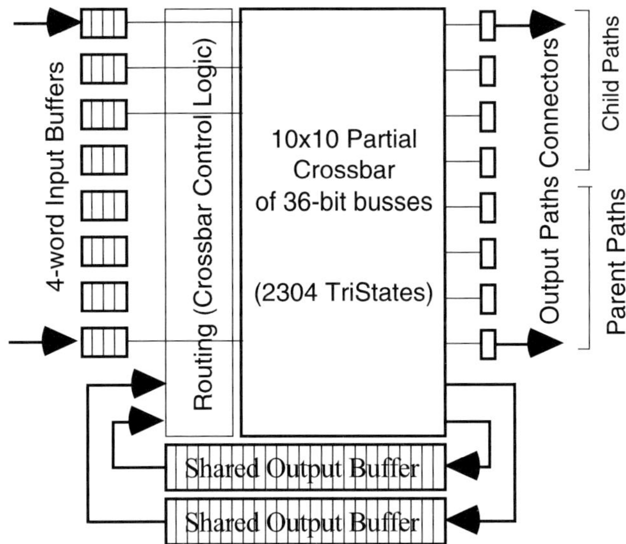

Fig. 6 Router datapaths synopsis

However, careful analysis of our topology shows that contention will mostly result from packets flowing down the tree towards child links, because alternate paths are only available for the father links, and also because those packets span more routers on average. We based our design (Fig. 6) on this property and the experience of the successful realization of the discrete router RCube [9]: Small (4-word) input buffers are necessary for hiding the delay of the control logic and the link latency. In case of output contention, child-bound packets can use two output buffers of 18 words each. They reduce cascaded contention by providing a longer side track for halted packets. It is more hardware-efficient than simply making all input buffers longer, because in most practical conditions only a couple of inputs will be subject to contention. The chosen size handles most efficiently packets shorter than 18 words, with payloads as large as typical bus bursts. The crossbar grows to 10×10 because of the output buffers, but it is not full because all routes are not possible.

We used the symbolic layout standard cell library of the ALLIANCE system to synthesize the router. The regular parts (buffers and crossbar) were placed manually using metal layers 4, 5, and 6 to interconnect the switch matrix. Figure 7 shows this placing. The area is $1\times0.8\,\mathrm{mm}^2$ in a $0.25\,\mu$ process. The control logic is synthesized, placed and routed by automatic tools in the empty space, using metal layers 2 and 3. From this geometry and from Table 1, we can forecast the network costs for future systems, which we summarized in Table 2.

The control logic can be pipelined to match virtually any speed the switching matrix may reach. Therefore the timing is constrained by the delay of the wires in the matrix. Electrical simulation taking into account crosstalk capacitances suggests

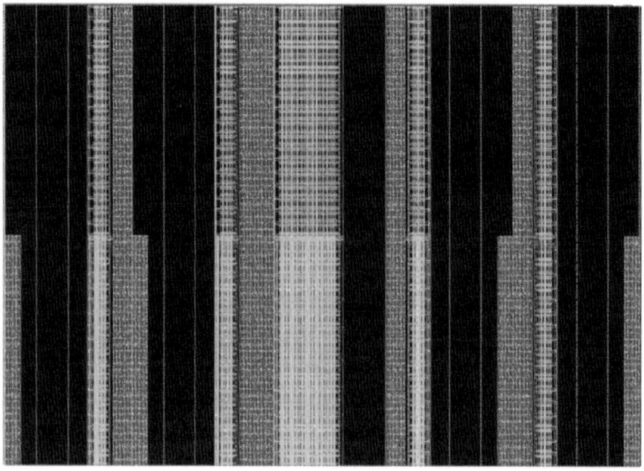

Fig. 7 Router floorplan

Table 2 Projected network costs

Process	Attached units	Peak bandwidth	Network area
.25 μ, 6ML	32	205 Gbit/s	<13 mm²
0.18 μ	64	568 Gbit/s	<20 mm²
0.13 μ	128	1.82 Tbit/s	<20 mm²

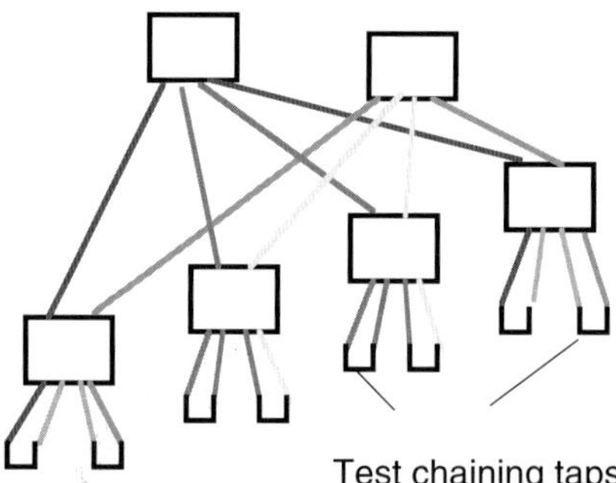

Fig. 8 Global test paths

a cycle time below 5 ns for a 0.25 μ CMOS process. Table 2 assumes this delay will scale in accordance with the SIA roadmap projections. A test silicon of the router macrocell is scheduled for the end of this year, which will enable accurate performance and power measurements.

Eventually, we paid careful attention to the network testability issues, as no off-the-shelf test method can handle a system comprising dozens of crossbars and hundreds of FIFOs. We use a graph property of the fat-tree shown on Fig. 8: the graph is *eulerian*, thus a common predefined connection scheme can be applied to all routers to create paths covering all links and buffers in the network. For the test mode, these paths are daisy-chained and fed with a pattern generator. All the global interconnect is tested by this method. Stalls and bubbles are included in the stream to stimulate the

FIFOs and test them without any scan path. In addition, an input–output loop-back test is applied to every switching matrix individually.

5 Performance Evaluation

A fast cycle-true, bit-true simulation model of the SPIN router was written in C language for the CASS system-level simulation tool [10]. This model allowed us to test large networks (up to 256 terminals) under a range of benchmark loads. The most common benchmark in use in the networking literature is the *uniform random distribution* of packet destinations. The results of this test are a *pessimistic estimate* of the practical performance of a network, because the load does not exhibit any locality that the tree clustering could exploit. Since all paths of the network are equally loaded, it also reduces the advantages of router adaptivity. Despite these artifacts, we tested the network under a random load to compare it against past realizations, while still getting a worst-case performance prediction.

Figure 9 summarizes the key data for a 32-terminal network (16 routers) with 20-word packets spread uniformly. The simulations were run for one million clock cycles. This is enough to provide ±1% accurate population counts. The *offered load* is the average proportion of cycles at which data are injected into unbounded buffers connected to each input terminal. These buffers become saturated when the network cannot absorb the offered load. Figure 9 shows how the packet latency grows as the network is loaded, until it reaches saturation. Note that this latency includes the time spent in the injection buffers before entering the network. The time actually spent *inside* the network is always lower.

The curves show the performance impact of the two shared output buffers. These simulations permitted to optimize the cost/performance of the realization presented above. The results are satisfactory by (discrete) networking standards, although

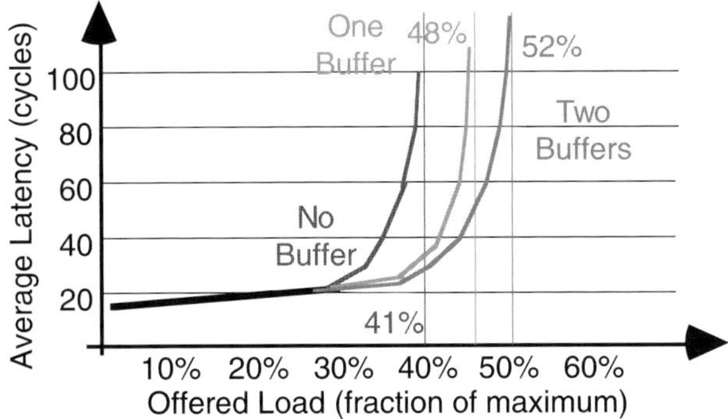

Fig. 9 Load benchmarks

Fig. 10 Latency distributions

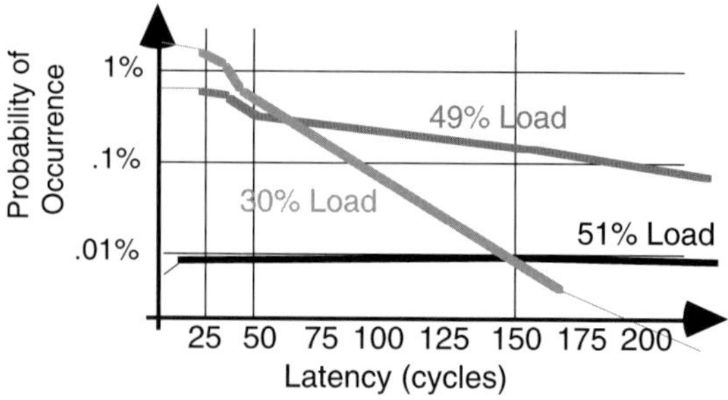

somewhat lower than state-of-the-art. This is remarkable given the very small depth of the input buffers, and it demonstrates the relevance of our topology-specific buffering policy. Simulations with larger networks have shown little performance degradation. We conclude that a network clocked at 200 Mhz would easily deliver 2 Gbit/s per terminal, and still scale nicely to aggregated throughputs up to 100 Gbit/s for 32 terminals.

Nevertheless, Fig. 10 shows an important drawback of the switched network: all latencies tend to occur, with an exponentially decreasing probability. This means that nearly all packets will be delivered in a small time. However, there *will be* some random hick-ups, all the more frequent as the network is heavily loaded. These simulations are for a 64-terminal network (48 routers), with a uniform load, where most packets have the maximum number of routing hops (i.e. 5 hops).

6 The Communication Protocols

This section presents results on the hardware implementation (i.e. wrappers) of the protocols enabling dataflow or address communications through the packet-switched network. The goal is to emulate these models using the message passing model. In addition, the two bandwidth-effective features of SPIN, adaptivity and output buffering, have the unpleasant property of swapping packets. Protocols must also enforce strict in-order delivery for dataflow and for some address-space transactions.

Regarding stream communications, we imagined a protocol based on our experience with sender-based protocols for adaptive networks [11]. The key is an *unambiguous chronological tagging* of every packet transmitted in a stream. Upon reception, packets are stored in the wrapper, which may deliver them in order to the dataflow processor. Buffer storage is reserved for the missing packets until they arrive. If a packet is very late, the processor stalls, the entire buffer becomes busy, and other incoming packets have to be rejected in the network, which is used as a delay line. An end-to-end credit-based traffic regulation bounds the amount of outstanding stream data and thus prevents rejection from causing a catastrophic network congestion. Both statistical modeling and cycle-true simulations show that this spurious traffic is negligible given the latency distribution of SPIN. The total bandwidth overhead introduced by the protocol is 31% of the stream throughput.

We synthesized a VHDL description of this hardware protocol stack, supporting four concurrent streams per terminal. The basic network access layer, necessary to plug in any service wrapper, test the network and check data integrity, represents about $0.15\,mm^2$ in a $.25\,\mu$ process. The stream traffic regulation layer is less than $0.1\,mm^2$. The stream protocol itself is only $0.1\,mm^2$, plus a dual-port SRAM of 1–4 kbits (0.1–$0.3\,mm^2$ with our symbolic SRAM generator). In any configuration, the full network interface is smaller than $0.6\,mm^2$ in this process, making it affordable for systems comprising up to a dozen stream processors, each with gigabit/s throughput.

Regarding address-space communications, we are defining a protocol for wrappers matching the "Virtual Component Interface" address-oriented standard [2]. The basic principle is to translate the VCI request/response packets into SPIN packets. However, for optimal performance, the initiator must take advantage of the VCI *split transactions*: because the delay of return-trips through the switched network is large, several transactions must be overlapped, as shown on Fig. 11, to use the full link bandwidth, e.g. for cache to memory traffic. Memory throughput can then be scaled by distributing it over several network terminals standing concurrent accesses, like in *Distributed Shared Memory* multi-computers.

Addressing the memory over a switching network raises a software compatibility concern, for those features that relied upon the snooping of a central bus, like semaphore synchronization and cache invalidation. DSM multicomputers have shown that a comprehensive support for cache consistency is possible but too complex for an embedded system [12]. Fortunately it is not needed because these systems are rather heterogeneous, asymmetric multiprocessors, with explicit invalidations and simple synchronization primitives. These can be supported more easily by simple protocols. We do not presently have a cost measurement of the VCI wrappers for SPIN. However, we believe they will be smaller than the dataflow wrappers, because they do not require any data buffering.

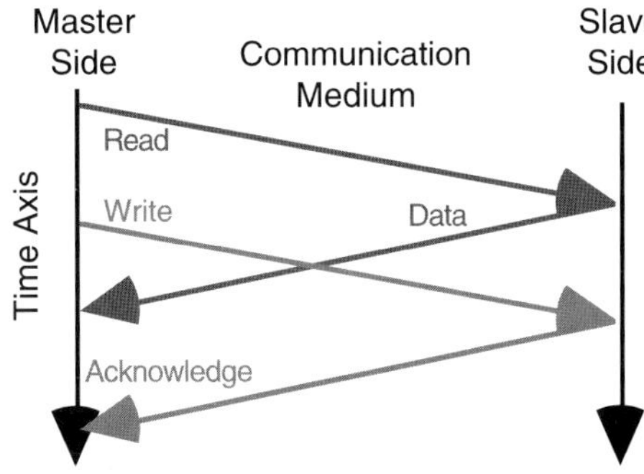

Fig. 11 The split transaction model

7   Present Limitations

A well-known criticism against our approach, is the complexity of switching
network concepts. The design space for networks is different and larger than for
busses, with new caveats and new refinements. Although all silicon engineers have
seamlessly used commodity networks (through telephones, NFS or Internet...),
they are unfamiliar and ill-at-ease with key network aspects like true concurrence
or statistical behavior prediction. This problem cannot be solved by improvements
to the proposed architecture, although protocols and CAD tools can help hiding its
internal intricacies. The true solution is the education of the designer community to
the broader perspectives opened by deep submicron systems [13].

8   Conclusions

We have shown a new architecture template for system-level interconnection, based
on switching networks. We assessed the cost and the performance of this template
through joint functional modeling and physical implementation of key parts of the
architecture. Our results demonstrate that it matches the throughput, latency and
area requirements of future systems-on-chip.

We acknowldged the need for legacy communication protocols to enable straightforward reuse of existing IPs in industry designs. Therefore, we have devised a

Table 3 The bus-versus-network arguments

Bus Pros and Cons			Network Pros and Cons
Every unit attached adds parasitic capacitance, therefore electrical performance degrades with growth	−	+	Only point-to-point one-way wires are used, for all network sizes
Bus timing is difficult in a deep submicron process	−	+	Network wires can be pipelined because the network protocol is globally asynchronous
Bus testability is problematic and slow	−	+	Dedicated BIST is fast and complete
Bus arbiter delay grows with the number of masters. The arbiter is also instance-specific	−	+	Routing decisions are distributed for all and the same router is reinstanciated, network sizes
Bandwidth is limited and shared by all units attached	−	+	Aggregated bandwidth scales with the network size
Bus latency is zero once arbiter has granted control	+	−	Internal network contention causes a small latency
The silicon cost of a bus is near zero	+	−	The network has a significant silicon area
Any bus is almost directly compatible with most available IPs, including software running on CPUs	+	−	Bus-oriented IPs need smart wrappers. in Software needs clean synchronization multiprocessor systems
The concepts are simple and well understood	+	−	System designers need *reeducation* for new concepts

layered protocol architecture, to provide a set of industry-standard communication mechanisms on top of a switching network. Results of such a mechanism for dataflow communication have been presented, including detailed silicon area evaluation. VCI-compliant wrappers for address-space communications are being actively investigated.

References

[1] K. Keutzer, *Chip Level Assembly (and not Integration of Synthesis and Physical) is the Key to DSM Design*, Proceedings of the ACM/IEEE International Workshop on Timing Issues in the Specification and Synthesis of Digital Systems (Tau'99), Monterey, CA, March 1999.

[2] Virtual Socket Interface Alliance, *On-Chip Bus Attributes* and *Virtual Component Interface – Draft Specification, v. 2.0.4*, http://www.vsia.com, September 1999 (document access may be limited to members only).

[3] C. Clos, *A Study of Nonblocking Switching Networks*, Bell System Technical Journal, vol. 32, no. 2, 406–424, 1953.

[4] J. Leijten et al. *Stream Communication between Real-Time Tasks in a High-Performance Multiprocessor*, Proceedings of the 1998 DATE Conference, Paris, France, March 1998.

[5] D. C. Chen, J. M. Rabaey, *A Reconfigurable Multiprocessor IC for Rapid Prototyping of Algorithmic-Specific High-Speed DSP Data Paths*, IEEE Journal of Solid-State Circuits, vol. 27, no. 12, 1895–1904, December 1992.

[6] W. Dally, C. Seitz, *Deadlock-free Message Routing in Multiprocessor Interconnection Networks*, IEEE Transactions on Computers, vol. C-36, no. 5, 547–553, May 1987.

[7] C. Leiserson, *Fat-Trees: Universal Networks for Hardware-Efficient Supercomputing*, IEEE Transactions on Computers, vol. C-34, no. 10, 892–901, October 1985

[8] M. Karol et al. *Input versus Output Queueing on a Space-Division Packet Switch*, IEEE Transactions on Communications, 1347–1356, December 1987.

[9] B. Zerrouk et al. *RCube: A Gigabit Serial Link Low Latency Adaptive Router*, Records of the IEEE Hot Interconnects IVth Symposium, Palo Alto, CA, August 1996.

[10] F. Pétrot et al. *Cycle-Precise Core Based Hardware/Software System Simulation with Predictable Event Propagation*, IEEE Computer Society Press, Proceedings of the 23rd Euromicro Conference, Budapest, Hungary, pp. 182–187, September 1997.

[11] F. Wajsbürt et al. *An Integrated PCI Component for IEEE 1355*, Proceedings of the 1997 EMMSEC Conference and Exhibition, Florence, Italy, November 1997.

[12] J. Hennessy, D. Patterson, *Computer Architecture, A Quantitative Approach – 2nd Edition*, Morgan Kaufmann, San Francisco, CA, 1996.

[13] H. de Man, *Education for the Deep Submicron Age: Business As Usual?* Proceedings of the 34th Design Automation Conference, Anaheim, CA, March 1997.

Trade-offs in the Design of a Router with Both Guaranteed and Best-Effort Services for Networks on Chip

E. Rijpkema, K.G.W. Goossens, A. Rădulescu, J. Dielissen,
J. van Meerbergen, P. Wielage, and E. Waterlander

Abstract Managing the complexity of designing chips containing billions of transistors requires decoupling computation from communication. For the communication, scalable, and compositional interconnects, such as networks on chip (NoC), must be used. In this paper we show that guaranteed services are essential in achieving this decoupling. Guarantees typically come at the cost of lower resource utilization. To avoid this, they must be used in combination with best-effort services. The key element of our NoC is a router consisting conceptually of two parts: the so-called guaranteed throughput (GT) and best-effort (BE) routers. We combine the GT and BE router architectures in an efficient implementation by sharing resources. We show the trade offs between hardware complexity and efficiency of the combined router, and motivate our choices. Our reasoning for the trade offs is validated with a prototype router implementation. We show a lay-out of an input-queued wormhole 5×5 router with an aggregate bandwidth of 80 Gbit/s. It occupies $0.26\,\mathrm{mm}^2$ in CMOS12. This shows that our router provides high performance at reasonable cost, bringing NoCs one step closer.

1 Introduction

Recent advances in technology raise the challenge of managing the complexity of designing chips containing billions of transistors. A key ingredient in tackling this challenge is *decoupling the computation from communication* [10, 14]. This decoupling allows IPs (the computation part), and the interconnect (the communication part) to be designed independently from each other.

In this paper, we focus on the communication part. Existing interconnects (e.g. buses) may no longer be feasible for chips with many IPs, because of the diverse and dynamic communication requirements. *Networks on a chip* (NoC) are emerging as an alternative to existing on-chip interconnects because they (a) structure and manage global wires in new deep-submicron technologies [2–5, 7], (b) share wires, lowering

Philips Research Laboratories, Eindhoven, The Netherlands

R. Lauwereins and J. Madsen (eds.), *Design, Automation, and Test
in Europe–The Most Influential Papers of 10 Years DATE*, 125–139
© Springer 2008

their number and increasing their utilization [5, 7], (c) can be energy efficient and reliable [3], and (d) are scalable when compared to traditional buses [8].

Decoupling the computation from communication requires that the *services* that IPs use to communicate are well defined, and hide the implementation details of the interconnect [10], Fig. 1a. NoCs help, because they are traditionally designed using layered protocol stacks [13], where each layer provides a well-defined interface which decouples service usage from service implementation [4, 14], see Fig. 1b.

In particular, *guaranteed services* are essential because they make the requirements on the NoC explicit, and limit the possible interactions of IPs with the communication environment. IPs can also be designed independently, because their use of guaranteed services is not affected by the interconnect or by other IPs. This is essential for a compositional construction (design and programming) of systems on chip (SoC). Moreover, failures are restricted to the IP configuration phase (a service request is either granted or denied by the NoC) which simplifies the IP programming model [7]. We view the guaranteed services to be offered by an interconnect as a requirement from the applications, see Fig. 1c.

The drawback of using guaranteed services is that they require resource reservations for worst-case scenarios. This is not acceptable in a SoC where cost constraints are typically very tight, see Fig. 1d. Therefore, we also provide *best-effort services* to exploit the network capacity that is left over, or reserved but unused. Guaranteed services are then used for the critical (e.g. real time) traffic, and best-effort services for non-critical communication.

The combination of guaranteed and best-effort classes is known from general computer network research [15], but not for on-chip networks. As on- and off-chip networks have different characteristics, the trade offs in their design are different.

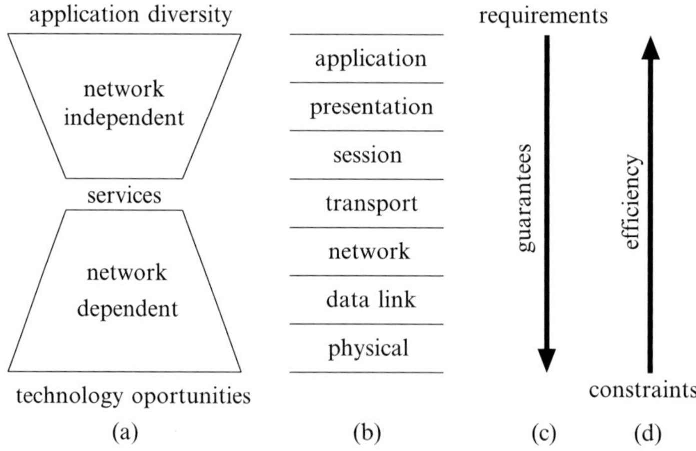

Fig. 1 Network services (a) hide the interconnect details and allow reusable components to be build on top of them, (b) are build using a layered approach, (c) are driven by the application requirements, and (d) their efficiency relies on technology and network organization

In this paper, we present the trade offs between hardware complexity and efficiency for networks on chip, and motivate our choices.

We present a prototype router architecture that reflects one particular set of design choices. It has an aggregate bandwidth of 80 Gbit/s, and its CMOS12 lay-out occupies 0.26 mm². We list other feasible variations that either increase performance, or lower the router cost.

In this paper, we first list a set of network-independent communication services that are essential in chip design (Section 2). Then, we show the trade-offs between efficiency and cost that we make in our NoC. In Section 3, we present some general network-related issues that are used in the sections to follow. In Section 4, we zoom into the internals of the key component of our NoC: a router that efficiently provides both guaranteed and best-effort services. In Section 5, we demonstrate the feasibility of our router design through a prototype implementation in CMOS12.

2 Services

The NoC services that we consider essential for chip design are: *data integrity*, meaning that data is delivered uncorrupted, *lossless data delivery*, which means no data is dropped in the interconnect, *in-order data delivery*, which specifies that the order in which data is delivered is the same order in which it has been sent, and *throughput* and *latency* services that offer time-related bounds. As shown in Section 1, guaranteed services are essential to simplify IP design and integration. With the current technology, we assume data integrity is solved at the data-link layer. All the other services can be guaranteed or not on request. In the next section, we describe briefly how these services are provided by our NoC, and in Section 4 we describe how our router architecture enables an efficient implementation of these services.

Guaranteed services require resource reservation for worst-case scenarios, which can be expensive. For example, guaranteeing throughput for a stream of data implies reserving bandwidth for its peak throughput, even when its average is much lower. As a consequence, when using guarantees, resources are often under-utilized.

Best-effort services do not reserve any resources, and hence, provide no guarantees. Best-effort services use resources well because they are typically designed for average-case scenarios instead of worst-case scenarios. They are also easy and fast to use, as they require no resource reservation. Their main disadvantage is their unpredictability: one cannot rely on a given performance (i.e., they do not offer guarantees). In the best case, if certain boundary conditions are assumed, a statistical performance can be derived.

The requirements for guaranteed services and the efficiency constraint (i.e. good resource utilization) are conflicting. Our approach to a predictable and low-cost interconnect is to integrate the guaranteed and best-effort services in the same interconnect. Guaranteed services would be used for critical traffic, and best-effort services for non-critical traffic. For example, a video processing IP will typically require a lossless, in-order video stream with guaranteed throughput, but possibly allows corrupted samples. Another example is cache updates which require uncorrupted, lossless, low-latency data transfer, but ordering and guaranteed throughput are less important. In Section 4.3 we show how integrated guaranteed and best-effort services efficiently can use common resources. In the remainder of this section we analyze the minimum level of abstraction at which the communication services must be offered to hide the network internals.

Traditionally, network services have been implemented and offered using a layered protocol stack, typically aligned to the ISO-OSI reference model [13], see Fig. 1b. NoCs also take this approach [3, 4, 7, 14], because it structures and decomposes the service implementation, and the protocol stack concepts aid positioning of services.

To achieve the decoupling of computation from communication, the communication services must be offered at least at the level of the transport layer in OSI reference model. It is the first layer that offers end-to-end services, hiding the network details; see Fig. 1a, b [4].

The lowest three layers in the protocol stack, namely physical, data-link, and network layers, are network specific. Therefore, these services should not be visible to the IPs when decoupling between computation from communication is desired. However, these layers are essential in implementing the services, because constructing guarantees without guarantees at the layer below is either very expensive, or even impossible. For example, implementing a lossless communication on top of a lossy service requires acknowledgment, data retransmission, and filtering duplicated data. This leads to an increase in traffic, and possibly larger buffer space requirements. Even worse, providing guarantees for time-related services is impossible if lower layers do not offer these guarantees. For example, latency can not be guaranteed if communication at a lower layer is lossy. As a consequence, guarantees can only be built *on top* of guarantees, see Fig. 1c. Similarly, a layer's efficiency is based on efficient implementations of the layers below it, see Fig. 1d.

3 Networks on Chip

General computer network research is a mature research field [15] which has many issues in common with NoCs. However, two significant differences between computer networks and on-chip networks make the trade offs in their design very different [5]. First, routers of a NoC are more resource constrained than those in computer network, in particular in the control complexity and in the amount of memory. Second, communication links of a NoC are relatively shorter than those in computer networks, allowing tight synchronization between routers.

We identify three important issues in the design of the router network architecture. These are: the *switching mode, contention resolution*, and *network flow control*. Equally important, *end-to-end flow control* and *congestion control* are handled in our NoC at the network edge instead of the routers; we therefore omit their discussion here. Similarly, we assume guaranteed data integrity at the link level and retain it at the network layer and and higher.

3.1 Switching Mode

The *switching mode* of a network specifies how data and control are related. We distinguish *circuit switching* and *packet switching*. In circuit switching data and control are separated. The control is provided to the network to *set up* a *connection*. This results in a *circuit* over which all subsequent data of the connection is transported. In *time-division circuit switching* bandwidth is shared by time-division multiplexing connections over circuits. Circuit-switched networks inherently offer time-related guaranteed services after resources are reserved during the connection set up.

In *packet switching* data is divided into *packets* and every packet is composed of a control part, the *header*, and a data part, the *payload*. Network routers inspect, and possibly modify, the headers of incoming packets to switch the packet to the appropriate output port. Since in packet switching the packets are self contained, there is no need for a set-up phase to allocate resources. Therefore, best-effort services are naturally provided by packet switching.

3.2 Contention Resolution

When a router attempts to send multiple data items over the same link at the same time *contention* is said to occur. As only one data item can be sent over a link at any point in time, a selection among the contending data must be made; this process is called contention resolution.

In circuit switching, contention resolution takes place at set up at the granularity of connections, so that data sent over different connections do not conflict. Thus, there is no contention during data transport, and time-related guarantees can be given.

In packet switching contention resolution takes place at the granularity of individual packets. Because packet arrival cannot be predicted contention can not be avoided. It is resolved dynamically by scheduling in which data items are sent in turn. This requires data storage in the router, see Section 4.2.1, and delays the data in a non predictable manner which complicates the provision of guarantees, see Section 4.1.1.

3.3 Network Flow Control

Network flow control, also called *routing mode*, addresses the limited amount of buffering in routers and data acceptance between routers. In circuit switching connections are set up. The data sent over these connections are always accepted by the routers and hence, no network flow control is needed. In packet switching, data items must be buffered at every router before they are sent on. Because routers have a limited amount of buffering they accept data only when they have enough space to store the incoming data.

There are three types of network flow control, namely *store-and-forward, virtual cut-through*, and *wormhole* routing. In store-and-forward routing, an incoming packet is received and stored in its entirety before it is forwarded to the next router. This requires storage for the complete packet, and implies a per-router latency of at least the time required for the router to receive the packet.

In virtual cut-through routing a packet is forwarded as soon as the next router guarantees that the complete packet will be accepted. When no guarantee is given, the router must be able to store the whole packet. Thus, virtual cut-trough routing requires buffer space for a complete packet, like store-and-forward routing, but allows lower-latency communication.

In wormhole routing packets are split in so-called *flits* (flow control digits). A flit is passed to the next router when the flit can be accepted, even when there is not enough buffer space for the complete packet. As soon as a flit of a packet is sent over an output port, that output port is reserved for flits of that packet only. When the first flit of a packet is blocked the trailing flits can therefore be spread over multiple routers, blocking the intermediate links. Wormhole routing requires the least buffering (buffer flits instead of packets) and also allows low-latency communication. However, it is more sensitive to deadlock and generally results in lower link utilization than virtual cut-through routing.

To allow low latency we consider both virtual-cut through and wormhole routing, which are both feasible in terms of buffer area, as shown in Section 5.

4   A Combined GT-BE Router

Section 2 defines our requirements for NoCs in terms of services that are to be offered, in particular, both guaranteed and best-effort services. Using the general network issues of the previous section we show in the following two subsections that the guaranteed and best-effort services can conceptually be described by two independent router architectures. The combination of these two router architectures is efficient and has a flexible programming model, as described in Section 4.3. Section 5 then shows a prototype implementation.

4.1   A GT Router Architecture

Our guaranteed-throughput (GT) router guarantees uncorrupted, lossless, and ordered data transfer, and both latency and throughput over a finite time interval.

As mentioned earlier, data integrity is solved at the data-link layer; we do not address it further. The GT router is lossless because we use a variant of circuit switching, described in the next section. Data is transported in fixed-size blocks. As only one block is stored per input in the GT router, data items remain ordered per connection. We now turn to the more challenging time-related guarantees, namely throughput and latency.

4.1.1 Time-related Guarantees

Latency is defined as the duration a packet is transported over the network. Guaranteeing latency, therefore, means that a worst-case upper bound must be given for this time. We define throughput for a given producer-consumer pair as the amount of data transported by the network over a finite, fixed time interval. Guaranteeing throughput means giving a lower bound.

We observe that guaranteeing latency even in a lossless router is difficult because contention requires scheduling and hence cause delays. Guaranteeing throughput is less problematic. Rate-based packet switching (for an overview see [16]) offers guaranteed throughput over a finite period, and hence a latency bound. This bound is very high, however, and the cost of buffering is also high. Deadline-based packet switching [12] offers preferential treatment for packets close to their deadline. This allows differential latency guarantees (under certain admissible traffic assumptions), but also at high buffer costs.

Circuit switching solves the contention at set up, so naturally providing guaranteed latency and throughput. Circuits can be pipelined to improve through-put [6], at the cost of additional buffering and latency. Time-division multiplexing connections over pipelined circuits additionally offers flexibility in bandwidth allocation. This requires a logical notion of router synchronicity, which is possible because a NoC is better controllable than a general network. We explain this variation in more detail in the next subsection. The associated programming model is described in Section 4.3.2.

4.1.2 Contention-free Routing

A router uses a *slot table* to (a) avoid contention on a link, (b) divide up bandwidth per link between connections, and (c) switch data to the correct output. Every slot table T has S time slots (rows), and N router outputs (columns). There is a logical notion of synchronicity: all routers in the network are in the same fixed-duration slot. In a slot s at most one *block* of data can be read/written per input/output port. In the next slot $(s + 1)\%S$, the read blocks are written to their appropriate output ports. Blocks thus propagate in a store and forward fashion. The latency a block incurs per router is equal to the duration of a slot and bandwidth is guaranteed in multiples of block size per S slots.

The entries of the slot table map outputs to inputs for every slot: $T(s, o) = i$. An entry is empty, when there is no reservation for that output in that slot. No contention arises because there is at most one input per output. Sending a single input to multiple outputs (multicast) is possible.

The slots reserved for a block along its path from source to destination increase by one (modulo S). If slot s is reserved in a router, slot $(s + 1)\%S$ must be reserved in the next router on the path. The assignment of slots to connections in the network is an optimization problem, and is described in Section 4.3.3. Section 4.3.2 explains how slots are reserved in our network.

4.2   A BE Router Architecture

Best-effort traffic can have a better *average* performance than offered by guaranteed services. This depends on boundary conditions, such as network load, that are unpredictable. Best-effort services thus fulfill our efficiency requirement, but without offering time-related guarantees. This section describes an architecture for a best-effort service with uncorrupted, lossless, in-order data transport.

The BE router cost and performance are largely dependent on the contention resolution scheme of the router. The contention resolution scheme has two components: buffering and scheduling. The main trade-off in Section 4.2.1 is between total buffer size, buffering strategy, and link utilization. Without taking global network requirements into account, no decisions will be made, rather we present a router that allows different instances, to trade off hardware complexity for link utilization at instantiation time. In Section 4.2.2 the trade-off is between link utilization and schedule complexity and we select an efficient scheduling algorithm that is easily specialized to the different instances.

4.2.1   Buffering Strategy

The buffering strategy determines the location of buffers inside the router. We distinguish *output queuing* and *input queuing*. In the following, N is the number of inputs, equal to the number of outputs, of our router. In output queuing N^2 queues are located at the outputs of the router as in Fig. 2a. From the inputs to the outputs there is a fully connected bipartite interconnect to allow every input to write to every output. Output queuing has the best performance among the buffering strategies, however, the interconnect will make the router wire dominated and expensive already for small values of N.

In input queuing the queues are at the input of the router. A scheduler determines at which times which queues are connected to which output ports such that no contention occurs. The scheduler derives contention-free connections, a switch matrix (crossbar switch) can be used to implement the connections. In traditional input queuing, or input queuing for short, there is a single queue per input, resulting in a buffer cost of N queues per router. However, due to the so-called *head-of-line blocking*, for large N, router utilization saturates at 59% [9]. Therefore, input queuing results in weak utilization of the links.

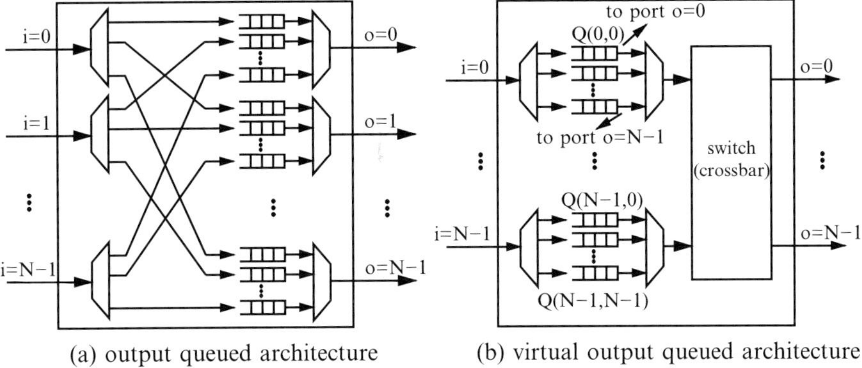

(a) output queued architecture   (b) virtual output queued architecture

Fig. 2 Schematics of two router architectures

Another version of input queuing is *virtual output queuing* (VOQ) [1]. VOQ combines the advantages of input queuing and output queuing. It has a switch like in input queuing and has the link utilization close to that of output queuing; 100% link utilization can still be achieved, when N is large [11]. As for output queuing, there are N^2 queues. For every input i there are N queues $Q(i, o)$, one for each output o, see Fig. 2b. Typically the set of N queues at each input port of a VOQ router are mapped onto a single RAM. However, for NoCs we strive at a small router and therefore we require the RAMs to have few addresses. But such RAMs have large overhead. Therefore, we use in-house developed dedicated fifos, which have almost no overhead, see Section 5.

The decision to select traditional input queuing or VOQ depends on system-level aspects like topology, network utilization, and global wiring cost, and is outside the scope of this paper. In Section 5 we show a prototype of an input queued router with dedicated hardware fifos and explain that VOQ is a valid option with minor additional cost.

4.2.2 Matrix Scheduling

The switch matrix, present in input queued architectures, is controlled by a contention resolution algorithm, known as matrix scheduling, to properly connects inputs to outputs.

The matrix scheduling problem can be modeled as a bipartite graph matching problem. Every input port i is modeled by a node u_i and every output port o by a node v_o. There is an edge between u_i and v_o if and only if queue $Q(i,o)$ is non-empty. A *match* is a subset of these edges such that every node is incident to at most one edge. For example, Fig. 3c is a match of Fig. 3a.

Matching can be done optimally, but because of time complexity and fairness, a nonoptimal algorithm is preferred [11].

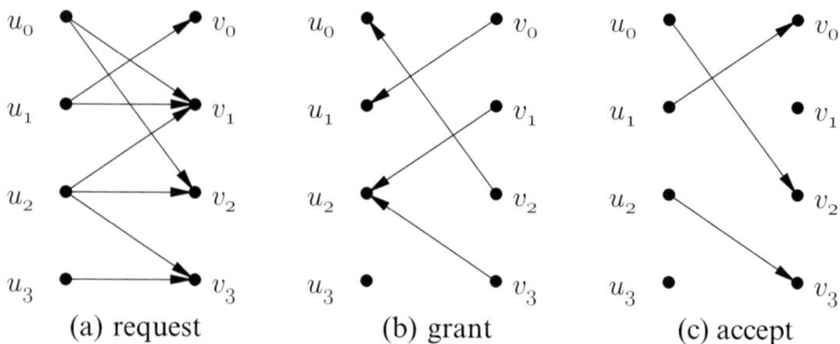

(a) request   (b) grant   (c) accept

Fig. 3 The three stages of a schedule iteration

Our matching algorithm is iterative and one iteration has three stages, illustrated by an example in Fig. 3 for $N = 4$. In the first stage, see Fig. 3a, every nonempty queue $Q(i,o)$ *requests* access to output port o from input port i. In the second stage, see Fig. 3b, every output port o *grants* one request, solving link contention at the output ports. In the third stage, see Fig. 3c, every input port i *accepts* one grant, to resolve memory contention at the input port. A next iteration then starts with the matching found so far. This scheme is used in various scheduling algorithms, including parallel iterative matching, round robin matching, and SLIP [11], and applies to both input queuing and VOQ. For input queuing, however, stage (c) in Fig. 3 is omitted since on contention on input ports can occur. To keep schedule latency as low as possible we use one iteration only.

4.3 Combining the GT and BE Routers

The GT and BE router architectures are combined to share resources, in particular the links and the switch. Moreover, best-effort traffic enables a packet-based programming model for the guaranteed traffic, as shown later, in Section 4.3.2.

The principal constraint for a combined router architecture is that guaranteed services are never affected by best-effort services. Fig. 4a shows that, conceptually, the combined router contains both router architectures (fat lines represent data, thin lines represent control). Incoming data is switched to either the GT or the BE router. The GT traffic, the traffic that is served by the GT router, has the higher priority, to maintain guarantees. This is ensured by the arbitration unit, which therefore affects the best-effort scheduling. Furthermore, best-effort packets can program the guaranteed router, as shown by the arrow labeled program. Thin lines going from the right to the left indicate network flow control, which is only required for best-effort packets because guaranteed blocks never encounter contention.

Figure 4b shows that the data path, consisting of buffers and switch matrix, is shared, and that the control paths of the BE and GT routers are separate, yet interrelated. Moreover, the arbitration unit of Figure 4a has been absorbed by the BE router. The following subsection shows how this can be done.

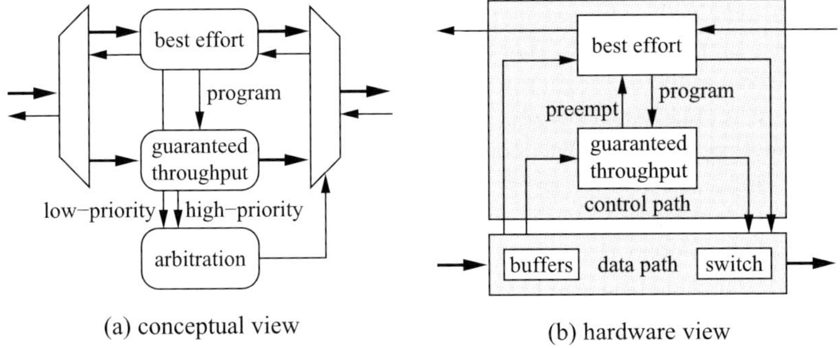

Fig. 4 Two views of the combined GT-BE router

4.3.1 Arbitration and Flit Size

When combining GT and BE traffic in a single network the impact on the network flow control scheme must be taken into account. Recall from Section 3.3 that a BE flit is the smallest unit at which flow control is performed. In other words, the BE scheduling can only react to GT blocks at flit granularity. To avoid alignment problems, the block size (B words) is a multiple of the flits (F words, $B = lF$) with l being constant. We prefer a small l to decrease the store-and-forward delay and reduce the buffer size for guaranteed traffic, and a small F for fine-grained switching and better statistical multiplexing.

The router architecture contains a data path and a control path, see Fig. 4b. The data path maximizes throughput for high link utilization, and the control path maximizes the rate of scheduling and switching. They can be designed and optimized independently. Given any combination of their operating frequencies, the router has both maximum throughput and switching rate by using the appropriate flit size F_{opt}. For $F > F_{opt}$, the control path is ready while data is still being transported, lowering the switching rate. For $F < F_{opt}$, flits have been transported before the control path finishes, wasting bandwidth.

We extend the schedule algorithm in Section 4.2.2, to handle the combination of GT and BE traffic. In this combination GT traffic always has priority over BE traffic. This is to ensure that guarantees are never corrupted.

4.3.2 Programming Model

In this section we show how GT connections are set up and torn down by means of BE packets to avoid introducing an additional communication infrastructure only to program the network. To ensure scalability, programming must not require a global view or centralized resources. Section 4.1.2 explains why our contention-free routing uses slot tables; we now see that they are distributed over routers for scalability.

Initially the slot table of every router is empty. There are three system packets: SetUp, TearDown, and AckSetUp. They are used to program the slot table in every router on their path. The SetUp packet creates a connection from a source to a destination, and travels in the direction of the data ("downstream"). When a SetUp packet arrives at the destination it is successful and is acknowledged by returning an AckSetUp. TearDown packets destroy (partial) connections, and can travel in either direction. SetUp packets contain the source of the data, the destination or a path to it, and a slot number. Every router along the path of the SetUp packet checks if the output to the next router in the path is free in the slot indicated by the packet. If it is free, the output is reserved in that slot, and the SetUp packet is forwarded with an incremented (modulo S) slot. Otherwise, the SetUp packet is discarded and a TearDown packet returns along the same path. Thus, every path must be reversible; this is the only assumption we make about the network topology. These upstream TearDown packets free the slot, and continue with a decremented slot. Downstream TearDown packets work similarly, and remove existing connections. A connection is successfully opened when an AckSetUp is received, else a TearDown is received.

The programming model is pipelined and concurrent (multiple-system packets can be active in the network simultaneously, also from the same source) and distributed (active in multiple routers). Given the distributed nature of the programming model, ensuring consistency and determinism is crucial. The outcome of programming may depend on the execution order of system packets, but is always consistent. The next section shows how to use this programming model.

4.3.3   Compile- and Run-time Slot Allocation

This section explains how to determine the slots specified in SetUp packets. A slot allocation for a single connection requires that, at every router along the path, the required output is free (not reserved by another connection) in the appropriate slot. Computing an optimal slot allocation for all connections requires a global network view and may be expensive. To reduce computational cost, heuristics can be used, possibly leading to nonoptimal solutions.

SetUp packets of different connections do not fail if connections are set up with conflict-free slots or paths. All execution orders of SetUp packets then give the same result, so that compile-time slot allocations can be recreated deterministically at run time.

Optimal run-time slot allocation is hard without a global (and central) slot table view, which is non-scalable and slows down programming. Distributed run-time slot allocation is scalable, but lacks a global view and is, therefore, perhaps suboptimal. Moreover, SetUp packets may interfere, making programming more involved, and perhaps nondeterministic. However, dynamic connection management at high rates will require distributed slot allocation. In a simple distributed greedy algorithm, all sources repeatedly generate random slot numbers for each set up until their connection succeeds. We conclude that our programming model allows both compile-time and run-time slot allocation. Computational complexity, deterministic results, and scalability can be balanced according to system requirements.

5   Current Results and Future Work

The previous section shows a prototype combined GT-BE architecture. We have synthesized an input-queued router using wormhole routing with arity 5, a queue depth of 8 flits of 3 words of 32 bits, and 256 slots in CMOS12 technology. The layout is shown in Figure 5. It has an aggregate bandwidth of $5 \times 500\,MHz \times 32\,bit = 80\,Gbit/s$. The area of the router is $0.26\,mm^2$.

The area of $0.26\,mm^2$ depends on the use of dedicated hardware fifos, labeled GQ and BQ in Fig. 5. The router would have been at least three times larger with register- or RAM-based fifos. The RAMs required for input queuing and VOQ in an on-chip router have few addresses so that their overhead makes them as large as (area-inefficient) register files. Decreasing the queue depths reduces the buffering area (with registers at least), but also degrades the router performance.

Dedicated hardware fifos enable both input and virtual output queuing strategies using wormhole routing because of the reasonable buffering cost. For example, VOQ with two-flit deep fifos is only moderately larger than the input queuing with fifos of depth 8 of Fig. 5. Virtual cut-through routing in combination with input queuing is also affordable now, because for packets of at most 8 flits, it has the same cost as the prototype.

The slot table (labeled STU in Fig. 5) occupies a significant part of the router, for two reasons. Logically the slot table is very large (256 slots). It is not worthwhile to reduce the number of slots because the RAM is very area inefficient. We are investigating more advanced slot table schemes and new memory architectures to reduce the size and area of the slot table. The cost of offering time-related guaranteed services is then lower.

We separately synthesized the data and control paths (cf. Fig. 4) with arities ranging from 3 to 13 to verify their speeds. With increasing arity, the speed of the data path reduces little. The speed of the control path decreases by a factor of two, corresponding to the complexity increase of the scheduling. For each arity, we balance the performance of the data and control paths by adjusting the flit size as needed, as shown in Section 4.3.1. The data and scheduling frequencies of the prototype router are $500\,MHz$ and $166\,MHz$, respectively, with a flit size of 3.

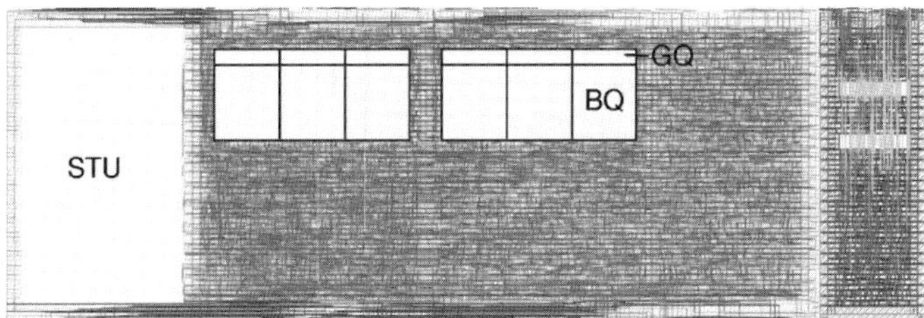

Fig. 5 Lay-out of a combined GT-BE router

Our results show that the cost and performance of the combined GT-BE router can make it the basis of a router-based network on chip. It further shows that dedicated hardware fifos significantly reduce buffering area and so enable both input queuing and VOQ, with wormhole and virtual-cut through routing.

6 Conclusions

In this paper we show that guaranteed services are essential to provide predictable interconnects that enable compositional system design and integration. However, guarantees typically utilize resources inefficiently. Best-effort services overcome this problem but provide no guarantees. So, integrating guaranteed and best-effort services allows efficient resource utilization, yet still providing guarantees for critical traffic.

Time-related guarantees, such as throughput and latency, can only be constructed on a NoC that intrinsically has these properties. We therefore define a router-based NoC architecture that combines guaranteed and best-effort services. The router architecture has conceptually two parts: the guaranteed-throughput (GT) and best-effort (BE) routers. Both offer data integrity, lossless data delivery, and in-order data delivery. Additionally, the GT router offers guaranteed throughput and latency services using pipelined circuit switching with time-division multiplexing. The BE router uses packet switching, virtual cut-through or wormhole routing, and input queuing or virtual output queuing.

We combine the GT and BE router architectures efficiently by sharing router resources. The guarantees are never affected by the BE traffic, and links are efficiently utilized because BE traffic uses all bandwidth left over by GT traffic. Connections are programmed using BE packets. The programming model is robust, concurrent, and distributed. It enables run-time and compile-time, deterministic and adaptive connection management.

For all our architecture choices, we show the trade offs between hardware complexity and efficiency. Our choices are motivated by a prototype router which has an area of $0.26\,\text{mm}^2$ in CMOS12 and offers 80 Gbit/s aggregate throughput. We use dedicated hardware fifos to significantly reduce the area of the data queues. With RAM- or register-based queues the router area would have been at least 3 times larger.

Dedicated hardware fifos enable (a) input queuing using both wormhole and virtual cut-through routing, and (b) virtual output queuing using wormhole routing. The buffer costs are too high, however, for virtual output queuing with virtual cut-through routing.

The cost of offering time-related guaranteed services is still high for our router. We are investigating how to reduce this cost.

An attractive feature of our router architecture is the ability to combine separately optimized data and control paths by adjusting the flit size.

In conclusion, we describe and motivate a choice of architectures for routers, which are an essential component in a NoC. They fulfill our NoC requirements by providing guaranteed services, and satisfy the efficiency constraint by offering best-effort services.

References

[1] M. Ali and M. Youssefi. The performance analysis of an input access scheme in a high-speed packet switch. In *INFOCOM*, 1991.

[2] J. Bainbridge and S. Furber. CHAIN: A delay-insensitive chip area interconnect. *IEEE Micro* (5), 2002.

[3] L. Benini and G. De Micheli. Powering networks on chips. In *ISSS*, 2001.

[4] L. Benini and G. De Micheli. Networks on chips: a new SoC paradigm. *IEEE Computer*, 35(1):70–80, 2002.

[5] W. J. Dally and B. Towles. Route packets, not wires: on-chip interconnection networks. In *DAC*, 2001.

[6] A. deHon. Robust, high-speed network design for large-scale multi-processing. TR 1445, MIT, AI Lab., 1993.

[7] K. Goossens et al. Networks on silicon: combining best-effort and guaranteed services. In *DATE*, 2002.

[8] P. Guerrier and A. Greiner. A generic architecture for on-chip packet-switched interconnections. In *DATE*, 2000.

[9] M. J. Karol et al. Input versus output queueing on a space-division packet switch. *IEEE Trans. Communications*, COM-35(12):1347–1356, 1987.

[10] K. Keutzer et al. System-level design: orthogonalization of concerns and platform-based design. *IEEE Trans. CAD of Integrated Circuits and Systems*, 19(12):1523–1543, 2000.

[11] N. McKeown. *Scheduling Algorithms for Input-Queued Cell Switches*. PhD thesis, University of California, Berkeley, 1995.

[12] J. Rexford. *Tailoring Router Architectures to Performance Requirements in Cut-Through Networks*. PhD thesis, University of Michigan, MI 1999.

[13] M. T. Rose. *The Open Book: A Practical Perspective on OSI*. 1990.

[14] M. Sgori et al. Addressing the system-on-a-chip interconnect woes through communication-based design. In *DAC*, 2001.

[15] A. S. Tanenbaum. *Computer Networks*. 1996.

[16] H. Zhang. Service disciplines for guaranteed performance service in packet-switching networks. *Proc. IEEE*, 83(10):1374–1396, 1995.

Exploiting the Routing Flexibility
for Energy/Performance-Aware Mapping
of Regular NoC Architectures*

Jingcao Hu and Radu Marculescu

Abstract In this paper, we present an algorithm which automatically maps the IPs onto a generic regular Network on Chip (NoC) architecture and constructs a deadlock-free deterministic routing function such that the total communication energy is minimized. At the same time, the performance of the resulting communication system is guaranteed to satisfy the specified constraints through bandwidth reservation. As the main contribution, we first formulate the problem of energy/ performance aware mapping, in a topological sense, and show how the routing flexibility can be exploited to expand the solution space and improve the solution quality. An efficient branch-and-bound algorithm is then described to solve this problem. Experimental results show that the proposed algorithm is very fast, and significant energy savings can be achieved. For instance, for a complex video/audio application, 51.7% energy savings have been observed, on average, compared to an adhoc implementation.

1 Introduction

Regular tile-based NoC architecture was recently proposed to mitigate complex on-chip communication problems [1,4,5]. As shown in the left of Fig. 1, such a chip consists of regular tiles where each tile can be a general-purpose processor, a DSP, a memory subsystem, etc. A router is embedded within each tile with the objective of connecting it to its neighboring tiles. Thus, instead of routing design-specific global on-chip wires, the inter-tile communication can be achieved by routing packets.

Given a target application described as a set of concurrent tasks which have been assigned and scheduled, to exploit such an architecture, the fundamental questions to answer are: (i) *which tile* each IP should be mapped to, (ii) *what*

---

*Research supported by NSF CCR-00-93104 and DARPA/Marco Gigascale Research Center (GSRC), and SRC 2001-HJ-898.

Department of Electrical and Computer Engineering Carnegie Mellon University Pittsburgh, PA 15213-3890, USA e-mail: {jingcao, radum} @ece.cmu.edu

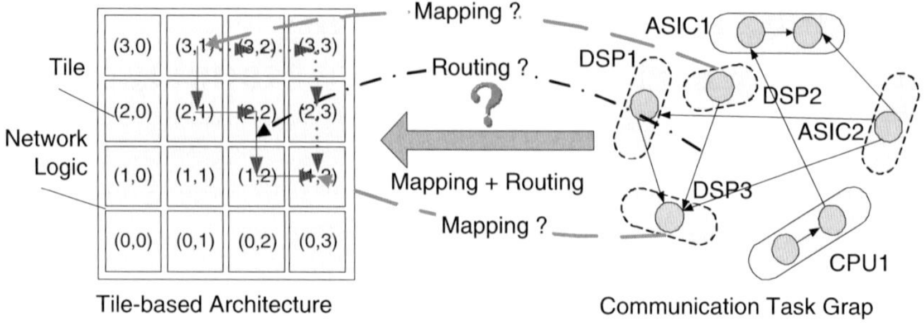

Fig. 1 Tile-based arch and mapping/routing problems

routing algorithm is suitable for directing the information among tiles, such that the metrics of interest are optimized. More precisely, in order to get the best power/performance tradeoff, the designer needs to determine the *topological placement* of these IPs onto different tiles. Referring to Fig. 1, this means to determine, for instance, onto which tile (e.g. (3,1), (1,3)) each IP (e.g. DSP2, DSP3) should be placed. Since there may exist multiple *minimal* routing paths, one also needs to select *one* qualified path for each communicating tile pair. For example, one has to determine which path (e.g. (3,1)→(2,1)→(2,2)→(1,2)→(1,3) or (3,1)→(3,2)→(3,3)→(2,3)→(1,3)) should the packets follow to send data from DSP2 to DSP3, if these two IPs are meant to be placed to tile (3,1) and (1,3), respectively.

While task assignment and scheduling problems have been addressed before [2], the *mapping* and *routing* problems described above represent a new challenge, especially in the context of the regular tile-based architecture, as this significantly impacts the energy and performance metrics of the system. In this paper, we address this very issue and propose an efficient algorithm to solve it. To this end, we first propose a suitable routing scheme (Section 3) and a new energy model (Section 4) for NoC. The problem of mapping and routing path allocation are then formulated in Section 5. Next, an efficient branch-and-bound algorithm is proposed to solve this problem under performance constraints (Section 6). Experimental results in Section 7 show that significant energy savings can be achieved, while guaranteeing the specified system performance. For instance, for a complex video/audio application, on average, 51.7% energy savings have been observed compared to an adhoc implementation.

2 Related Work

In [1,4,5], the on-chip interconnection networks are proposed to structure the top-level wires on a chip and facilitate modular design. While these papers discuss the overall advantages and challenges of regular NoC architecture, to the best of our knowledge, our work is the first to address the *mapping* and *routing path allocation* problems for tile-based architectures and provide an efficient way to solve them. Although routing (especially wormhole-based routing [3]) has been a hot research topic in the area of *direct* networks for parallel and distributed systems [6–8], the specifics of NoC force us to re-think standard network techniques and adapt them to the context of NoC architectures. In what follows, we address this issue by presenting a suitable routing technique for NoC together with an algorithm for automatic generation of the routing function.

3 The Proposed Routing Scheme for NoCs

The major differences between regular data-networks and NoCs are:

(A) **Buffering space**: To minimize the implementation cost, it is more reasonable to use *registers* as buffers for on-chip routers instead of huge memories (SRAM or DRAM) as is the case of regular data-networks. This also helps in decreasing access latency, which is critical for typical SoC applications.
(B) **Use of deterministic routing**: We believe that for NoC, deterministic routing is more suitable because:

- *Resource limitation and stringent latency requirements*

Compared to deterministic routers, implementing adaptive routers requires by far more resources. Moreover, since in adaptive routing the packets may arrive out of order, huge buffering space is needed to reorder them. This, together with its protocol overhead leads to prohibitive cost, extra delay and jitter.

- *Traffic predictability*

In contrast to typical regular data-networks, most NoCs need to support one application or, at most, a small class of applications. Consequently, the designer has a good understanding of the traffic characteristics and can use this information to avoid congestion by wisely mapping the IPs and allocating the routing paths. This further weakens the potential advantages of adaptive routing.

In summary, because of the aforementioned characteristics, together with consideration for deadlock problem, programmability, etc., we argue that the most appropriate routing technique for NoC should be *deterministic, deadlock-free, minimal, wormhole-based* [11].

4 Platform Description

In this section, we describe the regular tile-based architecture and the energy model for its communication network.

4.1   The Architecture

The system under consideration is composed of $n \times n$ tiles interconnected by a 2D mesh network.[1] Figure 2 shows an abstract view of a tile in this architecture.

Each tile in Fig. 2 is composed of a *processing core* and a *router*. The router is connected to the four neighboring tiles and its local processing core via channels (each consisting of two *one-directional point-to-point* links).

Due to the limited resources, buffers are implemented using registers (typically in the size of one or two flits each). A 5×5 crossbar switch is used as the switching fabric in the router. Based on the source/destination address, the routing table decides which output link the packet should be delivered to.

4.2   The Energy Model

Ye et al. [9] proposed a model for power consumption of network routers. The *bit energy* (E_{bit}) metric is defined as the energy consumed when one bit of data is transported through the router:

$$E_{bit} = E_{S_{bit}} + E_{B_{bit}} + E_{W_{bit}} \tag{1}$$

where $E_{S_{bit}}$, $E_{B_{bit}}$, and $E_{W_{bit}}$ represent the energy consumed by the switch, buffering and interconnection wires, respectively. (Ye et al. [9] assume the buffers are implemented in SRAM or DRAM.) However, the above power model is targeted

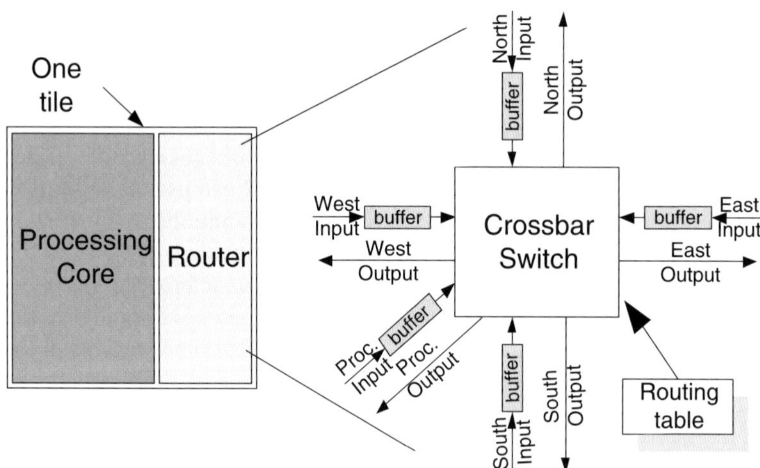

Fig. 2   The typical structure of a tile

[1] We use 2D mesh network simply because it naturally fits the tile-based architecture. However, our algorithm can be extended for other topologies.

for network routers where the *entire* chip is occupied by just *one* router. For tile-based NoC architectures, we need a new energy model because:

- In Eq. (1), $E_{B_{bit}}$ becomes dominant during congestion, as accessing and refreshing the memory are power expensive. This is no longer true for NoC where buffers are implemented with registers.
- In Eq. (1), $E_{W_{bit}}$ is the energy consumed on the wires inside the switch fabric. For NoC, the energy consumed *on the links* between tiles ($E_{L_{bit}}$) should also be included. Thus, the average energy consumed in sending one bit of data from a tile to its neighboring tile can be calculated as:

$$E_{bit} = E_{S_{bit}} + E_{B_{bit}} + E_{W_{bit}} + E_{L_{bit}} \qquad (2)$$

Since the length of a link is typically in the order of *mm*, the energy consumed by buffering ($E_{B_{bit}}$) and internal wires ($E_{W_{bit}}$) is negligible[2] compared to $E_{L_{bit}}$; Eq. (2) reduces to:

$$E_{bit} = E_{S_{bit}} + E_{L_{bit}} \qquad (3)$$

Consequently, in our new model, the average energy consumption of sending one bit of data from tile t_i to tile t_j is:

$$E_{bit}^{t_i,t_j} = n_{hops} \times E_{S_{bit}} + (n_{hops} - 1) \times E_{L_{bit}} \qquad (4)$$

where n_{hops} is the number of routers the bit passes.

For 2D mesh networks with *minimal* routing, Eq. (4) shows that the average energy consumption of sending one bit of data from t_i to t_j is determined by the *Manhattan* distance between them.

5 Problem Formulation

Simply stated, for a given application, we try to decide onto which tile should each IP be mapped and how should the packets be routed, such that the total communication energy consumption is minimized under performance constraints. To formulate this problem more formally, we define the following terms:

Definition 1 An *Application Characterization Graph* (APCG) $G = G(C,A)$ is a *directed* graph, where each vertex c_i represents one selected IP, and each directed arc $a_{i,j}$ characterizes the communication from c_i to c_j. Each $a_{i,j}$ has the following properties:

- $v(a_{i,j})$: arc volume from vertex c_i to c_j, which stands for the communication volume (*bits*) from c_i to c_j.

---

[2] We evaluated the energy consumption metrics using Spice for a $0.35 \propto m$ technology. The results show that $E_{Bbit} = 0.073pJ$, which is indeed negligible compared to E_{Lbit} (typically in the order of *pJ*).

- $b(a_{i,j})$: arc bandwidth requirement from vertex c_i to c_j, which stands for the minimum bandwidth (*bits/sec.*) that should be allocated by the network in order to meet the performance constraints.

Definition 2 An *Architecture Characterization Graph* (ARCG) $\mathcal{G}' = G(T,R)$ is a *directed* graph, where each vertex t_i represents one tile in the architecture, and each directed arc $r_{i,j}$ represents the routing from t_i to t_j. Each $r_{i,j}$ has the following properties:

- $P_{i,j}$: a set of candidate *minimal* paths from tile t_i to tile t_j. $\forall p_{i,j} \in P_{i,j}, L(p_{i,j})$ gives the set of links used by $p_{i,j}$.
- $e(r_{i,j})$: arc cost. This represents the average energy consumption (*joule*) of sending one bit of data from t_i to t_j, i.e. $E_{bit}^{ti,tj}$.

Definition 3 For an ARCG $\mathcal{G}' = G(T,R)$, a *routing function* R: $R{\rightarrow}P$ maps $r_{i,j}$ to *one* routing path $p_{i,j}$, where $p_{i,j} \in P_{i,j}$.

Using these definitions, the energy/performance aware mapping and routing path allocation problem can be formulated as:

Given an APCG and an ARCG that satisfy

$$size\,(APCG) \leq size\,(ARCG) \tag{5}$$

find a mapping function *map*() from APCG to ARCG and a *deadlock-free, minimal* routing function $\mathcal{R}$() which:

$$\min\{Energy = \sum_{\forall a_{i,j}} v(a_{i,j}) \times e(r_{map(c_i),map(c_j)})\} \tag{6}$$

such that:

$$\forall c_i \in C, \quad map(c_i) \in T \tag{7}$$

$$\forall c_i \neq c_j \in C, \quad map(c_i) \neq map(c_j) \tag{8}$$

$$\forall \text{link}\, l_k, B(l_k) \geq \sum_{\forall a_{i,j}} b(a_{i,j}) \times f(l_k \mathcal{R}(r_{map(c_i),map(c_j)})) \tag{9}$$

where $B(l_k)$ is the bandwidth of link l_k, and:

$$f(l_k, p_{m,n}) = \begin{cases} 0 : l_k \notin L(p_{m,n}) \\ 1 : l_k \in L(p_{m,n}) \end{cases}$$

In this formulation, conditions (7) and (8) mean that each IP should be mapped to exactly one tile and no tile can host more than one IP. Equation (9) guarantees that the communication traffic (load) of any link will not exceed its bandwidth.

We show in detail in [11] that the impact of mapping on the energy consumption is significant. In order to generate the best solution, routing paths have to be wisely allocated. In the following, we propose an efficient algorithm which can find nearly optimal solutions in reasonable run times.

6   Energy/Performance Aware Mapping and Routing Path Allocation

6.1   The Data Structure

Our approach is based on a *branch-and-bound* algorithm which efficiently walks through the *searching tree* that represents the solution space. Figure 3 shows a searching tree example for mapping a 4-IP application onto a 2×2-tile architecture.

Each node in the tree is either a *root* node, an *internal* node, or a *leaf* node. The root node corresponds to the state where *no* IP has been mapped and *no* routing path has been allocated. Each internal node represents a *partial* mapping. For example, the node labeled "23*xx*" represents a partial mapping where IP_0 and IP_1 are mapped to tile t_2 and tile t_3 respectively, while IP_2 and IP_3 are still unmapped. Each leaf node represents a *complete* mapping of the IPs to the tiles. Each node also has a *path allocation table* (PAT) which stores the routing paths for the traffic among its *occupied* tiles. For instance, the PAT of node "231*x*" in Fig. 3 shows that the traffic from t_1 to t_2 takes the path $t_1 \rightarrow t_0 \rightarrow t_2$, etc. The PAT of any node is automatically generated by the algorithm. When a child node is generated, the PAT of its parent node is inherited. Next, the routing paths for the traffic involving the newly occupied tile are allocated and added to its PAT. Caution must be taken to ensure freedom of deadlock.

To explain the algorithm, we define the following terms:

Definition 4 The *cost* of a node is the energy consumed by the communication among those IPs that have already been mapped.

Definition 5 Let $\mathcal{M}$ be the set of vertices in the APCG that have already have been mapped. A node is called a *legal* node if and only if it satisfies the following tow conditions:

- The routing paths specified by the PAT are deadlock free.
- $\forall l_k, B(l_k) \geq \Sigma_{\forall a_{i,j}, c_i c_j \in \mathcal{M}} b(a_{i,j}) \times f(l_k, p_{map(ci),map(cj)})$ where $p_{map(ci),map(cj)}$ is the routing path from tile $map(c_i)$ to $map(c_j)$ specified by the PAT.

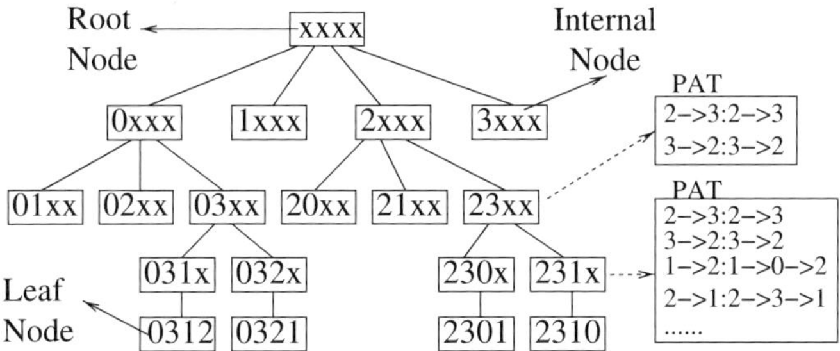

Fig. 3   An example search tree

6.2   The Branch-and-Bound Algorithm

Given the above definitions, finding the optimal solution of Eq. (6) is equivalent to finding the *legal leaf* node which has the *minimal* cost. To achieve this, our algorithm searches the optimal solution by alternating the following two steps:

Branch: An unexpanded node is selected and its next *unmapped* IP is enumeratively assigned to the remaining *unoccupied* tiles to generate the corresponding new child nodes. The PAT of each child node is also generated by first copying its parent node's PAT and then allocating the routing paths for the traffics between the newly occupied tile and the other occupied tiles. The routing paths specified by the PAT have to be deadlock-free.

Bound: Each of the newly generated child nodes is inspected to see if it is possible to generate the best leaf nodes later. A node can be trimmed away without further expansion if either its cost or its Lower Bound Cost (LBC) is higher than the *lowest* Upper Bound Cost (UBC) that has been found.[3]

How the algorithm allocates the routing paths is critical to its performance. Better routing path allocation helps balancing the traffic which leads to a better solution, but needs more time to compute. Next, we describe our routing path allocation heuristic which can find a good routing path within reasonably short computational times.

Routing path allocation has the objective to find *deadlock-free, minimal* routing paths, and at the same time, to balance the traffic across the links.

To be deadlock free, the routing algorithm needs to prohibit *at least one turn* in each of the possible routing cycles. In addition, it should *not* prohibit more turns than necessary to preserve the adaptiveness. Based on this, several deadlock-free adaptive routing algorithms have been proposed [7], including *west-first, north-last*, and *negative-first*. Chiu [8] proposed the *odd–even* turn model which restricts the locations where some types of turns can take place such that the algorithm remains deadlock-free.

In our algorithm, we convert the adaptiveness offered by the above algorithms into the flexibility for the routing path allocation by constructing a *Legal Turn Set* (LTS). A LTS is composed of and *only* of those turns allowed in the corresponding algorithm. Any path to be allocated can only employ turns from the LTS. The advantage of using LTS is two folds: first, since the turns are restricted to LTS, deadlock-free routing is guaranteed. Second, since only minimum turns are prohibited, the routing path allocation is still highly flexible.

Theoretically, the turns allowed in *any* of such deadlock-free adaptive routing algorithms can be used to construct a LTS. In this paper, we choose *west-first* and *odd–even* routing algorithm to build our LTS's.[4] Both of these two algorithms

---

[3]The computation of UBC/LBC is important for the performance of our algorithm. Please refer to [11] for a detailed description of the method that we use to compute UBC and LBC.

[4]*North-last* and *negative-first* routing are similar to *west-first* routing.

prohibit only $\frac{1}{4}$ of the total possible turns. However, the degree of the adaptiveness provided by *odd–even* routing is distributed more evenly than that provided by *west-first*.

Given an LTS, the following heuristic is used to allocate routing paths for a List of Communication Loads (LCL):

In Fig. 4, the Communication Load (CL) is first sorted by flexibility. The flexibility of a CL can be 1 (if there exists more than one *legal* path[5]) or 0 (otherwise). CLs with lower flexibility are given higher priorities for path allocation. If two CLs are tied in flexibility, the one with higher bandwidth requirement is given priority. Function *choose_link* returns the *least* loaded link allowed by LTS.

6.3   The Pseudo Code

Figure 5 gives the pseudo code of our algorithm. Two speedup techniques are employed to trim away more non-promising nodes, early in the search process:

- *IP ordering*: IPs are sorted by their communication demands ($\Sigma_{\forall j \neq i}\{v(a_{i,j})+v(a_{j,i})\}$ for IP c_i) so that the IPs with higher demand will be mapped earlier. Since the positions of the IPs with higher demands have larger impact on the overall communication energy consumption, fixing their positions earlier helps exposing those non-promising nodes earlier in the searching process; this reduces the number of nodes to be expanded.
- *Priority queue (PQ)*: The PQ sorts the nodes to be branched based on their cost. The lower the cost of the node, the higher the priority it has for branching, as expanding a node with lower cost will more likely decrease the minimum UBC.

Obviously, as the system size scales up, the run-time of the algorithm will also increase. Fortunately, we can trade off the solution quality with run time by limiting

```
Sort LCL by the flexibility of each CL
for each CL in LCL {
cur_tile = CL.source; dst_tile=CL.
destination
while(cur_tile ≠ dst_tile) {
    link = choose link(cur_tile, dst_tile);
cur tile = link.next tile();
link.add load(CL.bandwidth);  }}
```

Fig. 4 Routing path allocation

---

[5] A *legal* path has to be *minimal* and employ only those turns in the LTS.

```
Sort the IPs by communication demand
root_node = new node(NULL)
MUBC = +∞, best_mapping_cost = +∞
PQ.Insert(root_node)
while(!PQ.Empty()) {
    cur_node = PQ.Next()
    for each unoccupied tile t_i {
        generate child node n_new
        allocate routing paths
        if(n_new's mirror node exists in the PQ)
            continue
        if(n_new.LBC>MUBC) continue
        if(n_new.isLeafNode) {
            if(n_new.cost < best mapping cost) {
                best mapping cost = n_new.cost
                best mapping = n_new }}
        else {
            if(n_new.UBC<MUBC) MUBC = n_new.UBC
            PQ.insert(n_new) }}
}
```

Fig. 5 The pseudo code of the algorithm

the maximum length of PQ. When PQ's length reaches a threshold value, strict criteria are applied to select the child nodes for insertion into PQ.

7   Experimental Results

7.1   Evaluation on Random Applications

We first compare the run-time and the solution quality of our algorithm against a *simulated annealing* optimizer (SA). To make the comparison fair, SA was optimized by carefully selecting parameters such as number of moves per temperature, cooling schedule, etc. Since SA restricts itself to *XY* routing, for the first set of experiments, we also configure our algorithm to run without exploiting the routing flexibility by fixing it to just *XY* routing. In the following, we call this version EPAM-XY. (EPAM stands for Energy/Performance Aware Mapping.)

Four categories (I, II, III, IV) of random benchmarks were generated, each containing 10 applications with 9, 16, 25, and 36 IPs, respectively. EPAM-XY and

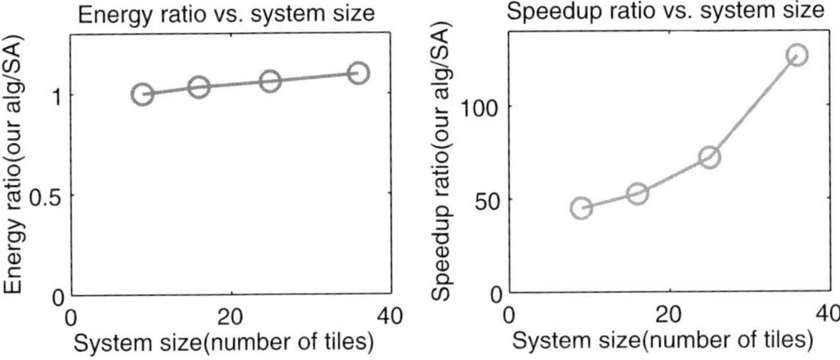

Fig. 6 Comparison between SA and EPAM-XY

SA were then applied to map these applications onto architectures with the same number of tiles. The results are shown in Fig. 6.

As shown in Fig. 6, for applications with 9 tiles, EPAM-XY runs **45** times faster than SA, on average. As the system size scales up, the speedup also increases dramatically. For applications with 36 tiles, the average speedup of EPAM-XY over SA increases to **127**. Meanwhile, the solutions produced by our algorithm remain very competitive. Indeed, the energy consumption of the solutions generated by EPAM-XY is only 3%, 6%, and 10% larger than the SA solutions for category II, III and IV, respectively. For category I, EPAM-XY even finds better solutions because it can walk through the whole search tree due to the small problem size.

To evaluate the benefits of exploiting the routing flexibility, we also applied our algorithm on benchmark applications using different LTS configurations. The comparison between the algorithms with LTS's based on *XY* routing (EPAM-XY), *Odd-Even* routing (EPAM-OE) and *West-First* (EPAM-WF) is shown in Fig. 7.

The left part of Fig. 7 shows how EPAM-XY, EPAM-OE and EPAM-WF perform for a typical 36-tile application as the link bandwidth constraints change. The results demonstrate two advantages of routing flexibility exploitation:

- First, it helps to find solutions for architectures with lower link bandwidth (which implies a lower implementation cost). For example, for this application, without exploiting the routing flexibility, the link bandwidth has to be 526 Mb/s (or higher) in order to find a solution which meets the performance constraints. By exploiting the routing flexibility, the link bandwidth requirement decreases to 476 Mb/s (EPAM-OE) and 500 Mb/s (EPAM-WF), respectively.
- Second, given the same architecture, exploiting routing flexibility leads to solutions with less energy consumption. The benefit becomes more significant as the links' loads reach their capacity. For instance, using the architecture where each link can provide 526 Mb/s bandwidth, the power consumption of the solution generated by EPAM-XY is 1.60 W. The power consumptions of the solutions generated by EPAM-OE and EPAM-WF are only 1.50 W and 1.04 W respectively. So significant power savings are achieved.

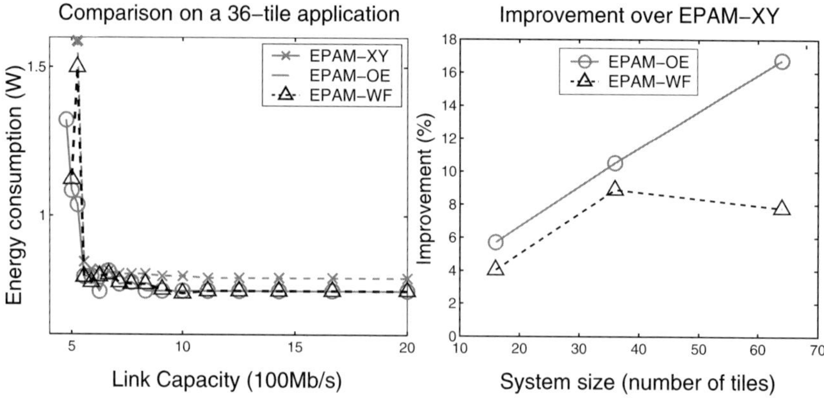

Fig. 7 The effectiveness of routing flexibility exploitation

The right part of Fig. 7 shows how the effectiveness of routing flexibility exploitation changes as the problem size scales. Let BW^{min} be the minimum link bandwidth needed by the corresponding algorithm to be able to find a solution which meets the performance constraints. The improvement (y-axis) is defined as $\dfrac{BW^{min}_{xy} - BW^{min}}{BW^{min}}$ which is a measure of how much the link bandwidth can be relaxed by exploiting routing flexibility compared to that using the EPAM-XY. As we can see, both EPAM-OE and EPAM-WF perform better than EPAM-XY. For applications with 16 tiles, EPAM-OE provides 5.69% improvement over EPAM-XY, on average. As the problem size scales up to 36 tiles and 64 tiles, the average improvement increases to 10.51% and 16.74%, respectively.

Overall, EPAM-OE performs *much better* than EPAM-WF, due to the fact that the *odd–even* routing provides more even adaptiveness than the *west-first* routing [8].

7.2 A Video/Audio Application

To evaluate the potential of our algorithm for real applications, we applied it to a generic MultiMedia System (MMS) [11]. MMS is an integrated video/audio system which includes an *H263* video encoder, an *H263* video decoder, an *MP3* audio encoder and an *MP3* audio decoder. We first partition MMS into 40 concurrent tasks and then assign/schedule these tasks onto 16 selected IPs available from industry [12]. These IPs range from DSP, generic processor, embedded DRAM to customized ASIC. We then use real video and audio clips as inputs to derive the communication patterns among these IPs.

Table 1 Comparison of ad hoc vs. EPAM-OE

Movie clips	Ad hoc(mW)	EPAM-OE(mW)	Savings
box/hand	171.2	82.8	51.6%
akiyo/cup	226.2	104.8	53.7%
man/phone	133.3	66.88	49.8%

Table 2 Comparison between SA and EPAM-OE

	SA	EPAM-OE	Improvement
Run time (s)	25.55	0.31	82.419%
Power (mW)	119.36	105.12	11.9%

Applying EPAM-OE to MMS, the solution is found in less than 0.5. CPU time. An ad hoc implementation was also developed to serve as reference. The results are shown in Table 1.

In Table 1, each row represents the power consumption of using two movie clips as simulation inputs, with one clip for the video/audio encoder and the other for the video/audio decoder. Compared to the ad-hoc solution, we observe around 51.7% energy savings, on average, which demonstrates the effectiveness of our algorithm.

To compare the performance between EPAM-OE and SA, both of them are applied to MMS for an architecture whose link bandwidth is fixed to 333 Mb/s. The results are shown in Table 2.

As shown in Table 2, our algorithm generates a better solution (about 12% less power) with significantly shorter run time compared to SA. We should point out that while the time of SA is affordable for this system (because it has only 4×4 tiles), the run time of SA increases dramatically as the system size scales up. For instance, for systems with 7×7 tiles, the average run time of SA increases to 2.2 h. For systems with 10×10 tiles (which may be available in the near future [5]), our algorithm needs just a few minutes to complete while the run time of SA becomes prohibitive (in our experiments, SA did not finish in 40 h of CPU time).

To show how much improvement our algorithm can achieve by exploiting routing flexibility, we applied EPAM-XY, EPAM-OE and EPAM-WF to MMS (Fig. 8). As shown in Fig. 8, EPAM-XY fails to find a solution when the link bandwidth decreases to 324 Mb/s (point A in Fig. 8). In contrast, both EPAM-OE and EPAM-WF can still find solution even the link bandwidth decreases to 307 Mb/s (point B in Fig. 8), thus suggesting a 5.5% improvement. Given the same architecture whose link bandwidth is fixed to 324 Mb/s, the power consumptions of the solutions found by EPAM-OE and EPAM-WF are both 109.2 mW, while the power

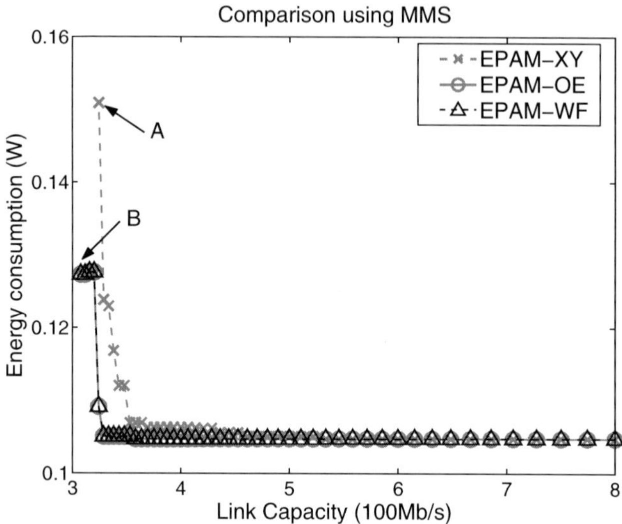

Fig. 8 Comparison using MMS application

consumption of the EPAM-XY solution is 150.9 mW; thus, **27.6%** of power savings are achieved. Again, this shows the advantage of exploiting the routing flexibility.

8   Conclusion and Future Work

In this paper, we proposed an efficient algorithm for solving the mapping and routing path allocation problems in regular tile-based NoC architectures. Although we focus on the architectures interconnected by 2D mesh networks, our algorithm can be adapted to other *regular* architectures with different network topologies. This remains to be done as future work.

References

[1] W. J. Dally, B. Towles, Route packets, not wires: on-chip interconnection networks, *Proc. DAC*, pp. 684–689, June 2001.

[2] J. Chang, M. Pedram, Codex-dp: co-design of communicating systems using dynamic programming, *IEEE Trans. CAD of Integrated Circuits and Systems*, vol. 19, no. 7, July 2002.

[3] W. J. Dally, C. L. Seitz, The torus routing chip, *Journal of Distributed computing*, vol. 1, no. 3, 187–196, 1986.

[4] A. Hemani, et al., Network on a chip: an architecture for billion transistor era, *Proc. IEEE NorChip Conf.*, Nov. 2000.

[5] S. Kumar, et al., A network on chip architecture and design methodology, *Proc. Symposium on VLSI*, pp. 117–124, April 2002.

[6] L. M. Ni, P. K. McKinley A survey of wormhole routing techniques in direct networks, *Computer*, vol. 26, no. 2, Feb. 1993.

[7] C. J. Glass and L. M. Ni, The turn model for adaptive routing, *Proc. ISCA*, May 1992.

[8] G. Chiu, The odd-even turn model for adaptive routing, *IEEE Trans. on Parallel and Distributed Systems*, vol. 11, no. 7, 729–738, July 2000.

[9] T. T. Ye, L. Benini, G. De Micheli, Analysis of power consumption on switch fabrics in network routers, *Proc. DAC*, June 2002.

[10] R. P. Dick, D. L. Rhodes, W. Wolf, TGFF: task graphs for free, *Proc. International Workshop on Hardware/Software Codesign*, March 1998.

[11] J. Hu, R. Marculescu, Exploiting the routing flexibility for energy/performance aware mapping of regular NoC architectures, *CSSI Technical Report*, Carnegie Mellon University, Sept. 2002. Available at http://www.ece.cmu.edu/~sld/publications.html

[12] http://www.mentor.com/inventra/cores/catalog/index.html

×pipesCompiler: A Tool for Instantiating Application-Specific Networks on Chip

Antoine Jalabert[1], Srinivasan Murali[2], Luca Benini[3],
and Giovanni De Micheli[4]

Abstract Future Systems on Chips (SoCs) will integrate a large number of processor and storage cores onto a single chip and require Networks on Chip (NoC) to support the heavy communication demands of the system. The individual components of the SoCs will be heterogeneous in nature with widely varying functionality and communication requirements. The communication infrastructure should optimally match communication patterns among these components accounting for the individual component needs. In this paper we present ×pipesCompiler, a tool for automatically instantiating an application-specific NoC for heterogeneous Multi-Processor SoCs. The ×pipesCompiler instantiates a network of building blocks from a library of composable soft macros (switches, network interfaces, and links) described in SystemC at the cycle-accurate level. The network components are optimized for that particular network and support reliable, latency-insensitive operation. Example systems with application-specific NoCs built using the ×pipesCompiler show large savings in area (factor of 6.5), power (factor of 2.4), and latency (factor of 1.42) when compared to a general-purpose mesh-based NoC architecture.

Keywords Systems on Chips, Networks on Chips, latency-insensitive design, application-specific, SystemC.

1   Introduction

With increasing transistor density, the number of cores on a chip and the communication demands between them is rapidly increasing. System interconnect scalability is limited for state-of-the-art SoC communication architectures based on

---

[1] CEA LETI-DSIS antoine.jalabert@cea.fr

[2] CSL Stanford University smurali@stanford.edu

[3] DEIS Univ of Bologna lbenini@deis.unibo.it

[4] CSL Stanford University nanni@stanford.edu

R. Lauwereins and J. Madsen (eds.), *Design, Automation, and Test
in Europe–The Most Influential Papers of 10 Years DATE*, 157–171
© Springer 2008

shared communication resources. *Networks on chip* (NoC) architectures have been proposed to address the scalability challenge [1–3, 8]. NoCs are scalable and compatible with design and reuse of cores, which is a critical feature required by SoC designers to meet tight time-to-market constraints [4].

An important design decision for NoCs is the choice of topology. Several researchers [10–13] envision NoCs as regular topologies (such as *mesh* networks and *fat trees*), which are suitable for interconnecting homogeneous cores in a chip multiprocessor (Figure 1a). However, many SoCs involve heterogeneous cores having varied functionality, size, and communication requirements. If a regular interconnect is designed to match the requirements of few communication-hungry components, it is bound to be largely overdesigned with respect to the needs of the remaining components. This is the main reason why most current SoCs use irregular topologies like bridged busses and/or dedicated point-to-point links [14].

As an example, consider the implementation of an *MPEG4 decoder* [5], depicted in Fig. 1b, where blocks are drawn roughly to scale and links represent inter-block communication. First, the embedded memory (SDRAM) is much larger than all other cores and it is a critical communication bottleneck. Block sizes are highly nonuniform and the floorplan does not match the regular, tile-based floorplan shown in Fig. 1a. Second, the total communication bandwidth to/from the embedded SDRAM is much larger than that required for communication among the other cores. Third, many neighboring blocks do not need to communicate. Even though it may be possible to implement MPEG4 onto a homogeneous fabric, there is a significant risk of either underutilizing many tiles and links, or, at the opposite extreme, of achieving poor performance because of local congestion. These factors motivate the use of an application-specific on-chip network [15].

With an application-specific network the designer is faced with the additional task of designing network components (e.g. switches) with different configurations

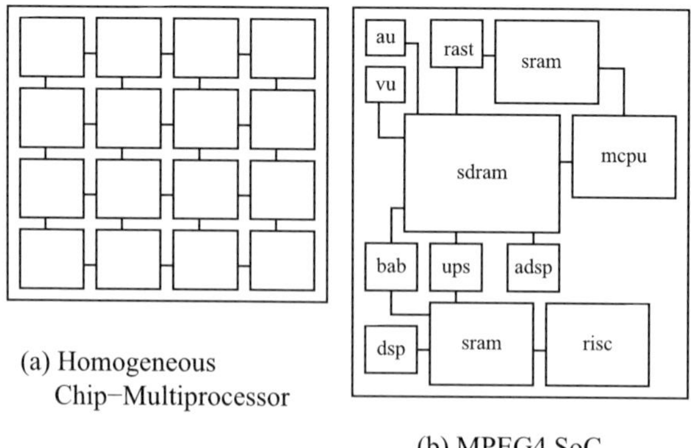

(a) Homogeneous
Chip–Multiprocessor

(b) MPEG4 SoC

Fig. 1 Homogeneous CMPs and heterogeneous SoC applications

(e.g. different I/Os, virtual channels, buffers) and interconnecting them with links of uneven length. These steps require significant design time and the need to verify network components and their communications for every design. In this paper we describe a tool that bridges the design gap for heterogeneous application-specific NoCs.

The xpipesCompiler automatically instantiates network components (routers, links, network interfaces) for a specific NoC topology, using the xpipes library of SystemC soft macros defined at the cycle-accurate and signal-accurate level [6]. The xpipes library is aggressively designed for high performance: links can be pipelined to an arbitrary degree to decouple clock cycle time from worst-case link delay (this mode of operation has been called *latency insensitive* in recent literature [9]). Latency insensitive link-level error control is fully supported, ensuring robustness against communication errors.

Even though xpipes can be instantiated as a high-performance regular NoC, its most innovative feature is that all its components are highly parameterized, and they can be tailored to the communication needs of a specific architecture. Thus, the xpipesCompiler can instantiate optimized NoCs. Significant improvements in area, power and latency are achieved with respect to regular NoC architectures, as demonstrated by several case studies detailed in the paper.

2   The xpipes Architecture

In this section we present a brief description of the architecture of switches, links, and network interfaces that form the xpipes library. We refer the reader to [6] for a detailed description of these components. A conceptual picture of the network architecture is shown in Fig. 2. It supports packet-switched communication, with source routing and wormhole flow-control. Source-based routing results in lightweight switch implementations when compared to dynamic routing and wormhole flow-control results in reduced buffering at each switch. Cores can be plugged into the network, provided they are OCP compliant [18] and the network communication protocols are completely hidden to the cores. The individual components are detailed below.

2.1   Network Interface

The *Network Interface* (NI) connects the core to the NoC. It converts the end-to-end OCP transactions into packets that are to be transmitted through the network. The NI builds the packet header using the routing information for the destination stored in a look-up table. The packet is broken down into header and payload flits and the flits are injected into the network at the rate of one flit every clock cycle. The NI synchronizes the network requests with the consuming rate of the core. To keep the interface complexity low, the NI supports only a single outstanding read

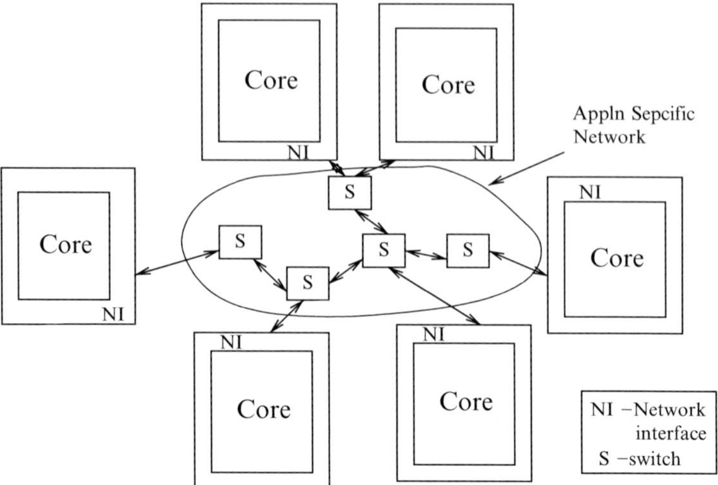

Fig. 2 NoC architecture block diagram

operation, but an arbitrary number of write transactions can be carried out after an outstanding read.

2.2 Switch Architecture

Switches in the ×pipes library are deeply pipelined to maximize the operating frequency. The pipeline structure for a single output module is shown in Fig. 3, and the structure is repeated for each output. Forward flow control is used and a flit is transmitted to the next switch only when adequate storage is available in that switch. Switches support multiple virtual channels and a physical link is assigned to different virtual channels on a flit-by-flit basis, thereby improving network throughput. The CRC decoders for error detection work in parallel with the switch operation, thereby hiding their impact on switch latency.

For latency insensitive operation, the switch has virtual channel registers to store $2N + M$ flits, where N is the number of link pipeline stages and M is an architecture dependent parameter (12 cycles in our design). The reason is that each transmitted flit has to be acknowledged before being discarded from the buffer. Before an ACK is received, the flit has to travel across the link (N cycles), an ACK/NACK decision has to be taken at the destination switch (a portion of M cycles), the ACK/NACK signal has to be propagated back (N cycles) and recognized by the source switch (remaining portion of M cycles). During this time, other $2N + M$ flits are transmitted but not yet ACKed.

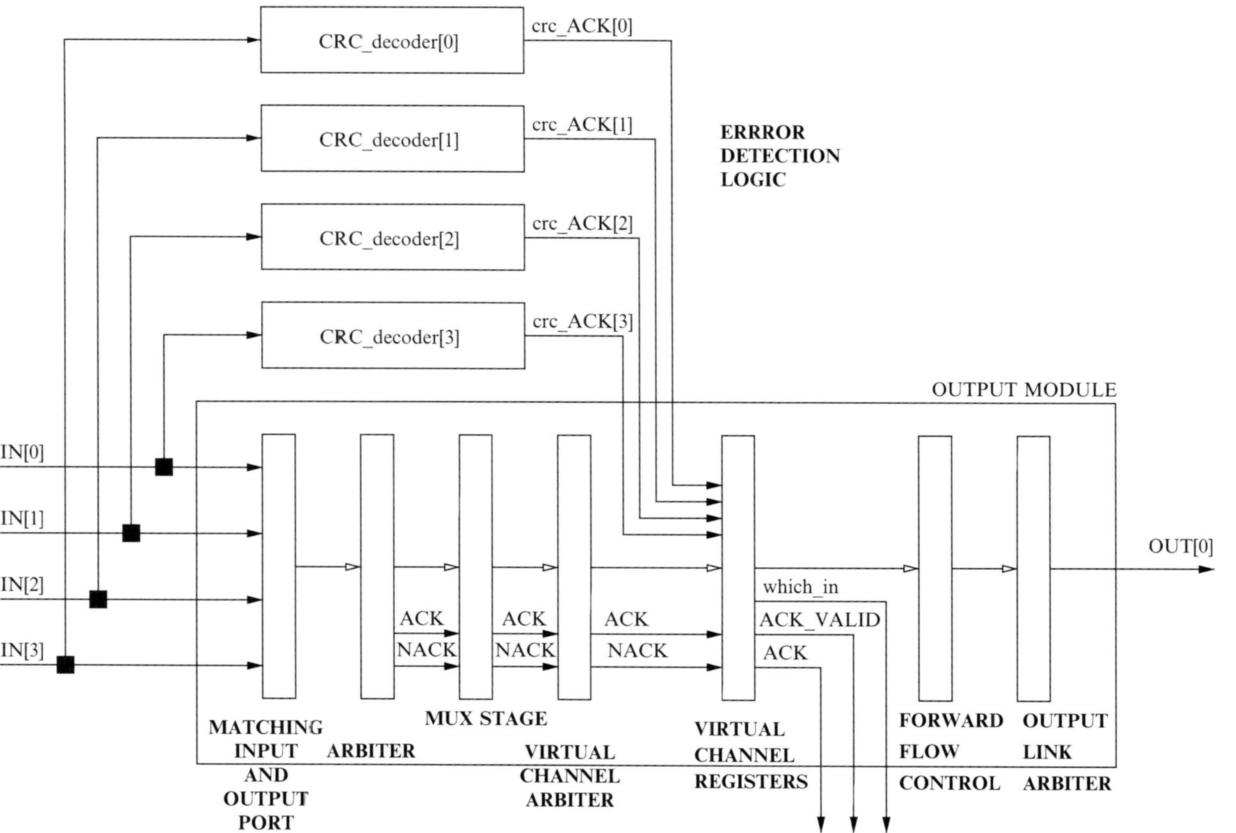

Fig. 3 Pipelined architecture of a switch

2.3   Link Architecture

The links of an irregular NoC are of varying lengths, and it takes a varying number of clock cycles to traverse the links. In order to maximize throughput, the links are subdivided into basic segments that require a single clock cycle for traversal, making the links *latency insensitive* by pipelining flits through them. This pipelining is applied to both data and control lines. As previously discussed, the switches and network interfaces have enough buffering resources and their functional correctness depends only on the flit arriving order and not on timing. This ensures a correct latency insensitive operation of the whole system.

2.4   Instantiation-time Parameters

All network components can be specialized through instantiation-time parameters. Some parameters are global and they affect all component instances. Global parameters are: flit size, address space of cores, degree of redundancy (minimum distance) of the CRC error-detection code to be used on the links, number of bits used for packet sequence count (for end-to-end flow control), maximum number of hops between any two network nodes (used for header sizing), number of flit types (to support special flits, for instance control flits with no payload), number of packet types (which can be used by higher-level protocols).

As for the local parameters affecting single network component instances, we have the following. For the network interface: number of data and address lines and maximum burst length in the OCP connection between NI and the core, type of interface (master/ initiator, slave or both), flit buffer size in the output port (which enables the network interface to continue packetization even when the network link is congested) and content of the routing table. For the switch: number of ports, number of virtual channels, link buffer size for each port (which relates to the number of pipeline stages in the corresponding link). For each link, the number of stages can obviously be specified.

3   The ×pipesCompiler

The complete NoC design flow is depicted in Fig. 4. From the specification of an application, the designer (or a high-level analysis and exploration tool) creates a high-level view of the SoC floorplan, including nodes (with their network interfaces), links and switches. Based on clock speed target and link routing, the number of pipeline stages for each link is also specified. The information on the network architecture is specified in an input file for the ×pipesCompiler. Routing tables for the network interfaces are also specified. The tool takes as additional input the SystemC library of soft components described in the previous section. The output is a SystemC hierarchical description, which includes all switches, links, network nodes, and interfaces and specifies their topological connectivity.

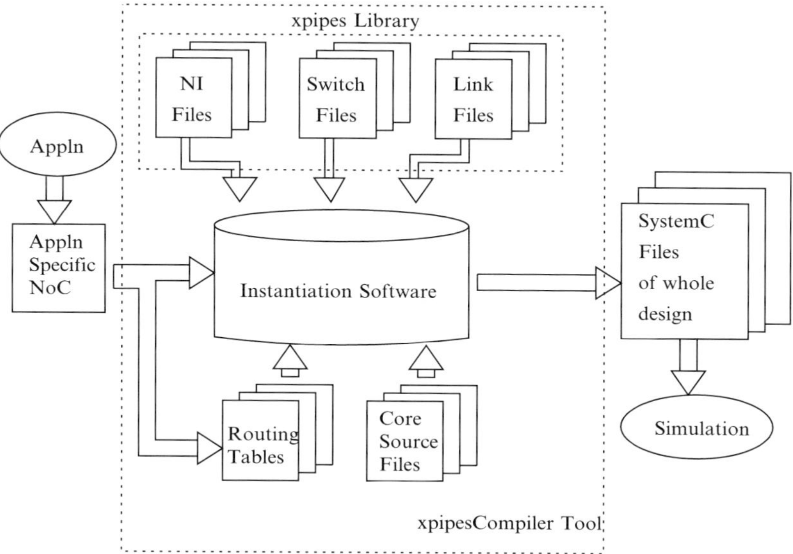

Fig. 4 NoC synthesis flow

The final description can then be compiled and simulated at the cycle-accurate and signal-accurate level. At this point, the description can be fed to back-end RTL synthesis tools for silicon implementation (the details of the back-end synthesis process are not covered in this paper).

In a nutshell, the xpipesCompiler generates a set of network component instances which are custom-tailored to the specification contained in its input network description file. Network instantiation follows a two-step procedure: first, the input file is parsed in to an internal data structure, then the structure is traversed and the output is generated. In the following subsections we describe these steps in more detail.

3.1 Input Specification and Parsing

The input file describes the cores, switches, links, and the relationships between them. From the designer's viewpoint, the implementation of the NI that connects a core to the NoC is transparent. The xpipesCompiler will instantiate the needed NIs according to the type of the core (Master, Slave, Master/Slave).

Example 1 *Part of a simple input specification is shown in Fig. 5. Attributes of a switch are: name (-NAME), number of I/O ports (-NPORT) and the number of virtual channels for each output port (-NVC). The attributes of a link (.module_lnk) are: name (-NAME), the name and port number of the source and destination switch for the link (-MAP), and the number of repeaters it includes (-NREP).*

While parsing the design description file, the xpipesCompiler dynamically allocates a tree data structure to store the necessary information for each object of the design.

```
.module_sw   -NAME=SW_0   -NPORT=3   -NVC=4
.module_sw   -NAME=SW_1   -NPORT=2   -NVC=2
.module_sw   -NAME=SW_2   -NPORT=3   -NVC=2
.module_sw   -NAME=SW_3   -NPORT=5   -NVC=3
.module_sw   -NAME=SW_4   -NPORT=4   -NVC=2
.module_lnk   -NAME=lnk_0_1   -MAP=SW_0,0;SW_1,2   -NREP=3
.module_lnk   -NAME=lnk_1_0   -MAP=SW_1,2;SW_0,0   -NREP=3
.module_lnk   -NAME=lnk_2_0   -MAP=SW_2,3;SW_0,1   -NREP=2
.module_lnk   -NAME=lnk_0_3   -MAP=SW_0,2;SW_3,0   -NREP=5
.module_lnk   -NAME=lnk_0_4   -MAP=SW_0,3;SW_4,1   -NREP=2
.module_lnk   -NAME=lnk_4_0   -MAP=SW_4,1;SW_0,3   -NREP=2
```

Fig. 5 Example input specification

Once the first step has been executed and no errors have been detected (like invalid parameters, parameters missing, global parameters not defined), the xpipesCompiler processes the collected information, as described in the following subsection.

3.2 Network Instantiation

The xpipesCompiler identifies the different types of switches, links, and network interfaces that are required in the design. Because all components are written in SystemC, a single class is created for each type, and multiple objects are instantiated whenever needed.

During network instantiation, the xpipesCompiler performs several optimizations to remove redundant logic from the generic library components. For example, if a switch has only an input link connected to a port, the logic and buffers of the missing output port will not be generated. In the case of a custom-made irregular design, this is a very valuable optimization that drastically reduces hardware complexity. The optimized network component classes are dynamically stored into arrays of structures.

For each object type, a recursive function processes the tree of files from the leaves to the main object file (the root), parsing each file and customizing it according to its type. It is also during this step that the routing tables for each NI are processed and converted into files that are to be included.

The top level (main. cc) of the design is then generated. This file instantiates all objects of the design at run-time. It also defines the signals needed to connect the objects according to the design description file. Here again, if possible, the xpipesCompiler will share the signals that are common to all objects. In order to automate tracing of signals, the *debug* command has been implemented, which enables monitoring of any signal in the design.

Example 2 *Referring to Example 1, Fig. 6 shows how the network is optimized by removal of redundant logic during instantiation. Because* port [1] *of SW_0 has only an input connection the* xpipesCompiler *will not instantiate a* PortOUT Block *that would be connected to this output. Similar optimization is performed for an output-only port*

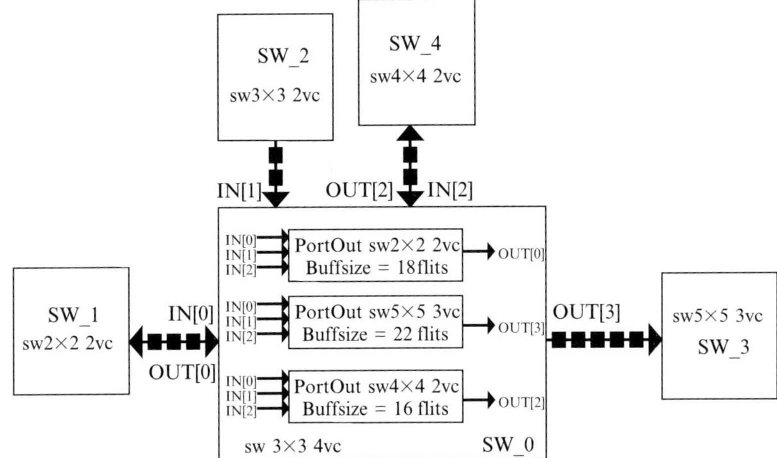

Fig. 6 Irregular n/w optimizations

like port [3]: *in each* PortOUT Block *generated, the signals related to this non-existing input are removed from the generic template files. The* xpipesCompiler *will also optimize the size of each output buffer according to the number of repeaters on each output link.*

The time of execution of the xpipesCompiler depends on the design characteristics. For example, a regular topology such as a 16×16 mesh can be generated faster than an application-specific topology with only few cores and switches. But in absolute terms execution time is not a major concern. For instance, for a custom design with 4 switches and 2 cores the execution time is about 2s on a Pentium running at 1.8 Ghz. For all our experiments, network generation was completed in few minutes.

4 Experimental Validation and Case Studies

We consider three video-processing applications: *Video Object Plane Decoder* (VOPD), *MPEG4 Decoder* (MPEG4) and *Multi-Window Displayer* (MWD), presented in [5, 7], that are mapped onto cores. The communication characteristics of these applications are shown in Fig. 7, with the edges annotated with the amount of data transferred between the cores in MB/s. We manually developed customized application-specific topologies that closely match the applications' communication characteristics. For comparison, we also developed a regular mesh NoC for each application. The application designs with different NoC configurations are shown in Figs. 8–10.[1]

---

[1]For clarity, we show pipelining only in Fig. 8 and we assume NI as part of core in the experiments.

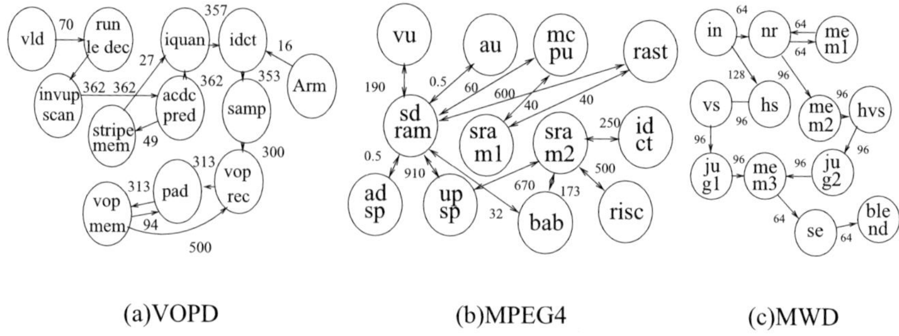

Fig. 7 Communication pattern of example designs

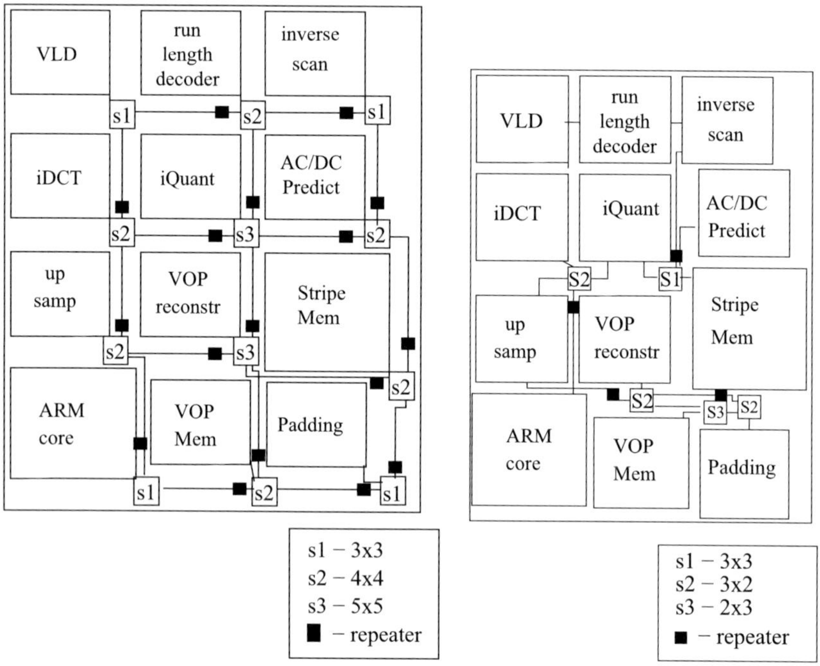

Fig. 8 Video object plane decoder

4.1 Application-Specific NoC Characterization

In the VOPD, about half the cores communicate to more than a single core. This
motivates the configuration of the custom NoC in Fig. 8b, having less than half the
number of switches than a regular mesh NoC. Also these switches are much

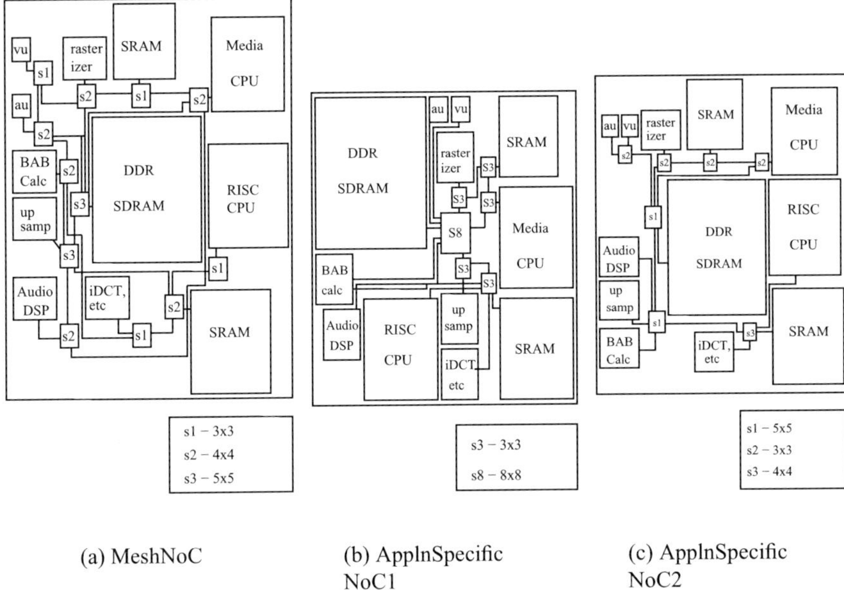

Fig. 9 MPEG4 decoder

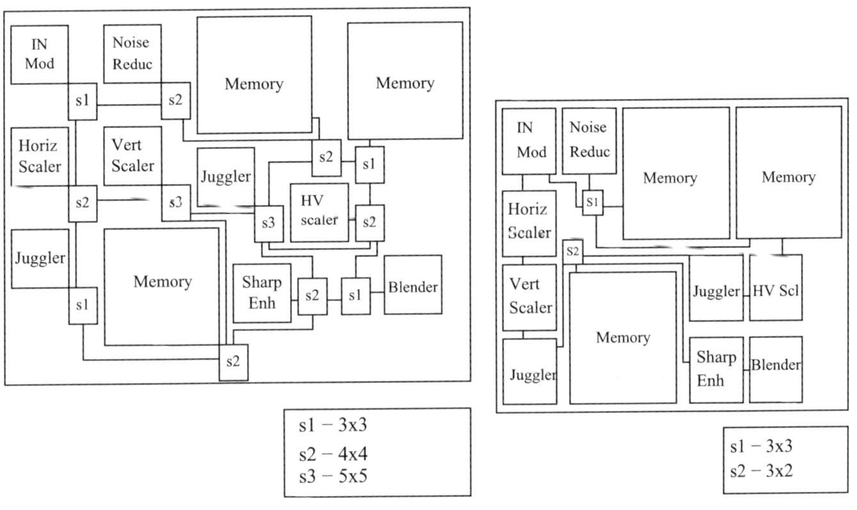

Fig. 10 Multi-window displayer

smaller than the mesh switches. In the MPEG4 design considered, many of the cores communicate with each other through the shared SDRAM. So a large switch is used for connecting the SDRAM with other cores (Fig. 9b) with smaller switches for other cores. We also consider an alternate custom NoC for MPEG4 (Fig. 9c) which is an optimized mesh network, with superfluous switches and switch I/Os removed. In the communication pattern of MWD, all but two cores communicate to only one other core. So the custom NoC for MWD (Fig. 10b) has only two switches.

4.2 Area-Power Analysis

We developed analytical models for estimating the switch areas. The area calculations include the crossbar area, buffer area, logic (including control) area. The models take into account the nuances of individual switch configurations and includes fine granularity of details (like accounting for pipeline registers, cross points, etc.).

The area estimates for the various NoC configurations for $0.1\,\mu$ technology is shown in Table 1. As custom VOPD and MWD NoCs (generated by the xpipes-Compiler) have relatively small number of switches, we obtain significant area improvement for the custom NoC. But for the MPEG4, as each core communicates to many other cores we have many switches and obtain only ×1.69 area improvement with the custom NoCs. We obtain an average of ×6.54 area savings for the custom NoCs when compared to the mesh NoC.

We used ORION [16], a power-modeling tool, for developing bit energy models for the switches. We use wiring parameters from [17] to estimate link power dissipation. We calculate the power dissipation for each NoC design, based on the average traffic (shown as edge annotations in Fig. 7) through each network component for a supply voltage of 1.8 V. The results of the power analysis is summarized in Table 2. For VOPD and MWD we obtain power savings of a factor of three, whereas for the MPEG4 the power savings are smaller. The reason for lower savings in MPEG4 is that most of the traffic traverses the bigger switches connected to the memories. As power dissipation on a switch increases nonlinearly with increase in switch size there is more power dissipation in the switches of custom NoC1 of MPEG4 (that has an 8×8 switch) as compared to the mesh NoC.

Table 1 Area estimates

Application	Type	Area mm²	Ratio mesh/cust
vopd	mesh	1.26	
	custom	0.22	5.73
mpeg4-1	mesh	1.31	
	custom	0.86	1.52
mpeg4-2	mesh	1.31	
	custom	0.71	1.85
mwd	mesh	1.22	
	custom	0.10	12.2

However most of the traffic traverses short links in this custom NoC (Fig. 9b), thereby giving marginal power savings for the whole design.

4.3   SystemC Simulation Results

We performed cycle-accurate simulation of the SystemC models of the NoCs generated by the xpipesCompiler. We used traffic generators to model the bursty nature of the application traffic, with average communication bandwidth matching

Table 2 Power estimates

Application	Topology	Power mW	Ratio mesh/custom
vopd	mesh	108.74	
	custom	40.08	2.71
mpeg4-1	mesh	114.36	
	custom	110.66	1.03
mpeg4-2	mesh	114.36	
	custom	93.66	1.22
mwd	mesh	25.9	
	custom	7.72	3.35

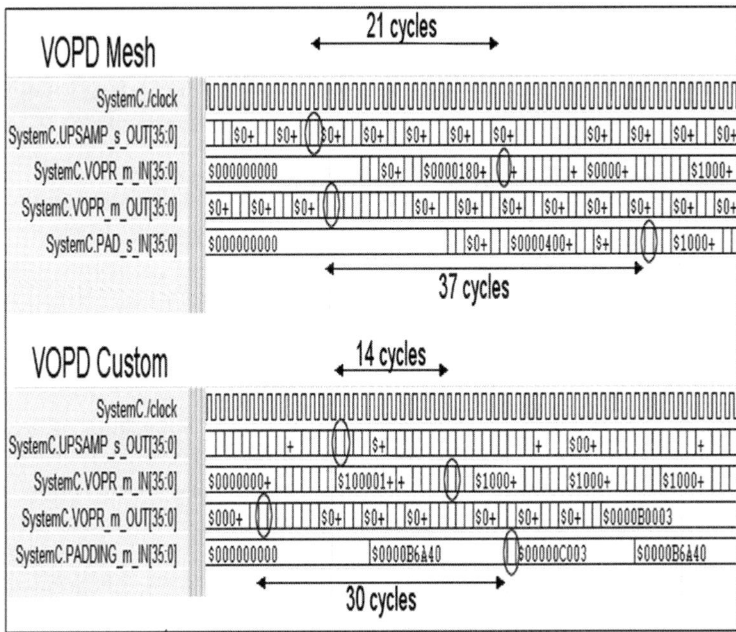

(a) SystemC Output

Fig. 11 Simulation results

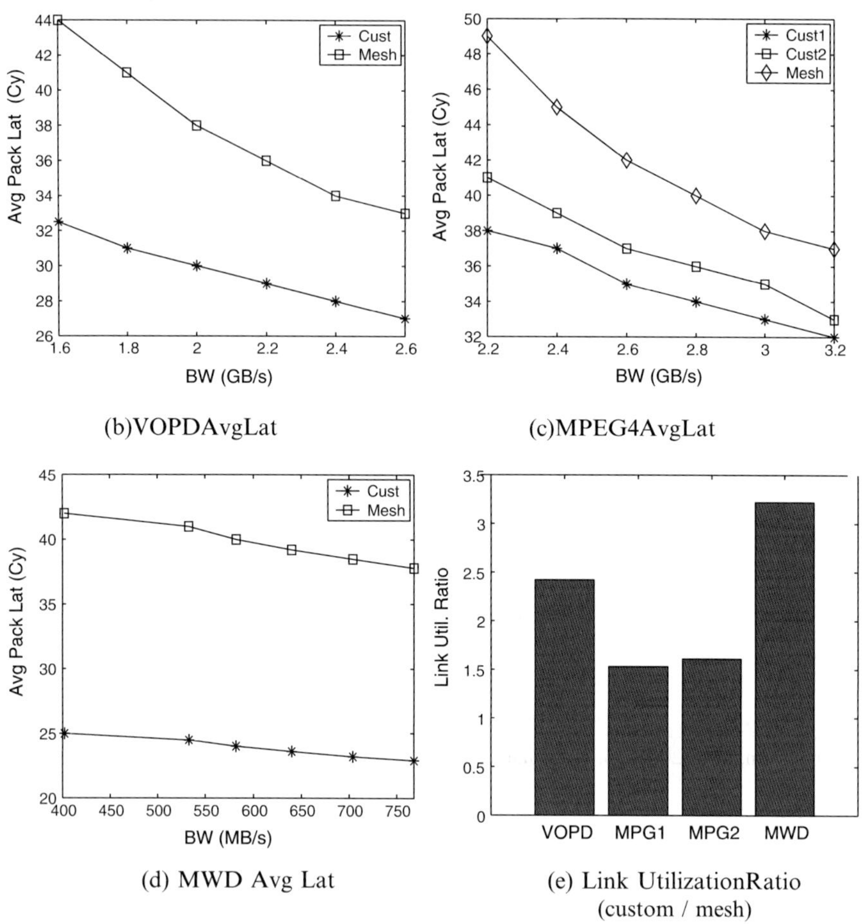

(b)VOPDAvgLat

(c)MPEG4AvgLat

(d) MWD Avg Lat

(e) Link UtilizationRatio
(custom / mesh)

Fig. 11 (continued)

the applications' average communication bandwidth. Snapshots of SystemC simulations of mesh and custom NoCs for some of the cores of VOPD are shown in Fig. 11a. The time between transmission of a flit and its reception, which includes the switch delay, link delay and contention delay, is marked in the figure. The variation of average packet latency (for 64B packets, 32 bit flits and 7 cycle switch delay) with link bandwidth is shown in Fig. 11b–d application-specific NoCs have lower packet latency as the average number of switch and link traversals is lower. Moreover, the latency increases more rapidly for the mesh NoCs with decrease in bandwidth. With the custom NoCs we achieve an average of 30% savings in latency (measured at the minimum plotted BW value). Also, custom NoCs have better link utilization as seen in Fig. 11e.

5 Conclusions and Future Work

In this paper we have presented xpipesCompiler, a tool that automatically instantiates application-specific NoCs. The tool bridges the design gap in building application-specific NoCs that optimally match the communication requirements of the system. The network components built are highly optimized for the particular NoC design, providing large savings in area, power and latency for example designs. In the future we plan to enhance the tool with automatic selection of network topology, thus providing a complete design flow for heterogeneous NoCs.

Acknowledgments   This research is supported by MARCO Gigascale Systems Research Center (GSRC) and NSF (under contract CCR-0305718).

References

[1] L. Benini, G.D. Micheli, Networks on Chips: A New SoC Paradigm, IEEE Computers, 70–78, Jan. 2002.

[2] A. Jantsch, H.Tenhunen, Networks on Chip, Kluwer Academic, 2003.

[3] E. Rijpkema et al. Trade-offs in the design of a router with both guaranteed and best-effort services for networks on chip, DATE 2003, pp. 350–355, Mar, 2003.

[4] W. Cesario et al. Component-based design approach for multi-core SoCs, DAC 2002, pp. 789–794, June, 2002.

[5] E.B. Van der Tol, E.G.T. Jaspers, Mapping of MPEG-4 decoding on a flexible architecture platform, SPIE 2002, pp. 1–13, Jan. 2002.

[6] M.Dallosso et al xpipes: a latency insensitive parameterized network-on-chip architecture for multi-processor SoCs, pp. 536–539, ICCD 2003.

[7] E.G.T. Jaspers, et al. Chip-set for video display of multimedia information, IEEE Trans. Consumer Electronics, Vol 45, No. 3, 707–716, Aug. 1999.

[8] F. Karim et al. An interconnect architecture for network systems on chips, IEEE Micro, Vol. 22, No. 5, 36–45, Sep. 2002.

[9] L.P. Carloni, K.L. McMillan, A.L. Sangiovanni-Vincentelli, Theory of latency-insensitive design, IEEE Trans. CAD of ICs and Systems, pp.1059–1076, Vol. 20, no. 9, Sept. 2001.

[10] P. Guerrier, A. Greiner, A generic architecture for on-chip packet switched interconnections, Proc. DATE, pp. 250–256, March 2000.

[11] S. Kumar et al. A network on chip architecture and design methodology, ISVLSI 2002, pp. 105–112, 2002.

[12] S.J. Lee et al. An 800MHz star-connected on-chip network for application to systems on a chip, ISSCC 2003, Feb. 2003.

[13] J. Hu, R. Marculescu, Energy-aware mapping for tile-based NOC architectures under performance constraints, ASP-DAC 2003.

[14] H.Yamauchi et al. A 0.8 W HDTV video processor with simultaneous decoding of two MPEG2 MP@HL streams and capable of 30 frames/s reverse playback, ISSCC, Vol.1, 473–474, Feb. 2002

[15] H. Zhang et al. A IV Heterogeneous reconfigurable DSP IC for wireless baseband digital signal processing, IEEE Journal of SSC, 1697–1704, Vol. 35, no. 11, Nov. 2000.

[16] H.S Wang et al. Orion: a power-performance simulator for interconnection networks, MICRO, Nov. 2002.

[17] R. Ho, K. Mai, and M. Horowitz, The Future of Wires, Proceedings of the IEEE, pp. 490–504, April 2001.

[18] http://www.ocpip.org/home OCP specification

A Network Traffic Generator Model for Fast Network-on-Chip Simulation

Shankar Mahadevan[1], Federico Angiolini[2], Jens Sparsø[1], Michael Storgaard[1], Jan Madsen[1], and Rasmus Grøndahl Olsen[1]

Abstract For Systems-on-Chip (SoCs) development, a predominant part of the design time is the simulation time. Performance evaluation and design space exploration of such systems in bit- and cycle-true fashion is becoming prohibitive. We propose a traffic generation (TG) model that provides a fast and effective Network-on-Chip (NoC) development and debugging environment. By capturing the type and the timestamp of communication events at the boundary of an IP core in a reference environment, the TG can subsequently emulate the core's communication behavior in different environments. Access patterns and resource contention in a system are dependent on the interconnect architecture, and our TG is designed to capture the resulting reactiveness. The regenerated traffic, which represents a realistic workload, can thus be used to undertake faster architectural exploration of interconnection alternatives, effectively decoupling simulation of IP cores and of interconnect fabrics. The results with the TG on an AMBA interconnect show a simulation time speedup above a factor of 2 over a complete system simulation, with close to 100% accuracy.

1 Introduction

An important step in the design of a complex System-on-Chip is to select the optimal architecture for the on-chip network (NoC). In order to do so, it is imperative to analyze and understand network traffic patterns through simulation. This can be accomplished at various stages in the design flow, from abstract transaction level models (TLM) to bit- and cycle-true models. In many cases, only the most detailed models prove capable of capturing important aspects of communication performance, e.g. the latency associated with resource contention. The obvious drawback of these approaches is slower simulation speed.

[1]Informatics and Mathematical Modelling (IMM) Technical University of Denmark (DTU) Richard Petersens Plads, 2800 Lyngby, Denmark e-mail: {sm, -,-,jsp, jan}@imm.dtu.dk

[2]Dipartimento di Elettronica, Informatica e Sistemistica (DEIS) University of Bologna Viale Risorgimento, 2 40136 Bologna, Italy e-mail: {fangiolini}@deis.unibo.it

R. Lauwereins and J. Madsen (eds.), *Design, Automation, and Test in Europe–The Most Influential Papers of 10 Years DATE*, 173–184
© Springer 2008

">

In this paper, we focus on enabling the exploration of different NoC architectures at the bit- and cycle-true level by increasing the speed of the complete SoC simulation. This is a key advantage, since architectural exploration typically involves carrying out the same set of simulations for each design alternative, and simulations may consist of millions of clock cycles each.

We assume as a requirement the availability of a reference SoC design, consisting of IP cores and of a NoC, and of an application partitioned and compiled onto the various IP cores. This application might either be software executing on programmable IP cores, or code synthesized into dedicated hardware. This reference system will be used to collect an initial trace of the IP cores' behavior. In order to increase simulation speed for subsequent design space exploration, we propose to replace the IP cores with Traffic Generators (TG) which emulate their communication at the interface with the network, as illustrated in Fig. 1.

The goal is to perform only one reference simulation using bit- and cycle-true simulation models of the IP cores running the target application, and to speed up subsequent variants of that simulation using traffic generators coupled with accurate models of the alternative interconnects only. While the internal processing of IP cores does not need thorough replication by the generators, and can often be modeled by waiting for an amount of cycles between network transactions, the unpredictability of network latency of different NoC architectures may lead to changes in the number and relative ordering of transactions. Thus, traffic generators should have at least some reactive capabilities, as will be explained in Section 3.

In order to capture reactive behavior, we propose a TG implementation as a very simple instruction set processor. Our approach is significantly different from a purely behavioral encapsulation of application code into a simulation device, in analogy with TLM modeling. The TG model we propose is aimed at faithfully replicating traffic patterns generated by a *processor running an application*, not just by the application; this includes e.g. accurate modeling of cache refills and of latencies between accesses, allowing for cycle-true simulations. At the same time, this approach allows a straightforward path towards deployment of the TG device on a silicon NoC test chip.

To evaluate the TG concept, we have integrated the proposed TG model into MPARM [8], a homogeneous multiprocessor SoC simulation platform, which

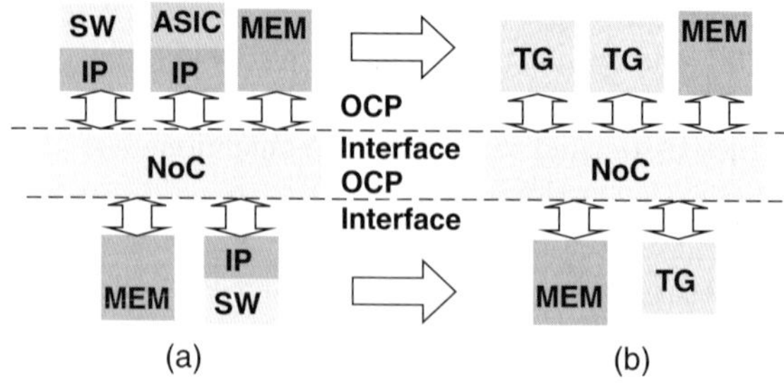

Fig. 1 Simulation environment with bit-and cycle-true: (a) IP-cores, (b) TG model

provides a bit-and cycle-true SoC simulation environment. The current version of MPARM supports several NoC architectures, e.g. AMBA [8], STBus and the xpipes [3], and leverages ARMv7 processors as IP cores. The use of the OCP [1] protocol at the interfaces between the cores and the interconnect allows for easy exchange of IP cores for TGs, as indicated in Fig. 1.

The rest of the paper is organized as follows. Section 2 introduces related work, and is followed by a discussion of the requirements for modelling traffic patterns in Section 3. Section 4 details the TG implementation, and Section 5 describes how communication traces are extracted and turned into programs executing on the TG. Section 6 presents initial simulation results which show the potential of our TG approach. Finally, Section 7 provides conclusions.

2 Related Work

The use of traffic generators to speed up simulation is not new, and several traffic generator approaches and models have been proposed.

In [6], a stochastic model is used for NoC exploration. Traffic behavior is statistically represented by means of uniform, Gaussian, or Poisson distributions. Such distributions assume a degree of correlation within the communication transactions which is unlikely in a SoC environment. Traffic patterns in SoC systems have shown to be reactive and bursty [2, 7]. The simplicity and simulation speed of stochastic models may make them valuable during preliminary stages of NoC development, but, since the characteristics (functionality and timing) of the IP core are not captured, such models are unreliable for optimizing NoC features.

A modeling technique which adds functional accuracy and causality is transaction-level modeling (TLM), which has been widely used for NoC and SoC design [4, 5, 9–11]. In [9, 10], TLM has been used for bus architecture exploration. The communication is modeled as read and write transactions which are implemented within the bus model. Depending on the required accuracy of the simulation results, timing information such as bus arbitration delay is annotated within the bus model. In [10] an additional layer called "cycle count accurate at transaction boundary" is presented. Here, the transactions are issued at the same cycle as that observed in bus-cycle-accurate models, thus, intratransaction visibility is traded-off for simulation speedup. While modeling the entire system at higher abstraction i.e. TLM, both [9] and [10] present a methodology for preserving accuracy with gain in simulation speed.

We would like to underline that our approach is dual with respect to TLM. While transaction-level models usually represent interconnects as a collection of available services and emphasize local processing on IP cores, the platform we describe is composed of accurate models for the interconnect, while processing resources are abstracted away. Simulation speed is gained like in TLM models, but the purpose of this gain is enabling accurate assessment of interconnect performance, not of core or application performance. The above methods are suitable for feature exploration once the NoC architecture has been chosen, but are not thought for NoC exploration itself.

3 Traffic Modeling Requirements

The generation of a traffic pattern emulating that of a real IP core can be faced at varying degrees of accuracy.

At the most basic level, a trace with time stamps can be collected in the reference system and then be independently replayed, an approach that we might call "cloning". This approach is clearly inadequate when the variance of network latency is taken into account; whenever a transaction is delayed, either due to hardware design or congestion, the effect should propagate to subsequent transactions, which would also be delayed in real systems. A simple example of such critical blocking is a cache refill request.

This observation leads to the deployment of "timeshifting" traffic generators: adjacent transactions are tied to each other, and are issued at times which are a function of the delay elapsed before receiving responses to previous transactions. This implicitly means that the trace collection mechanism must include not only timestamps for processor-generated commands, but also for network responses. However, even this model fails when multi-core systems come under scrutiny: the arbitration for resources in such designs is timing-, and thus, architecture dependent. Therefore, very different transaction patterns may be observed as a function of the chosen interconnection design. To make an example, checks for a shared resource done by polling generate different amounts of traffic depending on the relative ordering of accesses to the resource.

As a consequence, the need for "reactive" TG models is justified. Such models must have some knowledge about the system architecture and about the application behavior to correctly generate (and not just duplicate) traffic patterns across different underlying networks. A TG should be able to mimic the behavior of an IP core even when facing unpredictable network performance, e.g. due to resource contention, packet collisions, arbitration, and routing policies.

To illustrate the requirements driving the development of our TG model, we will now describe how two typical transactions occurring in the MPARM modelling and simulation environment can be reproduced. MPARM features in-order, single-pipeline ARM cores as system masters and two types of memory as system slaves: private (only accessible by one master) or shared (visible by all masters in the system). Figure 2 shows examples of the two types of communication: (a) processor-initiated communication towards an exclusively owned slave peripheral, and (b) processor-initiated communication towards a system-shared slave peripheral. We call network latency (t_{nwk}) the time taken for the communication to traverse from the master OCP interface to the slave OCP interface and vice versa; this latency depends both on the chosen architecture and on the network congestion at the time of communication.

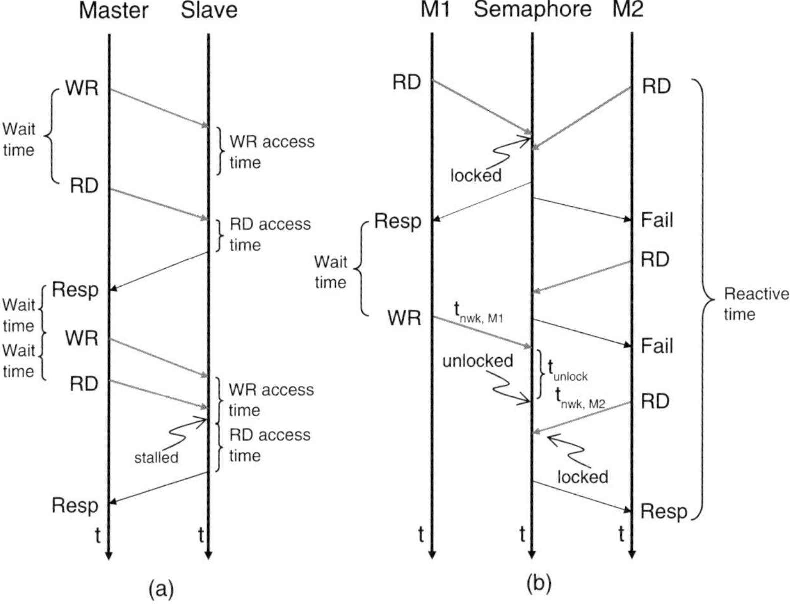

Fig. 2 Two typical MPARM transactions

In Fig. 2a, the first two master transactions are a write (WR) and a read (RD). The time to service the WR transaction, which is a posted write, is just the network latency plus the slave access time. The RD, which in MPARM uses blocking semantics, pays an additional penalty because the response has to make its way back to the master. From the TG point of view, this pattern is easily recordable: network latency and slave access time are unimportant factors, and the essential point to capture is just the delay between WR assertion and RD assertion, and between RD response and the following command. In a subsequent simulation with traffic generators replacing cores, these delays will be modeled by explicit idle waits in the TG, while the network latency will be dependent on the NoC model to simulate. In the next set of transactions, where a RD closely follows a WR, the RD command reaches the slave before the latter has finished servicing the WR, and is thus stalled at the slave interface. This stalling behavior does not need to be explicitly captured in a TG model, since, from a processor perspective, it simply appears to be part of the slave response time. This simplistic example of a master accessing a private slave proves that if the type and the time stamp of the communication events are captured, the behavior of the master can be emulated via nonpreemptive sequential communication transactions interleaved with an appropriate amount of idle wait cycles.

In Fig. 2b, two master devices (M1 and M2) attempt to gain access to a single hardware semaphore. M1 arrives first and locks the resource; the attempt by M2 thus fails. In MPARM, semaphore checking is performed by polling, i.e. M2 regularly issues read events until eventually the semaphore is granted to it. Since the transactions occur over a shared network fabric, the unlock event (WR) issued by M1 and the success of the next request (RD) event by M2 are dependent. Only if the M2 RD event is issued at least $t_{\text{nwk,M1}} + t_{\text{unlock,S}} - t_{\text{nwk,M2}}$ after the unlocking by M1, then M2 will be granted the semaphore and additional polling events will not be required. Therefore, depending on network properties, a variable amount of transactions might be observed at the OCP interfaces of M1 and M2. This is the *reactive behavior* that needs to be captured by the TG model: both M1 and M2 need to react to accommodate the network latency. Thus, the simplistic model which could be applied to transactions towards a privately owned slave now needs to be extended with additional information about the master process execution, about system properties, and about input/output data. In detail, the TG must be able to recognize polling accesses (i.e. a knowledge of what addressing ranges represent pollable resources) and must add support for recording of actual data transfers (e.g. writing a "1" or a "0" to a shared memory location might be the difference between locking or unlocking a resource).

We take the above discussion as a requirement to implement accurate TG models. The examples in Fig. 2 demonstrate that traces collected at the IP-NoC interface are sufficient to accurately reproduce the IP's communication, provided that the reactive behavior of the master IP cores is taken into account. These traces should collect sequences of communication transactions, comprising of requests and responses, separated by time intervals with no communication, i.e. idle time. A simulation of the entire system should produce several traces, one per IP core interface.

4 Implementation of the Traffic Generators

In this section we describe in some detail the implementation of our traffic generators. As mentioned before, our proposed TG model is designed as a simulation tool, but allows future deployment as a hardware device. Within the simulation environment for NoC exploration, the emphasis is on simulation speedup, while within a hardware instance the emphasis is on ease of (re)programmability and a small silicon footprint in order to support implementation of test chips containing NOC prototypes.

Conceptually, three different TG entities might be needed: (1) A TG emulating a processor (an OCP master). This TG must be able to issue conditional sequences of traces composed of communication transactions separated by idle/wait-periods; (2) A TG emulating a shared memory (an OCP slave). This TG must contain a data structure modeling an actual shared memory (since the values read by the masters may affect the sequence of transactions seen at the master IP cores); and (3) A TG emulating a slave memory (an OCP slave). This TG must be able to respond, possibly with dummy values, to communication transactions issued by a master. Only the first is actually required for deployment in a simulation environment, which already provides its own system slaves, thus only this entity will be described in the present paper. On the other hand, it is important to notice that both slave TG

Table 1 OCP-master TG instruction set

Instructions	Description
OCP Instructions:	
Read(addr)	Read from an address
Write(addr, data)	Write to an address
BurstRead(addr, count)	Burst read a range of addr.
BurstWrite(addr, data, count)	Burst write an address set
Other Instructions:	
If(arg1, arg2, operand)	Branch on condition
Jump(location)	Branch direct
SetRegister(reg, value)	Set register (load immediate)
Idle(counter)	Wait for given no of cycles

modules are much simpler in design with respect to the master TG, as their logic basically just involves a small state machine to handle OCP transactions.

For the processor TG, we have implemented and modeled a multicycle processor with a very simple instruction set as listed in Table 1. The processor has an instruction memory and a register file, but no data memory. The instruction set consists of a group of instructions which issue OCP transactions (whose arguments are set up in registers) and a group of instructions allowing the programming of conditional sequencing and parameterized waits such that the required traces can be implemented/programmed. The process for deriving TG programs from traces obtained in the reference simulation is explained in Section 5.

5 The TG Simulation Flow

In order to use the traffic generators, a user must first perform a reference simulation using bit-true and cycle-true IP models. It is interesting to note that, at this stage, the interconnect does not yet need to be accurately modeled, allowing for time savings. During this simulation, traces are collected from all OCP interfaces in the system. For this purpose, the OCP interface modules within the MPARM platform (the network interfaces in the case of the xpipes interconnect, the bus master in the case of AMBA AHB) were adapted to collect traces of OCP request and response communication events into a predefined file format (*.trc*). The address and (if any) data fields of the transactions are also observed. Trace entries are single or burst read/write transactions. Figure 3a shows an example trace.

The next step is to convert the traces into corresponding TG programs (*.tgp*). A translator outputs symbolic code; Fig. 3b shows the TG program derived for traces in Fig. 3a. Finally, an assembler is used to convert the symbolic TG program into a binary image (*.bin*) which can be loaded into the TG instruction memory and executed. Execution might be within a simulation model (which is the approach presented in this paper) or in hardware on a NoC test-chip. Validation of the trace collection and processing mechanism can be achieved by collecting traces with IP

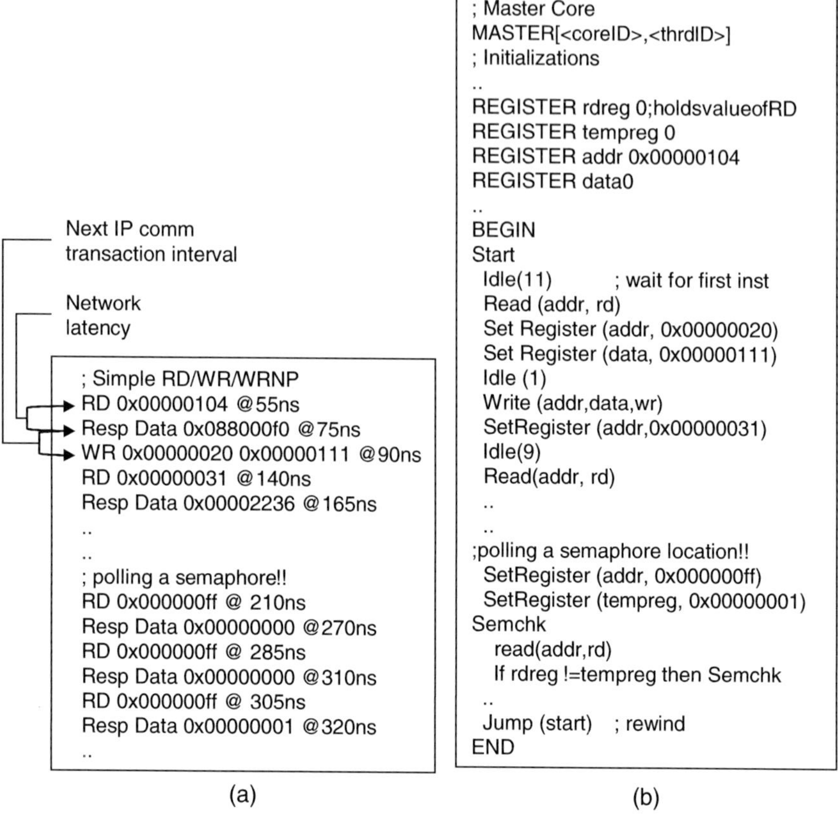

(a)   (b)

Fig. 3 (a) MPARM trace to (b) TG program

cores running on different interconnects, and verifying the resulting *.tgp* and *.bin* programs to match. The conversion process is fully automated and the time taken for this process is discussed in Section 6.

As seen in Fig. 3b, the TG program starts with a header describing the type of core and its identifier. The next few statements express initialization of the register file. Register rdreg is defined as special register where the value of RD transactions is stored.

By looking at the code in Fig. 3a, it is possible to notice that the first communication events in the trace occur at time 55, 75, and 90ns. We assume each TG cycle to take 5ns, the same as the IP core for which the trace is collected. At the beginning of the simulation, the TG has no instruction to perform until the 11th (55/5) cycle, so an Idle wait is observed. The trace of the RD event is followed by a response, at a time which is dependent on the network latency. The IP core is blocked until this response arrives. A WR event occurs three ((90-75)/5) cycles after the response is received; these cycles are partially spent for TG internal operations (data and address register setting), and an ensuing Idle wait is added to fill the gap. Then the following RD instruction is translated into the corresponding Read program call after 10 cycles, one of which is taken to set up the RD address. This is blocking until a response is received, 5 cycles later.

Now, consider the trace entries from time 210–320 ns. By identifying the address as belonging to a semaphore location and knowing the polling behavior of the MPARM IP core, the translator inserts the Semchk label and an If conditional statement. This statement checks whether the read value is equal to "1", which reflects an unblocked semaphore. This loop effectively represents the semaphore polling behavior. All master devices attempting to access this address incorporate the same routine in their TG program, thus, capturing the system dynamics.

At this stage, additional simulations can be run on a platform with traffic generators and a variety of interconnect fabrics, thereby evaluating performance of NoC design alternatives. Compared with the reference setup, where the interconnect fabric could be modeled at a high level, the target NoC should now be simulated at the cycle- and bit-true level to carefully assess its performance. Validation of the TG model can be achieved by coupling the TG with the same interconnect used for tracing with IP cores, and checking the accuracy of the IP core emulation. Results for this validation, and for tests on different interconnects than the reference one, will be presented in the next section.

6 Results

We simulated within the MPARM framework, using the AMBA NoC, and four benchmarks. The first benchmark was a single-processor application (SP matrix manipulation), with the purpose of assessing accuracy and speedup in the simplest environment. The second benchmark (Cacheloop) was a test performing idle loops within the processors' cache, and only minimal bus interaction; this allowed an assessment of the speedup provided by the TG model when scaling the number of processors in the system up to twelve. Finally, the remaining two benchmarks (MP matrix manipulation and DES encryption/decryption) were multiprocessor tests stressing synchronization and resource contention with traffic patterns as discussed in Section 3, and were used mainly to ascertain the accuracy of the whole design flow when stressed by complex transactions.

In the first experiment we aimed at validating the trace collection/processing environment. We ran the same benchmarks over AMBA and ×pipes, noticing very different execution times due to different latency and scalability features. However, after translation, a check across *.tgp* programs showed no difference at all. This result demonstrates the feasibility of an approach which decouples simulation of the IP cores and of the underlying interconnect fabric.

Table 2 summarizes the results of simulations done on the AMBA AHB interconnect with ARM processors and then with TGs. The left columns report the

Table 2 TG vs. ARM performance with AMBA

#IPs	Cumulative Execution Time			Simulation Time		
	ARM	TG	Error	ARM	TG	Gain
SP matrix:						
IP	6610680	6610659	0.00%	73 s	34 s	×2.15
Cacheloop:						
2P	2500903	2500913	0.00%	47 s	14 s	×3.36
4P	2501760	2501701	0.00%	87 s	22 s	×3.95
6P	2502558	2502640	0.00%	127 s	29 s	×4.38
8P	2503404	2503522	0.00%	163 s	37 s	×4.41
10P	2504250	2504404	0.01%	197 s	42 s	×4.69
12P	2505096	2505286	0.01%	239 s	51 s	×4.69
MP matrix:						
2P	3276505	3276030	0.01%	66 s	25 s	×2.64
4P	3528038	3530759	0.08%	128 s	42 s	×3.05
6P	3691454	3697854	0.17%	195 s	61 s	×3.20
8P	3997878	4058812	1.52%	260 s	82 s	×3.17
10P	4881007	4902806	0.45%	334 s	106 s	×3.15
12P	5901290	5901131	0.00%	432 s	143 s	×3.02
DES:						
3P	978080	980098	0.21%	26 s	10 s	×2.60
4P	1054839	1057944	0.29%	34 s	11 s	×3.09
6P	1491570	1492274	0.05%	53 s	20 s	×2.65
8P	1959755	1960575	0.04%	73 s	30 s	×2.43
10P	2441026	2441743	0.03%	95 s	42 s	×2.26
12P	2927359	2927218	0.00%	125 s	62 s	×2.02

number of simulated cycles, while the right ones illustrate simulation time.[1] The column "Error" is a measure of the accuracy of replacing IP cores with TGs, based upon the difference in simulated cycles, while the column "Gain" describes the improvement in simulation time.

The table shows that replacing ARM processors with TGs yields excellent accuracy, close to 100% for small numbers of processors, while guaranteeing a speedup factor of 2–4. This speedup is mostly due to the drastic simplification in the amount of logic needed to generate communication transactions, compounded in small part with the elimination of any adaptation layer in the system since the TG is natively implemented with an OCP interface. This speedup compares favorably to previous work in the area (a speedup of ×1.55 is reported in [10]), and must be evaluated by taking into account the fact that it involves no shift in the level of abstraction of the simulations.

Inaccuracies in execution time can be explained as follows. In Cacheloop, and until about 6–8 processors in MP matrix and four in DES, the TG platform shows an

---

[1] Benchmarks taken on a multiprocessor Xeon 1.5 GHz with 12 GB of RAM, eliminating any disk-swapping effect. Especially for benchmarks with a short duration, time measurements were taken by averaging over multiple runs and care was put in minimizing disk loading effects.

acceptable accuracy degradation. This is due to the compounding of minimal timing mismatches caused by the conversion from traces to TG programs. When adding even more processors, however, accuracy improves again, because the AMBA bus starts to saturate, causing the processors to idle wait for bus arbitration for long amounts of time. The congestion is serious enough to dominate the effect of the timing mismatches. Since TGs cannot save simulation complexity if the replaced processors are in idle state, this is also the reason causing the speedup to get smaller with large numbers of processors in MP matrix and DES. For Cache-loop, which always executes from the local caches without any bus traffic, this phenomenon does not appear. Thus, the reduced speedup is not a property of the TG.

The impact of trace collection is small, and is incurred only once. For example, when running the MP matrix benchmark on the AMBA interconnect with four ARM processors, a plain benchmark run takes 128 s; the benchmark run with TG tracing enabled takes 147 s, and subsequent parsing and elaboration requires an additional 145 s for a 20 MB trace file.[1] Only one such iteration is needed to be able to take advantage of ×2–×4 speedups in subsequent design space exploration. Additionally, since processed TG programs are identical regardless of the reference interconnect in which raw traces were collected, such collection could be performed on top of a transactional fabric model, further reducing the impact of the reference simulation.

7 Conclusions

Experimental results prove the viability of a TG-based approach which decouples simulation of IP cores and of interconnect fabrics. Even in presence of unpredictable contention for shared resources in a multiprocessor environment, our TG model proved capable of delivering speedups in the order of ×2–×4 when run on AMBA while keeping a remarkable accuracy.

The TG model we propose provides a wide range of features, with a simple but powerful instruction set allowing for sophisticated flow control and therefore a variety of communication patterns. It is very useful for fast and accurate verification and exploration of different NoC architectures, which is the motivation of this work. While this paper was focused on simulation speedup, the TG may also be used as a flexible tool in a variety of platforms. The TG might be used in association with manually written programs to generate traffic patterns typical of IP cores still in the design phase, helping in the tuning of the communication performance between the underlying NoC and that IP core.

Our future work includes synthesis of the TG device, and support for processors allowing out-of-order transactions. Research will also include analysis of the behavior of a system in which multiple tasks run on a single processor and are dynamically scheduled by an OS, either based upon timeslices (preemptive multi-tasking) or upon transition to a sleep state followed by awakening on interrupt receipt. Context switching-related issues will need to be modeled or predicted.

Acknowledgments The work of Shankar Mahadevan is partially funded by SoC-Mobinet, Nokia, and the Thomas B. Thrige Foundation. The work of Federico Angiolini is partially funded by ARTIST.

References

[1] Open Core Protocol Specification, Release 2.0 http://www.ocpip.org, 2003.

[2] E. Bolotin, I. Cidon, R. Ginosar, and A. Kolodny. QNoC: QoS architecture and design process for network on chip. In *Journal of Systems Architecture*. Elsevier, 2004.

[3] M. Dall'Osso, G. Biccari, L. Giovannini, D. Bertozzi, and L. Benini. xpipes: a latency insensitive parameterized Network-on-Chip architecture for multi-processor SoCs. In *Proceedings of 21st International Conference on Computer Design*, p. 536–539. IEEE Computer Society, 2003.

[4] F. Fummi, P. Gallo, S. Martini, G. Perbellini, M. Poncino, and F. Ricciato. A timing-accurate modeling and simulation environment for networked embedded systems. In *Proceedings of the 42th Design Automation Conference*, p. 42–47, 2003.

[5] T. Grötker, S. Liao, G. Martin, and S. Swan. *System Design with SystemC*. Kluwer Academic, 2002.

[6] K. Lahiri, A. Raghunathan, and S. Dey. Evaluation of the traffic-performance characteristics of System-on-Chip communication architectures. In *Proceedings of the 14th International Conference on VLSI Design*, p. 29–35, 2001.

[7] K. Lahiri, A. Raghunathan, G. Lakshminarayana, and S. Dey. Communication architecture tuners: a methodology for the design of high-performance communication architectures for System-on-Chips. In *Proceedings of the 2000 Design Automation Conference, DAC'00*, p. 513–518, 2000.

[8] M. Loghi, F. Angiolini, D. Bertozzi, L. Benini, and R. Zafalon. Analyzing on-chip communication in a MPSoC environment. In *Proceedings of the 2004 Design, Automation and Test in Europe Conference (DATE'04)*. IEEE, 2004.

[9] O. Ogawa, S. B. de Noyer, P. Chauvet, K. Shinohara, Y. Watanabe, H. Niizuma, T. Sasaki, and Y. Takai. A practical approach for bus architecture optimization at transaction level. In *Proceedings of Design, Automation and Testing in Europe Conference 2004 (DATE03)*. IEEE, March 2003.

[10] S. Pasricha, N. Dutt, and M. Ben-Romdhane. Extending the transaction level modeling approach for fast communication architecture exploration. In *Proceedings of 38th Design Automation Conference (DAC'04)*, p. 113–118. ACM, 2004.

[11] M. Sgroi, M. Sheets, A. Mihal, K. Keutzer, S. Malik, J. Rabaey, and A. Sangiovanni-Vincentelli. Addressing the System-on-Chip interconnect woes through communication-based design. In *Proceedings of the 38th Design Automation Conference (DAC'01)*, p. 667–672, June 2001.

Part III
Modeling, Simulation and Run-Time Management

Modeling, Simulation and Run-Time Management

Enrico Macii

The continuous scaling of electronic technologies has changed quite substantially the way integrated circuits and systems are designed. On the one hand, increased complexity calls for more powerful methods to execute the traditional CAD tasks, such as modeling, simulation, and optimization. On the other hand, new tools have become necessary to support design, as modern electronic systems differ significantly from devices of the "old times", where the entire functionality was implemented as silicon resources. Hardware capabilities are complemented by software support, including low-level primitives and application-level services, in order to guarantee high-performance, low-power, and reliable operation of embedded systems that are at the core of most mobile and consumer electronics products.

To comply and adapt to this changed design scenario, the CAD landscape has witnessed an evolution along different axes, driven by technology, architecture, and application shifts.

From the side of traditional modeling, there are two major issues that captured the attention of the researchers in the last few years: (1) The need of representing systems at high levels of abstraction, where implementation details are hidden and thus designers can focus on the definition, simulation and validation of the overall system behavior. (2) The need of representing systems that contain heterogeneous components, both from the hardware and the software perspectives, and to offer a unified interface that enables designers to evaluate different design options, architectural configurations and software solutions.

The specific problem of model-based design of embedded software has been addressed from different angles. One approach that has reached a significant amount of consensus relies on the usage of Domain-Specific Modeling Languages (DSMLs), i.e. languages targeting a well-defined domain that offer the users abstractions and notations that are customized towards the application classes of their interest. DSMLs differ from mainstream modeling efforts that focus on creating a language for a wide range of applications, such as UML. One of the shortcomings of existing model-based design frameworks is their limited capability of explicitly and unambiguously defining the model behavioral semantics.

Politecnico di Torino

R. Lauwereins and J. Madsen (eds.), *Design, Automation, and Test in Europe–The Most Influential Papers of 10 Years DATE*, 187–193
© Springer 2008

The problem of making DSML semantically precise is addressed in [Chen-DATE07]. The work takes the MIC tool-suite [Sztipanovits97] and its extensions (e.g. [Chen05]) as the reference framework that incorporates some basic infrastructure for "semantic anchoring" of DSMLs. The Abstract State Machine (ASM) [Boerger03] formalism is at the basis of the MIC framework. ASM is in use since the mid-1980s for different kinds of applications, ranging from hardware modeling to the description of complex algorithms, from database models to distributed systems modeling. Key contribution of [Chen-DATE07] is the enhancement of the anchoring approach to heterogeneous behaviors; this is achieved by defining structural and behavioral composition rules that enable the combination of primary semantic units. The flexibility of the extended framework is demonstrated by utilizing it for the generation of the behavioral semantics of the EFSM, a modeling language for vehicle motion control behavioral specification developed by GM Research [Wang06].

As mentioned earlier, embedded systems (possibly networked) represent the enabling technology for some of the fastest growing segments of the electronic business, including mobile, infotainment, and consumer products. From a more far-reaching perspective, growing interest is attracted by distributed embedded systems, as they provide an innovative, fresh look at the most traditional networking concepts (both from the architectural and the protocol points of view). One example, in this respect, is given by the Wireless Sensor Networks (WSNs) [Estrin99].

A WSN is a wireless network consisting of spatially distributed autonomous devices using sensors to cooperatively monitor physical or environmental conditions, such as temperature, sound, vibration, pressure, motion or pollutants, at different locations. In addition to one or more sensors, each node in a sensor network is typically equipped with a radio transceiver or other wireless communications device, a small microcontroller, and an energy source, usually a battery. The size of a single sensor node can vary from shoebox-sized nodes down to devices of the size of a grain of dust. The cost of sensor nodes is similarly variable, ranging from hundreds of dollars to a few cents, depending on the size of the sensor network and the complexity required of individual sensor nodes. Size and cost constraints on sensor nodes result in corresponding constraints on resources, such as energy, memory, computational speed and bandwidth [Rabaey00]. Energy minimization in WSNs can be achieved in various ways, including the design of the communication protocols [Heinzelman00] and the implementation of adequate dynamic power management capabilities [Sinha01].

An important step towards the transition of the WSN technology from research to industrial application is represented by the definition of the IEEE 802.15.4 standard, which specifies the physical and MAC layers of the protocol and addresses particularly low data rates, very long battery life (months or even years) and very low complexity. The work in [Bougard-DATE05] offers a detailed assessment of the energy efficiency of the standard in the context of a dense WSN. The modeling work maps to realistic conditions, as it assumes an implementation of the network made through commercially available components; it is enhanced by the study of an energy-aware radio management (i.e. activation and deactivation) policy, which enables power reduction and reliability improvement. The capability of the model to provide detailed breakdown of the power consumed by the various packet transmission stages and architecture components is demonstrated by applying it to the power-managed WSN.

Accurate and compact modeling is instrumental for the implementation of efficient simulation capabilities, which find utilization at any level of design abstraction and for different purposes, ranging from design exploration to validation and sign-off. An area of raising concern in circuit design for nanometer technologies is the effect of process variation on reliability and correct operation [Borkar05]. In fact, scaling of feature sizes has not been accompanied by a suitable scaling of geometric tolerances. In addition, phenomena such as edge or surface roughness, or the fluctuation of the number of doping atoms within the channels are becoming significant. As a result, production yields, as well as figures of merit of a circuit such as performance and power have become extremely sensitive to uncontrollable statistical process variations.

The problem of predicting the effects of process variation is commonly approached using statistical techniques [Mutlu05]. In particular, Monte Carlo simulation [Fishman05] is, as of today, the solution of choice adopted by most statistical simulators in charge of estimating timing, power, yield and other quantities for nanometer circuits and devices. Monte Carlo simulation is a method for iteratively evaluating a deterministic model using sets of random inputs. The method is often used when the model is complex, nonlinear, or involves several uncertain parameters. Monte Carlo simulation is categorized as a sampling method because the inputs are randomly generated from probability distributions to simulate the process of sampling from an actual population.

Monte Carlo techniques are, by construction, very good at sampling the statistically likely cases. In other terms, they are conceived so that rare (thus, presumably unimportant) events are excluded from the simulation sample, therefore guaranteeing faster convergence and more accurate results. Unfortunately, for elementary devices that tend to be replicated many times in large circuits (e.g. SRAM cells, flip-flops), ignoring statistically rare events during the simulation may result in reliability estimates for the entire design which are poor, sometimes unacceptable.

The work in [Singhee-DATE07] addresses specifically the issue of being able to sample only the statistically rare events. Existing solutions considered analytical or semi-analytical techniques in replacement of Monte Carlo sampling, e.g. [Mahmoodi05]. The drawback of the analytical approach stands in the approximations that are needed in the construction of the models in order to make the complexity of the problem manageable. The [Singhee-DATE07] paper proposes a Monte Carlo method that allows filtering out from the tail distribution to be used for the simulation unwanted samples that are not rare enough. Extreme Value Theory and data mining concepts are the building bricks of the proposed method, which has been validated on a number of test cases, including a 90 nm SRAM cell, a 45 nm flip-flop and a 64 bit, 90 nm SRAM column.

Complex, heterogeneous systems are difficult to model and to simulate, as discussed earlier in this chapter. On the positive side, they offer great opportunities for trading off enhanced service provision and optimization of metrics such as performance and power consumption. The trade off can be explored statically, i.e. at design time. More interesting, however, is the possibility of implementing systems that support different operation modes, whose selection is made at run-time, based on the actual workload, the environment and the occurrence of external events.

Systems that are able to support run-time (i.e. dynamic) resource management to trade performance for power minimization are now very common, even at the commercial level. This is particularly true in mobile, battery-operated devices.

The underlying assumption of a power-managed system is that the activity of its components is event-driven, that is, it is triggered by external events and it is often interleaved with long periods of quiescence. An intuitive way of reducing the average power dissipated by the whole system consists of shutting down (or reducing the performance of) the components during their periods of inactivity (or under-utilization). In other words, one can adopt a dynamic power management (DPM) policy that dictates how and when the various components should be powered (or slowed) down in order to minimize the total system power budget under some performance/throughput constraints [Benini97].

A component of an event-driven computing system can be modeled through a Finite State Machine (FSM) that, in the simplest case, has two states: *Active* and *Idle*. When the component is idle, it is desirable to shut it down (or to move it to a low-power state) by lowering its power supply or by turning off its clock; in this way, its power dissipation can be drastically reduced [Benini00]. The simplest dynamic power management policies that have been devised are timeout-based: a component is put in its power-down mode only T time units after its finite state machine model has entered the *Idle* state. This is because it is assumed that there will be a very high chance for the component to be idle for a much longer time if it has been in the *Idle* state for at least T time units.

Timeout policies can be inefficient for three reasons: first, the assumption that, if the component is idle for more than T time units, it will be so for much longer may not be true in many cases. Second, whenever the component enters the *Idle* state, it stays powered for at least T time units, wasting a considerable amount of power in that period. Third, speed and power degradations due to shutdowns performed at inappropriate times are not taken into account; in fact, it should be kept in mind that the transition from power-down to fully functional mode has an overhead: it takes some time to bring the system up to speed, and it may also take more power than the average, steady-state power.

To overcome the limitations of timeout-based policies, more complex strategies have been developed. Predictive [Srivastava96], adaptive [Zen05], and stochastic algorithms [Benini99] [Qiu99] are a few examples of effective policies; they are based on complex mathematical models and they all try to exploit the past history of the active and idle intervals to predict the length of the next idle period, and thus decide whether it may be convenient to power down a system component.

Beside component idleness, also component under-utilization can be successfully exploited to reduce power consumption. For example, if the time a processing unit requires to complete a given task is shorter than the actual performance constraint, execution time can be increased by making the processor running slower; this can be achieved by either reducing the clock frequency, or by lowering the supply voltage, or both, provided that the hardware and the software enable this kind of operation. In all cases, substantial power savings are guaranteed with almost no performance penalty. Policies for dynamic voltage scaling (DVS) and dynamic frequency scaling (DFS) are of increasing popularity in electronic products that feature DPM capabilities [Horowitz94].

The application of more sophisticated DPM policies, possibly including also DVS/DFS, calls for the availability of finer power models. The two-state FSM mentioned earlier can be extended to the more general case of Power FSM (PSM), which associates the workload values with a set of resources relevant to its power behavior [Benini98]. At any given point in time, each resource is in a specific power state, which consumes a discrete amount of power.

The power savings and performance impact of several DPM policies are quantitatively compared in paper [Lu-DATE00]. A total of ten policies are benchmarked on some realistic applications, including hard disks in desktop and notebook computers, X-servers and network applications. The results of the analysis are put in perspective and conclusions are drawn on the effectiveness of the different classes of policies.

In addition to policy development, there are several other issues that must be considered when implementing DPM in practice. These include the need of a model for power managed systems, the choice of an implementation style for the power manager that must guarantee accuracy in measuring inter-arrival times and service times for the system components, flexibility in monitoring multiple types of components, low perturbation and marginal impact on component usage. Also, key is the choice of the style for monitoring component activity; options here are off-line (traces of events are dumped for later analysis) and on-line (traces of events are analyzed while the system is running; statistics related to component usage are constantly updated) monitoring.

The work in [Chung-DATE99] introduces some DPM policies that focus specifically on the modeling and characterization of the user (i.e. the service requestor) in order to capture the nonstationarity of its behavior. The paper presents two techniques based on sliding windows to detect the time-varying parameters of the stochastic model of the workload source. In addition, it shows how policies for dynamic power management under nonstationary service request models can be determined by interpolating optimal policies computed under the assumption that user requests are stationary. The proposed DPM algorithms are tested, validated, and their performance assessed on a laptop computer with a hard disk. Results demonstrate the feasibility of the policy in a real system environment.

References

[Chen-DATE07] K. Chen, J. Sztipanovits, S. Neem, Compositional Specification of Behavioral Semantics, *DATE-07: IEEE Design Automation and Test in Europe*, pp. 906–911, Nice, France, April 2007.
[Sztipanovits97] J. Sztipanovits, G. Karsai, Model-Integrated Computing, *IEEE Computer*, Vol. 30, No. 4, 110–112, April 1997.

[Chen05] K. Chen, J. Sztipanovits, S. Neem, Toward a Semantic Anchoring Infrastructure for Domain-Specific Modeling Languages, *EMSOFT-05: ACM International Conference on Embedded Software*, pp. 35–44, Jersey City, NY, September 2005.

[Boerger03] E. Boerger, R. Staerk, *Abstract State Machines: A Method for High Level System Design and Analysis*, Springer, 2003.

[Wang06] S. Wang, S. Birla, S. Neem, A Modeling Language for Vehicle Motion Control Behavioral Specification, *International Workshop on Software Engineering for Automotive Systems*, pp. 53–60, Shanghai, China, May 2006.

[Estrin99] D. Estrin, R. Govindan, J. Heidemann, S. Kumar, Next Century Challenges: Scalable Coordination in Sensor Networks, *MobiCom-99: Fifth Annual ACM International Conference on Mobile Computing and Networking*, pp. 263–270, Seattle, WA, August 1999.

[Rabaey00] J. M. Rabaey, M. J. Ammer, J. L. da Silva Jr., D. Patel. S. Roundy, PicoRadio Supports Ad Hoc Ultra-Low Power Wireless Networking, *IEEE Computer*, Vol. 33, No. 7, 42–48, July 2000.

[Heinzelman00] W. Heinzelman, A. Sinha, A. Wang, A. Chandrakasann, Energy-Scalable Algorithms and Protocols for Wireless Microsensor Networks, *ICASSP-00: IEEE International Conference on Acoustics, Speech, and Signal Processing*, pp. 3722–3725, Istanbul, Turkey, June 2000.

[Sinha01] A. Sinha, A. Chandrakasan, Dynamic Power Management in Wireless Sensor Networks, *IEEE Design & Test of Computers*, Vol. 18, No. 2, 62–74, March–April 2001

[Bougard-DATE05] B. Bougard, F. Catthoor, D. C. Daly, A. Chandrakasan, W. Dehaene, Energy Efficiency of the IEEE 802.15. 4 Standard in Dense Wireless Microsensor Networks: Modeling and Improvement Perspectives, *DATE-05: IEEE Design Automation and Test in Europe*, pp. 196–201, Munich, Germany, March 2005.

[Borkar05] S. Borkar, Designing Reliable Systems from Unreliable Components: The Challenges of Transistor Variability and Degradation, *IEEE Micro*, Vol. 25, No. 6, 10–16, November/ December 2005.

[Mutlu05] A. Mutlu, M. Rahman, "Statistical Methods for the Estimation of Process Variation Effects on Circuit Operation," *IEEE Transactions on Electronics Packaging Manufacturing*, Vol. 28, No. 4, 364–375, October 2005.

[Fishman05] G. S. Fishman, *A First Course in Monte Carlo*, Duxbury Press, October 2005.

[Singhee-DATE07] Amith Singhee, Rob A. Rutenbar, Statistical Blockade: A Novel Method for Very Fast Monte Carlo Simulation of Rare Circuit Events and its Application, *DATE-07: IEEE Design Automation and Test in Europe*, pp. 1379–1384, Nice, France, April 2007.

[Mahmoodi05] H. Mahmoodi, S. Mukhopadhyay, K. Roy, "Estimation of Delay Variations due to Random Dopant Fluctuations in Nanoscale CMOS Circuits," *IEEE Journal of Solid-State Circuits*, Vol. 40, No. 9, 1787–1796, September 2005.

[Benini97] L. Benini, G. De Micheli, *Dynamic Power Management of Circuits and Systems: Design Techniques and CAD Tools*, Kluwer Academic 1997.

[Benini00] L. Benini, A. Bogliolo, G. De Micheli, A Survey of Design Techniques for System-Level Dynamic Power Management, *IEEE Transactions on VLSI Systems*, Vol. 8, No. 3, 299–316, June 2000.

[Srivastava96] M. B. Srivastava, A. P. Chandrakasan, R. W. Brodersen, Predictive System Shutdown and Other Architectural Techniques for Energy Efficient Programmable Computation, *IEEE Transactions on VLSI Systems*, Vol. 4, No. 1, 42–55, March 1996.

[Zen05] Z. Ren, B. H. Krogh, R. Marculescu, Hierarchical Adaptive Dynamic Power Management, *IEEE Transactions on Computers*, Vol. 54, No. 4, 409–420, April 2005.

[Benini99] L. Benini, G. Paleologo, A. Bogliolo, G. De Micheli, Policy Optimization for Dynamic Power Management,, *IEEE Transactions on Computer-Aided Design*, Vol. 18, No. 6, 813–833, June 1999.

[Qiu99] Q. Qiu, M. Pedram, Dynamic Power Management Based on Continuous-Time Markov Decision Processes, *DAC-36: ACM/IEEE Design Automation Conference*, pp. 555–561, New Orleans, LA, June 1999.

[Horowitz94] M. Horowitz, T. Indermaur, R. Gonzalez, Low Power Digital Design, *IEEE Symposium on Low Power Electronics*, pp. 8–11, San Diego, CA, October 1994.

[Benini98] L. Benini, R. Hodgson, P. Siegel, System-Level Power Estimation and Optimization, *ISLPED-98: ACM/IEEE InternationalS Symposium on Low Power Electronics and Design*, pp. 173–178, Monterey, CA, August 1998.

[Lu-DATE00] Y. Lu, E. Y. Chung, T. Simunic, L. Benini, G. De Micheli, Quantitative Comparison of Power Management Algorithms, *DATE-00: IEEE Design Automation and Test in Europe*, pp. 20–26, Paris, France, March 2000.

[Chung-DATE99] E. Y. Chung, L. Benini, A. Bogiolo, G. De Micheli, Dynamic Power Management for Non-Stationary Service Requests, *DATE-99: IEEE Design Automation and Test in Europe*, pp. 77–81, Munich, Germany, March 1999.

Dynamic Power Management for Nonstationary Service Requests[*]

Eui-Young Chung[1], Luca Benini[2], Alessandro Bogliolo[2],
and Giovanni De Micheli[1]

Abstract Dynamic Power Management is a design methodology aiming at reducing power consumption of electronic systems, by performing selective shutdown of the idle system resources. The effectiveness of a power-management scheme depends critically on an accurate modeling of the environment, and on the computation of the control policy. This paper presents two methods for characterizing non-stationary service requests by means of a prediction scheme based on sliding windows. Moreover, it describes how control policies for nonstationary models can be derived.

1 Introduction

Design methodologies for energy-efficient system-level design are receiving increasingly larger attention [2, 6, 7, 10], because of the widespread use of portable electronic appliances (e.g. cellular phones, laptop computers, etc.) and of the concerns about the environmental impact of electronic systems.

Dynamic power management (DPM) [1] is a flexible and general design methodology aiming at controlling performance and power levels of digital circuits

---

[*]This work was supported in part by MARCO and ARPA.

[1]Eui-Young Chung was supported by Samsung Electronics. Co. LTD.

[1]Stanford University Computer System Laboratory Stanford, CA, 94305, USA.

[2]Università di Bologna Dip. Informatica, Elettronica, Sistemistica Bologna, ITALY 30165.

eychung@stanford.edu

nanni@galileo.stanford.edu

lbenini@deis.unibo.it

abogliolo@deis.unibo.it

and systems, by exploiting the idleness of their components. A system is provided with a *power manager* that monitors the overall system and component states and controls the state transitions. The control procedure is called power management *policy*.

Srivastava et al. [5] proposed heuristic policies to shut down a system during idle periods. The basic idea in [5] is to predict the length of idle periods and shut down the system when the predicted idle period is long enough to amortize the cost (in latency and power) of shutting down and later reactivating the system. A shortcoming of the predictive shutdown approaches proposed by Srivastava is that they are based on off-line analysis of usage traces, hence they are not suitable to nonstationary request streams whose statistical properties are not known a priori. This shortcoming is addressed by Golding et al. [8, 9] and Hwang and Wu [4] proposed on-line methods that dynamically adapt the shutdown policy based on the distribution of past user requests.

All predictive shutdown techniques share a few limitations. First, they do not deal with resources with multiple states of operation (instead of just active and sleep). Second they cannot accurately trade off performance losses (caused by transition delays between states of operation) with power savings. Third, they do not deal with general system models where incoming requests can be enqueued while waiting for service.

These limitations are addressed in [3] where general systems (with multiple states and queueing) and user request are modeled as Markov chains. The Markov model enables a rigorous formulation of the search for optimal power management policies as a constrained minimization problem whose exact solution can be found in polynomial time. Unfortunately a basic assumption in [3] is that the Markov model is stationary. This assumption clearly does not hold if the system experiences nonstationary load conditions.

The contribution of this paper is twofold. We propose prediction schemes for the service requester (e.g. user) that captures the nonstationary property of its behavior. We present two schemes based on sliding windows to capture the time-varying parameters of the stochastic processing modeling the requester. Next we show how policies for dynamic power management under nonstationary service request models can be determined by interpolating optimal policies computed under an assumption of stationariety.

2 System Modeling

In this section, we briefly review the system model introduced in [3]. The overall system model for DPM is shown in Fig. 1. An electronic system is modeled as a unit providing a service, called *service provider* (SP) while receiving requests from another unit, called *service requester* (SR). A *queue* buffers incoming unserviced requests. The service provider can be in one of several states (e.g. *active, sleep, idle,* etc.). Each state is characterized by the ability/inability of providing a service and

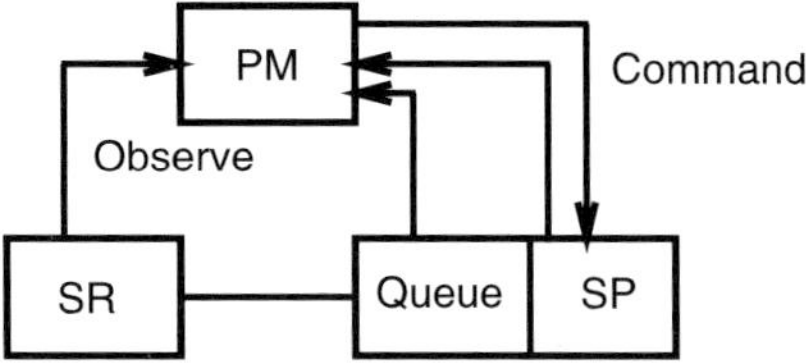

Fig. 1 Overall system model for DPM

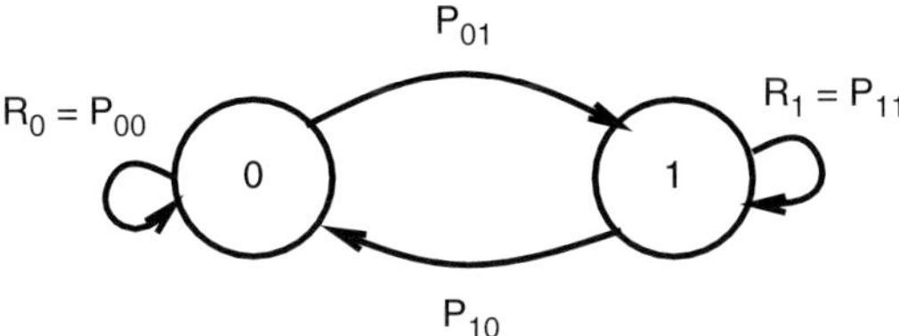

Fig. 2 An example of a Markov chain

by a power consumption level. Transitions among states may have a performance penalty (e.g. latency in reactivating a unit) and a power penalty (e.g. power loss in spinning up a hard disk).

A *power manager* (PM) is a control unit that controls the transitions among states. The power consumption of the *PM* is negligible with respect to the overall power dissipation. At given points in time, the power manager evaluates the overall state of the system (provider, queue and requester) and decides to issue a command to stimulate a state transition. A *control policy* is a sequence of decisions. For our purposes, a policy can be thought of as a table, that associates a probability of issuing a command with any system state.

We model the system components as Markov chains [3]. In particular, we use a controlled Marcov chain model for the system provider, so that the transition probabilities can be made dependent on the command issued by the power manager. We use a discrete time setting, i.e. we assume that time is divided into time slices.

whereas *SP* behaviors can be modeled as a stationary process, because the *SP* response to stimuli does not change over time, the workload bounce can be highly nonstationary, and that it is appropriate to model *SR* as a nonstationary process. In this work, we focus on nonstationary service requesters. A generic requester can have s states. For the sake of simplicity, we will discuss the case of $s = 2$, as shown in Fig. 2. When in state 0, no request is issued. When in state 1, one request per time slice is issued. The corresponding transition matrix is denoted by P. We call the diagonal elements of P *user request probabilities* and we denote them by R_i, $i = 0, 1$. The probabilities $R_i = Prob(s(t+1) = i | s(t) = i)$, $i = 0,1$ (and the entire transition matrix P) are fixed for the stationary model [3]. For capturing the nonstationariety of SR, we assume that R_i can change over time and are initially unknown.

3 User Request Prediction

3.1 Single Window Approach

A sliding window is adopted to keep the recent user request history and this information is used to predict future user requests. A sliding window denoted as W, consists of WS slots and each slot, $W(i)$, $i = 0,1,\ldots WS - 1$, stores one previous user request, i.e. $s \in 0,1,\ldots,S - 1$. The basic window operation is to shift one slot constantly every time slice. An example for a window operation for two-state user requests is shown in Fig. 3. As shown in Fig. 3, at each time point, $W(i + 1) \leftarrow W(i)$, $i = 0,1,\ldots,WS - 1$ and $W(0)$ stores a new user request from SR. The user request prediction for the next time point is done as follows. Let $l = \Sigma_{k=1}^{WS-1} (W(k) = i)$. Then,

$$P(i,j) = \begin{cases} 1/l \sum_{k=1}^{WS_{-1}} [(W(k) = i) \wedge (W(k-1)] & \text{if } l \neq 0 \\ 0 & \text{if } l = 0, \text{ and } \quad i = j \\ 1/(S-1) & \text{otherwise} \end{cases}$$

(1)

where, "=" is the equivalence operation with a Boolean output, (i.e. it yields "1" when the two arguments are same, otherwise returns "0"), and where "$\wedge$" is the "*conjunction*" operation. Thus, a policy applied at a given time point should be optimized for the P predicted at the previous time point. This can be achieved through policy table look-up method described in Section 4.

3.2 Multi-Window Approach

The basic structure for multi-window approach is shown in Fig. 4. There are as many windows as the number of states S of SR and their sizes are same (WS). At a time point, the user request of the previous time point is stored in the Previous

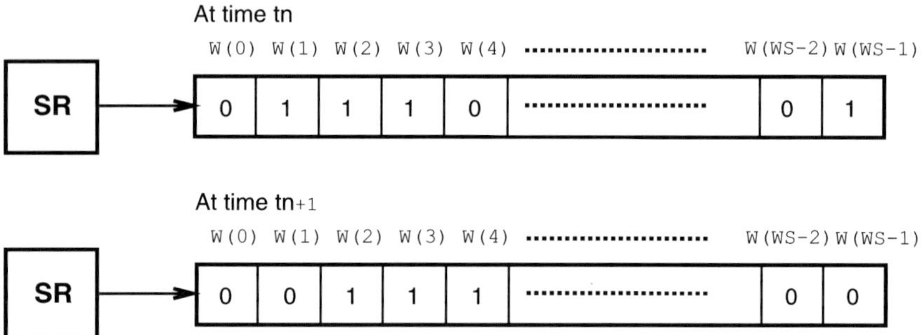

Fig. 3 Single window operation for two-state user requests

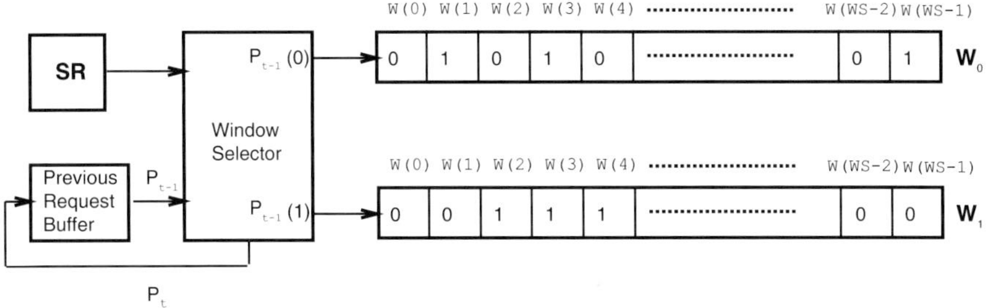

Fig. 4 Multi-window operation for two-state user requests

Request Buffer (PRB) and this buffer controls the window selector to choose a window in which the current user request is stored. For convenience, the window selected when PRB stores state s of a user request is represented as W_s. Each window slice stores the state of a user request. Note that only the selected window performs the shift operation, while the other windows stay constant. Thus, each window W_s stores $W S$ previous user requests and plays a role to predict the transition probabilities from state i to any other states. Each row of P is mapped to the window corresponding to the state which is source of the transition and $P(i,j)$ can be easily calculated as follows.

$$P(i, j) = \frac{\sum_{k=0}^{WS-1}(W_i(k)=j)}{WS} \text{ for all } i, j \qquad (2)$$

One important factor in a window approach is the window size, $W S$. It highly contributes to the accuracy of prediction because WS determines the precision of the predicted user request probability. If $W S$ is too small, a small change in user requests causes a large effect to user request prediction. Conversely, if $W S$ is too large, a large change in user requests gives only a small amount of effect to the prediction. Extensive experimental work for window size selection is described in later section.

4 Dynamic Power Management

4.1 Policy Table Construction

A look-up table stores the data sampled from a n-dimensional function which has n input variables. Sampling points of each variable are mapped to the table indices of the corresponding dimension of the table. For two-state service requester, input variables are R_0 and R_1. By denoting the number of sampling points of each variable as k_i, i=0,1, the sampling points for R_0 and R_1 can be be generally denoted as

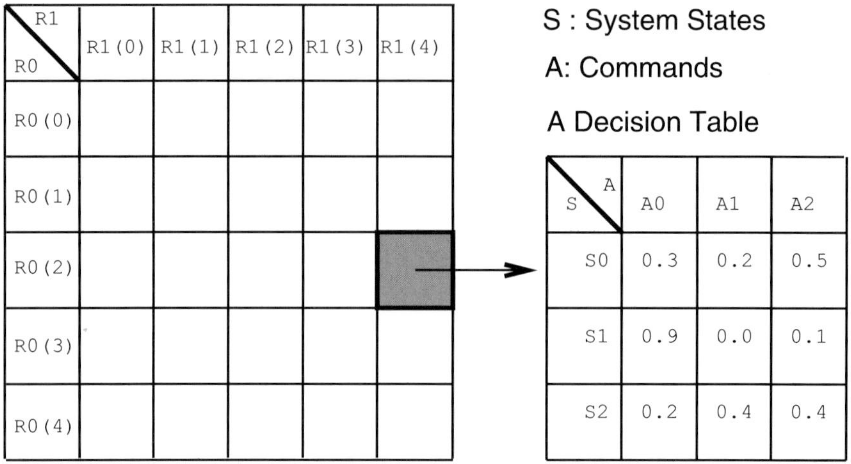

Fig. 5 An example of a 2D policy table

$R_O(i)(i = 0,1,\ldots,k_0 - 1)$, $R_1(j)(j = 0,1,\ldots,k_1 - 1)$, respectively. An example of a two-dimensional policy table constructed for $k_0 = k_1 = 5$ is shown in Fig. 5. Each cell of a policy table is also a two-dimensional table, which we call a decision table. A decision table is a matrix with as many rows as the total system states and as many columns as the command issued by the PM to SP. Each cell of a policy table can be indexed as a pair $(R_0(i), R_1(j))$. For each pair $(R_0(i), R_1(j))$, a policy optimization is performed to get the decision table and the obtained decision table is stored to a cell of the policy table with the corresponding index. The overall table is constructed once for all and its size is $k_0 \times k_1$ times the size of the table used in [3]. Apart from the larger storage overhead, there is no performance penalty in accessing this table.

4.2 Decision Using Interpolation

The predicted user request can be used to select cells in the policy table containing optimal policy at a given time. If the predicted user request is one of the sampling points used in policy optimization, we can choose a decision from the cell in the corresponding table index according to the system state denoted as *CS*. But there may be a gap between the predicted user request and sampling points because predicted user request is a value calculated in a continuous interval (from 0 to 1), while the sampling points are discretely sampled. We use a two-dimensional linear interpolation/extrapolation technique to compute the decision tables for (R_0, R_1) value pairs that do not correspond to sampled values in the policy table. The pseudo-code of the interpolation/extrapolation procedure is shown in Fig. 6. Extrapolation is used if the values of any of the R_i is either larger than the largest sampled value or smaller than the smallest sampled value. In all other cases, the interpolated value is computed

```
2DInterpolation (CS, cell,PR₀,PR₁,K₀,K₁)
for (i = 0; i < 2; i++){
if (PRᵢ ≤ Rᵢ(0)) { /* extrapolation */
id⁰ᵢ= 0;
id¹ᵢ = 1;
}else if (PRᵢ ≥ Rᵢ(Kᵢ-1)){ /* extrapolation */
id⁰ᵢ = Kᵢ - 2;
id¹ᵢ = Kᵢ - 1;
}else /* interpolation */
id⁰ᵢ = j s.t. Rᵢ(j) ≤ PRᵢ ≤ Rᵢ(j+1);
id¹ᵢ = j+1;
for (i = 0; i < 2; i++){
for (j = 0; j < 2; j++)    {
Select a decision d2i;+j from cell(idⁱ₀, idʲ₁)
for State CS;
 }
}
foreach (command)
d₅ = OneDimInterp(d₀,d₁);
d₆ = OneDimInterp(d₃,d₄);
d₇ = OneDimInterp(d₅,d₆);
return( );
```

Fig. 6 2-dimensional interpolation

as three successive one-dimensional linear interpolations on the table entries corresponding to (R_0, R_1) pairs surrounding the predicted values of R_0 and R_1.

5 Experimental Results

We applied the proposed prediction schemes to a Hard Disk Drive with highly nonstationary workload. The HDD (i.e. the SP) consumes 3W in *active* state and 0W in *sleep* state. The transition time from active to sleep (and vice versa) takes in average 10 time slices. Requests can be stored in a queue with length 2, and require in average 1.25 time slices to be served. The Markov model of the system (SP, Queue and SR) has 14 states. The PM can issue 2 commands: GO_ACTIVE and GO_SLEEP. The decision table from the policy optimization is 14×2 matrix, where the row is the total number of states and the column is the number of commands [3]. We tested our adaptive policy on a worst-case, highly nonstationary workload, constructed by concatenating 26 workloads with different R_0, R_1, and different duration (between 30,000 and 50,000 time slices). With a perfect workload estimation scheme, each workload would be immediately identified, the optimal policy computed and used until the change to a new workload. We call this policy *Best-Adaptive*. We can also define another ideal policy, called the *Best-Oracle*, that assumes perfect knowledge not only of workload parameters, but also of the duration of every single idle period in the request stream. In this case, we can analyze the stream to find idle periods which are longer than the sum of transition time from active state to sleep state and vice versa. These idle periods can be used for system shutdown without performance degradation. Obviously, both the best-adaptive and best-oracle policies are theoretical limits, which cannot be achieved in practice because they require the availability of pefect predictors of future events (they can be tested in simulation by examining in advance the entire input trace that drives the simulator). They will be used for comparison with our adaptive policies. We extended the optimizer and simulator introduced in [3] to generate the policy table and simulate user work-load with the proposed prediction schemes. We computed minimum power performance-constrained policies. Performance constraints are expressed by limiting the average time spent by a request in the queue (W_p) and the probability of an incoming request to find the queue full (we call this event *request loss* L_p). Notice that the constraints are not equivalent: L_p is usually tighter for workloads with high number of requests (because of the limited queue lengh), while W_p dominates for light workloads. For the first set of experiments, L_p was chosen to be 10% more than the loss experienced without power management, and W_p was set to increase expected waiting time by 1 time slice. Optimization and simulation were performed on SUN Ultra2 SPARC workstation (200 MHz, 520 MB main memory) to find out the optimal window size (*WS*). The computation time for policy table construction (10×10 table) was approximatively 5 min. We compared the power and performance of our adaptive approaches with the best-adaptive policy. We also compare with the nonadaptive approach [3] which considers the overall workload as a stationary Markov chain with constant R_0 an R_1. Figure 7a, b report respectively power and average waiting time as a function of sliding window width. The nonadaptive approach produces higher power consumption than our adaptive approaches. Also, constraints are now well matched (even though, in this case, the nonadaptive policy is conservative). The graphs also show that both window approaches are very close to the best-adaptive policy when the window size is reasonably large. Notice that double window approach is more stable and accurate than the single window approach for the same window size. The appropriate window size is over 40 for the double window approach and is over 90 for the single window approach. Not only the overall trade-off between the performance and power, but also the local trade-off is important because it represents how well the policy exploits idleness on a short time scale. Figure 8 reports on the x axis the index of the 26 different workloads of the nonstationary trace. Power obtained with best-adaptive, best-oracle and our adaptive policies (with window size 50) is reported on the

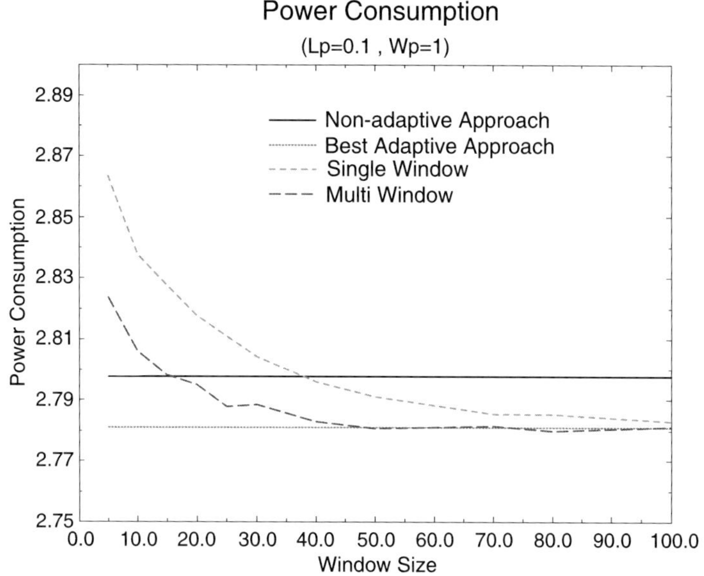

(a)

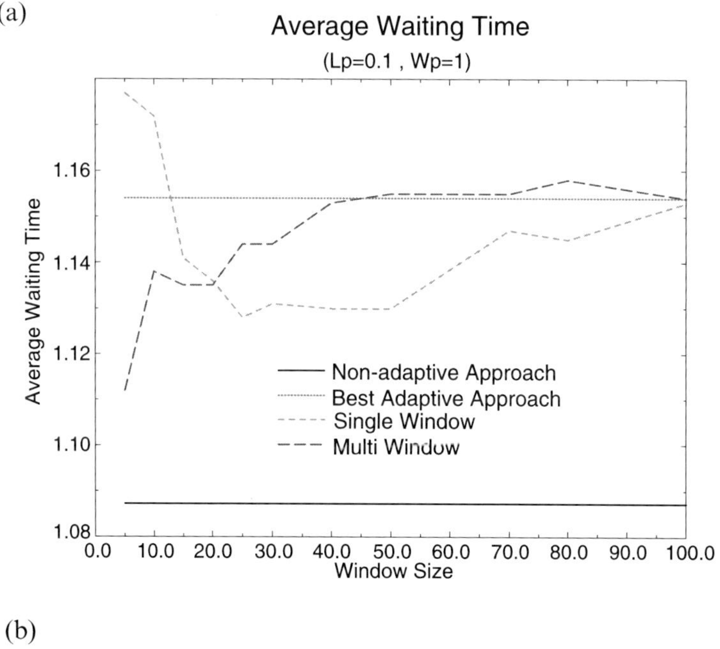

(b)

Fig. 7 (a) Power comparison (b) average waiting time comparison

y axis. When the performance constraint is tight ($L_c = 0.1$, $W_p = 1$) as shown in Fig. 8a, the window approaches are still comparable to the best-adaptive approach, but even the best-adaptive policy does worse than the best-oracle for light workloads

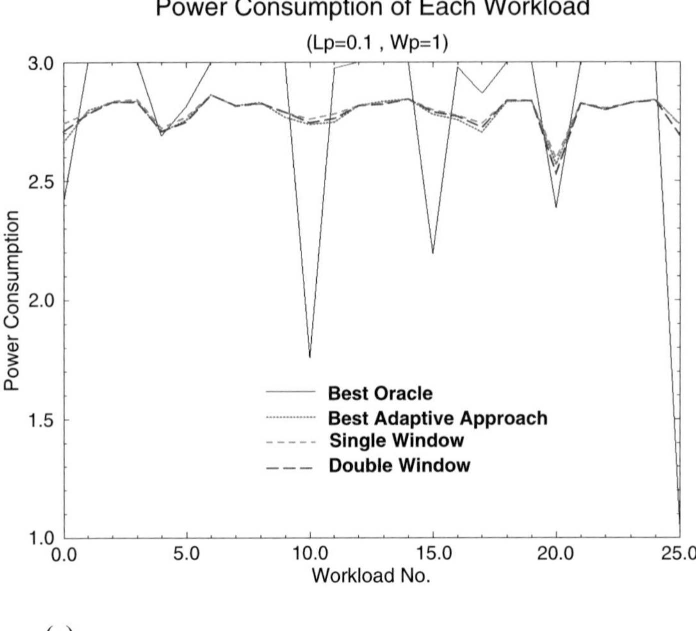

(a)

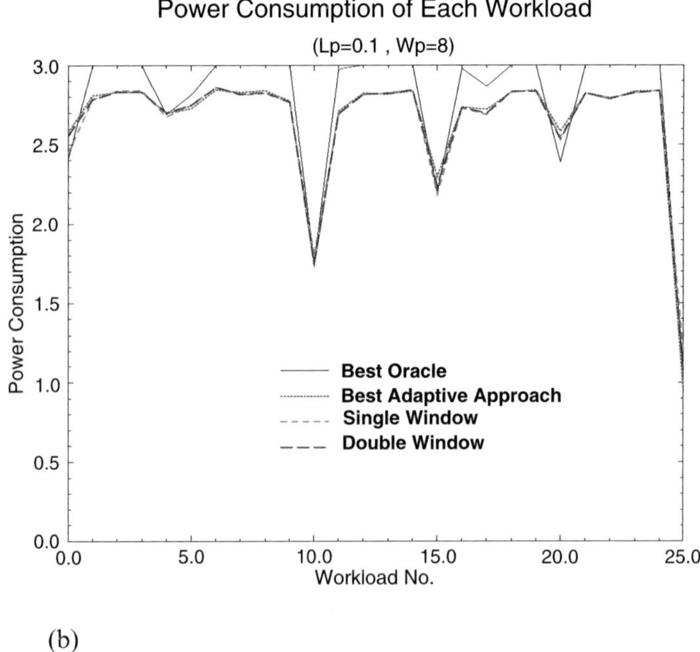

(b)

Fig. 8 (a) Power comparison of each workload ($W_p = 1$); (b) power comparison of each workload ($W_p = 8$)

with tight performance constraints. When the waiting time constraint, W_p is relaxed without change of L_p ($L_p = 0.1$, $W_p = 8$), the best-adaptive approach and both window approaches are comparable to the best-oracle as shown in Figure 8 (b). Notice that relaxing the constraint does not imply that the waiting time increases proportionally: for heavy workloads the request loss constraint dominates and keeps the waiting time low. To confirm this intuitive fact, we observed that the waiting time increase on the entire trace when $W_p = 8$ is only 10% with respect to the waiting time when $W_p = 1$. It is also interesting to observe that for some workloads the adaptive policies save even more power than the best-oracle. This is possible because some performance is traded off for power, while this is not allowed in the best-oracle policy, where performance is constrained to be the same as without power management. Finally, we applied the window approaches to a real trace of time-stamped disk accesses [3]. We performed two experiments with different constraints, namely $L_p = 0.1$, $W_p = 1$ and with $L_p = 0.1$, $W_p = 8$ and compared the power consumption with that of the best-oracle policy. Results are shown in Fig. 9. Since the trace is for a relatively light workload, the power consumption of the adaptive policies with tight W_p constraints is high with respect to the best-oracle policy. On the contrary, by relaxing the W_p constraint ($W_p = 8$), we obtain policies that are even more aggressive than the best-oracle with a measured waiting time increase of just 6% compared to the average waiting time when $W_p =$

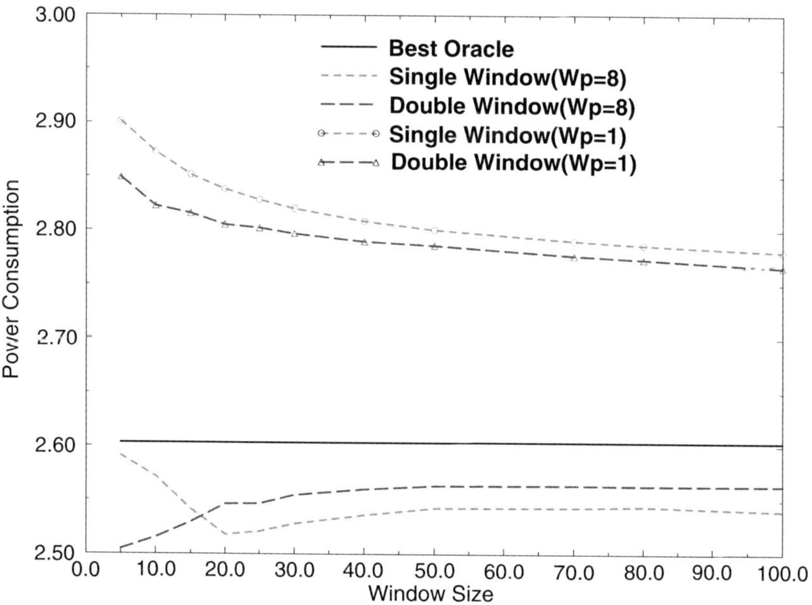

Fig. 9 Power consumption for real trace

1. The adaptive policies compare favourably with the nonadaptive approach for both tight and loose W_p: average power is 2.81 W for the nonadaptive approach (for both W_p values) while it is 2.78 W (with tight constraint) and 2.57 W (with loose constraint) for the double-window adaptive policy ($WS = 100$).

6   Conclusions

In this paper, we described adaptive power management policies for nonstationary workloads. Our adaptive approach is based on sliding windows and interpolation to find an optimal policy from a precharacterized optimal policy table. The proposed approach deals effectively with highly non-stationary workloads: good overall and instantaneous accuracy is achieved. Moreover, our adaptive method offers the possibility of trading off power for performance in a controlled fashion. Simulation results show that our method is comparable to ideal predictive policies which cannot be implemented in practice and outperforms nonadaptive policies.

References

[1] L. Benini and G. De Micheli. *Dynamic Power Management: Design Techniques and CAD Tools*. Kluwer, 1997.

[2] A. Chandrakasan and R. Brodersen. *Low-Power Digital CMOS Design*. Kluwer, 1995.

[3] G. Paleologo, L. Benini, A. Bogliolo and G. D. Micheli. Policy optimization for dynamic power management. *DAC - Proceedings of the Design Automation Conference*, pp. 182–187, 1998.

[4] C.-H. Hwang and A. Wu. A predictive system shutdown method for energy saving of event-driven computation. *Proceedings of the International Conference on Computer Aided Design*, pp. 28–32, 1997.

[5] M. Srivastava, A. Chandrakasan and R. Brodersen. Predictive system shutdown and other architectural techniques for energy efficient programmable computation. *IEEE Transactions on VLSI Systems*, 4(1), 42–55, March 1996.

[6] J. Monteiro and S. Devadas. *Computer-Aided Techniques for Low-Power Sequential Logic Circuits*. Kluwer, 1997.

[7] W. Nebel and J. Mermet (Eds.). *Low-Power Design in Deep Submicron Electronics*. Kluwer, 1997.

[8] R. Golding, P. Bosh and J. Wilkes. Idleness is not sloth. *Proceedings of Winter USENIX Technical Conference*, pp. 201–212, 1995.

[9] R. Golding, P. Bosh and J. Wilkes. Idleness is not sloth. *HP Laboratories Technical Report HPL-96–140*, 1996.

[10] J. M. Rabaey and M. Pedram (Eds.). *Low-Power Design Methodologies*. Kluwer, 1996.

Quantitative Comparison of Power Management Algorithms

Yung-Hsiang Lu, Eui-Young Chung, Tajana Šimunić,
Luca Benini[1], and Giovanni De Micheli

Abstract Dynamic power management saves power by shutting down idle devices. Several management algorithms have been proposed and demonstrated effective in certain applications. We quantitatively compare the power saving and performance impact of these algorithms on hard disks of a desktop and a notebook computers. This paper has three contributions. First, we build a framework in Windows NT to implement power managers running realistic workloads and directly interacting with users. Second, we define performance degradation that reflects user perception. Finally, we compare power saving and performance of existing algorithms and analyze the difference.

1 Introduction

Dynamic power management (DPM) is an effective approach to reduce power consumption without significantly degrading performance [2]. DPM shuts down devices when they are not being used and wakes them up when necessary. When a device is not used, it is called *idle*; otherwise, it is called *busy*. DPM algorithms observe request patterns and predict the length of idle periods. Idle periods can be defined in different ways [8]. In this paper, we consider an idle period as "time with no requests waiting for service". The device is in a *working* state when it can serve requests with higher power consumption, P_w (Table 1 summarizes all symbols.). The device is *sleeping* when it consumes less power, P_s ($P_s < P_w$), and cannot serve requests. Shutting down and waking up a device usually cause performance degradation and require extra energy. Therefore, DPM algorithms shut down a device only when an idle period is long enough to justify performance degradation and state-transition energy.

This paper compares DPM algorithms for controlling the power states of hard disk drives on a desktop and a notebook computers. Hard disks are of particular interest for power management due to three reasons. First, hard disks may consume up to 20% of total energy in a computer [7]. Recent studies find

Computer Systems Laboratory, Stanford University, USA {luyung, eychung, tajana, nanni}@
stanford.edu [1]Dip. Informatica, Elettronica, Sistemistica, Università di Bologna, Italy [1]lbenini@
deis.unibo.it

R. Lauwereins and J. Madsen (eds.), *Design, Automation, and Test
in Europe–The Most Influential Papers of 10 Years DATE*, 207–219
© Springer 2008

Table 1 Symbols and meanings

Symbol	Meaning
T_{sd}	Shutdown delay
T_{wu}	Wakeup delay
T_{be}	Break-even time for an idle period
T_{ms}	Minimum sleeping time to save energy
T_{bs}	Time before shutdown
T_{ss}	Average time in sleeping state
$T_{idle}[i]$	Current idle period, candidate for shutdown
$t_{idle}[i]$	Predicted value of $T_{idle}[i]$
$T_{busy}[i]$	Busy period before $T_{idle}[i]$
$W_S[i]$	Starting time of a waiting period
$W_E[i]$	Ending time of a waiting period
τ	Timeout value
E_{sd}	Energy to shut down
E_{wu}	Energy to wake up
P_s	Power in sleeping state
P_w	Power in working state
N_{sd}	Number of shutdowns
N_{wd}	Number of wrong shutdowns

that hard disks will remain major power consumers in the near future [10, 12]. Second, hard disks have large performance and power overhead because of mechanical inertia. Spinning down or up disk plates takes several seconds, equivalent to hundreds of millions of instructions in modern computers. Finally, hard disks are not always needed when computers are running if the physical memory contains all the information needed; for example, caching may avoid unnecessary spin-ups [5].

Our study has three major contributions: a framework to implement power managers (PM), definition of user-perceived performance, and quantitative comparison of algorithms. In the past, DPM algorithms were evaluated mainly by simulation. In contrast, we build a framework for comparing DPM algorithms running realistic workloads on a commercial operating system, Windows NT, and directly interacting with users. Consequently, users can perceive performance degradation while power is saved. Although some algorithms support multiple sleeping states, we use only one sleeping state in the implementation as a common denominator for fair comparison.

2 Foundation for Algorithm Comparison

2.1 *Idle Periods*

Figure 1 shows requests and the power states of a device. An idle period ($T_{idle}[i]$) occurs between two periods with requests, also called busy periods. The device is

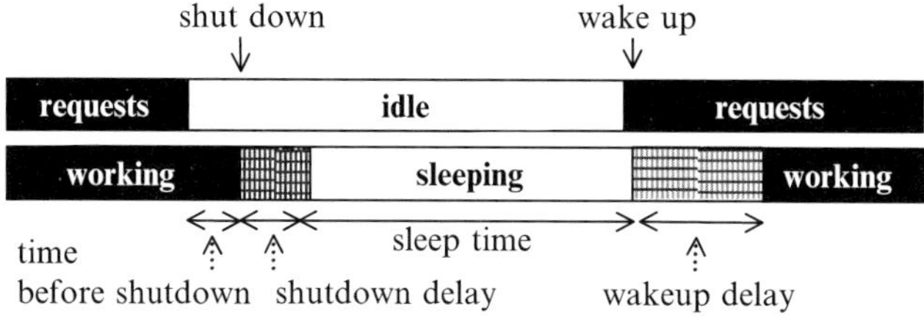

Fig. 1 State transitions

shut down after it enters $T_{idle}[i]$; it does not enter the sleeping state immediately due to shutdown delay (T_{sd}). Some DPM algorithms do not shut down a device immediately after an idle period starts; instead, they wait until they are confident that $T_{idle}[i]$ is long enough to save power. This waiting time is called "time before shutdown" (T_{bs}). Later, when requests arrive, the device is woken up and enters the working state after wakeup delay (T_{wu}). The energy consumed during the shutdown and wakeup delay are denoted as E_{sd} and E_{wu}.

2.2 Break-even Time

As we explained earlier, in order to compensate the shutdown and wakeup overhead, a device has to stay in the sleeping state long enough. We call this minimum duration the *minimum sleeping time* (T_{ms}). Since $E_{sd} + E_{wu} + P_s \bullet T_{ms} = P_w \bullet (T_{ms} + T_{sd} + T_{wu})$, we can find T_{ms} in Eq. (1). The minimum length of an idle period to save energy is the *break-even time* for idle period (T_{be}). It includes the shutdown and wakeup delay in addition to T_{ms}, so $T_{be} = T_{ms} + T_{sd} + T_{wu}$. A shutdown command saves power only if $T_{idle}[i] > T_{bs} + T_{be}$.

$$T_{ms} = \frac{E_{sd} + E_{wu} - P_w \cdot (T_{sd} + T_{wu})}{P_w - P_s} \tag{1}$$

$$T_{be} = \frac{E_{sd} + E_{wu} - P_s \cdot (T_{sd} + T_{wu})}{P_w - P_s} \tag{2}$$

2.3 Performance Metrics

DPM trades off between power and performance. Several ways were proposed to quantify performance, such as total or average waiting time; however, total or average waiting time can be misleading. Using these metrics, a system that causes a total of 50 s of waiting in 5 h is better than one that causes 60 s of waiting in 5 h.

However, if the former requires a user to wait for 40 s in a 1-min period while the latter requires at most 10 s in 1 min, most users think the second system has better performance. Traditional total waiting time does not reflect this discrepancy. Studies in psychology show that waiting time can affect user behavior [18]. In this paper, we use two objective performance metrics: long waiting or repetitive waiting "within short time intervals" to reflect the perception of performance degradation.

Figure 2 gives an example of four waiting periods. A waiting period (W) starts at W_S and ends at W_E. Our first performance metric is the largest total waiting time in a duration of length d called W_d. It is obtained by finding the sum of waiting time in a sliding window of size d. A window may contain multiple waiting periods, such as W [3] and W [4] in the third window; a window can also contain part of a waiting period, such as W [3] by the second window. In general, W_d can be calculated by Eq. (3).

$$W_d = \max_{t} \sum_{\substack{i\,\text{such that} \\ W_s[i] \geq t \\ W_E[i] \leq t+d}} (W_E[i] - W_s[i]) \tag{3}$$

This equation finds all waiting periods inside a window of size d and calculates the sum of these periods. By adjusting t, the starting point of a window, it then finds the window that contains the longest total waiting time. If a waiting period is partially covered, we consider only the part within the window; for simplicity, Eq (3) does not include partially covered waiting periods.

The second performance metric is the longest shutdown sequence in which the time between two adjacent shutdowns is less than a threshold th. This metrics measures the number of consecutive waiting periods that are close and cause delay repetitively. It is the largest m for which there is a sequence $W[i]$, $W[i + 1]$, ... $W[i + m]$ such that the following conditions hold:

$$\begin{aligned}
W_s[K+1] \quad & -W_E[k] \leq th \\
W_s[i] \quad & -W_E[i-1] > th \\
W_s[i+m+1] \quad & -W_E[i+m] > th
\end{aligned} \tag{4}$$

where $k \in [i, i + m - 1]$

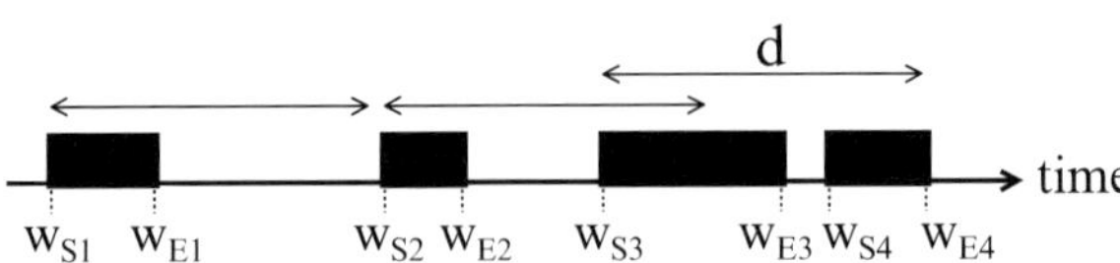

Fig. 2 Waiting due to power management

3 Algorithms and Parameter Selection

We compare algorithms originally designed for various applications, such as X-servers and hard disks; these algorithms are listed in Table 2. They assume different characteristics of applications. For example, T_{sd} and T_{wu} are fairly small for X-server but large for hard disks due to mechanic inertia. We compare them in the same environment and study the deviations from their originally intended applications. In this section, we briefly explain these algorithm and the parameters recommended by their authors.

TIMEOUT: Timeout algorithms are simple and widely used; they assume that if a device is idle longer than τ, it will remain idle for a long time ($T_{idle} > \tau \Rightarrow T_{idle} > \tau + T_{be}$) with up to 95% confidence level [8]. Timeout algorithms wait for τ before shutdown, so $T_{bs} = \tau$. Microsoft Windows Control Panel allows users to set τ as small as 60 s. We use a filter device driver [14] and can choose any τ value; we use 30 s and 2 min in this study.

The first adaptive timeout (ATO1) algorithm adjusts τ by considering the value of $\dfrac{T_{idle}[i-1]}{T_{wu}}$ [6]. When the ratio is small, τ increases; when the ratio is large, τ decreases. We start with 30 s for τ and use $(\alpha_m, \beta_m, \rho) = (1.5, 0.5, 0.1)$ because they produced better results. Golding suggests updating τ asymmetrically: increasing τ by 1 s or decreasing by half [8] (ATO2). In our experiments for [8], τ is limited to 1–120 s. Another approach adjusts τ according to $T_{busy}[i]$ [13] (ATO3). If $T_{busy}[i]$ is small, τ decreases; if $T_{busy}[i]$ is large, τ increases. We use 120 s for the initial value of τ with 1 Hz sampling and 2 s for the adjustment factor.

L-SHAPE: If a short busy period is frequently followed by a long idle period, their scatter plot will form an "L-shape" (LS) as shown in Fig. 3. If a busy period is short enough, the following idle period is likely to be long ($T_{idle}[i] \geq T_{be} \Leftrightarrow T_{busy}[i] \leq T_{threshold}$). Consequently, when request patterns form an L-shape, the device should be shut down after a short busy period [17].

Table 2 DPM algorithms compared

Algorithm	Features	Applications
[6] (ATO1)	Adaptive timeout	Hard disk
[8] (ATO2)	Adaptive timeout	Hard disk
[13] (ATO3)	Adaptive timeout	Hard disk
[17] (LS)	L-shape	X-server
[9] (EA)	Exponential average	Telnet
[15] (DM)	Discrete-time Markov	Hard disk
[3] (SW)	Sliding windows	Hard disk
[16] (CM)	Continuous time Markov	Real-time inputs
[20] (SM)	Time-index semi-Markov	Hard disk
[11] (CA)	Competitive	Spin-block
[4] (LT)	Learning tree	Hard disk

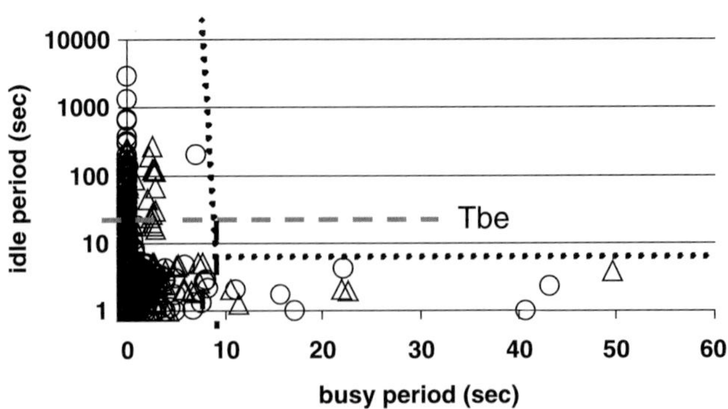

Fig. 3 $T_{busy}[i]$ vs. $T_{idle}[i]$ of two different users in our experiments

EXPONENTIAL AVERAGE: Hwang and Wu [9] use the predicted and the actual lengths of the previous idle period to predict the length of the current idle period. They use exponential average (EA) for predicting T_{idle} by $t_{idle}[i] = a \bullet T_{idle}[i - 1] + (1 - a) \bullet t_{idle}[i - 1]$, which is equivalent to

$$t_{idle}[i] = (1 - a)^{i+1} t_{idle}[0] + \sum_{k=0}^{i} a(1 - a)^k T_{idle}[i - k]$$

where $t_{idle}[i]$ is the prediction of $T_{idle}[i]$. It is an average of previous idle periods with exponential weights. If $t_{idle}[i] > T_{be}$, the device is shut down. This algorithm limits $t_{idle}[i]$ such that it cannot exceed $c \bullet t_{idle}[i - 1]$ where c is a constant greater than one. We use 0.5 for a and 2 for c; we ignore all $T_{idle}[i - 1] < 0.1$ s so that t_{idle} are not affected by short idle periods. This is an on-line filtering because $T_{idle}[i - 1]$ is known before computing $t_{idle}[i]$. We use 0.1 s so that our implementation has only negligible difference from the original algorithm.

STOCHASTIC MODEL: Power management can be solved as an optimization problem when devices and requests are modeled as stochastic processes. This approach formulates power management as a constrained optimization problem; it provides the flexibility to trade off between power and performance. For example, power management can be modeled as a discrete-time Markov decision process (DM) [15]. The algorithm assumes stationary geometric distribution of request arrivals. It is extended to handle non-stationary requests in [3]. Nonstationarity is captured by sliding windows (SW); the algorithm interpolates preoptimized look-up tables for shutdown decisions. The shutdown decision is evaluated each period, even when the device is in the sleeping state, thus causing computation overhead. For example, for a 10 W processor, SW could waste as much as 1800 J of energy during a 30-min idle period just due to reevaluating shutdown decisions.

By modeling a device as a continuous-time Markov process, PM can change power states upon event occurrences, such as "request queue empty" events, instead of at discrete time intervals [16] (CM). State transitions are assumed to follow exponential distributions. This approach makes a decision as soon as certain events happen. Our measurements show no significant power saving since the algorithm tends to shut down too soon, thus incurring large wakeup costs and sometimes even missing shutting down on long idle periods.

Both discrete and continuous-time approaches model request arrivals and the power state transitions using memoryless distributions, which is not accurate in real situations. Semi-Markov approach [19] models transition times between power states with uniform distributions. As the request arrivals are better modeled using Pareto distribution, the semi-Markov approach is further generalized with time-indexed semi-Markov models in [20] (SM). Optimal power saving based on this model reevaluates the decisions during idle periods until either an request occurs or the device is shut down. When the device is sleeping, no decision evaluation is needed. Therefore, this algorithm has low computation overhead and shows the largest power saving in our study.

COMPETITIVE ALGORITHM: A "c-competitive" on-line algorithm (CA) can find a solution with cost less than c times the cost generated by an optimal off-line algorithm. It can be proven that 2-competitive power saving is achievable if $\tau = T_{be}$ [11]. In other words, this algorithm consumes at most twice the minimum power consumed by an off-line PM.

LEARNING TREE: Adaptive learning trees (LT) transform sequences of idle periods into discrete events and store them into tree nodes [4]. This algorithm predicts idle periods using finite-state machines similar to branch prediction used in microprocessors and selects a path which resembles previous idle periods. At the beginning of an idle period, it determines an appropriate sleeping state; this algorithm is capable of controlling multiple sleeping states.

4 Experiment Results

4.1 Experiment Environment

We use an environment built specifically for implementing and evaluating DPM algorithms [14]. It consists of two ACPI-compliant [1] computers. The first is a Pentium II desktop computer with an IBM DTTA 350640 hard disk; the other is a Pentium II notebook computer with a Fujitsu MHF 2043AT hard disk. Both are running a beta version of Microsoft Windows NT V5. We implemented filter drivers for each algorithm to control the power states of the hard disks, to record disk accesses and to analyze the performance impact and the power management overhead of each algorithm. Table 3 shows the parameters of the disks. T_{be} is 17.6 and 5.43 s for the IBM and Fujitsu disks respectively.

Table 3 Disk parameters

Model	P_s Watt	P_w Watt	T_{sd} s	E_{sd} J	T_{wu} s	E_{wu} J
IBM	0.75	3.48	0.51	1.08	6.97	52.5
Fujitsu	0.13	0.95	0.67	0.36	1.61	4.39

4.2   PM Overhead and Power Consumption

We recorded disk accesses for two users, one developing C programs and the other making presentation slides. These traces include disk accesses from user requests and operating system activities. Then the traces are replayed taking approximately 11 h for each algorithm. We find that all algorithms spend less than 1% of computation time on power managers; hence, power management pays off when it is able to effectively reduce power consumption. These algorithms are compared by five figures:

- Power consumption (P), unit: Watt.
- Number of shutdowns (N_{sd})
- Number of wrong shutdowns (N_{wd}) that have sleeping time shorter than T_{ms} and actually waste energy
- Average time in sleeping state (T_{ss}), unit: second
- Average time before shutdown (T_{bs}), unit: second

Table 4 orders the algorithms by power consumption. In this table, smaller values are better except T_{ss}. The first row contains the minimum power consumption without performance degradation; it is generated off-line with full knowledge about future requests. The last row shows the power consumption if no power management is applied. This table shows that SM, CA, and SW can save nearly 50% of power on both platforms. Even though they have close power consumption on the mobile disk, they differ significantly in performance. CA and SM have more than twice wrong shutdowns (N_{wd}) compared with SW. For algorithms with similar power consumption, performance is an important factor for evaluation. The total waiting time is approximately proportional to the total number of shutdowns (N_{sd}). Users may notice substantially different performance degradation even for two algorithms that have similar values of N_{sd} if some wrong shutdowns occur repetitively within a short time interval.

4.3   Performance Measurement

Figure 4 (next page) draws the worst waiting time (W_d) for d between 1 to 10 min. It shows that, in the worst case, CA requires users to wait for 98 seconds in a 10-min duration on the desktop hard disk. The bottom of the figure is the waiting time by percentage. When the window size increases the percentage of waiting time decreases for all algorithms. When the window size is small, such as 1 min, some algorithms may require users to wait for more than 50% of the time. This demonstrates the importance to measure the worst-case performance for small d. Traditional performance metric using the total waiting time cannot provide enough information for determining user perception of performance degradation. This figure has several "jumps" as d increases because the worst-case waiting time may change from one window to another. Figure 4 also shows that the waiting time is considerably shorter on the mobile hard disk. Figure 5 plots the longest shutdown sequence in which the time between two adjacent shutdowns is shorter than a

Table 4 Algorithm comparison

Algorithm	P	N_{sd}	N_{wd}	T_{ss}	T_{bs}
			desktop		
off-line	1.64	164	0	166	0
SM	1.92	156	25	147	18.2
CA	1.94	160	15	142	17.6
SW	1.97	168	26	134	18.7
$\tau = 30$	2.05	147	18	142	30.0
LT	2.07	379	232	62	5.7
ATO3	2.09	147	26	138	29.9
ATO1	2.19	141	37	135	27.6
ATO2	2.22	595	430	41	4.1
$\tau = 120$	2.52	55	3	238	120.0
DM	2.60	105	39	130	48.9
EA	2.99	595	503	30	7.6
always-on	3.48	–	–	–	–
			notebook		
off-line	0.33	250	0	118	0
SM	0.40	326	76	81	8.0
SW	0.43	191	28	127	13.4
CA	0.44	323	64	79	5.4
LT	0.46	437	217	56	6.1
ATO1	0.47	273	73	88	12.4
EA	0.50	623	427	37	3.0
$\tau = 30$	0.51	139	7	157	30.0
ATO3	0.52	196	48	109	24.5
DM	0.62	173	54	102	35.2
ATO2	0.64	881	644	19	2.3
$\tau = 120$	0.67	55	0	255	120.0
always-on	0.95	–	–	–	–

threshold. In this figure, the x-axis is the threshold value and the y-axis is the lengths of sequences. The arrow indicates that EA has a sequence of 27 waiting periods with less than one minute between two shutdowns. Users perceive delays every minute or even more frequently for 27 times.

5   Discussion

5.1   Correlation of Adjacent Busy and Idle Periods

LS uses the length of the previous busy period to predict the length of the current idle period by "L-shape" approximation. Figure 3 shows two traces we collected; the dash line encloses most requests into an L shape. If we consider only idle periods longer than 10 s and redraw the distribution in Fig. 6, we find this

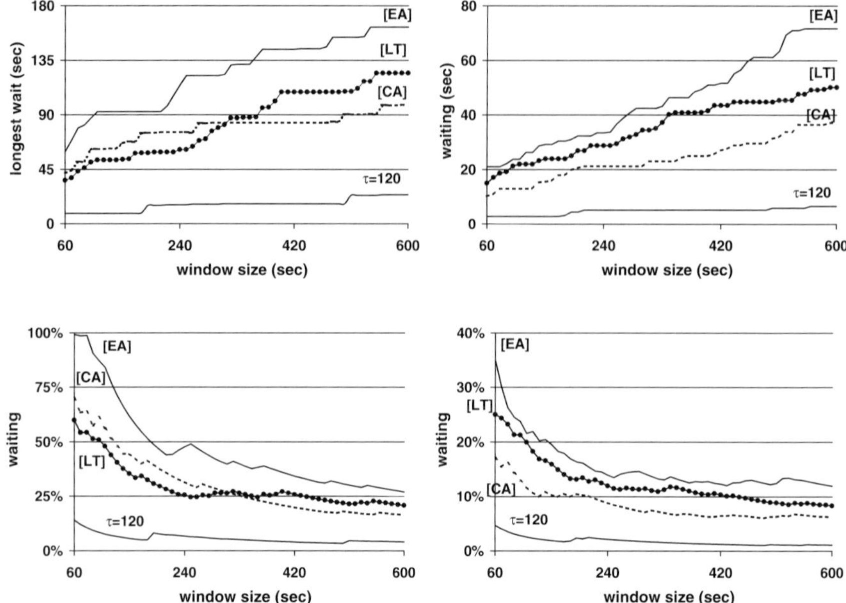

Fig. 4 W_d on desktop (left) and mobile (right) disks

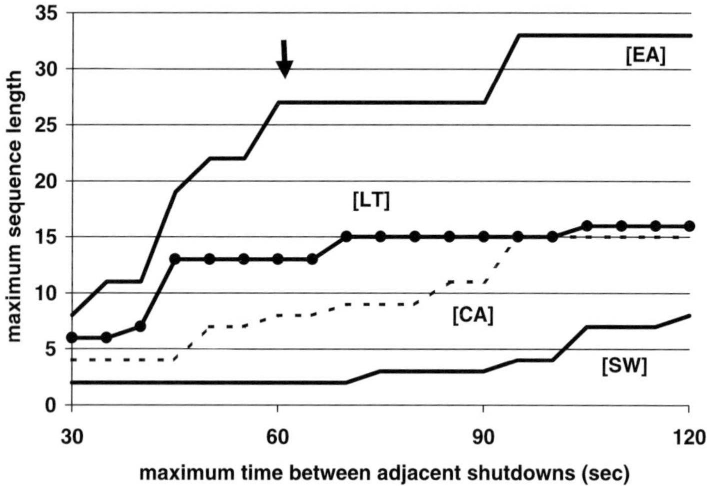

Fig. 5 Maximum length of shutdown sequences

approximation requires refinement. One long idle period (197 s, pointed by the arrow) follows a long busy period; L-shape algorithm will keep a hard disk in the working state during this period. Furthermore, a large group of short idle periods follow short busy periods enclosed by the circle. L-shape approximation is not sufficient to determine the length of an idle period.

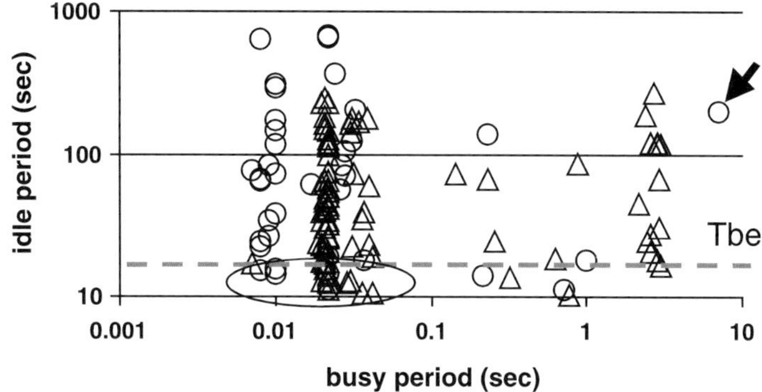

Fig. 6 $T_{busy}[i]$ vs. $T_{idle}[i]$ (Zoomed In)

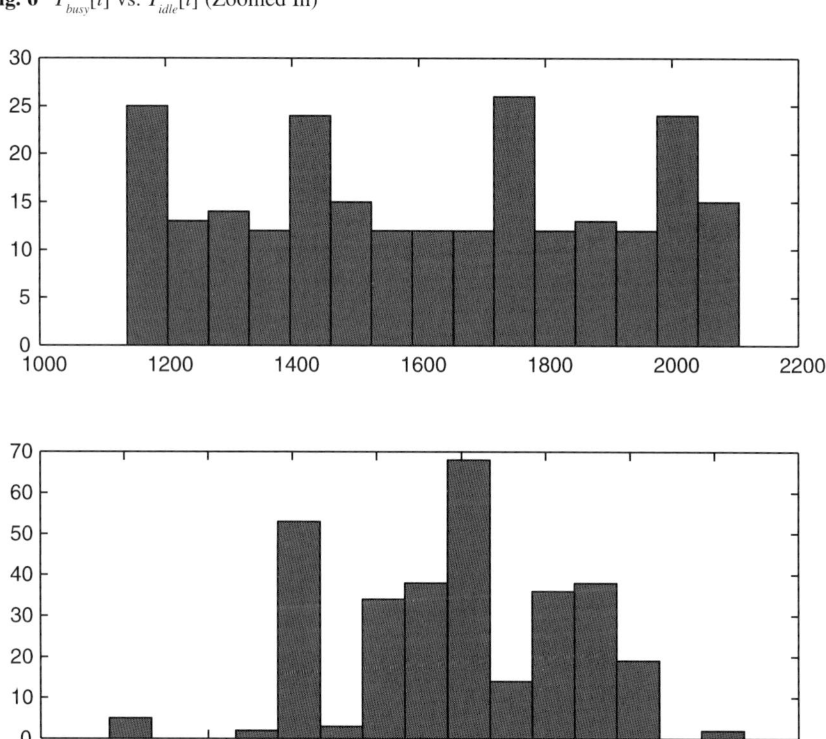

Fig. 7 T_{wu} (ms) for mobile (top) and desktop disks

5.2 State Transition Time

Table 3 lists the average of T_{sd} and T_{wu}. Most algorithms treat them as constants for simulation; in reality, they are not constants. Figure 7 shows significant variation of T_{wu}. For the desktop hard disk, the average of T_{wu} is 6.97 s and the standard deviation is 0.65 s or 9.25%. For the mobile disk, T_{wu} is even more widely distributed; it is inappropriate to use a single value for T_{wu}. Markov models in DM are inexact approximations because our experiments show that T_{wu} is not exponentially distributed.

5.3   Successive Wrong Shutdowns After a Long T_{idle}

EA assumes that a long idle period is followed by other long idle periods and a short idle period is followed by other short idle periods. When a long idle period is followed by short idle periods, however, the performance deteriorates considerably and generates long sequences in Fig. 5. Consider an example, a user leaves the office and creates an idle periods of 1 h. When the user returns, the hard disk will be shut down repetitively and made unusable during the first minute. In order to remedy this problem, a desirable algorithm should change its prediction sooner once successive wrong shutdowns happen. We do not use predictive wakeup because it consumes 96% more energy on the mobile disk without significant performance improvement.

5.4   Algorithm Ordering and Device Parameters

Compared to the desktop hard disk, $T_{sd} + T_{wu}$ and $E_{sd} + E_{wu}$ are 70% and 91% less on the mobile hard disk. As a result, DPM algorithms can be more aggressive to shut down the mobile disk in order to save energy since the penalty is considerably less. The ordering on the desktop computer in Table 4 is similar to the ordering on the notebook computer except EA where it consumes less power than most other algorithms. This algorithm was originally designed for X-server and telnet, which have small shutdown and wakeup overhead because no mechanic device is involved. This example shows the importance to tune an algorithm for its intended application.

6   Conclusions

We built a framework to compare power management algorithms. To our knowledge, this is the first time DPM algorithms are compared while running realistic workloads and interacting with users. Our experience shows that waiting in a short duration and consecutive waiting, instead of total waiting time, directly affect user perception of performance. We quantitatively define performance metrics that reflect user experience. Our study concentrates on comparing power management algorithms for control the power states of computer hard disks and discusses several key issues in designing DPM algorithms.

Acknowledgments This work is supported by MARCO, ARPA, GSRC, and NSF (CCR-9901190).

References

[1] ACPI. http://www.teleport.com/~acpi.

[2] L. Benini, A. Bogliolo, and G. D. Micheli. A Survey of Design Techniques for System-Level Dynamic Power Management. *IEEE Transactions on VLSI Systems*, March 2000.

[3] E.-Y. Chung, L. Benini, A. Bogliolo, and G. D. Micheli. Dynamic Power Management for Non-Stationary Service Requests. In *Design Automation and Test in Europe*, pp. 77–81, 1999.

[4] E.-Y. Chung, L. Benini, and G. D. Micheli. Dynamic power management using adaptive learning tree. In *International Conference on Computer-Aided Design*, pp. 274–279, 1999.

[5] F. Douglis, R. Cáceres, F. Kaashoek, K. Li, B. Marsh, and J. A. Tauber. Storage Alternatives for Mobile Computers. In *USENIX Symposium on Operating Systems Design and Implementation*, pp. 25–37, 1994.

[6] F. Douglis, P. Krishnan, and B. Bershad. Adaptive Disk Spin-Down Policies for Mobile Computers. In *Computing Systems*, 8, 381–413, 1995.

[7] F. Douglis, P. Krishnan, and B. Marsh. Thwarting the Power-Hungry Disk. In *USENIX Winter Conference*, pp. 293–306, 1994.

[8] R. Golding, P. Bosch, and J. Wilkes. Idleness is not Sloth. In *USENIX Winter Conference*, pp. 201–212, 1995.

[9] C.-H. Hwang and A. C. Wu. A Predictive System Shutdown Method for Energy Saving of Event-Driven Computation. In *International Conference on Computer-Aided Design*, pp. 28–32, 1997.

[10] Intel. Mobile Power Guide '99.

[11] A. Karlin, M. Manasse, L. McGeoch, and S. Owicki. Competitive Randomized Algorithms for Nonuniform Problems. *Algorithmica*, 11(6):542–571, June 1994.

[12] J. R. Lorch and A. J. Smith. Software Strategies for Portable Computer Energy Management. *IEEE Personal Communications*, 5(3):60–73, June 1998.

[13] Y.-H. Lu and G. D. Micheli. Adaptive Hard Disk Power Management on Personal Computers. In *Great Lakes Symposium on VLSI*, pp. 50–53, 1999.

[14] Y.-H. Lu, T. Šimunić, and G. D. Micheli. Software Controlled Power Management. In *International Workshop on Hardware/Software Codesign*, pp. 157–161, 1999.

[15] G. A. Paleologo, L. Benini, A. Bogliolo, and G. D. Micheli. Policy Optimization for Dynamic Power Management. In *Design Automation Conference*, pp. 182–187, 1998.

[16] Q. Qiu and M. Pedram. Dynamic Power Management Based on Continuous-Time Markov Decision Processes. In *Design Automation Conference*, pp. 555–561, 1999.

[17] M. B. Srivastava, A. P. Chandrakasan, and R. W. Brodersen. Predictive System Shutdown and Other Architecture Techniques for Energy Efficient Programmable Computation. *IEEE Transactions on VLSI Systems*, 4(1):42–55, March 1996.

[18] M. T. Stokes. *Time in Human-Computer Interaction*. PhD. Thesis, Psychology Department, Texas Tech University, 1991.

[19] T. Šimunić, L. Benini, and G. D. Micheli. Event-Driven Power Management of Portable Systems. In *International Symposium on System Synthesis*, pp. 18–23, 1999.

[20] T. Šimunić, L. Benini, and G. D. Micheli. Dynamic Power Management of Laptop Hard Disk. In *Design Automation and Test in Europe*, 2000.

Energy Efficiency of the IEEE 802.15.4 Standard in Dense Wireless Microsensor Networks: Modeling and Improvement Perspectives

Bruno Bougard[1,3], Francky Catthoor[1,3], Denis C. Daly[2],
Anantha Chandrakasan[2], and Wim Dehaene[3]

Abstract Wireless microsensor networks, which have been the topic of intensive research in recent years, are now emerging in industrial applications. An important milestone in this transition has been the release of the IEEE 802.15.4 standard that specifies interoperable wireless physical and medium access control layers targeted to sensor node radios. In this paper, we evaluate the potential of an 802.15.4 radio for use in an ultralow power sensor node operating in a dense network. Starting from measurements carried out on the off-the-shelf radio, effective radio activation and link adaptation policies are derived. It is shown that, in a typical sensor network scenario, the average power per node can be reduced down to $211\,\mu W$. Next, the energy consumption breakdown between the different phases of a packet transmission is presented, indicating which part of the transceiver architecture can most effectively be optimized in order to further reduce the radio power, enabling self-powered wireless microsensor networks.

1   Introduction

Wireless microsensor networks are autonomous networks for monitoring purposes, ranging from short-range, potentially *in vivo* health monitoring [1] to wide-range environmental surveillance [2]. Thanks to the tremendous range of applications they will enable, wireless microsensor networks have received a great deal of attention in recent years. Designing such a network and more specifically the protocols to support its functioning, is a challenging task. Despite the wide variety of applications, all sensors networks face similar constraints [3]:

---

[1] IMEC, Leuven Belgium, {bougardb, catthoor}@imec.be

[2] Department of EECS, Massachusetts Institute of Technology, Cambridge, MA 02139, {ddaly, anantha}@mit.edu

[3] Department of EE, K.U. Leuven, Belgium, wim.dehaene@esat.kuleuven.ac.be

R. Lauwereins and J. Madsen (eds.), *Design, Automation, and Test in Europe–The Most Influential Papers of 10 Years DATE*, 221–234

Density High-end microsensor networks are expected to have a density of approximately 20 nodes/m^3. Hence, the medium access control layer (MAC) should be able to accommodate several hundred to thousand nodes.

Distributed traffic Due to their high node density, wireless sensor networks must have a high capacity. However, the data rate requirements per node are low (<10 kbps). This results in a very low radio duty cycle.

Energy Microsensor nodes are required to be small and autonomous. Their small form factor limits the amount of energy that can be stored in batteries. Furthermore, the density of the network as well as the environment where nodes are deployed often prohibits periodic replacement of the batteries. An existing goal is for a microsensor node to have an average power on the order of 100 µW, which would allow the device to obtain its power from the environment by energy scavenging [4].

Wireless microsensor network research in recent years has strived to design radio circuitry and transmission protocols to meet these novel constraints [5, 6] and it is expected that results from this research will soon emerge in industrial applications. An important milestone in this transition has been the release of the IEEE 802.15.4 standard [7] that specifies interoperable physical and medium access control layers targeted to sensor node radios.

In [8], the performance of the 802.15.4 standard in terms of throughput and energy efficiency is assessed based on simulation. However, this work focuses on a scenario with few nodes and low load, which diverges significantly from the conditions encountered in wireless microsensor networks. In this paper, we evaluate the potential of the standard for use in an ultra low power sensor node operating in the aforementioned dense network conditions.

Section 2 provides a short description of the 802.15.4 standard and outlines a commercial radio implementing the standard, the Chipcon CC2420 [9]. Measurement results carried out on the off-the-shelf component are presented in Section 3. Next, in Section 4, we develop an energy-aware activation policy for the radio and model the resulting average power consumption, as well as the corresponding transmission reliability. In Section 5, it is shown how the model can be used to optimize the energy efficiency in a dense microsensor network scenario. Finally, the energy consumption breakdowns between the different phases of a packet transmission and the different states of the radio are presented, indicating which part of the transceiver architecture can most effectively be optimized in order to further reduce the radio power consumption and thus enable self-sustained wireless microsensor networks.

2   The IEEE 802.15.4 Standard and its Implementation

2.1   Physical Layer

An IEEE 802.15.4 compliant radio can operate in 16 channels in the 2450 MHz ISM band, 10 channels in the 915 MHz band (only in the US) and 1 channel in the 868 MHz band (EU and Japan). The 2450 MHz band allows higher datarate and offers more channels than the other bands and thus is well suited for sensor networks with high network load. Signaling in the 2450 MHz band is based on orthogonal quadrature phase shift keying (O-QPSK) and direct sequence spread spectrum (DSSS). The chip rate equals 2 Mchip/s. One 4-bit symbol is mapped into a 32-chip PN sequence, resulting in a symbol period T_S of 16μs, and a throughput of 250 kbps corresponding to a byte period T_B of 32μs. The CC2420 IC implements the 2450 GHz PHY and supports MAC functionalities. The transmitter and receiver have respectively a direct up-conversion and low IF I/Q architecture. The transmit power can be programmed from −15 to 0 dBm in 8 steps.

2.2 Medium Access Control Layer

Many possible network topologies can be built based on the 802.15.4 MAC. We focus only on 1-hop star networks (Fig. 1a) where a *network coordinator* is elected. In a wireless microsensor network, the network coordinator can be the base-station. Communication from nodes to coordinator (uplink), from coordinator to node (downlink) or from node to node (ad hoc) is possible. In the following, we model uplink communication, which occurs more often than downlink or ad hoc communication in a network that gathers information from the environment and forwards it to the base-station.

In a star network, the beacon mode appears to allow for the greatest energy efficiency. Indeed, it allows the transceiver to be completely switched off up to 15/16 of the time when nothing is transmitted/received while still allowing the transceiver to be associated to the network and able to transmit or receive a packet at any time [10]. The beacon mode introduces a so-called superframe structure (Fig. 2). The superframe starts with the beacon, which is a small synchronization packet sent by the network co-ordinator, carrying service information for the network maintenance and notifying nodes about pending data in the downlink. The inter-beacon

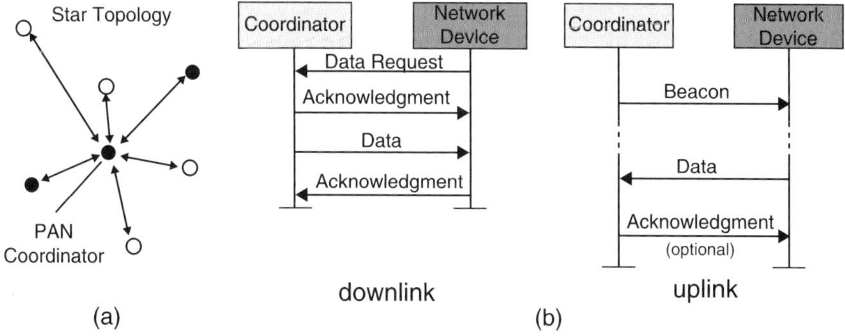

Fig. 1 (a) Star topology in beacon mode. (b) Indirect transmission is used in the downlink while slotted CSMA/CA is used for the uplink

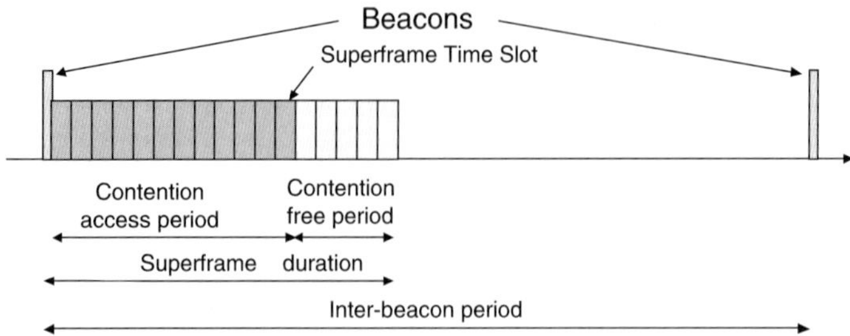

Fig. 2 Superframe structure in beacon mode

period is partially or entirely occupied by the superframe, which is divided in 16 slots. A number of slots at the tail of the superframe may be used as *guaranteed time slot (GTS)*, i.e. they are dedicated to specific nodes. This functionality targets very low latency applications but does not fit well in a dense sensor network since the number of dedicated slots would not be sufficient to accommodate several hundreds of node. In such conditions, it is better to use the contention access mode where the sparse data is statistically multiplexed.

In the contention access period, distributed channel accesses in the uplink are coordinated by a slotted carrier sense multiple access—collision avoidance (CSMA/CA) mechanism while indirect transmission is used in the downlink (Fig. 1b). As we will see later, the CSMA/CA mechanism has a significant impact on the overall energy and performance of the uplink.

According to the slotted CSMA/CA algorithm [7], a node must sense the channel free at least twice before being able to transmit, this corresponds to the decrement of the so-called contention windows (CW). The first sense must be delayed by a random delay chosen between 0 and 2^{BE}-1, where BE is the backoff exponent. This randomness serves to reduce the probability of collision when two nodes simultaneously sense the channel, assess it free and decide to transmit at the same time. When the channel is sensed busy, transmission may not occur and the next channel sense is scheduled after a new random delay computed with an incremented backoff exponent. If the latter has been incremented twice and the channel is not sensed to be free, a transmission failure is notified and the procedure is aborted. When a packet collides or is corrupted, it can be retransmitted after a new contention procedure.

The contention procedure starts immediately after the end of the beacon transmission. All channel senses or transmissions must be aligned with the CSMA slot boundaries that are separated by a fixed period of $T_{slot} = 20 \times T_s$.

As we will see later, the contention procedure introduces a significant overhead in energy consumption. Therefore, a *Battery Life Extension* mode, where the backoff exponent is limited to 0–2 is supported by the 802.15.4 standard. However, in dense network conditions, this mode would results into an excessive collision rate. Hence, we are not using this feature in our experiment.

3 Characterization of the Radio

To be able to assess the average power consumption of an 802.15.4 node in a network we must characterize the instantaneous power consumption of the transceiver when operating in and switching between states. The CC2420 transceiver supports four states:

1. *Shutdown*: The clock is switched off and the chip is completely deactivated waiting for a startup strobe
2. *Idle*: The clock is turned on and the chip can receive commands (for example, to turn on the radio circuitry)
3. *Transmit*
4. *Receive*

In the context of wireless microsensor networks, which are characterized by a very low transmission duty cycle, it has been shown that the transient energy when switching from one mode to another significantly impacts the total power consumption [11, 12]. As we will see later, when considering the MAC, this effect becomes more significant. Hence, it is important to precisely characterize the transition time and energy between the transceiver states. Steady state power, transient time and energy measurement have been made through the use of the Chipcon CC2420EM/EB evaluation board and the *SmartRF Studio* software [9]. Measurement results are summarized in Fig. 3. The state transition energy has been evaluated by multiplying the transition time by the power in the arrival state. This is a worst-case assumption. Notice that the idle state power of 712 µW is already 7 times higher that the average power goal of 100 µW. To achieve lower power, the

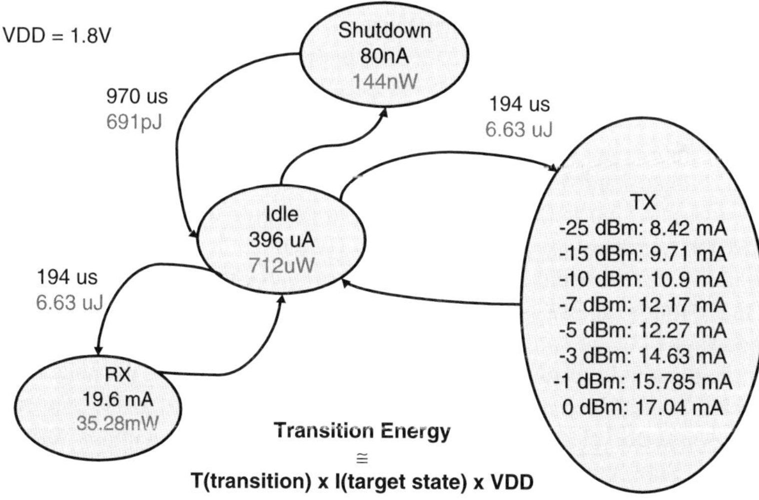

Fig. 3 Steady state and transient power and energy measurement results

transceiver must enter the shutdown state when no action is required during a superframe. This is handicapped by the relatively long transition between shutdown and idle states (~1 ms). To account for this delay, the chip must be preemptively turned on 1 ms before the beacon. Fortunately, this transition requires a relatively low energy (691 pJ). However, additional hardware is required to stay synchronized with the superframe, as the CC2420's clock is turned off in the shutdown state.

After characterizing the energy behavior of the transceiver, the next step is to examine the overall link performance. The bit error probability has been estimated on a testbench composed of a CC2420 transmitter wired to a second CC2420 in receive mode, through a set a calibrated attenuators. Using a wired channel allows one to precisely control the received power. The conditions of an additive white gaussian noise channel (AWGN) are reproduced to assess the packet error probability as a function of the received power. The assumption of an AWGN channel is valid as long as the channel is coherent during the transmission of a packet (slow fading). With the maximum packet size of 123 bytes transmitted at the gross rate of 250 kbps, the packet transmission takes 4 ms, which is smaller than the coherence time encountered in the 2450 GHz band without mobility issues [13]. The estimated bit error probability is plotted in Fig. 4. An exponential regression is done leading to Eq. (1) where Pr_{bit} is the bit-error probability and P_{Rx} the received power, equal to the transmit power P_{Tx} minus the path-loss A.

$$\mathrm{Pr}_{bit} = 2.35.10^{-30}.e^{-0.659.P_{Rx}} \tag{1}$$

$$P_{Rx} = P_{Tx} - A \tag{2}$$

4 Radio Activation Policy, Link Adaptation, and Average Power Consumption

4.1 Transmission Procedure and Radio Activation

The data of Figs. 3 and 4 are sufficient to characterize the performance and energy of the physical layer but they are not sufficient to compute the average power and the transmission reliability of a node in network conditions. Indeed, as sketched for the uplink in Fig. 5, the medium access control procedures introduce a significant overhead. In the following, we assume that a node will attempt to transmit a single packet per superframe. To do so, it will first listen the beacon, after having preemptively turned on its radio in receive mode. After the beacon is received, the node can enter idle mode. As explained in Section 2, the contention procedure requires at least two channel senses for clear channel assessment (CCA), which requires turning the receiver on. Between the CCAs, the receiver can return to the idle state. The node must stay in idle rather than shutdown because of the 1 ms delay to recover from the shutdown state. Once the channel is assessed clear twice, the transmission can start.

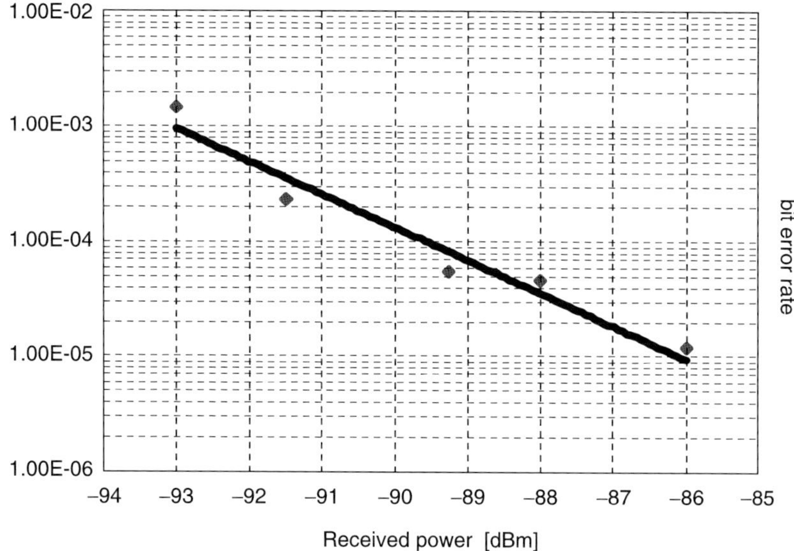

Fig. 4 Bit error probability estimation results

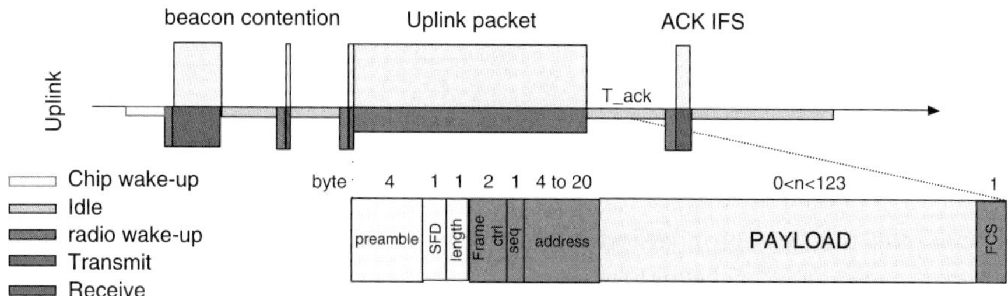

Fig. 5 MAC overheads for the uplink

If the channel coherence time is sufficiently large (larger than a few packet transmissions), the transmit power can be selected as a function of the path loss measured during the reception of the beacon. Since the data rate is fixed, the best link adaptation policy at this level is channel inversion, i.e. keeping the receiver signal to noise ratio constant by compensating for the channel fading by increasing transmit power. In Section 5, we compute the thresholds to switch from one power level to the other.

In addition to the modulated payload data, the transmitted packet consists of a preamble sequence to ease synchronization (corresponding to 4 bytes of data), a frame delimiter (1 byte) and 1 and 8 bytes of PHY and MAC service data, respectively. We assume that short (4 bytes) addresses are used. Let $L_o = 13$ be the total overhead in byte introduced by the PHY and the MAC. The time needed to transmit a packet is given by (3).

$$T_{packet} = (L_0 + L) \times T_B \tag{3}$$

As aforementioned, despite the CSMA/CA procedure, there exists a probability the packet collides with the transmission of another node. Also, the packet can be corrupted by bit errors due to noise. Therefore, a packet acknowledgment mechanism is implemented. If the packet is well received, a short acknowledgement packet is fed back to the transmitter after a minimum time $t^-_{ack} = 192\mu s$. The transmitter waits for such an acknowledgement for maximum $t^+_{ack} = 864\mu s$. If nothing is received, the transmitter repeats the transmission. The node can enter idle mode during the t^-_{ack} period but must be in receiving mode between the end of t^-_{ack} and the reception of the acknowledgement or until the end of t^+_{ack}. Since we assume that the node transmits only one packet per frame, it can shutdown after receiving the acknowledgement.

To compute both the average power consumption of a node and the probability a transaction fails, we still have to characterize the average duration of the contention procedure ($\overline{T}_{cont}$), the average number of CCAs ($\overline{N}_{CCA}$) done during this period, the residual probability of collision (Pr_{col}) and the probability a channel access failure is reported (Pr_{cf}). These quantities depend mainly of the network load (λ)—defined as the aggregate data rate relative to the maximum bandwidth—and the packet duration (T_{packet}). We have characterized those relations empirically by Monte-Carlo simulation of the contention procedure. Results for a network of 100 nodes per channel are depicted in Fig. 6.

4.2 Average Power Consumption

To compute the average node power, we have to determine how long the node occupies each state. To account for the state transition energy, we add the transition delay ($T_{si} = 1\,ms$: transition time between shutdown and idle; $T_{ia} = 194\,\mu s$: transition time between idle and transmit/receive) to the corresponding active time. The expressions of the average time the node is in idle, transmit, and receive modes when following the proposed activation policy are given in (4–6).

$$T_{idle} = T_{si} + Pr_{ef} \times \overline{T}_{cont} + (1 - Pr_{ef}) \times \left[\left(\sum_{i=1}^{N_{max}} P_{tr}(i) \times i + P_{tr}(> N_{max}) \times N_{max} \right) \times (\overline{T}_{cont} + t^+_{ack}) \right] \tag{4}$$

$$T_{Tx} = (1 - Pr_{cf}) \times \left(\sum_{i=1}^{N_{max}} P_{tr}(i) \times i + P_{tr}(> N_{max}) \times N_{max} \right) \times T_{packet} \tag{5}$$

$$T_{Rx} = T_{ia} + T_{beacon} + Pr_{cf} \times \overline{N}_{CCA} \times T_{ia}$$

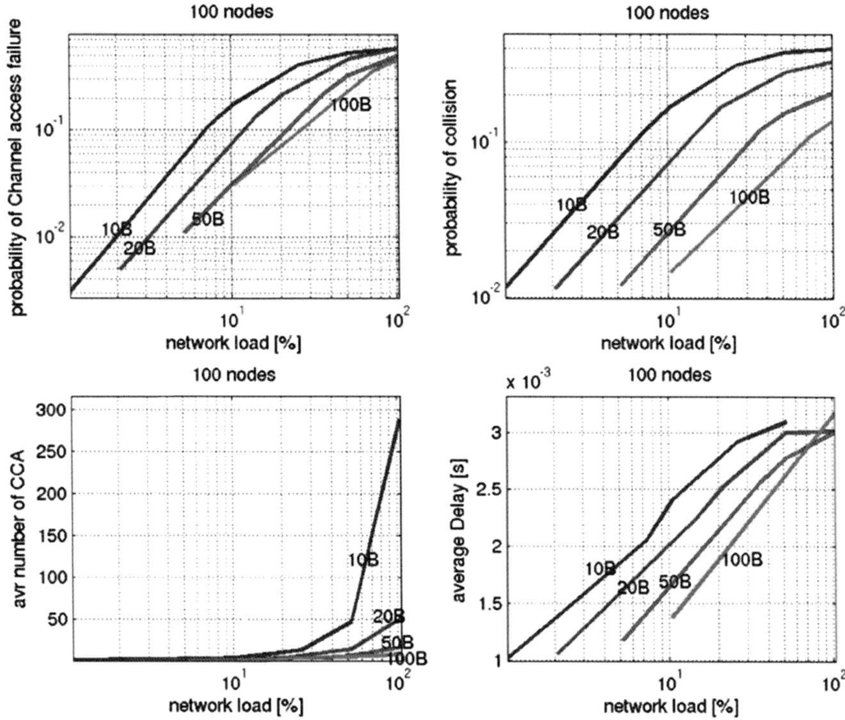

Fig. 6 Behavior of the slotted CSMA/CA algorithm for different packet sizes (10 bytes, 20 bytes, 50 bytes, and 100 bytes)

$$+(1-\text{Pr}_{cf})\left[\left(\sum_{i=1}^{N_{\max}}P_{tr}(i)\times i+P_{tr}(>N_{\max})\times N_{\max}\right)\times \bar{N}_{CCA}\times T_{ia}\\ +\sum_{i=2}^{N_{\max}}P_{tr}(i)\times(i-1)\times t_{ack}^{+}+(1-P_{tr}(>N_{\max}))\times T_{ack}\right] \quad (6)$$

$P_{tr}(i)$ is the probability i transmissions are required to transmit a packet. The maximum number of transmissions, N_{max}, is limited to 5 in our investigation. $P_{tr}(i)$ can be computed by (7, 8) where Pr_{tf} is the probability of transmission failure (9) combining the probability of collision Pr_{col} and the probability of transmission error Pr_{e}, which is computed as a function of the bit error probability (Pr_{bit}) and the total packet size (L_{packet}) minus the synchronization preamble (10).

$$P_{tr}(i) = \text{Pr}_{tf}^{i-1}\times(1-\text{Pr}_{tf}) \quad (7)$$

$$P_{tr}(>N_{\max}) = 1-\sum_{i=1}^{N_{\max}}P_{tr}(i) \quad (8)$$

$$\mathrm{Pr}_{tf} = 1 - (1 - \mathrm{Pr}_{col}) \times (1 - \mathrm{Pr}_e) \tag{9}$$

$$\mathrm{Pr}_e = 1 - (1 - \mathrm{Pr}_{bit})^{(L_{packet}-4)\times 8} \tag{10}$$

To compute the average power, T_{idle}, T_{Rx}, and T_{Tx} must be multiplied by the steady state power in the corresponding mode (P_{idle}, P_{Rx}, and P_{Tx}) and divided by the inter-beacon period (11). The leakage power when the chip is shutdown is neglected. The inter-beacon period (T_{ib}) is computed as a function of the minimum superframe duration ($T_{ib}^{min} = 15.36\,\mathrm{ms}$) and the so-called beacon order (BO), which can be chosen between 0 and 15 (12).

$$P_{avr} = \frac{P_{idle} \times T_{idle} + P_{Tx} \times T_{Tx} + P_{Rx} \times T_{Rx}}{T_{ib}} \tag{11}$$

$$T_{ib} = T_{ib}^{min} \times 2^{BO} \tag{12}$$

4.3 *Probability of Transmission Failure*

In addition to the average power consumption, the total probability of transmission failure can be computed. A transmission failure can be due to a channel access failure, which occurs with a probability Pr_{cf}, or if the packet cannot be transmitted after N_{max} trials (probability: $P_{tr}(>N_{max})$). The transmission failure probability can be computed as:

$$\mathrm{Pr}_{fail} = 1 - (1 - \mathrm{Pr}_{cf}) \times (1 - P_{tr}(> N_{max})) \tag{13}$$

5 Case study

With the model developed in Section 4, it is now possible to study the energy efficiency of the IEEE 802.15.4 standard in the context of dense microsensor networks. We consider a scenario where 1,600 nodes are uniformly distributed in a circular area around a base-station. Since 16 channels are available, 100 nodes are sharing the same channel. We will assume that each node attempts to transmit 1 byte of data every 8 ms, resulting in an effective data rate of 1 kbps per node and 100 kbps per channel.

5.1 *Link Adaptation*

We assume that all the nodes are within communication range of the base-station, i.e. the pass loss between one node and the base-station is such that the received power when 0 dBm are transmitted is above the receiver sensitivity. Since the different nodes experience different path losses, to achieve maximum energy efficiency, they have to adapt their transmit power. To determine the energy-optimal thresholds to switch between transmit power levels, the total energy per transmitted bit is computed for the full range of path loss. We assume that if a transmission fails in a given superframe, the application will retry transmission in the next superframe. The average transmission delay and average energy per bit are hence, given by:

$$delay = T_{ib} \times \frac{1}{1 - \mathrm{Pr}_{fail}} \tag{14}$$

$$energy = \frac{P_{avr} \times delay}{L_{data} \times 8} \tag{15}$$

The results for a packet size of 120 bytes are depicted in Fig. 7. Power level thresholds (circles) correspond to the crossing of energy-path-loss curves for the different transmit power levels. It can be seen that the thresholds are independent of the network load. The transmission is efficient for path losses up to 88 dB. The energy per bit ranges from 135 nJ/bit for a pathloss lower than 55 dB to 220 nJ/bit for a pathloss of 88 dB. Hence, adaptation of the transmit power can save up to 40% of the total energy.

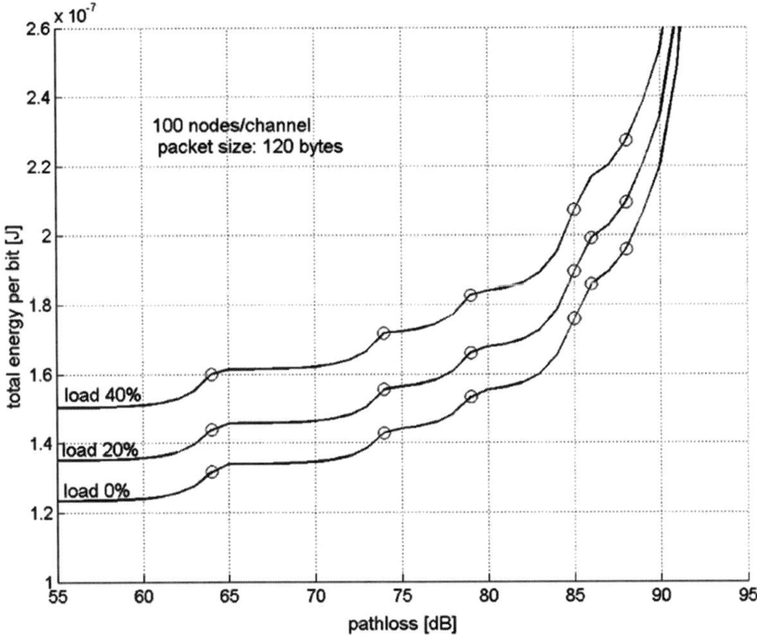

Fig. 7 Optimal energy per bit for different path loss and network loads

5.2 *Packet Size*

When the optimal transmit power is known, it is interesting to determine which packet size leads to the minimum energy per bit. On one hand, small packets require the same MAC overhead as large packets, which increases their energy per useful bit. However, large packets are more subject to transmission error, and hence, require retransmission more often. In addition, when network load is high, large packets will increase the channel access failure probability. Intuitively a tradeoff is expected. However, as depicted in Fig. 8, the energy per bit decreases monotonically up to a packet payload size of 123 bytes, which is the maximum possible in 802.15.4. Reaching the optimum requires a larger packet size.

We hence, choose for this case study a packet size of 120 bytes. Gathered data is buffered until 120 bytes are accumulated. Thus, each node attempts to transmit a packet every 960 ms. We set the beacon order to 6, so that one packet per node is transmitted during each superframe. This corresponds to a load of 42% in each channel. Assuming that the path loss is distributed uniformly between 55 and 95 dB, one can compute using the model presented in section 4 that the average power equals 211 μW with a delivery delay of 1.45 s and a probability of transmission failure of 16%.

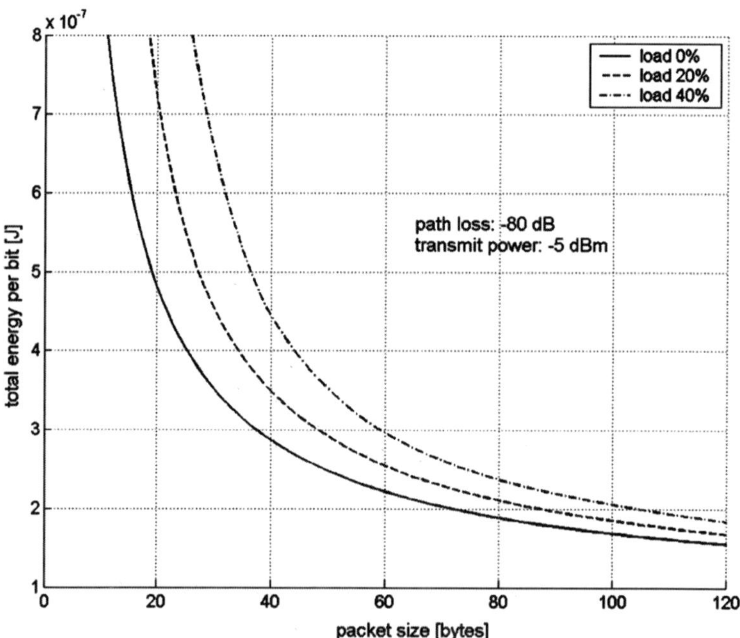

Fig. 8 Impact of the MAC overhead at different network loads

5.3 *Power Breakdown*

Interestingly, the calculated power consumption is close but still over the existing $100\,\mu\text{W}$ constraint of energy scavenging. To lower power consumption in future designs, it is valuable to know the energy breakdown of the node. Figure 9 presents the energy breakdown between the different phases of the protocol in our scenario. We notice that the effective transmission uses less than 50% of the total energy. Twenty-five percent of the energy is spent during contention. This is due to the multiplicative effect of the CSMA/CA mechanism to the transceiver start-up energy. The acknowledgement mechanism uses 15% of the energy, mainly because of the necessity of activating the receiver during the acknowledgment waiting-time. Twenty percent of the energy is spent for listening for the beacon.

Based on the energy breakdown for the transceiver, one can see several key ways to improve the overall energy efficiency of sensor networks. Specific methods include reducing the transition time between states and designing a scalable receiver. Reducing the transition time between states by a factor two would decrease the total average power by 12%. Furthermore, a scalable receiver that offers a low power mode for sensing the channel and waiting for an acknowledgment frame has the potential of reducing the total average power by an additional 15%.

6 Conclusions and Perspectives

The release of the IEEE 802.15.4 standard has been a major milestone in the transition of wireless microsensor networks from the research world to industrial applications. In this paper, we have studied how this standard can be used to support communication in dense, data-gathering networks. An energy-aware radio activation policy has been proposed and the corresponding average power consumption and transmission reliability have been analyzed as a function of important network parameters. The resulting model has been used to optimize the physical and medium access control layers parameters in a dense sensor network scenario. We

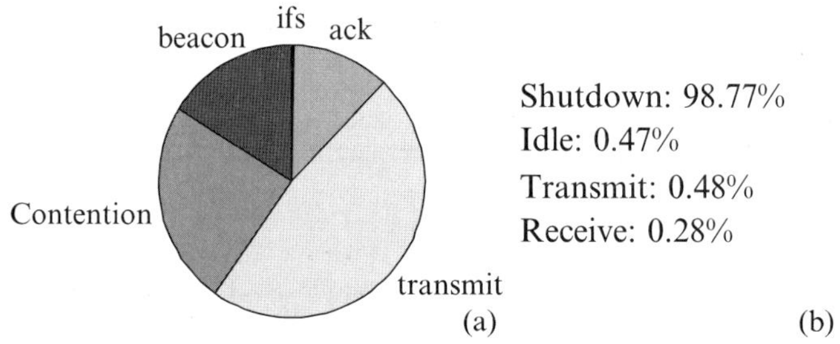

Fig. 9 Breakdown of the energy per bit (a) and time (b) spent in the different phases of the protocol

have analyzed a sensor network of 1,600 nodes transmitting 1 byte every 8 ms and calculated the average power consumption to be 211 µW. To achieve this power figure, buffering is necessary in order to use the largest packet size allowed by the standard. Indeed, the energy per bit decreases monotonically with the packet size up to the maximum allowed size. Allowing larger packets would allow further energy efficiency improvement, at the cost of increased latency. It has been shown that in the considered scenario, less than 50% of the energy is used for actual data transmission. A significant percentage of energy is consumed during the contention procedure (25%) and waiting for an acknowledgment (15%). The overhead of the contention is mainly due to the receiver start-up energy when doing clear channel assessment. The acknowledgement overhead results from the receiver power consumption when waiting for an acknowledgment. Based on the energy breakdown, several ways to improve the overall energy efficiency are proposed. These physical level improvements combined with continued MAC optimizations will allow for energy efficient, self-powered sensor networks.

Acknowledgment This work is the result of the collaboration between the Digital Integrated Circuits and Systems Lab at MIT and the Wireless Research group at IMEC, thanks to special credit from the Belgian National Science Foundation (FWO). Bruno Bougard is Research Assistant at the FWO. The author would like to thank Chipcon Inc. for donating the evaluation boards used in this experiment.

References

[1] R.L. Ashok and D.P. Agrawal, Next-generation wearable networks, *Computer*, Vol. 36, No. 11, Nov. 2003, 31–39.

[2] D. Estrin and R. Govindan, Next century challenges: Scalable coordination in sensor networks, in *Proc. Mobicom*, 1999, pp. 263–270.

[3] A.P. Chandrakasan et al. Design considerations for distributed micro sensor systems, in *Proc. CICC*, 1999, pp. 279–286.

[4] S. Roundy at al. Energy Scavenging for Wireless Sensor Networks, with Special Focus on Vibration, Springer, 2003

[5] R. Min et al. Energy-centric enabling technologies for wireless sensor networks, IEEE Wireless Communications, Vol. 9, No. 4, 28–39, Aug. 2002.

[6] J.M. Rabaey et al. PicoRadio supports ad hoc ultra low power wireless networking, *IEEE Computer*, Vol. 33, 42–48, 2000

[7] http://standards.ieee.org/getieee802/download/802.15.4-2003.pdf

[8] G. Lu et al. Performance Evaluation of the IEEE 802.15.4 MAC for Low-Rate Low-Power Wireless Networks, in *Proc. EWCN'04*, April 2004

[9] http://www.chipcon.com/files/CC2420_Data_Sheet_1_2.pdf

[10] J. Zheng and M.J. Lee, Will IEEE 802.15.4 Make Ubiquitous Networking a Reality? IEEE *Communication Mag.*, Jun. 2004

[11] Rex Min et al. A Frame work for Energy-Scalable Communication in High-Density Wireless Networks, in *Proc. ACM ISLPED*, pp. 36–41, Monterey, Aug. 2002

[12] E. Shih et al. Design Consideration for Energy-Efficient Radios in Wireless Microsensor Networks, in *Journal of VLSI Signal Processing*, Vol. 37, 77–94, 2004

[13] S. Thoen et al. Channel Time-Variance for Fixed Wireless Communications: Modeling and Impact, in *Proc. IASTED Wireless and Optical Communications Conference*, Banff, June 2002

Statistical Blockade: A Novel Method for Very Fast Monte Carlo Simulation of Rare Circuit Events, and its Application

Amith Singhee and Rob A. Rutenbar

Abstract Circuit reliability under statistical process variation is an area of growing concern. For highly replicated circuits such as SRAMs and flip-flops, a rare statistical event for one circuit may induce a not-so-rare system failure. Existing techniques perform poorly when tasked to generate both efficient sampling and sound statistics for these rare events. *Statistical Blockade* is a novel Monte Carlo technique that allows us to efficiently filter—to *block*—unwanted samples insufficiently rare in the tail distributions we seek. The method synthesizes ideas from data mining and Extreme Value Theory, and shows speed-ups of $\times 10$–$\times 100$ over standard Monte Carlo.

1 Introduction

Circuit reliability under statistical process variation is an area of growing concern. Designs that add excess safety margin, or rely on simplistic assumptions about "worst case" corners no longer suffice. Worse, for critical circuits such as SRAMs and flip flops, replicated across 10K–10 M instances on a large design, we have the new problem that statistically rare events are magnified by the sheer number of these elements. In such scenarios, an exceedingly rare event for one circuit may induce a not-so-rare failure for the entire system.

Monte Carlo analysis remains the gold standard for the required statistical modeling. Standard Monte Carlo techniques are, by construction, most efficient at sampling the statistically likely cases. Indeed, classical modifications such as *Importance Sampling* [1] allow Monte Carlo methods to avoid sampling these unlikely (i.e. unimportant) events. Ours is the mirror image problem: how can we efficiently sample *only* the statistically rare events? How can we *model* the statistics in the tails of these heavy-tailed distributions? Importance Sampling also gives us some help to sample in the tails [2], but changes the statistics of these rare samples. Unfortunately, we need both samples and rigorous statistics to determine the reliability

Department of ECE, Carnegie Mellon University, Pittsburgh, Pennsylvania, 15213 USA, {asinghee,rutenbar}@ece.cmu.edu

R. Lauwereins and J. Madsen (eds.), *Design, Automation, and Test in Europe–The Most Influential Papers of 10 Years DATE*, 235–251
© Springer 2008

of critical circuits like large SRAMs, or flips-flop in aggressively clocked designs operating with small setup slack. Standard Monte Carlo methods are poorly suited to this important problem.

One avenue of attack is to abandon Monte Carlo. Several analytical and semi-analytical approaches have been suggested to model the behavior of SRAM cells [3–5] and digital circuits [6] in the presence of process variations. All suffer from approximations necessary to make the problem tractable. [4] and [6] assume a linear relationship between the statistical variables and the performance metrics (e.g. static noise margin), and assume that the statistical process parameters and resulting performance metrics are normally distributed. This can result in gross errors, especially while modeling rare events, as we shall show later. When the distribution varies significantly from Gaussian, [4] chooses an F-distribution in an ad hoc manner. Bhavnagarwala et al. [3] presents a complex analytical model limited to a specific transistor model (the transregional model) and further limited to only static noise margin analysis for the 6T SRAM cell. Calhoun and Chandrakasan [5] again models only the static noise margin (SNM) for SRAM cells under assumptions of independence and identical distribution of the upper and lower SNM, which may not always be valid.

A different avenue of attack is to modify the Monte Carlo strategy. Hocevar et al. [2] show how Importance Sampling can be used to predict failure probabilities. Recently, Kanj et al. [7] applied an efficient formulation of these ideas for modeling rare failure events for single 6T SRAM cells, based on the concept of *Mixture Importance Sampling* from Hesterberg [8]. The approach uses real SPICE simulations with no approximating equations. However, the method only estimates the exceedence probability of a *single* value of the performance metric. A rerun is needed to obtain probability estimates for another value. No complete model of the tail of the distribution is computed. The method also combines all performance metrics to compute a failure probability, given *fixed* thresholds. Hence, there is no way to obtain separate probability estimates for each metric, other than a separate run per metric. Furthermore, given that Hocevar et al. [2] advice against importance sampling in high dimensions, it is unclear if this approach will scale efficiently to large circuits with many statistical parameters.

In this paper, we present a novel, general and efficient Monte Carlo method that addresses both problems previously described: very fast generation of samples—rare events—with sound models of the tail statistics for any performance metric. The method imposes almost no *a priori* limitations on the form of the statistics for the process parameters, device models, or performance metrics. The method is conceptually simple, and it exploits ideas from two rather nontraditional sources.

To obtain both samples and statistics for rare events, we may need to generate and evaluate an intractable number of Monte Carlo samples. *Generating* each sample is neither challenging nor expensive: we are merely creating the parameters for a circuit. *Evaluating* the sample is expensive, because we simulate it. What if we could quickly *filter* these samples, and *block* those that are unlikely to fall in the low-probability tails we seek? Many samples could be generated, but very few simulated. We show how to exploit ideas from *data mining* [9] to build *classifier* structures, from a small set of Monte Carlo training samples, to create the necessary blocking filter. Given these samples, we show how to use the rigorous mathematics of *Extreme Value Theory* (EVT [10], the theory of the limiting behavior of sampled maxima and minima) to build sound models of these tail distributions. The essential "blocking" activity of the filter gives the technique its name: *Statistical Blockade*.

Statistical blockade has been tested on both SRAM and flip-flop designs, including a complete 64-cell SRAM column (a 403-parameter problem), accounting for both local and global variations. (In contrast to several prior studies [5–6, 9] we shall see that simulating only one cell does not correctly estimate the critical tail statistics.) However, statistical blockade allows us to generate both samples and accurate statistics, with speedups of ×10–×100 over standard Monte Carlo.

This paper is organized as follows. Section 2 reviews the mathematics for modeling rare event tail distributions, derived from EVT. Section 3 develops our tail distribution fitting strategy, based on probability-weighted moments, and our method for probability prediction, once the model has been built. Section 4 develops the core of the statistical blockade method: an efficient tail sampling strategy, using a classifier-based blocking filter. Section 5 presents experimental results. Section 6 offers concluding remarks.

2   Extreme Value Theory

EVT provides us with mathematical tools to build models of the tails of distributions. It has been used extensively in climatology and risk management, among other applications: wherever the probability of extreme and rare events needs to be modeled. Here we introduce the mathematical concepts from EVT that our approach relies on.

Suppose we define a threshold t for some random variable (e.g. the SNM of an SRAM cell) with cumulative distribution (CDF) $F(x)$: All values above t constitute the *tail* of the distribution. Throughout this paper, we are considering only the upper tail: this is without loss of generality, since a simple sign change converts a lower tail to the upper tail. Now define the conditional CDF of excesses above t as

$$F_t(x) = P\{X - t \geq x \mid X \geq t\} = \frac{F(x+t) - F(t)}{1 - F(t)} \text{ for } x \geq 0 \tag{1}$$

An important distribution in the theory of extreme values is the *Generalized Pareto Distribution* (GPD), which has the following CDF:

$$G_{a,k}(x) = \begin{cases} 1 - (1 + kx/a)^{1/k}, & k \neq 0 \\ 1 - e^{-x/a}, & k = 0 \end{cases} \tag{2}$$

The seminal result we exploit is from Balkema and de Haan [11] and Pickands [12] (referred to as *BdP*) who proved that

$$\lim_{t\to\infty} \sup_{x\geq 0} \left| F_t(x) - G_{a,k}(x) \right| = 0 \tag{3}$$

if and only if F is in the *maximum domain of attraction* (MDA) of the *Generalized Extreme Value* distribution (GEV): $F \in MDA(H_\eta)$. This means that when the distribution F satisfies the given condition ($F \in MDA(H_\eta)$), the conditional CDF of F tends, as we move the threshold farther and farther out on the tail, towards a particularly tractable analytical form. Let us look at the condition in more detail.

The GEV CDF is as follows

$$H_\eta(x) = \begin{cases} e^{-(1+\eta x)^{-1/\eta}}, & \eta \neq 0 \\ & \quad\quad \text{where } 1+\eta x>0 \\ e^{-e^{-x}}, & \eta = 0 \end{cases} \tag{4}$$

It combines three simpler distributions into one unified form:

- for $\eta = 0$ we get the *Gumbel-type* (or Type I) distribution

$$\Lambda(x) = e^{-e^{-x}} \tag{5}$$

- for $\eta > 0$, we get the *Fréchet-type* (or Type II) distribution

$$\Theta_\alpha(x) = e^{-x^{-\alpha}} \text{ for } x > 0 \tag{6}$$

- for $\eta < 0$, we get the *Weibull-type* (or Type III) distribution

$$\Psi_\alpha(x) - e^{-|x|^\alpha} \text{ for } x < 0 \tag{7}$$

Let us now look at what the "maximum domain of attraction" means. Consider the maxima (M_n) of n i.i.d. random variables.

Suppose there exist normalizing constants a_n and b_n, such that

$$P\{(M_n - b_n)/ a_n \leq x\} = F^n(a_n x+b_n) \to H(x) \text{ as } n\to\infty \tag{8}$$

for some non-degenerate $H(x)$. Then we say that F is "in the maximum domain of attraction" of H. In other words, the maxima of n i.i.d. random variables with CDF F, when properly normalized, converge in distribution to a random variable with the distribution H. Fisher and Tippett [13] and Gnedenko [14] showed that for a large class of distributions,

$$F \in MDA(H) \Rightarrow H \text{ is of type } H_\eta \tag{9}$$

For example McNeil [10], MDA(Λ) includes the normal, exponential, gamma and lognormal distributions; MDA(Θ_α) includes the Pareto, Burr, log-gamma, Cauchy and t-distributions; MDA(Ψ_α) includes finite-tailed distributions like the uniform and beta distributions. Hence, for a large class of distributions, the *BdP* theorem holds true. In other words, if we can generate enough points in the tail of a distribution ($x \geq t$), in most cases, we can fit a GPD to the data and make predictions further out in the tail.

This is a remarkably practical and useful result for the rare circuit event scenarios we seek to model. In particular, it shows that most prior ad hoc fitting strategies are at best suboptimal, and at worst, simply wrong. Let us next consider how to use these results.

3   Model Fitting and Prediction

Assuming we can generate points in the tail, there remains the problem of fitting a GPD form to the conditional CDF. Several options are available here [15]: moment matching, maximum likelihood estimation (MLE) and probability weighted moments (PWM) [16]. We have chosen PWM because it seems to have lower bias [15] and does not have the convergence problems of MLE.

The PWMs of a continuous random variable x with CDF F are the quantities

$$M_{p,r,s} = E[x^p \{F(x)\}^r \{1 - F(x)\}^s] \tag{10}$$

which often have simpler relationships with the distribution parameters than conventional moments $M_{p,0,0}$. For the GPD it is convenient to use these particular PWMs

$$\alpha_s = M_{1,0,s} = E[x\{1 - F(x)\}^s] = \frac{a}{(s+1)(s+1+k)} \tag{11}$$

which exist for $k > -1$: this is true for most cases of interest [15]. The GPD parameters are then given by

$$a = \frac{1\alpha_0\alpha_1}{\alpha_1 - 2\alpha_1}, k = \frac{\alpha_0}{\alpha_0 - 2\alpha_1} - 2 \tag{12}$$

where the PWMs are estimated from the samples as

$$\tilde{\alpha}_i = n^{-1} \sum_{j=1}^{n} (1 - p_{j|n})^i x_{j|n} \tag{13}$$

where $x_{1|n} \leq \ldots \leq x_{n|n}$ are the ordered samples and $p_{j|n} = (j+\gamma)/(n+\delta) \cdot \gamma = -0.35$ and $\delta = 0$ are as suggested in [15].

Given the ability to fit the GPD form, now let us consider the problem of predicting useful probabilities. Once we have a GPD model of the conditional CDF above a threshold t, we can predict the exceedence probability—the *failure* probability—for any value x_f:

$$P(X > x_f) = [1 - P(X \le t)][1 - F_t(x_f - t)] \tag{14}$$

Here, $P(X \le t)$ can be computed using empirical data obtained from standard Monte Carlo, or more sophisticated variance reduction techniques, for example, mixture importance sampling [7]. $F_t(x_f - t)$ is just the prediction by the GPD model. Hence, we can write (14) as

$$P(X > x_f) = [1 - F(t)][1 - G_{a,k}(x_f - t)] \tag{15}$$

4   Statistical Blockade: Classification-based Sampling

Even with all the useful theory presented above, we still need a way to efficiently generate samples in the tail of the distribution of the performance metric of a circuit. Standard Monte Carlo (MC) is very unsuited to this job, because it generates samples that follow the complete distribution. The problem is severe for rare event statistics: if our threshold t is the 99% point of the distribution, only one out of 100 simulations will be useful for building the tail model.

Our approach is to build a so-called *classifier* to filter out candidate MC points that will not generate a performance value in the tail. Then, we simulate only those MC points that will generate points in the tail. For clarity, we shall refer to this structure as the *blockade filter*, and its action as *blockade filtering*. We borrow ideas from the data mining community [9] to build the filter. A *classifier* is an indicator function that allows us to determine set membership for complex, high-dimensional, nonlinear data. Given a data point, the classifier reports true or false on the membership of this point in some arbitrary set. For statistical blockade, this is the set of parameter values *not* in the extremes of the distributional tail we seek. The classifier is built from a relatively small set of representative sample data, and as we shall see, need not be perfectly accurate to be effective.

Let us look at this filter, and its construction. Suppose the statistical parameters (V_t, t_{ox}, etc.) in a circuit are denoted by s_i, and the performance metric being measured is y. Our sampling strategy tries to simulate only those points $\{s_i\}$, that result in values of $y \ge t$. This is accomplished in three steps (shown in Fig. 1):

1. *Perform initial sampling* to generate data to build a classifier. This initial sampling is also used for estimating $F(t)$ in (15), and could be standard Monte Carlo or importance sampling.
2. *Build a classifier* using a classification threshold t_c. To minimize false negatives (tail points classified as non-tail points), choose $t_c < t$.
3. *Generate more samples* using MC, following the CDF F, but simulate only those that are classified as tail points.

Using the tail points generated by the blockade-filtered sampling, we can then build a conditional CDF model for the tail, using the tools of Sections 2 and 3. As long as the number of false negatives is acceptably low, the simulated tail points are

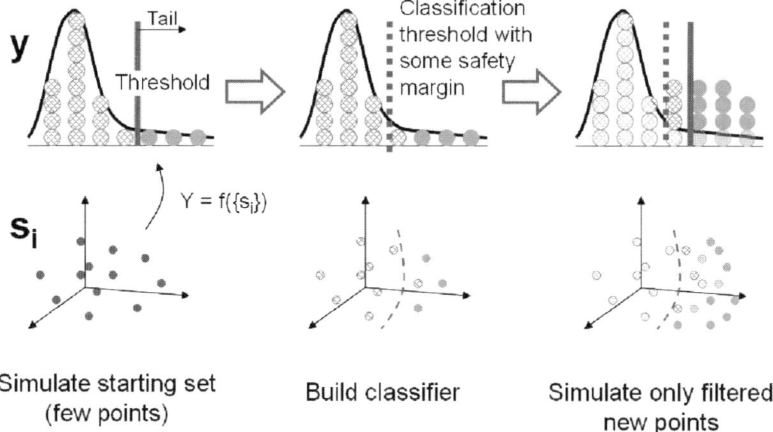

Fig. 1 Classification-based sampling

true to the actual distribution. Hence, there is no need to unbias the estimates. Note that the approach is reminiscent of acceptance–rejection sampling [1].

In this work, the classifier used is a *Support Vector Machine* (SVM, [17]). The time for model building and classification is negligible compared to the total simulation time. Apart from this practical consideration, there is no restriction on the type of classifier that can be used. Classification is a rich and active field of research in the data mining community and there are many options for choosing a classifier [9]. SVMs are a popular, well researched classifier strategy, and optimized implementations are readily available [17].

5 Experimental Results

We now apply the statistical blockade method to three testcases: a single 90 nm SRAM cell, a 45 nm master-slave flip-flop and a full 64-bit 90 nm SRAM column. The initial sampling to construct each blockade filter was a standard MC run of 1000 points. An SVM classifier was built using the 97% point (of each relevant performance metric) as the classification threshold t_c. The tail threshold t was defined as the 99% point.

One technical point to note about the SVM construction: since the sample set is biased with many more points in the body of the distribution than in the tail, we need to unbias the classification error [18]. Suppose that, of the 1000 simulated training points, $T \ll 1000$ actually fall into the tail we seek. Since the two classification sets (true/false) have an unbalanced number of points, the SVM classifier will be biased toward the body (1000-T points). Even if all T of the tail points are misclassified, the error rate is quite low if the body is classified correctly. Hence, classification error in the tail is penalized more—by a weighting factor of roughly T—than errors in the body, to try to avoid missing tail points. We use a weight value of 30 for these results.

5.1   Single 6-T SRAM Cell

The first testcase is shown in Fig. 2: a 6-T SRAM cell, with bit-lines connected to a column multiplexor and a non-restoring write driver. The metric being measured is the write time τ_w: the time between the wordline going high to the non-driven cell node (node 2) transitioning. Here, "going high" and "transitioning" imply crossing 50% of the full voltage change. The device models used are from the Cadence 90 nm Generic PDK library. There are 9 statistical parameters: 8 V_t variations to model random dopant fluctuation (RDF, [19]) effects in the transistors named in the figure, and 1 global gate-oxide variation. All variations are assumed to be normally distributed about the nominal value. The V_t standard deviation is

$$\sigma(V_t) = \frac{5mV}{\sqrt{WL}} \text{ where W,L are in } \mu m \tag{16}$$

This variation is too large for the 90nm process, but is in the expected range for more scaled technologies; this creates a good stress test for the method. The gate-oxide standard deviation is taken as 2%.

   100,000 MC points were blockade-filtered through the classifier, generating 4,379 tail candidates. After simulating these 4,379 points, 978 "true" tail points were obtained. The tail model obtained from these points is compared with the

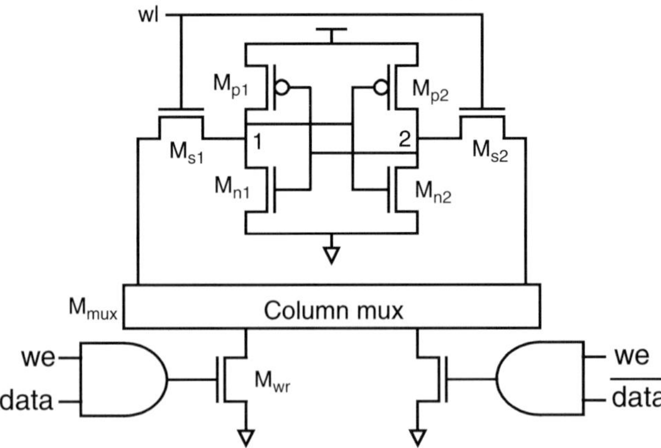

Fig. 2 6-T SRAM cell with column mux and write drivers. V_t variation on named devices and global t_{ox} variation

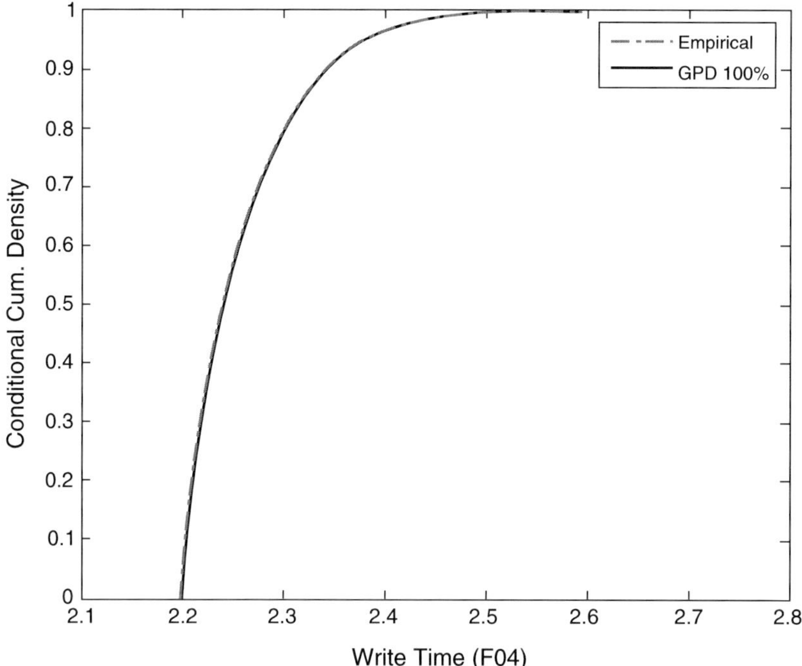

Fig. 3 Comparison of tail model CDF (5379 simulations) with empirical tail CDF (1 million simulations)

empirical tail conditional CDF obtained after simulating 1 million MC points, in Fig. 3. Table 1 shows a comparison of the failure probability predictions for different values of τ_w, expressed as equivalent sigma points:

$$x_\sigma = \Phi^{-1}(1 - p(X > x_f)) \tag{17}$$

where Φ is the standard normal CDF. This is the equivalent point on a standard normal that would have the same cumulative probability. For example, $x_\sigma = 3$ implies a cumulative probability of 0.99865 and a failure probability of 0.00135. The delays are expressed as multiples of the fanout-of-four (FO4) delay of the process. The table also shows x_σ predictions from an accurate tail model built using the 1 million MC points, without any filtering. The empirical prediction fails beyond 2.7 FO4 because there are simply *no* points generated by the MC run so far out in the tail (beyond 4.8σ).

The table shows two important advantages of our approach:

- Even without any filtering, the GPD tail model is better than Monte Carlo, since it can be used to predict probabilities far out in the tail, even when there are no points that far out.

Table 1 Comparison of predictions by Monte Carlo, Monte Carlo with tail modeling and statistical blockade filtering, for single SRAM cell. The number of simulations includes the 1000 training samples

τ_w	Standard MC (1M sims)	GPD No Blockade Filter (1M sims)	GPD With Blockade Filter (5,379 sims)
2.4	3.404	3.408	3.379
2.5	3.886	3.886	3.868
2.6	4.526	4.354	4.352
2.7	∞	4.821	4.845
2.8	∞	5.297	5.356
2.9	∞	5.789	5.899
3.0	∞	6.310	6.493

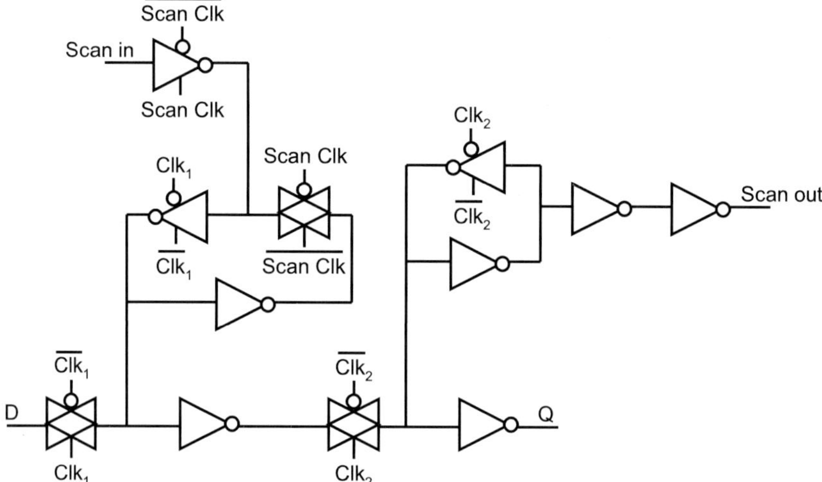

Fig. 4 Master-Slave Flip-Flop with Scan Chain

- Using blockade filtering, coupled with the tail model, we can drastically reduce the number of simulations (from 1 million to 5,379) and still generate a reliable tail model.

5.2 *Master-Slave Flip-Flop with Scan Chain*

A large chip can have tens of thousands of instances of the same flip-flop. Typically, these flip-flops are in a scan chain to enable rigorous testing. Random threshold variation in the scan chain transistors can also impact the performance of the flip-flop. Hence, our next testcase (Fig. 4) is a commonly seen Master-Slave Flip-Flop with scan chain (MSFF).

The design has been implemented using the 45 nm CMOS Predictive Technology Models from [20]. Variations considered include RDF for all transistors in the circuit and one global gate-oxide variation. Threshold variation is modeled as normally distributed V_t variation:

$$\sigma(V_t) = 0.0135 \frac{V_{t0}}{\sqrt{WL}} \text{ where W,L are in } \mu\text{m} \tag{18}$$

V_{t0} is the nominal threshold voltage. This results in 30% standard deviation for a minimum-sized transistor. The t_{ox} standard deviation is taken as 2%. The metric being measured is the clock-output delay, τ_{cq} in terms of the FO4 delay. A GPD model was built using 692 true tail points, obtained from 7,785 candidates blockade filtered from 100,000 MC samples. Figure 5 compares this model with (1) the empirical CDF from 500,000 standard MC simulations, and (2) a GPD model built from after blockade filtering these 500,000 points. The discrepancy of the models can be explained by looking at the empirical PDF of the delay in Fig. 6. Due to the heavy tail, slight variations in the tail samples chosen can cause large variations in the model. Our method is still able to generate an acceptably accurate model, as is

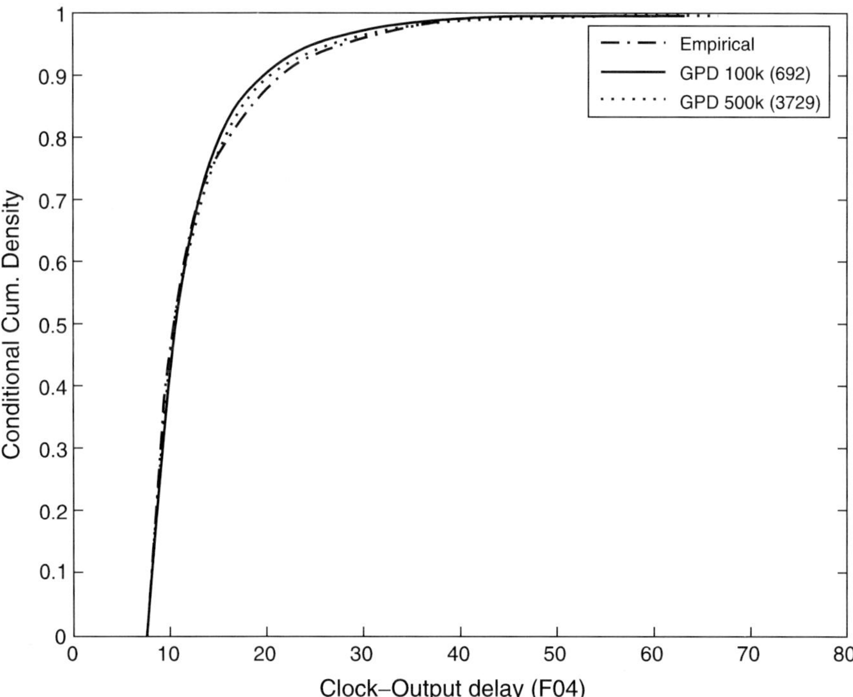

Fig. 5 Tail model for MSFF (1692 and 4729 simulations) compared with empirical model (500,000 simulations)

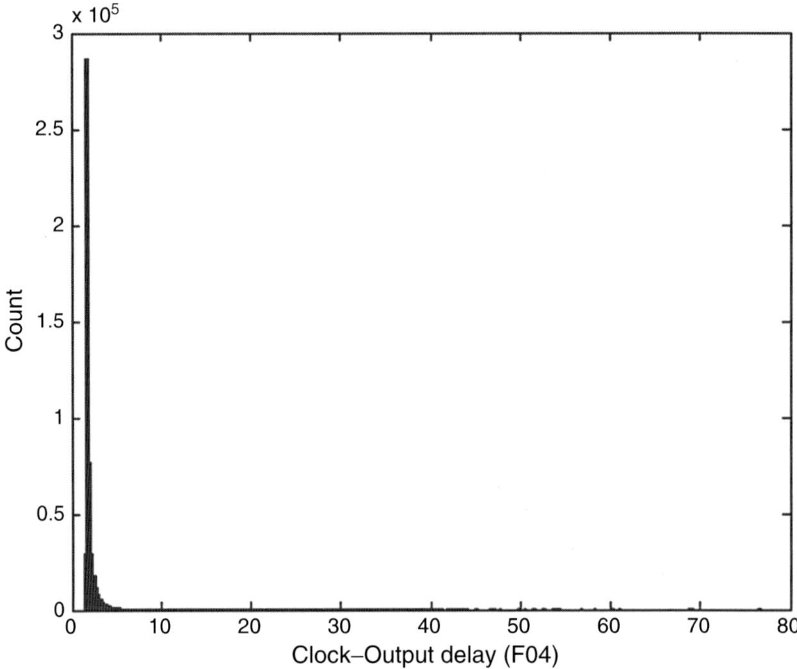

Fig. 6 Probability density plot for Clock-Output delay of the MSFF, showing a long, heavy tail

evident by the comparison of x_σ in Table 2. Standard MC starts under-estimating the failure probability (overestimating x_σ) far out in the tail (from row 3 on). The tail model has much better predictive power (column 2): $x_\sigma = 4.283$ implies a failure probability of 9.2 ppm. Even with blockade filtering, the tail model is still quite accurate. The table also shows the estimates from a standard Gaussian distribution fit to 20,000 MC points: it is obvious that such a simplifying assumption *severely* underestimates the failure probability.

5.3 64-bit SRAM Column

The next testcase is a 64-bit SRAM column, with nonrestoring write driver and column multiplexor (Fig. 7). Only one cell is being accessed, while all the other wordlines are turned off. Random threshold variation on all 402 devices (including the write driver and column mux) are considered, along with a global gate-oxide variation. The device and variation models are the same 90 nm technology as the single SRAM cell (Section 5.1). In scaled technologies, leakage is no longer negligible. Hence, process variations on devices that are meant to be inaccessible can also

Table 2 Comparison of predictions by MC, MC with GPD modeling, blockade filtered GPD modeling, and standard Gaussian approximation, for MSFF. The number of simulations includes the 1000 training samples

τ_{cq}	Standard MC (500K sims)	GPD No Blockade Filter (500K sims)	GPD With Blockade Filter (8,785 sims)	Gaussian Tail Approximately (20K sims)
30	3.424	3.466	3.431	22.127
40	3.724	3.686	3.661	30.05
50	4.008	3.854	3.837	37.974
60	4.219	3.990	3.978	45.898
70	4.607	4.102	4.095	53.821
80	∞	4.199	4.195	61.745
90	∞	4.283	4.282	69.669

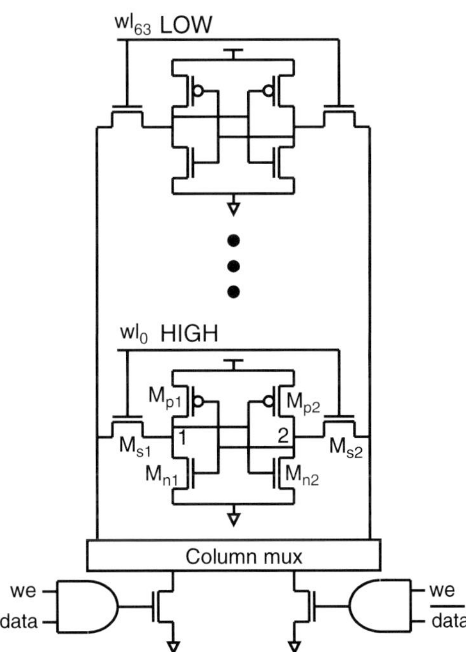

Fig. 7 64-bit SRAM column with column mux and write drivers. V_t variation on all devices and global t_{ox} variation

impact the overall behavior of a circuit. This testcase allows us to see the impact of leakage through the 63 off cells, along with variations in the write driver.

The metric measured is the write time (τ_w), from wl$_0$ to node 2. The number of statistical parameters is 403 in this case. Building a classifier with only 1000 points in 403-dimensional space is nearly impossible. Hence, the dimensionality is reduced by choosing only those parameters that significantly affect the output. We employ standard statistical sensitivity techniques. We measure this significance with Spearman's Rank Correlation Coefficient [21], r_s. Suppose R_i and S_i are the ranks of corresponding values of two variables in a dataset, then their rank correlation is given as:

$$r_S = \frac{\sum_i (R_i - \bar{R})(S_i - \bar{S})}{\sqrt{\sum_i (R_i - \bar{R})^2}\sqrt{\sum_i (S_i - \bar{S})^2}} \tag{19}$$

This measure of correlation is more robust than a linear Pearson's correlation, in the presence of non-linear relationships in the data. Figure 8 shows the sorted magnitudes of the 403 rank correlation values, computed between the statistical parameters and the output τ_w. For classification, only the parameters with $|r_s| > 0.1$ were chosen. This reduced the dimensionality to only 11: the devices chosen by this method were the pull-down and output devices in the active AND gate, the column mux device, the bitline pull-down devices and all devices in the 6-T cell, except for M_{p2} (since node 2 is being pulled down in this case). This selection coincides with a designer's intuition of the devices that would have the most impact on the write time in this testcase.

The empirical CDF from 100,000 MC samples is compared with the tail model obtained by blockade filtering 20,000 MC samples (218 true tail points from 1046 filtered candidates) in Fig. 9. Also shown is the tail model obtained by blockade filtering the 100,000 MC samples. Table 3 compares the following: the x_σ predictions

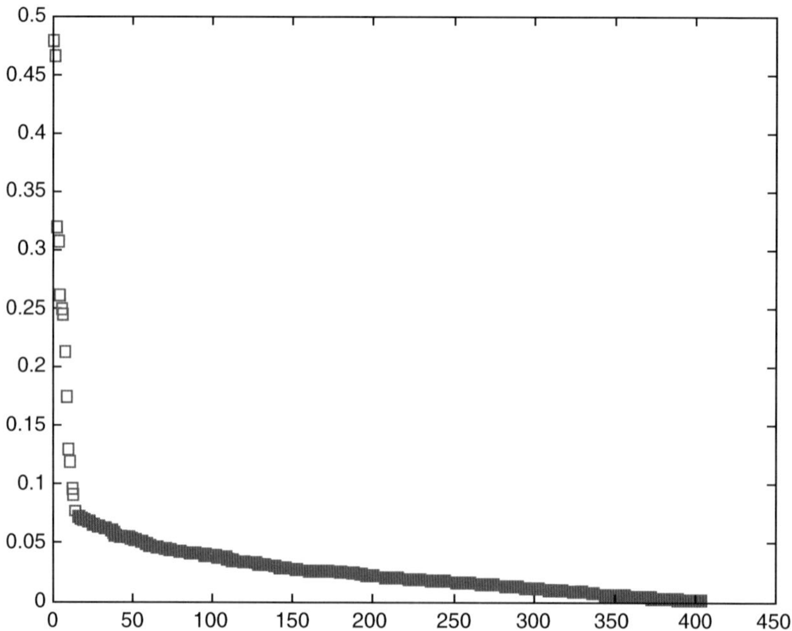

Fig. 8 Absolute values of rank correlation between the statistical parameters and write time of the SRAM column

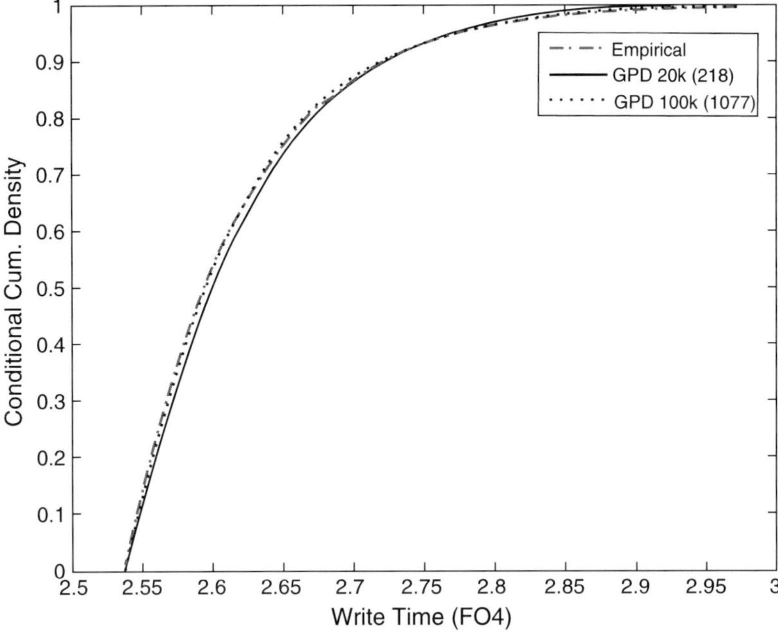

Fig. 9 Tail model for SRAM column (2046 and 6314 simulations) compared with empirical model (100,000 simulations)

Table 3 Comparison of predictions by MC, MC with tail modeling, blockade filtered tail modeling and Gaussian approximation, for SRAM column. The number of simulations includes the 1,000 training samples

τ_w	Standard MC (100K sims)	GPD No Blockade Filter (100K sims)	GPD w/ Blockade Filter (20K pts 2,046 sims)	GPD w/ Blockade Filter (100K pts 6,314 sims)	Standard Gaussian Approximately (20K sims)
2.7	2.966	2.986	2.990	3.010	3.364
2.8	3.367	3.373	3.425	3.390	3.898
2.9	3.808	3.743	3.900	3.747	4.432
3.0	∞	4.101	4.448	4.088	4.966
3.1	∞	4.452	5.138	4.416	5.499
3.2	∞	4.799	6.180	4.736	6.033
3.3	∞	5.147	–	5.049	6.567
3.4	∞	5.496	–	5.357	7.100

from standard MC; a GPD tail model with no filtering; two *different* GPD tail models with filtering of 20,000 and 100,000 points, respectively; and a standard Gaussian fit to 20,000 points. We can see that the 218 true tail points obtained by blockade filtering only 20,000 MC samples is not enough to build a reliable tail model. However, we get much better results using the 1,077 true tail points obtained by blockade filtering 100,000 MC samples (5314 simulations). The Gaussian again underestimates the failure probability.

Comparing with Table 1, we can see that simulating variations in a single cell, without modeling variation in the environment circuitry (other cells in the column and the write driver itself), can lead to large underestimation of the delay spread: 3.0 FO4 delay is estimated as a 6.3σ point (Table 1), while it is actually a 4.1σ point (Table 3).

Before concluding, we mention two points. First, across all three test cases, we see significant improvements in accuracy over simple Gaussian fits, and similar improvements in fitting if we use our GPD model and simple MC sampling. However, we also see significant speedups over simple Monte Carlo, ranging from roughly one to two orders of magnitude.

Finally, we mention an obvious extension to these ideas. The test cases shown herein all measure a single performance metric. Our method is, however, flexible enough to accommodate multiple metrics: multiple classifiers can be trained from the same training set, one for each metric. Each classifier would then identify potential tail points for its corresponding metric, which can be simulated and used to build a tail model for every metric. In the worst case, the tail samples of two metrics might be mutually exclusive, resulting in approximately twice the number of simulations as compared to the case of a single metric. In the best case, the tail samples of the metrics would overlap and there would not be any significant increase in the number of simulations.

6 Conclusions

Statistical blockade is a novel, efficient and flexible framework for (1) generating samples in the tails of distributions of circuit performance metrics, and (2) deriving sound statistical models of these tails. This enables us to make predictions of failure probabilities given thresholds far out in the tails. This capability has become critical for reliable and efficient design of high-replication circuits, such as SRAMs, as transistor sizes move deeply into the nanometer regime. Our methods offer both significantly higher accuracy than standard Monte Carlo, and speedups of one to two orders of magnitude across a range of realistic circuit testcases and variations.

Acknowledgments This work was supported by the MARCO/DARPA Focus Research Center for Circuit and System Solutions (C2S2) and the Semiconductor Research Corp.

References

[1] G.S. Fishman, A First Course in Monte Carlo, Duxbury Press, Oct. 2005.

[2] D.E. Hocevar, M.R. Lightner, T.N. Trick, A Study of Variance Reduction Techniques for Estimating Circuit Yields, *IEEE Trans. CAD*, 2(3), July, 1983.

[3] A.J. Bhavnagarwala, X. Tang, J.D. Meindl, The Impact of Intrinsic Device Fluctuations on CMOS SRAM Cell Stability, *J. Solid State Circ*, 26(4), 658–665, Apr. 2001.

[4] S. Mukhopadhyay, H. Mahmoodi, K. Roy, Statistical Design and Optimization of SRAM Cell for Yield Enhancement, *Proc. ICCAD*, 2004.

[5] B.H. Calhoun, A. Chandrakasan, Analyzing Static Noise Margin for Sub-threshold SRAM in 65nm CMOS, *Proc. ESSCIRC*, 2005.

[6] H. Mahmoodi, S. Mukhopadhyay, K. Roy, Estimation of Delay Variations due to Random-Dopant Fluctuations in Nanoscale CMOS Circuits, *J. Solid State Circuits*, 40(3), 1787–1796, Sep. 2005.

[7] R. Kanj, R. Joshi, S. Nassif, Mixture Importance Sampling and its Application to the Analysis of SRAM Designs in the Presence of Rare Failure Events, *Proc. DAC*, 2006.

[8] T.C. Hesterberg, Advances in Importance Sampling, PhD Dissertation, Dept. of Statistics, Stanford University, 1988, 2003.

[9] T. Hastie, R. Tibshirani, J. Friedman, The Elements of Statistical Learning, Springer, 2003.

[10] A.J. McNeil, Estimating the Tails of Loss Severity Distributions using Extreme Value Theory, *ASTIN Bull.* 27(1), 117–137, 1997.

[11] A. Balkema, L. de Haan, Residual Life Time at Great Age, *Ann. Probab*, 2(5), 792–804, 1974

[12] J. Pickands III, Statistical Inference Using Extreme Order Statistics, *Ann. Stat.* 3(1), 119–131, Jan. 1975.

[13] R. Fisher, L. Tippett, Limiting Forms of the Frequency Distribution of the Largest or Smallest Member of a Sample, *Proc. Cambridge Phil. Soc.*, 24, pp 180–190, 1928.

[14] B. Gnedenko, Sur La Distribution Limite du Terme Maximum D'Une Serie Aleatoire, *Ann. Math.* 44(3), Jul. 1943.

[15] J.R.M. Hosking, J.R. Wallis, Parameter and Quantile Estimation for the Generalized Pareto Distribution, *Technometrics*, 29(3), 339–349, Aug. 1987

[16] J.A. Greenwood, J.M. Landwehr, N.C. Matalas, J.R. Wallis, Probability Weighted Moments: Definition and Relation to Parameters of Several Distributions Expressable in Inverse Form, *Water Resour. Res.* 15, 1049–1054, 1979.

[17] T. Joachims, Making Large-Scale SVM Learning Practical, LS8-Report, 24, Universität Dortmund, 1998.

[18] K. Morik, P. Brockhausen, T. Joachims, Combining Statistical Learning with a Knowledge-based Approach – A Case Study in Intensive Care Monitoring, *Proc. 16th Int'l Conf. on Machine Learning*, 1999.

[19] D.J. Frank, Y. Taur, M. Ieong, H.P. Wong, Monte Carlo Modeling of Threshold Variation due to Dopant Fluctuations, *Symp. VLSI Technology*, 1999.

[20] http://www.eas.asu.edu/~ptm/

[21] G.E. Noether, Introduction to Statistics: The Nonparametric Way Springer, 1990.

Compositional Specification of Behavioral Semantics

Kai Chen[1], Janos Sztipanovits[2], and Sandeep Neema[3]

Abstract An emerging common trend in model-based design of embedded software and systems is the adoption of Domain-Specific Modeling Languages (DSMLs). While abstract syntax metamodeling enables the rapid and inexpensive development of DSMLs, the specification of DSML semantics is still a hard problem. In previous work, we have developed methods and tools for the semantic anchoring of DSMLs. Semantic anchoring introduces a set of reusable "semantic units" that provide reference semantics for basic behavioral categories using the Abstract State Machine (ASM) framework. In this paper, we extend the semantic anchoring framework to heterogeneous behaviors by developing a method for the composition of semantic units. Semantic unit composition reduces the required effort from DSML designers and improves the quality of the specification. The proposed method is demonstrated through a case study.

1   Introduction

Model-based design of embedded software uses formal, composable and manipulable models in the design, implementation and system integration process. An emerging common trend in model-based software and systems design is that modeling languages are domain-specific: they offer software/system developers abstractions and notations that are tailored to characteristics of their application domains.

Model analysis and model-based code generation require the precise specification of DSMLs. This is partly achieved by metamodeling languages and metamodels describing the abstract syntax of DSMLs [4]. While abstract syntax metamodeling has been an important step in model-based design and been used in various model-based design frameworks, explicit and formal specification of behavioral semantics has not received much attention. For instance, the UML SPT [1] does not have

[1]Motorola Labs Schaumburg, IL, 60646, USA Kai.Chen@motorola.com

[2]ISIS, Vanderbilt University Nashville, TN, 37203, USA Sztipaj@isis.vanderbilt.edu

[3]ISIS, Vanderbilt University Nashville, TN, 37203, USA Sandeep@isis.vanderbilt.edu

R. Lauwereins and J. Madsen (eds.), *Design, Automation, and Test in Europe–The Most Influential Papers of 10 Years DATE*, 253–265
© Springer 2008

precisely defined semantics [2], which creates possibility for semantic mismatch between design models and modeling languages of analysis tools. This is particularly problematic in safety critical real-time and embedded systems domain, where semantic ambiguities may produce conflicting results across different tools.

We started addressing these problems by extending our Model Integrated Computing (MIC) tool suite [6] with a semantic anchoring infrastructure for DSMLs. The Semantic Anchoring infrastructure [7, 8] includes a set of well-defined "semantic units" that capture the behavioral semantics of basic dynamic behavior categories. The semantics of a DSML is defined by specifying the transformation rules between the abstract syntax metamodel of the DSML and that of a selected semantic unit. In this paper we build on the previous results and address the impact of system heterogeneity by developing a method to specify DSML semantics as the composition of semantic units.

The organization of this paper is the following: Section 2 provides a short overview of the concepts of semantic anchoring and semantic units. The core idea for semantic unit composition is explained in Section 3. In Section 4, we demonstrate our approach using a simple case study. Section 5 is our conclusion.

2   Semantic Anchoring and Semantic Units

A DSML can be formally defined as a 5-tuple $L = <A, C, S, M_S, M_C>$ consisting of abstract syntax (A), concrete syntax (C), syntactic mapping (M_C), semantic domain (S) and semantic mapping (M_S) [7]. The abstract syntax A defines the language concepts, their relationships, and well-formedness rules available in the language. The concrete syntax C defines the specific notations used to express models, which may be graphical, textual, or mixed. The syntactic mapping, $M_C: C{\rightarrow}A$, assigns syntactic constructs to elements in the abstract syntax. The DSML semantics are defined in two parts: a semantic domain S and a semantic mapping $M_S: A{\rightarrow}S$. The semantic domain S is usually defined in some formal, mathematical framework, in terms of which the meaning of the models is explained. The semantic mapping relates syntactic concepts to those of the semantic domain.

Although DSMLs use many different modeling concepts and notations for accommodating needs of domains and user communities, the scope of well understood behavioral abstractions are more limited. Broad categories of component behaviors can be represented by a finite set of *basic behavioral categories*, such as Finite State Machine (FSM), Timed Automaton (TA), and Hybrid Automaton (HA). Similarly, analyzability requirements and the need for correct-by-construction system composition have led to the emergence of a set of *basic component interaction categories* expressed as Model of Computations such as Synchronous Data Flow (SDF), and Process Networks (PN) [5]. This observation led us to propose a semantic anchoring infrastructure for defining behavioral semantics of DSMLs. The development and use of the semantic anchoring infrastructure includes the following tasks [7]:

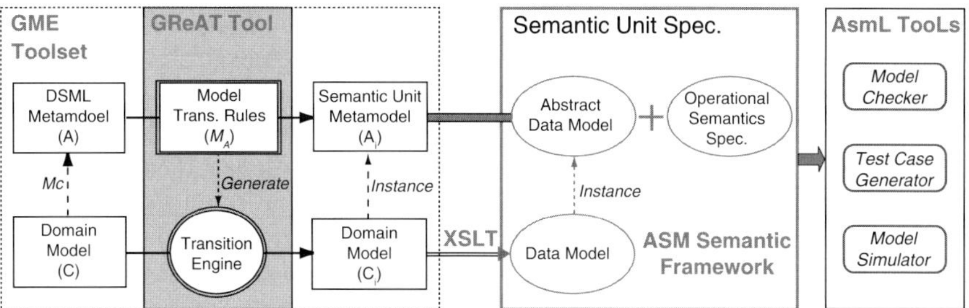

Fig. 1 Semantic anchoring tool suite

1. Definition of a set of modeling languages $\{L_i\}$ for capturing semantics of the basic behavioral abstractions and development of the precise specifications for all components of $L_i = <C_i, A_i, S_i, M_{Si}, M_{Ci}>$. We use the term *semantic units* to describe these basic modeling languages.

2. Definition of the behavioral semantics of an arbitrary DSML, $L = <C, A, S, M_S, M_C>$, is accomplished by specifying the mapping, $M_A: A \rightarrow A_i$, to a predefined semantic unit L_i. The semantic mapping, $M_S: A \rightarrow S$, of L is then defined by the composition $M_S = M_{Si} \circ M_A$, which indicates that the semantics of L is *anchored* to the S_i semantic domain of the L_i modeling language.

Figure 1 shows our semantic anchoring tool suite. It comprises (1) the ASM-based AsmL tool suite [3, 9] for specifying semantic units and (2) the MIC modeling (GME) and model transformation (GReAT) tool suites [6] that support the specification of transformation rules between the DSML metamodels and the Abstract Data Models defined in the semantic units. The behavioral semantics of semantic units are specified as a Control State ASMs [3] using AsmL. Microsoft Research has developed a set of tools to support the simulation, test case generation and model checking for AsmL specifications.

3   Semantic Unit Composition

In the semantic anchoring infrastructure, we define a finite set of semantic units, which capture the semantics of basic behavioral and interaction categories. If the semantics of a DSML can be directly anchored to one of these basic categories, its semantics can be defined by simply specifying the model transformation rules between the metamodel of the DSML and the Abstract Data Model of the semantic unit [7]. However, in heterogeneous systems, the semantics is not always fully captured by a predefined semantic unit. If the semantics is specified from scratch (which is the typical solution if it is done at all) it is not only expensive but we loose the advantages of anchoring the semantics to (a set of) common and well-established semantic units. This is not only loosing reusability of previous efforts, but has negative consequences on our ability to relate semantics of DSMLs to each other and to guide language designers to use well understood and safe behavioral and interaction semantic "building blocks" as well.

Our proposed solution is to define semantics for heterogeneous DSMLs as the composition of semantic units. If the composed semantics specifies a behavior which is frequently used in system design (for example, the composition of SDF interaction semantics with FSM behavioral semantics defines semantics for modeling signal processing systems [5]), the resulting semantics can be considered a *derived semantic unit*, which is built on *primary semantic units*, and could be offered up as one of the set of semantic units for future anchoring efforts. Note that *primary semantic units* refer to the semantic units that capture the semantics of the *basic behavioral categories*, such as FSM, TA, and HA. The composition approach we describe in the rest of the paper is strongly influenced by Gossler and Sifakis [11] framework for composition by clearly separating behavior and interaction. In the following we provide a brief overview of the composition approach that will be followed by a case study.

Mathematically, a semantic unit specification can be represented as a 2-tuple $<A, R>$, where A is an Abstract Data Model specifying the abstract syntax of the semantic unit and R represents a set of Operations and Transition Rules (*updates*, in the ASM terminology). We use $M = I(A)$ to denote the set of all instances of A. Then, each $m \in M$ is a well-formed Data Model defined by the Abstract Data Model A and R specifies the behavior of each $m \in M$. The behavior in ASM is modeled by a sequence of *steps*, where a Step in a given state includes the execution *simultaneously* of all Rules whose guard conditions are true (and the updates are consistent) [3]. Since ASM states are mathematical structures (*sets* with basic operations and predicates), it is easy to integrate Abstract Data Models and Rules. The integrated tool suite ensures that the behavior of domain models defined in a DSML is simulated according to their "reference semantics" by automatically transforming them into AsmL Data Models using the transformation rules.

We model semantic unit composition as *structural* and *behavioral compositions* (see Fig. 2). An ASM instance includes an m data model, the R rule set, and the S dynamic state variables updated during runs. The structural composition defines relationships among selected elements of Abstract Data Models using partial maps. In Fig. 2, we demonstrate semantic composition with two semantic units, SU1, and SU2. The composed semantics is also represented as a 2-tuple $<A, R>$. The structural composition yields the composed Abstract Data Model $A = <A_C, A_{SU1}, A_{SU2}, g_1, g_2>$, where g_1, g_2 are the partial maps between concepts in $A_C, A_{SU1},$ and A_{SU2}.

Behavioral composition is completed by the R_C set of rules that together with R_{SU1} and R_{SU2} form the R rule set for the composed semantics. The role of the R_C set of rules is to receive the possible sets of actions that can be offered by the embedded semantic units using the Get(…) rules, to restrict these sets according to the interactions created by the structural composition and to send back selected subset of actions through the Run(…) rules to complete their next step. The executable actions are represented as partial orders above the set of actions. (This will be shown in detail in the Section 4.)

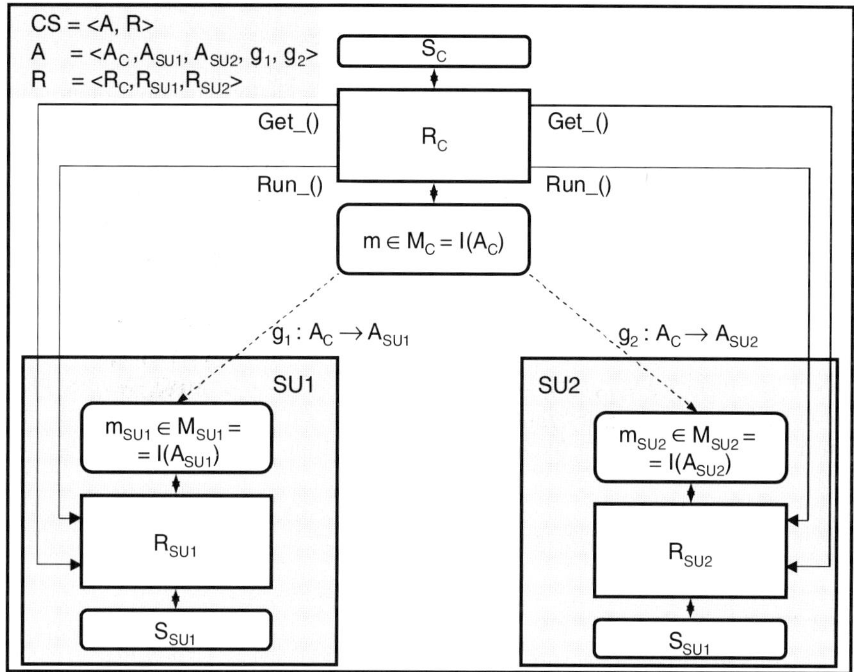

Fig. 2 A graphical representation for the semantic unit composition

In fact, the behavioral composition specifies a controller, which restricts the executions of actions. Since the behavior of the embedded semantic units can be described as partial orders on the sets of actions they can perform, the behavioral composition can be modeled mathematically as a composition of the partial orders.

4   Case Study: Semantic Specification for EFSM

EFSM was developed by General Motors Research to specify vehicle motion control software [12]. As a case study, we defined the behavioral semantics of EFSM as the composition of two primary semantic units, Finite State Machine (FSM-SU) and Synchronous Dataflow (SDF-SU). The full semantic specification includes two composition steps: (1) the semantics of EFSM Components are defined as the composition of FSM-SU and SDF-SU; (2) the semantics of SEFSM Systems are then defined as the composition of the semantics of EFSM Components, which can also be considered as a derived semantic unit, called Action Automaton Semantic Unit, and SDF-SU. Due to space limitations, we can only briefly describe the first composition step. The full semantic specifications can be downloaded from [10].

4.1 EFSM Components

An EFSM model is a synchronous reactive system including a set of components communicating through event channels and data channels. An EFSM component is an FSM-based model. We use a simple component model shown in Fig. 3 as an example to explain the structure and the behavior of EFSM components. The component communicates with other components through *ports*, including a single *input event port* (IEP), an *output event port* (OEP), two *input data ports* (IDP1 and IDP2) and two *output data ports* (ODP1 and ODP2). A component also includes an FSM, where *transitions* are labeled with a *trigger event*, a *guard*, an *output event*, and set of *actions*. Guards and actions are *computational functions* within the component and receive their input data through *input data ports*. The execution of an action (a function) may produce new data, while the execution of a guard only returns a Boolean value for the true or false evaluation.

4.2 Primary Semantic Units Used

4.2.1 FSM-SU Specification

The specification contains two parts: an Abstract Data Model $A_{FSM\text{-}SU}$ and Operations and Transformation Rules $R_{FSM\text{-}SU}$ on the data structures defined in A. The AsmL abstract class *FSM* prescribes the top-level structure of a FSM. All the abstract members of *FSM* are further specified by a concrete FSM, which is an instance of the Abstract State Model. (see detailed examples in [7])

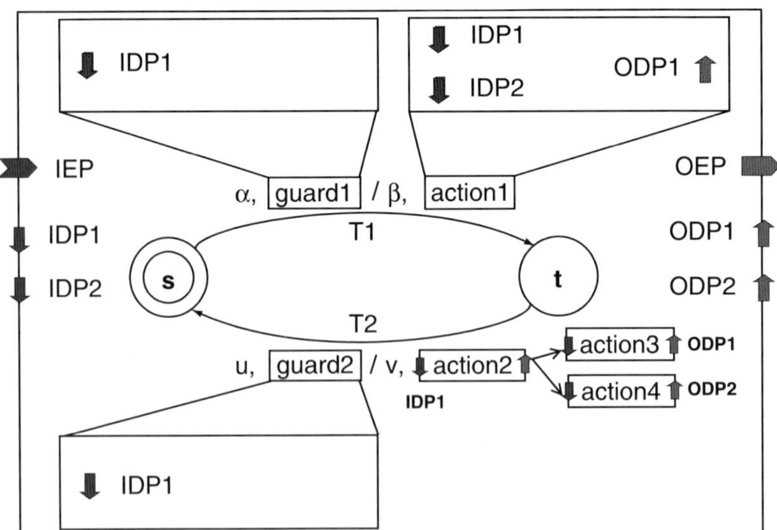

Fig. 3 A simple EFSM component model

```
structure Event
   eventType as String
class State
   initial as Boolean
   var active as Boolean = false
class Transition
abstract class FSM
   abstract property states                           as set of state
      get
   abstract property transitions                      as Set of Transition
      get
   abstract property outTransitions                   as Map of
      <State, Set of Transition>
      get
   abstract property dstState as Map of <Transition, State>
      get
   abstract property triggerEventType                 as Map of
      <Transition, String>
      get
   abstract property outputEventType                  as Map of
      <Transition, String>
      get
```

The behavioral semantics of FSM-SU is specified as a set of AsmL rules. The rule *Run* specifies the top-level system reaction of a FSM when it receives an event. Note that the '?' modifier after *Event* means the return from the *Run* rule may be either an event or an AsmL *null* value.

```
abstract class FSM
   Run (e as Event) as Event?
      step
         let CS as State = GetCurrentState ()
      step
         let enabledTs as Set of Transition = {t | t in
         outTransitions (CS) where e.eventType = triggerEventType(t)}
      step
         if Size (enabledTs) >= 1 then
            choose t in enabledTs
               step
                  CS.active := false
               step
                  dstState(t).active := true
               step
                  if t in me.outputEventType then
                     return Event(outputEventType(t))
                  else
                     return null
```

> **else**
>> **return null**

4.2.2 SDF-SU Specification

The AsmL specification of the Abstract Data Model $A_{SDF\text{-}SU}$ is shown below. *Token* is defined as an AsmL structure to package data using the AsmL construct *case*. *Port* and *Channel* are defined as first-class types. The Boolean attribute *exist* of a port indicates whether the port has a valid data token. When all the input ports of a node have valid data tokens, the node is enabled to fire. In the specification, *Fire* is an abstract function. A concrete node will override the abstract function *Fire* with a computational function. The AsmL abstract class *SDF* captures the top-level structure of a model. The abstract property *inputPorts* contains a sequence of the SDF model's input ports that do not belong to any internal nodes. The abstract property *outputPorts* expresses the similar meaning.

```
structure Value
   case IntValue
      v as Integer
   case DoubleValue
      v as Double
   case BoolValue
      v as Boolean
structure Token
   value as Value?
class Port
   var token as Token = Token (null)
   var exist as Boolean = false
class Channel
   srcPort as Port
   dstPort as Port
abstract class Node
   abstract property inputPorts
      as Seq of Port get
   abstract property outputPorts
      as Seq of Port get
   abstract Fire ()
abstract class SDF
   abstract property nodes
      as Set of Node get
   abstract property channels
      as Set of Channel get
   abstract property inputPorts
      as Seq of Port get
   abstract property outputPorts
      as Seq of Port get
```

The operational rule *Run* specifies the steps it takes to execute a set of nodes. This rule can be considered as a composition interface for SDF-SU. In the beginning, some of the nodes in the set may not be enabled, but they are supposed to be enabled by the execution of already enabled ones. The rule non-deterministically chooses an enabled node from the set of enabled nodes (returned by the operational rule *GetEnabledNodes*) and fires it. The execution of a node consumes the data tokens in all input ports of the node and produce tokens to all output ports as well. An error is reported if there are no enabled nodes in the set while the set is not empty.

> **abstract class** SDF
>> Run (ns **as** Set **of** Node)
>>> **step** while Size(ns) <> 0
>>>> **choose** n **in** ns **where** n **in** GetEnabledNodes ()
>>>>> **remove** n **from** ns
>>>>> Fire (n)
>>>> **ifnone**
>>>>> error (*"Some Nodes are not enabled to fire."*)

4.3 Compositional Semantic Specification for EFSM Components

The behaviors of individual EFSM components can be divided into two different behavioral aspects: the FSM-based behaviors expressing reactions to events and the SDF-based behaviors controlling the execution of computational functions (actions and guards). In this section, we formally specify the behavioral semantics of EFSM components as the composition of two primary semantic units: FSM-SU and SDF-SU. The compositional semantics specification consists of two parts: (1) an Abstract Data Model defining the structural composition $<A_C, A_{FSM-SU}, A_{SDF-SU}, g_1, g_2>$, where $g_1: A_C \rightarrow A_{FSM-SU}$ and $g_2: A_C \rightarrow A_{SDF-SU}$ are structural relation maps; and (2) Operations and Transformation Rules specifying the behavioral composition $<R_C, R_{FSM-SU}, R_{SDF-SU}>$.

4.3.1 Structural Composition

The structural composition defines mapping from elements in the Abstract Data Model of the composed semantic unit to elements in FSM-SU and those in SDF-SU. Figure 4 shows the role of FSM-SU and SDF-SU in the EFSM component model by restructuring the example in Fig. 3. In the modified structure, the FSM model controls the event-related behaviors, while the SDF model takes charge of the data-related computations. Comparing Figs. 3 and 4, we can find that the overall structure of the FSM model closely matches that of the original EFSM component, except for events, guards and actions. The trigger events and the output events in

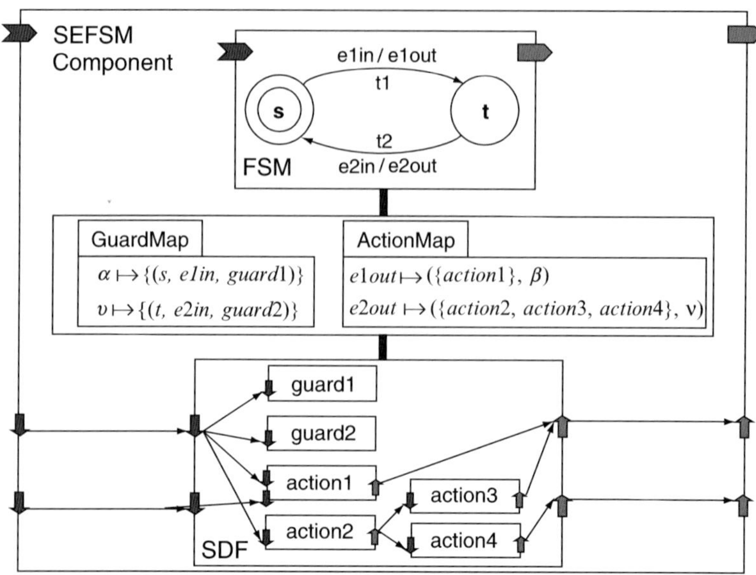

Fig. 4 A compositional structure of the EFSM component originally shown in Fig. 3

the FSM model are renamed. The guards and actions are represented as nodes in the SDF model. The relationships between the FSM model and the SDF model are specified by two maps: *GuardMap* and *ActionMap*. In this section, we only briefly explain how these two maps help to relate the FSM model with the SDF model. More details will be introduced in the following behavioral composition section.

The new compositional structure is built in a way that each transition in the original component is decomposed into three parts: a transition in the FSM model, a node representing the guard and a node representing the action in the SDF model. In the original component, a transition can be unambiguously located by the combination of the source state, the trigger event, and the guard. In the compositional structure, the information can be expressed by a 3-tuple (s, e, n), where s refers a state in the FSM model; e is a local trigger event in the FSM model; and n represents a node in the SDF model. When a component receives an event, this event is a global event and will not be directly forwarded to the FSM model. The *GuardMap* maps this global event to a set of 3-tuples, each tuple referring to a transition in the original component whose trigger event matches this global event. Using the example in Fig. 3 again, the event α is the trigger event only for the transition *T1*. In the compositional structure as shown in Fig. 4, the *T1* transition is decomposed into the *t1* transition in the FSM model, whose source state is s and trigger event is *e1in*, and the *guard1* and *action1* node in the SDF model. As a result, *GuardMap* assigns the event α to the set $\{(s, e1in, guard1)\}$.

```
class EventPort
   var evnt as Event = Event ("")
   var exist as Boolean = false
abstract class Component
   abstract property inPort       as EventPort
      get
   abstract property outPort      as EventPort
      get
   abstract property GuardMap as Map of <String,
      Set of (String, String, Node?)>
      get
   abstract property ActionMap as Map of <String,
      (Set of Node, String?)>
      get
   abstract property fsm as FSM
      get
   abstract property sdf as SDF
      get
```

4.3.2 Behavioral Composition

In essence, the behavioral composition specifies the rules R_C, which is akin to a component-level controller (or scheduler) that orchestrates the executions and interactions of the FSM model and the SDF model.

The execution of a transition in the original EFSM component can be decomposed into a three-step process: (1) the evaluation of the guard functions for all outgoing transitions from the current state as nodes in the SDF model; (2) the selection of an enabled transition in the FSM model; and (3) the execution of actions of the transition as nodes in the SDF model. The three steps are related to each other by the maps *GuardMap* and *ActionMap*. The output event produced by the execution of a transition in the FSM model is a local event. *ActionMap* maps it to a 2-tuple *({n}, e)*, where *{n}* refers to a set of nodes (actions) in the SDF model and *e* refers to a global output event that will be propagated out of the component. For instance, the execution of the *t2* transition of the FSM model in Fig. 4 generates a local event *e2out*. Since the *t2* transition corresponds to the *T2* transition in the original component (Fig. 3), which is attached with actions, *action2*, *action3* and *action4*, and an output event *v*, the *ActionMap* maps the local event *e1out* to a 2-tuple *({action2, action3, action4}, v)* accordingly.

The rules verbalized above are specified in AsmL as Operations and Transition Rules. The operational rule *Run* of *Component* specifies the top-level component operations as a sequence of *updates*. The AsmL construct *require* asserts that the component's input event port must have a valid event. The rule first consumes the event in the port and checks whether this event triggers further updates in the component. If the event does, the rule *MapToLocalInputEvent* returns the corresponding local event used to trigger the FSM model; if not, a *null* value is returned and the reaction is completed. If a valid local event is returned, it activates the FSM model. The reaction of the FSM model returns a local output event. If the EFSM component produces an output event in this reaction, the rule *MapToGlobalOutputEvent* maps the local event to the global output even, which is then stored in the output port of the component.

```
abstract class Component
  Run ()
    require inPort.exist
    step
      inPort.exist := false
      let localEvent as Event? =
        MapToLocalInputEvent (inPort.evnt)
    step
      if localEvent <> null then
        step let e as Event? = fsm.React (localEvent)
        step
          let global Event as Event?=MapToGlobalOutputEvent(e)
        step
          if globalEvent <> null then
            outPort.evnt := globalEvent
            outPort.exist : = true
```

Furthermore, we observe that this behavioral semantics specification is not limited to the EFSM components. It actually specifies the semantics of a common behavioral category that captures the reactive computation behaviors. Therefore, we can consider the semantic specification for EFSM components as a new derived semantic unit, called Action Automaton Semantic Unit (AA-SU). We can leverage this AA-SU to compositionally specify the semantics of EFSM systems. (Please refer to [10] for the full specification.)

5   Conclusion

Compositional semantic specification is a necessary step for making DSMLs semantically precise and practical. The proposed approach builds on a large body of work on ASM [3], semantics of composition [5], and on our earlier work on semantic anchoring [7] [8]. As a future step we will continue the construction of a library of primary semantic units and will move toward increased automation in semantic unit composition.

References

[1] Object Management Group. *UML™ Profile for Modeling and Analysis of Real-Time and Embedded systems*. realtime/05-02-06.

[2] Susan Graph, Ileana Ober. How Useful is the UML profile SPT Without Semantics? In *Workshop on the Usage of the UML Profile for Scheduling, Performance and Time (SIVOES '04)*, Toronto Canada, 2004.

[3] E. Boerger, R. Staerk. *Abstract State Machines: A Method for HighLevel System Design and Analysis*. Springer, 2003.

[4] G. Karsai, J. Sztipanovits, A. Ledeczi, T. Bapty. Modelintegrated Development of Embedded Software. *Proceedings of the IEEE*, vol. 91, 2003.

[5] Johan Eker, Jörn W. Janneck, Edward A. Lee, Jie Liu, Xiaojun Liu, Jozsef Ludvig, Stephen Neuendorffer, Sonia Sachs, Yuhong Xiong. Taming Heterogeneity The Ptolemy Approach. *Proceedings of the IEEE*, vol. 91, p. 127–144, 2003.

[6] The MIC Tool Suite. http://www.escherinstitute.org/Plone/tools/suites/mic.

[7] K. Chen, J. Sztipanovits, S. Neema, M. Emerson, S. Abdelwahed, Toward a Semantic Anchoring Infrastructure for Domain-Specific Modeling Languages, In *Proceedings of the Fifth ACM International Conference on Embedded Software (EMSOFT '05)*, p. 35–44, September 2005.

[8] K. Chen, J. Sztipanovits, S. Abdelwahed, E. Jackson, Semantic Anchoring with Model Transformations. In *Proceedings of European Conference on Model Driven Architecture - Foundations and Applications (ECMDA-FA)*, November 2005. LNCS, vol. 3748. p. 115–129, Springer 2005.

[9] The Abstract State Machine Language. www.research.microsoft.com/fse/asml.

[10] The Semantic Anchoring Tool Suite. www.isis.vanderbilt.edu/SAT.

[11] G. Gossler, J. Sifakis, Composition for Component-Based Modeling. *Science of Computer Programming*, vol. 55, 2005.

[12] S. Birla, S. Wang, S. Neema, T. Saxena. Addressing Cross-tool Semantic Ambiguities in Behavior Modeling for Vehicle Motion Control. In *Automotive Software Workshop 2006*, April 2006.

Design Technology for Advanced Digital Systems in CMOS and Beyond

Hugo De Man

1 Introduction

The period 1997–2007 is characterized by a real paradigm shift in semiconductor technology with a deep impact on the design of digital systems. Let us mention just a few major changes:

- Technology scaling evolved from 350 to 45 nm minimum feature size. As a result the number of transistors per chip increased by two orders of magnitude leading to Systems-on-Chip (SoC) with over 100 million transistors.
- Supply voltage VDD decreased from 3.3 to 0.9 V but threshold voltage VT cannot be decreased further due to increased device leakage power. The latter is now becoming comparable to dynamic power. VDD, VT, and clock frequency have now become design variables in order to control dynamic as well as static energy/task consumption at run time.
- As we are reaching the atomic scale and are writing features smaller than the wavelength of light used in litho, no two identically designed devices are alike anymore. Although systematic variability can be circumvented at design and test time, coping with stochastic variability needs novel run-time methodologies for the design of reliable systems from unreliable components.
- As a result, scaling no longer happens in the traditional way but by the introduction of a plethora of new materials and device architectures.
- The main application domain shifts from desktop to mobile low power, cheap consumer products with ever shorter time-to-market requirements and coping with rapidly changing de facto standards and feature requirements at a time where the nonrecurring design engineering (NRE) cost exceeds US $50 million at the 90 nm level.
- In 10–15 years from now, CMOS scaling as such will most likely come to an end and today device physicists are looking intensively for alternatives but they live too much in isolation from the manufacturing and design community.

Senior Fellow IMEC, Belgium; Professor Emeritus K.U. Leuven

R. Lauwereins and J. Madsen (eds.), *Design, Automation, and Test in Europe–The Most Influential Papers of 10 Years DATE*, 269–273

All of these changes are cumulative and concurrent. This creates an enormous challenge to the design community as all these changes at the physical level have to ripple through the complete design trajectory from systems to atoms. This makes R&D for new design technology more exciting than ever. At the same time, we see a movement "back to the future" as the link between systems and physics can no longer be neglected and new abstraction models are definitely needed.

More in particular the above changes give rise to number of emerging trends in the design area such as:

1. Exponentially increasing NRE cost favors embedded software programmable rather than ASIC architectures. In this way NRE cost can be amortized over a much larger production volume by reusing the hardware design for a larger number of emerging applications and standards.
2. On the other hand, mapping software applications on a general purpose instruction set processor is orders of magnitude less power efficient due to intense data and instruction traffic between memory and processor. Hence, power optimization becomes a crucial aspect.
3. This power optimization can be done in many ways:

    (a) By using multicore architectures whereby each core operates at the optimum voltage and clock frequency for minimizing the energy consumption within the real time constraints of the application
    (b) By using a custom heterogeneous multicore architecture whereby each core is just flexible enough to do the jobs requested
    (c) By minimizing and localizing instruction and data traffic between memory hierarchy and cores
    (d) By optimizing the memory hierarchy accordingly
    (e) By encoding the data and instructions on parallel busses or interconnect structures in order to minimize power and correct storage and communication errors
    (f) By making use of asynchronous optimized packet- and circuit switching networks-on-chip connecting locally synchronous cores

4. Increased variability leads to the need to design reliable systems from unreliable components and interconnect. In the end we will even have to live with failing or defect components either at first use or during use. Fault and variability resilient design therefore will be crucial for ultimately scaled CMOS, but also for novel nanoscaled technologies that will resemble more and more biological systems that "survive" in spite of numerous defects.
5. A definite need arises to establish a "back to the future" movement linking the physics of emerging devices to new abstractions to design circuits and systems with them.

Many of these trends can be found back in the papers from the last 10 years of the DATE conference. In the area of digital system design, four papers have been selected on the basis of citations and/or peer review evaluation. Below, we briefly discuss them in chronological order and discuss their relevance in the light of the above trends and paradigm shifts.

1. Address Bus Encoding Techniques for System-Level Power Optimization by L. Benini, G. De Micheli, E. Macii, D. Sciuto, and C. Silvano (1998)

This paper refers to the trends under 3a, c, and especially e. As scaling continues, the energy consumption of instruction and data transfer by far exceeds energy consumption in computation. The paper proposes a substantial reduction in bus switching activity by proposing three novel so called mixed bus encoding techniques for the address streams of on-chip cores.

The authors consider both instruction and data address streams as well as multiplexed address streams and report power savings on bus switching activity ranging from 20–35%.

They also pay attention to the power dissipated in the encoding and decoding logic showing that, especially for on-chip address buses, a careful trade-off must be respected between the complexity of the encoding and the corresponding reduction in switching activity of the bus.

The results of this paper have been referred to in 241 subsequent papers discussing power reduction in SoC's and hence can be considered as a true seminal paper in this domain.

2. MOCSYN: Multiobjective Core-Based Single-Chip System Synthesis by R.P. Dick and N.K. Jha (1999)

This paper refers to nearly all points mentioned in trends 1, 2, and 3 above. It is one of the first papers that covers the complete concurrent hardware-software design trajectory for heterogeneous multi-core architectures all the way from multiple multirate periodic real-time task graph specifications to the core placement as well as to the automated synthesis of an asynchronous custom bus interconnect network and optimal clock frequency selection for the synchronous cores. New is the use of a multiobjective optimization technique based on genetic algorithms. This leads not only to a single architecture optimized for a single objective, but to a set of Pareto optimal architectures, which trade off price, area and average power consumption. This paper has triggered an enormous amount of research on automated synthesis methods for low-power embedded SoC systems at the major design conferences such as DAC, ASP-DAC, ICCAD and DATE.

3. Minimum Energy Fixed-Priority Scheduling for Variable Voltage Processors by G. Quan and X.S. Hu (2002)

This paper focuses on the important trend 3a, i.e. finding the VDD and clock frequency schedule that minimizes dynamic energy consumption for executing a set of independent real-time tasks on a single processor according to a fixed-priority preemptive policy. Up to 2002 several authors had proposed heuristic methods for this problem and the optimality of the resulting schedules were not known. The great contribution of this paper is that it proposes two algorithms that provide the theoretical inherent limit to optimally schedule fixed-priority real-time tasks on variable voltage processors that now become available as IP blocks on most SoC's. Of course this is at the expense of a high computational complexity but at the same time it allows to evaluate better existing or new emerging heuristic methods to solve this important problem in low power SoC design. It has influenced most subsequent research in this area of dynamic voltage scaling. For example, more recent work[1,2] tackles the fact that leakage power increases fivefold for each new technology generation. This has a great impact on this issue as the minimization of the total energy now will require a trade off between higher speed (and thus dynamic power) resulting in a longer idle slack and leakage power that can then be cut off in idle periods by appropriate circuit techniques.

4. Synthesis and Optimization of Threshold Logic Networks with Application to Nanotechnologies by R. Zhang, P. Gupta, L. Zhong, and N. K. Jha (2004)

This is an interesting example of a "back to the future" paper trying to bridge the gap between the emerging device physics and the circuit and system design community. Around 2015–2020 we will reach the ultimate scaling limits of CMOS around the 10 nm level. Today there is an intensive research activity to investigate new technologies beyond CMOS that can possibly lead to further continuation of Moore's law. Emerging technologies such as the use of hetero-juncton transistors combined with single electron resonant tunneling diodes as well as quantum cellular automata offer the possibility to design threshold logic gates. Although they still operate on Boolean variables, their output switches when a weighted sum or difference of the Boolean inputs exceeds a threshold T. In this way an equivalent Boolean function can be synthesized with a smaller number of threshold logic gates than when using traditional CMOS gates. In addition, their switching speed and energy efficiency are in principle higher than for CMOS. Unfortunately, at present, such circuits only operate reliably at a few degrees Kelvin and therefore do not satisfy yet the "system-ability" criteria of the ITRS roadmap.

However, it is not excluded that more system-able Threshold Logic devices result form the intensive research for the "new post CMOS switch" with nanoscale dimensions.

This paper is quite significant in the sense that it presents the first comprehensive and formal methodology for multilevel multioutput threshold network synthesis based on an abstraction of the physical model. Based on real benchmarks, the paper shows that threshold logic synthesis on average leads to 52% reduction in gate count compared to classical CMOS gates. The paper paves the way to a smoother transition towards logic design using emerging nano-technologies making use of the threshold logic formalism.

---

[1] G. Quant et al. Fixed Priority Scheduling for Reducing Overall Energy on Variable Voltage Processors, Proc. 25th IEEE International Real-Time Systems Symposium, pp. 309–318

[2] R. Jejurikar et al. Leakage Aware Dynamic Voltage Scaling for Real-Time Embedded Systems, Proc. DAC'04, June 4–7, 2004, pp. 275–280

Epilog

CMOS technology scaling will probably continue for another 10–15 years. However, it will not be business as usual as discussed in the introduction. We also have to face the fact that, although a new technology node may occur every 2–3 years, history shows that the introduction of a new design paradigm takes 7–10 years from concept to acceptance by the design community. This clearly indicates that the EDA R&D community in principle should be looking at the challenges of designing scaled CMOS architectures at or below the 22 nm level. Such chips may contain a network of 1000+ cores which causes huge challenges to the parallel computing community in addition to the other challenges at the nano-physics level mentioned above.

In addition, advanced research should be devoted to establish a dialogue between the EDA and the emerging device research communities to make sure that devices are "system-able" in the end, i.e. that the link can be made to designing meaningful systems out of them.

Looking back to the challenges mentioned above, we see that during the last 10 years the EDA community focused mainly on the trends 1–3 mentioned above. Very little attention however, is paid to the issue of coping with variability and fault resilient design methods and establishing the link back to emerging nanoscale devices. Thereby one should be aware that perhaps the latter may play a more important role in "More than Moore" systems such as smart sensors, actuators, and biosystems, rather than in "More Moore," i.e. trying to extend computational power beyond what ultimate CMOS will be capable to achieve around 2020. We believe that it is time to encourage and fund a revival of out-of-the-box thinking in the EDA world in order to be ready to face the exciting world of nanoelectronic systems of the future.

Address Bus Encoding Techniques
for System-Level Power Optimization

Luca Benini[3], Giovanni De Micheli[3], Enrico Macii[2], Donatella Sciuto[1],
and Cristina Silvano[4]

Abstract The power dissipated by system-level buses is the largest contribution
to the global power of complex VLSI circuits. Therefore, the minimization of the
switching activity at the I/O interfaces can provide significant savings on the overall
power budget. This paper presents innovative encoding techniques suitable for minimizing the switching activity of system-level address buses. In particular, the schemes
illustrated here target the reduction of the average number of bus line transitions per
clock cycle. Experimental results, conducted on address streams generated by a real
microprocessor, have demonstrated the effectiveness of the proposed methods.

1   Introduction

The increasing performance requirements and the existing gap between the speed
of current microprocessors and the speed of the systems interfaces has pushed the
designers to increase the bandwidth of the data transfers. For example, the PowerPC
620 processor can be configured with either a 64 or a 128-bit data bus to support
the transfer of the 64-byte cache block within four clock cycles. Moreover, modern
software applications span a very large address space: The latest versions of
existing processors, such as the DEC Alpha AXP and the PowerPC 620, support a
64-bit address space. Hence, both data and address buses have become very wide
for modern microprocessor-based systems.

Due to the intrinsic capacitances of the bus lines, a considerable amount of power
is required at the I/O pins of a microprocessor when data have to be transmitted over
the bus. More specifically, it has been estimated that the capacitance driven by the I/O
nodes is usually much larger (up to three orders of magnitude [1]) than the one seen
by the internal nodes of the microprocessor. As a consequence, dramatic optimizations

---

[1]Politecnico di Milano Dip. di Elettronica e Informazione Milano, ITALY 20133

[2]Politecnico di Torino Dip. di Automatica e Informatica Torino, ITALY 10129

[3]Stanford University Computer Systems Laboratory Stanford, CA 94305

[4]Università di Brescia Dip. di Elettronica per l'Automazione Brescia, ITALY 25123

R. Lauwereins and J. Madsen (eds.), *Design, Automation, and Test*
in Europe–The Most Influential Papers of 10 Years DATE, 275–289
© Springer 2008

of the average power consumption can be achieved by minimizing the number of transitions (i.e. the switching activity) on system-level buses.

Encoding paradigms for reducing the switching activity on the bus lines have been recently investigated. Stan and Burleson [2] have proposed the use of a redundant encoding scheme, called the *bus-invert code*, to limit the average power. The bus-invert method performs well when patterns to be transmitted are randomly distributed in time and no information about pattern correlation is available. Therefore, the method seems to be appropriate for encoding the information traveling on data buses.

Concerning address buses, other techniques have been explored. They all rely on the well-known fact that the addresses generated by a microprocessor are often consecutive, due to the *spatial locality* principle [3], which states that the memory elements whose addresses are close to each other tend to be referenced during successive clock cycles.

To exploit this peculiar characteristics of the address buses, Su, Tsui, and Despain [4] have proposed to encode the patterns with the *Gray code*, since it guarantees a single bit transition when consecutive addresses are produced. The issue of modifying the Gray code so as to preserve the one-transition property for consecutive addresses of byte-addressable machines is discussed in [5].

The Gray code achieves the minimum switching activity among irredundant codes; however, better performance can be obtained by using redundant codes. We [6] have proposed an encoding technique, called *T0 code*, which is based on the idea of avoiding the transfer of consecutive addresses on the bus by adding a redundant line, *INC*, to communicate to the receiving memory subsystem the information on the sequentiality of the addresses. The increments between consecutive patterns can be parametric, reflecting the addressability scheme adopted in the given architecture. The T0 code outperforms the Gray code in the case of streams of consecutive addresses of limited and unlimited lengths. Furthermore, savings are not offset by the power absorbed by the additional encoding/decoding circuitry. The exploitation of the sequential behavior of addresses has also been approached, at a high level of abstraction, by Panda and Dutt [1]. That work introduces techniques for determining a mapping of the data to the physical memory which reduces the total switching activity on the address buses.

An additional contribution to the area of address bus encoding is provided in [7]. The target of the *Beach code* is that of reducing the bus activity in the cases where the percentage of in-sequence addresses is limited. In this situation, it may be possible to exploit other types of temporal correlations than arithmetic sequentiality that may exist between the patterns that are being transmitted over the address bus. More specifically, it has been noted that time-adjacent addresses usually show remarkably high block correlations. The encoding strategy is then determined depending on the particular stream being transmitted. Therefore, the Beach code best performs on special purpose systems, where a dedicated processor (e.g. core, DSP, microcontroller) repeatedly executes the same portion of embedded code.

In this paper, we propose innovative encoding techniques targeting the reduction of the average number of transitions on the address buses of microprocessor-based systems. Here, the data and address buses are at the core of the interface between the processor and the memory and I/O subsystems. In particular, they are used, on-processor, to access the first and the second-level caches, as well as, off-processor, to access the external lower level caches, the main memory (usually through a memory controller), and the I/O subsystems (to support direct memory accesses from the I/O controllers). Our goal is to avoid any modification to the standard memory components, hence adding the encoding circuitry inside the processor, and the decoding logic inside the memory and the I/O controllers. In the next section, we summarize the main characteristics of currently available codes, and we compare their performance through analytical and experimental analysis. In Section 3 we propose new encoding schemes that combine the properties of existing approaches. This with the ultimate objective of exploiting the advantages offered by each technique, and thus, identifying the best method for the specific microprocessor, the MIPS RISC [8], we have used for the experiments. Power and timing-efficient implementations of the encoding/decoding logic for such encoding technique are also presented.

2 Previous Work

2.1 Bus-Invert Encoding

Stan and Burleson [2] have proposed a redundant code, called the *bus-invert* code, to decrease the average power. A redundant bus line, called *INV*, is needed to signal to the receiving end of the bus which polarity is used for the transmission of the incoming pattern. The encoding depends on the Hamming distance (i.e. the number of bit differences) between the value of the encoded bus lines at time $t - 1$ (also counting the redundant line at time $t - 1$) and the value of the address bus lines at time t. The Hamming distance is compared to $N/2$, where N is the bus width (assuming N even without loss of generality). If the Hamming distance between two successive patterns is larger than $N/2$, the current address is transmitted with inverted polarity and the redundant line is asserted; otherwise, the current address is transmitted *as is*, and the *INV* line is de-asserted. The bus-invert method can be expressed by the following equation:

$$(\mathbf{B}^{(t)}, \mathrm{INV}^{(t)}) = \begin{cases} (\mathbf{b}^{(t)}, 0) & \text{if } \mathrm{H}^{(t)} \le N/2 \\ (\overline{\mathbf{b}}^{(t)}, 1) & \text{if } \mathrm{H}^{(t)} > N/2 \end{cases} \tag{1}$$

where $\mathbf{B}^{(t)}$ is the value of the encoded bus lines at time t, $INV^{(t)}$ is the additional bus line, $\mathbf{b}^{(t)}$ is the address value at time t, $\mathrm{H}^{(t)} = (\mathbf{B}^{(t-1)}|INV^{(t-1)}, \mathbf{b}^{(t)}|0)$ is the Hamming distance, and N is the bus width of $\mathbf{b}^{(t)}$.

The corresponding decoding scheme is simply defined as:

$$b^{(t)} = \begin{cases} (B^{(t)} \cdot \text{if INV} = 0 \\ (\overline{B}^{(t)} \text{ if INV} = 1 \end{cases} \tag{2}$$

2.2 Asymptotic Zero-Transition Encoding

As mentioned in Section 1, the Gray code achieves its asymptotic best performance of a single transition per emitted address when infinite streams of consecutive addresses are considered. However, the code is optimum only in the class of irredundant codes, that is, codes that employ exactly N-bit patterns to encode a maximum of 2^N data words. Adding redundancy to the code, better performance can be achieved by adopting the T0 code [6]. The T0 method requires a redundant line, *INC*, to signal with value one that a consecutive stream of addresses is output on the bus. If *INC* is high, all other lines on the bus are frozen, to avoid unnecessary switchings. The new address is computed directly by the receiver. On the other hand, when two addresses are not consecutive, the *INC* line is low and the remaining bus lines are used as standard binary codes for the new addresses.

Obviously, this redundant code outperforms the Gray code on unlimited streams of consecutive addresses. Since all addresses are consecutive, the *INC* line is always asserted, and the bus lines never switch. As a consequence, the asymptotic performance of the T0 code is zero transitions per emitted consecutive address. The increments between consecutive patterns can be parametric, reflecting the addressability scheme adopted in the given architecture. In this respect, our code has the same capabilities of the Gray schemes reported in [5].

The T0 encoding scheme can be formally specified as follows:

$$(B^{(t)}, \text{INC}^{(t)}) = \begin{cases} (B^{(t-1)}, 1) & \text{if } b^{(t)} = b^{(t-1)} + S \\ (b^{(t)}, 0) & \text{otherwise} \end{cases} \tag{3}$$

where $\mathbf{B}^{(t)}$ is the value on the encoded bus lines at time t, *INC*$^{(t)}$ is the additional bus line, $\mathbf{b}^{(t)}$ is the address value at time t and $\mathbf{S}$ is a constant power of 2, called *stride*.

The corresponding decoding scheme can be formally defined as follows:

$$b^{(t)} = \begin{cases} (b^{(t-1)} + S) & \text{if INC} = 1 \\ B^{(t)} & \text{if INC} = 0 \end{cases} \tag{4}$$

2.3 Analytical Performance Comparison

Let us compare the two bus-encoding techniques discussed so far, considering the binary code as the reference for the comparison. For the analysis we consider a stream of unlimited length whose addresses have a random uniform distribution, and a stream of unlimited length whose addresses are consecutive. Table 1 reports the results of the comparison, where N is the address bus width. Notice that, in

Table 1 Analytical performance comparison

Stream Type	Code	Avg. Trans. per Clock	Avg. Trans. per Clock per Line	Avg. I/O Pow. Diss.
Out-of-Seq Addresses	Binary	$N/2$	0.5	1
	T0	$N/2$	0.5	1
	Bus-Inv	T_N	T_N/N	
In-Seq Addresses	Binary	$N/2$	0.5	1
	T0	0	0	0
	Bus-Inv	T_N	T_N/N	$T_N/N/_2$

the case of the bus-invert code, the average number of transitions per clock cycle, T_N, is given by:

$$T_N = \frac{1}{2^N} \cdot \sum_{k=0}^{N/2} k C_{N+1}^k \tag{5}$$

where C_N^k is the binomial coefficient.

2.4 Experimental Performance Comparison

In this section, we first analyze the behavior of the addresses generated by a RISC microprocessor, then we provide experimental results for comparing the above-discussed encoding techniques. The number of transitions occurring on the address bus during the execution of benchmark programs, derived from different application domains, has been used as the metric to estimate the power consumption.

The addresses generated by a microprocessor obey to the spatial locality principle, that is, they are often in-sequence and, in general, they present a very high correlation. However, the things are quite different if the addresses correspond to instructions or data. In fact, addresses corresponding to instructions usually show a more sequential behavior with respect to data addresses. Instructions are usually stored in adjacent locations of the memory space, while only structured data such as arrays are generally stored in consecutive memory locations to achieve better locality. Clearly, there are exceptions to this behavior: control-flow instructions cause interruptions in the sequence of consecutive addresses on the instruction flow, and data not stored in arrays are often accessed without any regular pattern. Nevertheless, sequential addressing dominates on average.

These considerations are supported by the data contained in Tables 2–4, where we report the percentage of in-sequence addresses measured on real address streams occurring on the 32-bit address bus of the MIPS microprocessor, when different benchmark programs are executed. Three distinct cases have been considered:

- The instruction address bus
- The data address bus
- The instruction/data multiplexed address bus

The tables report also the number of transitions on the address bus, for the above mentioned classes of codes, and the percentage of transitions saved with respect to binary encoding.

The average percentage of sequential addresses in the benchmark streams is higher for instructions addresses (63.04%) than for data address streams (11.39%). This behavior can be due to the fact that references to automatic variables such as loop counters destroy the sequentiality of the address streams even if array data structures are accessed sequentially. As expected, when the probability of consecutive addresses is high, as for instruction addresses, a scheme such as T0 can be effective to reduce the transitions count, offering an average savings of 35.52% (see Table 2), while the bus-invert approach does not introduce any savings.

On the other hand, when the probability of in-sequence addresses is very low, as in the case of the data addresses reported in Table 3, the T0 code can provide only marginal advantages with respect to binary encoding (3.37%). The binary encoding can thus represent a good alternative for data addresses, since it does not require any redundancy and therefore no encoding and decoding circuitry. Among redundant codes, another alternative for data address buses can be represented by the bus-invert code, that is advantageous for the minimization of the average number of transitions (10.78% savings on average).

When the address bus is multiplexed (i.e. mixed instructions and data addresses), as in the MIPS architecture, the sequential behavior is often interrupted, when the selection signal switches from instruction to data and vice versa. Hence, the multiplexed address bus shows an intermediate behavior (48.64%) for the data in Table 4. As a matter of fact, the sequentiality of the addresses on the bus is somewhat reduced by the time multiplexing and by the inherent randomness of the data addresses.

3   Mixed Bus-Encoding Techniques

The different effects of the encoding schemes discussed in Section 2 on the number of bus transitions have suggested us the use of combined techniques which may exploit the best properties of each method to further improve the overall system-level power budget. In this section, we introduce some novel encoding options, and we present transition count results collected during the execution of the usual benchmark programs.

3.1   T0_BI Encoding

For architectures based on a single address bus used to transmit both instruction and data addresses, as in the case of external second-level unified data and instruction caches, a new encoding can be applied to exploit the advantages provided by the T0 and bus-invert approaches. The combined code, called *T0_BI code*, reduces the

Table 2 Experimental comparison of existing encoding schemes for instruction address streams

Benchmark	Stream length	In-seq addr	Binary trans	T0 Trans	T0 Savings	Bus-Invert Trans	Bus-Invert Savings
gzip	119102	60.16%	232587	140845	39.44%	232586	0.00%
gunzip	63884	64.94%	118409	81433	31.23%	118094	0.27%
ghostview	404595	57.28%	678295	470949	30.57%	678236	0.01%
espresso	1751673	60.17%	3014854	2239400	25.72%	3014649	0.01%
nova	544994	52.99%	959912	836708	12.83%	959334	0.00%
jedi	14690249	53.96%	23145174	18521606	19.98%	23145174	0.00%
latex	700317	71.42%	1400540	200490	85.68%	1400540	0.00%
matlab	6400326	76.56%	12000527	5800465	51.66%	12000527	0.00%
oracle	500326	69.99%	880547	680485	22.72%	880547	0.00%
Average		63.04%			35.52%		0.03%

Table 3 Experimental comparison of existing encoding schemes for data address streams

Benchmark	Stream length	In-seq addr.	Binary trans.	T0 Trans.	T0 Savings	Bus-Invert Trans.	Bus-Invert Savings
gzip	34393	52.47%	131651	97157	26.20%	118184	10.23%
gunzip	14602	10.59%	92349	91739	0.66%	81953	11.26%
ghostview	112689	10.74%	554499	542323	2.20%	507987	8.39%
espresso	570750	5.56%	3956543	3945485	0.28%	3426699	13.39%
nova	220050	10.56%	1381975	1368763	0.95%	1204694	12.82%
jedi	6145049	12.67%	47268786	47156168	0.24%	41114510	13.02%
latex	200109	0.01%	763	771	−1.05%	558	26.87%
matlab	1204110	0.00%	5400375	5400385	0.00%	5351262	0.91%
oracle	120111	0.01%	466606	466616	0.00%	466039	0.12%
Average		11.39%			3.37%		10.78%

Table 4 Experimental comparison of existing encoding schemes for multiplexed address streams

Benchmark	Stream length	In-seq addr.	Binary trans.	T0 Trans.	T0 Savings	Bus-Invert Trans.	Bus-Invert Savings
gzip	153495	57.43%	833799	738693	11.41%	809194	2.95%
gunzip	78486	52.81%	482093	427145	11.40%	406744	15.63%
ghostview	517284	58.25%	3067635	2795525	8.87%	2911494	5.09%
espresso	2322423	54.39%	15147444	13958602	7.85%	13404439	11.51%
nova	765045	56.63%	5526484	5325856	3.63%	5412030	2.07%
jedi	20835298	57.88%	152427486	142947798	6.22%	137950083	9.50%
latex	900426	55.55%	7002828	6202750	11.43%	5202521	25.71%
matlab	7604436	65.37%	42405967	34005893	19.81%	39474425	6.91%
oracle	620437	59.13%	3939551	3479461	11.68%	3594330	8.76%
Average		57.62%			10.25%		9.79%

average power and requires two redundant lines, *INC* and *INV*. The formal definition of the code is the following:

$$(\mathbf{B}^{(t)}, INC^{(t)}, INV^{(t)}) = \begin{cases} (\mathbf{B}^{(t-1)},1,0) & \text{if } b^{(t)} = b^{(t-1)} + S \\ (b^{(t)},0,0) & \text{if } b^{(t)} \neq b^{(t-1)} + S \wedge \\ & \quad H^{(t)} \leq (N+2)/2 \\ (\overline{b}^{(t)},0,1) & \text{if } b^{(t)} \neq b^{(t-1)} + S \wedge \\ & \quad H^{(t)} > (N+2)/2 \end{cases} \qquad (6)$$

where $\mathbf{B}^{(t)}$ is the value of the encoded bus lines at time t, $INV^{(t)}$ and $INC^{(t)}$ are the additional bus lines, $b^{(t)}$ is the address value at time t, $H^{(t)} = (\mathbf{B}^{(t-1)}|INC^{(t-1)}|INV^{(t-1)}, b^{(t)}|0|0)$ is the Hamming distance, and N is the bus width of $b^{(t)}$.

The corresponding decoding scheme is given by:

$$b^{(t)} = \begin{cases} \mathbf{B}^{(t)} & \text{if } INC=0 \wedge INV=0 \\ \overline{\mathbf{B}}^{(t)} & \text{if } INC=0 \wedge INV=1 \\ b^{(t-1)} + S & \text{if } INC=1 \end{cases} \qquad (7)$$

3.2 *Dual_T0 Encoding*

For architectures based on multiplexed address buses, such as the one of the MIPS microprocessor, two address streams α and β, with quite different behavior, are time-multiplexed on the same address bus. Stream α, corresponding to instruction addresses, has high probability of having two consecutive addresses on the bus in two successive clock cycles; on the contrary, stream β, corresponding to data addresses, has almost no in-sequence patterns. The control signal, *SEL*, available in the standard bus interface to de-multiplex the bus at the receiver side, is asserted when stream α is transmitted; otherwise, *SEL* is de-asserted. An extension of the T0 code, called *dual_ T0 code*, can be effective in these cases, providing the application of the T0 code and the updating of the encoding/decoding registers whenever *SEL* is asserted; otherwise, the binary code is applied and the encoding/decoding registers hold.

Analytically, the dual_T0 encoding can be expressed as:

$$(\mathbf{B}^{(t)}, INC^{(t)}) = \begin{cases} (\mathbf{B}^{(t-1)},1) & \text{if } SEL=1 \wedge b^{(t)} = \tilde{b}^{(t)} + S \\ (b^{(t)},0) & \text{otherwise} \end{cases} \qquad (8)$$

where $\mathbf{B}^{(t)}$ is the value on the encoded bus lines at time t, $INC^{(t)}$ is the additional bus line, $b^{(t)}$ is the address value at time t, $\mathbf{S}$ is the stride, *SEL* is the selection signal, and $\tilde{b}^{(t)}$ is given by:

$$\tilde{b}^{(t)} = \begin{cases} \tilde{b}^{(t-1)} & \text{if } SEL=0 \\ b^{(t-1)} & \text{if } SEL=1 \end{cases} \qquad (9)$$

The corresponding decoding scheme follows:

$$b^{(t)} = \begin{cases} (\tilde{b}^{(t)} + S) & \text{if INC} = 1 \\ B^{(t)} & \text{if INC} = 0 \end{cases} \qquad (10)$$

3.3 Dual_T0_BI Encoding

Similarly to what we have done for the T0_BI code, we can define the the *dual_T0_BI code*. With this scheme, which is expected to work well on multiplexed buses, the overall switching activity can be reduced by resorting the T0 code anytime stream α is transmitted on the bus, while the bus-invert is used for stream β. As for the T0_BI code, the dual_T0_BI can offer significant savings for the average power. The dual_T0_BI encoding is defined as:

$$(B^{(t)}, \text{INCV}^{(t)}) = \begin{cases} (B^{(t-1)}, 1) & \text{if SEL} = 1 \wedge b^{(t)} = \tilde{b}^{(t)} + S \\ (\overline{b}^{(t)}, 1) & \text{if SEL} = 0 \wedge H^{(t)} > N/2 \\ (b^{(t)}, 0) & \text{otherwise} \end{cases} \qquad (11)$$

where $\mathbf{B}^{(t)}$ is the value on the encoded bus lines at time t, $INCV^{(t)}$ is the additional bus line, $\mathbf{b}^{(t)}$ is the address value at time t, $\mathbf{S}$ is the stride, SEL is the selection signal, $\mathbf{H}^{(t)} = (\mathbf{B}^{(t-1)}|INCV^{(t-1)}, \mathbf{b}^{(t)}|0)$ is the Hamming distance, N is the bus width of $\mathbf{b}^{(t)}$, and $\tilde{\mathbf{b}}^{(t)}$ is defined as before.

The decoding scheme can be summarized with the following equation:

$$b^{(t)} = \begin{cases} (\tilde{b}^{(t)} + S) & \text{if INCV} = 1 \wedge \text{SEL} = 1 \\ \overline{B}^{(t)} & \text{if INCV} = 1 \wedge \text{SEL} = 1 \\ B^{(t)} & \text{if INCV} = 0 \end{cases} \qquad (12)$$

3.4 Experimental Performance Comparison

In order to test out the performance of the newly proposed encoding techniques, we have used again the MIPS microprocessor as reference architecture on which measuring the number of address bus transitions required by the execution of the usual set of benchmark programs.

Tables 5–7 summarize the results, in terms of number of transitions and transitions savings with respect to the pure binary encoding.

When the percentage of in-sequence addresses is high, as in the case of instruction address streams (see Table 5), the savings provided by the T0_BI, the dual_T0 and the dual_T0_BI codes are very remarkable (35.52%, on average, with respect to pure binary). However, the same savings have been obtained by using the simple

Table 5 Experimental comparison of mixed encoding schemes for instruction address streams

Bench	Stream	In-Seq	Binary	T0_BI		Dual_T0		Dual_T0_BI	
mark	Length	Addr.	Trans.	Trans.	Savings	Trans.	Savings	Trans.	Savings
gzip	119102	60.16%	232587	140845	39.44%	140845	39.44%	140845	39.44%
gunzip	63884	64.94%	118409	84003	29.06%	81433	31.23%	81433	31.23%
ghostview	404595	57.27%	678295	481845	28.96%	470949	30.57%	470949	30.57%
espresso	1751673	60.17%	3014854	2286402	24.16%	2239400	25.72%	2239400	25.72%
nova	544994	52.99%	959912	838481	12.65%	836708	12.83%	836708	12.83%
jedi	14690249	53.96%	23145174	18521606	19.98%	18521606	19.98%	18521606	19.98%
latex	700317	71.42%	1400540	200490	85.68%	200490	85.68%	200490	85.68%
matlab	6400326	76.56%	12000527	5800465	51.66%	5800465	51.66%	5800465	51.66%
oracle	500326	69.98%	880547	680485	22.72%	680485	22.72%	680485	22.72%
Average		63.05%			34.92%		35.52%		35.52%

Table 6 Experimental comparison of mixed encoding schemes for data address streams

Bench	Stream	In-Seq	Binary	T0_BI		Dual_T0		Dual_T0_BI	
mark	Length	Addr.	Trans.	Trans.	Savings	Trans.	Savings	Trans.	Savings
gzip	34393	52.47%	131651	83540	36.54%	131651	0.00%	118184	10.23%
gunzip	14602	10.59%	92349	82928	10.20%	92349	0.00%	81953	11.26%
ghostview	112689	10.74%	554499	500750	9.69%	554499	0.00%	507987	8.39%
espresso	570750	5.56%	3952944	3471146	12.27%	3956543	0.00%	3426699	13.39%
nova	220050	10.56%	1381975	1233051	10.77%	1381975	0.00%	1218763	11.81%
jedi	6145049	12.67%	47268786	42116937	10.90%	47268786	0.00%	41114510	13.02%
latex	200109	0.01%	763	577	24.38%	763	0.00%	558	26.87%
matlab	1204110	0.00%	5400375	5369926	0.56%	5400375	0.00%	5351262	0.91%
oracle	120111	0.01%	466606	466293	0.07%	466606	0.00%	466039	0.12%
Average		11.40%			12.82%		0.00%		10.66%

Table 7 Experimental comparison of mixed encoding schemes for multiplexed address streams

Benchmark	Stream Length	In-Seq Addr.	Binary Trans.	T0_BI		Dual_T0		Dual_T0_BI	
				Trans.	Savings	Trans.	Savings	Trans.	Savings
gzip	153495	57.44%	833799	715717	14.16%	736373	11.68%	709769	14.88%
gunzip	78486	52.81%	482093	356057	26.14%	426705	11.49%	343335	28.78%
ghostview	517284	58.25%	3067635	2661381	13.24%	2785341	9.20%	2785341	14.29%
espresso	2322423	54.39%	15147444	12349832	18.47%	13586402	10.31%	11883056	21.55%
nova	765045	56.63%	5526484	5218195	5.57%	5254072	4.92%	5054617	8.53%
jedi	20835298	57.88%	152427486	131617956	13.65%	141103112	7.43%	126132486	17.25%
latex	900426	55.55%	7002828	4202462	39.99%	6202718	11.43%	4002340	42.85%
matlab	7604436	65.37%	42405967	32257681	23.93%	34005845	19.81%	30775139	27.43%
oracle	620437	59.13%	3939551	3116685	20.89%	3479391	11.68%	2964637	24.75%
Average		57.62%			19.56%		12.15%		22.25%

T0 code. Thus, the latter seems to be the most preferable solution in this case, due to the relatively low cost of the T0 encoding/decoding circuitry.

Regarding the data address streams, shown in Table 6, no savings are provided by the dual_T0 code, while the T0_BI and the dual_T0_BI offer some advantages with respect to binary encoding (12.82% and 10.66%, respectively). By comparing these results to the ones provided by the bus-invert method, we can claim that the T0_BI represents the most effective solution for average power minimization.

If we consider multiplexed address buses (see Table 7), the dual_T0_BI code shows the best savings (22.25% on average) if compared to the T0_BI and dual_T0 codes (19.56% and 12.15% over pure binary, respectively). Furthermore, the dual_ T0_BI provides the *absolute* best savings, thus being the most effective code, concerning the average power, for the address bus of the MIPS microprocessor.

4   Encoding and Decoding Logic

The results of Section 3.4 show that, for a muxed bus, the dual_T0_BI code is the most effective in terms of transition count. It is now necessary to evaluate if the power savings achievable through activity reduction is not offset by the circuitry required to implement the code on a system bus. After describing the the basic encoder/decoder architecture, we analyze their power consumption when the encoding scheme is used to minimize the power of on-chip and off-chip buses.

4.1   Architectures

The encoder architecture can be derived directly from Eq. 11. It consists of a section for the T0 encoding which generates the *INC* signal, a section for the bus-invert logic providing the *INV* signal, and the output multiplexor, which is controlled by the input signal *SEL* and by signal $INCV = INC + INV$. While the architecture of the T0 section was fully described in [6], the bus-invert section has been realized by a Hamming distance evaluator of the encoded bus lines at time $t - 1$ concatenated with the *INCV* signal and the address value at the present time t, followed by a majority voter to decide if the computed Hamming distance is greater than half of the bus width. Besides the encoded bus lines, the circuit outputs the *INCV* signal (the *SEL* signal, needed by the decoder, is already present on the bus). The circuit has been synthesized using Synopsys Design Compiler, and implemented onto a 0.35µm, 3.3 V library from SGS-Thomson. Its critical path is 5.36 ns, and it is through the bus-invert section and the output mux.

Concerning the decoder, its architecture can be obtained easily from Eq. 12, and it is reminiscent of the T0 decoder detailed in [6] (*SEL* and *INCV* are the control signals of the multiplexer).

4.2   Power Analysis: On-Chip Busses

The analysis is carried out for three codes: Binary, T0, and dual_T0_BI. The binary encoder and decoder consist only of internal buffers, the T0 circuitry is the one presented in [6], and the dual_T0_BI encoding/decoding logic has the structure described in Section 4.1.

The same reference input switching activities (derived from the benchmark address streams) are applied to the inputs of the three encoders, and the power consumption is estimated using the probabilistic mode of Synopsys Design Power at a frequency rate of 100 MHz.

Concerning the decoders, the reference switching activities derived from the benchmark streams cannot be used to estimate the power of the T0 and the dual_T0_BI decoders; this is because the circuits receive, as input, the encoded address streams whose switching activities are reduced. Therefore, the switching activities used for the estimation are derived from such streams. Table 8 reports the total power consumption results. The power value of the dual_T0_BI encoder is approximately one order of magnitude worse than the one of the T0 encoder for on-chip loads up to 0.4 *pF*, while for higher values the difference is reduced. On the other hand, the power values of the decoders for the T0 and dual_T0_BI codes are comparable, due to the similarity in their architectures.

Table 8 Enc/Dec power consumption for on-chip loads

Load (pF)	Binary Enc/Dec (mW)	T0 Encoder (mW)	T0 Decoder (mW)	Dual_T0_BI Encoder (mW)	Dual_T0_BI Decoder (mW)
0.01	0.1319	6.2891	3.1422	60.3139	3.6518
0.02	0.1660	6.3786	3.2043	60.4658	3.7238
0.04	0.2342	6.5579	3.3284	60.6372	3.8687
0.06	0.3023	6.7393	3.4525	60.8119	4.0176
0.08	0.3704	6.9219	3.5766	60.9895	4.1665
0.10	0.4381	7.1048	3.7088	61.1708	4.3163
0.16	0.6398	7.6538	4.1084	61.7393	4.7664
0.20	0.7742	8.0198	4.3772	62.1226	5.0682
0.30	1.1104	8.9376	5.0523	63.1054	5.8232
0.40	1.4465	9.8562	5.7274	64.1040	6.5785
0.50	1.7827	10.7749	6.4025	65.1125	7.3337
0.60	2.1188	11.6935	7.0776	66.1232	8.0889
0.70	2.4549	12.6122	7.7527	67.1388	8.8441
0.80	2.7911	13.5309	8.4278	68.1622	9.5993
0.90	3.1272	14.4496	9.1029	69.2003	10.3545
1.00	3.4633	15.3682	9.7780	70.2499	11.1098
2.00	6.8246	24.5550	16.5289	80.8256	18.6619
3.00	10.1859	33.7418	23.2798	91.5777	26.2141
4.00	13.5472	42.9286	30.0308	102.4174	33.7663
5.00	16.9085	52.1154	36.7817	113.2670	41.3184
10.00	33.7150	98.0493	70.5364	167.7545	79.0793

4.3   Power Analysis: Off-Chip Busses

When a off-chip system bus is considered, it is necessary to introduce input and output pads at the chip interface. Pads usually represent the most power consuming part of the entire chip. The binary encoder is thus constituted only by the output pads driving typical output loads. This same structure must be added at the outputs of the T0 and dual_T0_BI encoders. Concerning the decoding circuitry, only the power dissipated by the logic is considered in the analysis, since the power consumption due to the input pads has shown to be negligible with respect to the one consumed by the output pads seen by the encoders. The reference input switching activities are applied to the three encoders. In the binary encoder, the output pads commute at the reference switching activities. Conversely, the T0 and the dual_T0_BI encoders receive, as input, the reference switching activities, which are reduced at their outputs in the same manner as for the on-chip case. The encoders outputs drive the typical input capacitances of the output pads (0.01 *pF* for a 8 *mA* output pad). The reduced encoders output switching activities constitutes the switching activities applied to the output pads, which drive huge external loads. Therefore, the major power gains of the proposed codes derive from such reduction in the switching activities applied to the output pads.

The T0 and the dual_T0_BI decoders receive, as input, the reduced switching activities resulting at the corresponding output pads, and drive typical on-chip capacitances.

Power results are reported in Table 9. By looking at the data, we can see that the use of the T0 code is convenient for loads between 20 and 100 *pF*, while for larger values the use of the dual_T0_BI code is recommended.

Table 9 Enc/Dec power consumption for off-chip loads

Load (pF)	Binary Global (mW)	T0 Pads (mW)	T0 Global (mW)	Dual_T0_BI Pads (mW)	Dual_T0_BI Global (mW)
5	43.4593	38.3835	47.8148	31.9732	95.9389
10	60.1896	53.1597	62.5910	44.2817	108.2474
15	76.9198	67.9360	77.3673	56.5902	120.5559
20	93.6501	82.7123	92.1436	68.8988	132.8645
25	110.3804	97.4885	106.9198	81.2073	145.1730
30	127.1107	112.2648	121.6961	93.5158	157.4815
40	160.5713	141.8174	151.2487	118.1329	182.0986
50	194.0318	171.3699	180.8012	142.7500	206.7157
60	227.4924	200.9224	210.3537	167.3670	231.3327
70	260.9530	230.4750	239.9063	191.9841	255.9498
80	294.4135	260.0275	269.4588	216.6012	280.5669
90	327.8741	289.5801	299.0144	241.2182	305.1839
100	361.3347	319.1326	328.5639	265.8353	329.8010
110	394.9327	348.8066	358.2379	290.5535	354.5192
120	428.5308	378.4806	387.9119	315.2718	379.2375
130	462.1288	408.1546	417.5859	339.9900	403.9557
140	495.7270	437.8285	447.2598	364.7082	428.6739
150	529.3250	467.5026	476.9339	389.4265	453.3922

5 Conclusions and Future Work

A comparative analysis of existing low-power bus encoding techniques, such as the T0 and the bus-invert codes, has allowed us to come up with mixed encoding schemes which exploit the best characteristics of the codes above. More specifically, we have proposed the T0_BI, the dual_T0, and the dual_T0_BI codes, and we have discussed their performance concerning the reduction in switching activity obtained on the multiplexed address bus (i.e. instruction and data addresses travel on the same bus) of the MIPS microprocessor when real programs are executed. The dual_T0_BI has shown to be the most effective scheme, since it has produced a 22.25% savings over the pure binary encoding, while the T0 code—the best approach known so far—has only given a 10.25% switching activity reduction.

We have implemented the dual_T0_BI encoding/decoding logic, and we have compared its power performance to the ones of the binary and T0 encoders and decoders when on-chip and off-chip buses with different capacitive loads have to be driven.

Concerning future work, we are now looking into the problem of identifying the most appropriate encoding schemes for different types of memory hierarchies (e.g. main memory, L1 and L2 caches), address bus connections and I/O subsystems. As a first step in this direction, we are thus working on the characterization of existing microprocessors (e.g. MIPS, SPARC, PowerPC, DEC-Alpha, PA-RISC, Intel) with respect to these architectural options.

References

[1] P. R. Panda, N. D. Dutt, Reducing Address Bus Transitions for Low Power Memory Mapping, *EDTC-96: IEEE European Design and test Conference*, pp. 63–67, Paris, France, March 1996.

[2] M. R. Stan, W. P. Burleson, Bus-Invert Coding for Low-Power I/O, *IEEE Transactions on Very Large Scale Integration (VLSI) Systems*, Vol. 3, No. 1, 49–58, March 1995.

[3] J. L. Hennessy, D. A. Patterson, *Computer Architecture - A Quantitative Approach*, II Edition, Morgan Kaufmann Publishers, 1996

[4] C. L. Su, C. Y. Tsui, A. M. Despain, Saving Power in the Control Path of Embedded Processors, *IEEE Design and Test of Computers*, Vol. 11, No. 4, 24–30, Winter 1994.

[5] H. Mehta, R. M. Owens, M. J. Irwin, Some Issues in Gray Code Addressing, *GLS-VLSI-96: IEEE 6th Great Lakes Symposium on VLSI*, pp. 178–180, Ames, IA, March 1996.

[6] L. Benini, G. De Micheli, E. Macii, D. Sciuto, C. Silvano, Asymptotic Zero-Transition Activity Encoding for Address Busses in Low-Power Microprocessor-Based Systems, *GLS-VLSI-97: IEEE 7th Great Lakes Symposium on VLSI*, pp. 77–82, Urbana-Champaign, IL, March 1997.

[7] L. Benini, G. De Micheli, E. Macii, M. Poncino, S. Quer, System-Level Power Optimization of Special Purpose Applications: The Beach Solution, *ISLPED-97: IEEE/ACM International Symposium on Low Power Electronics and Design*, pp. 24–29, Monterey, CA, August 1997.

[8] G. Kane and J. Heinrich, *MIPS RISC Architecture*, Prentice Hall, 1994.

MOCSYN: Multiobjective Core-Based Single-Chip System Synthesis

Robert P. Dick and Niraj K. Jha

Abstract   In this paper, we present a system synthesis algorithm, called MOCSYN, which partitions and schedules embedded system specifications to intellectual property cores in an integrated circuit. Given a system specification consisting of multiple periodic task graphs as well as a database of core and integrated circuit characteristics, MOCSYN synthesizes real-time heterogeneous single-chip hardware–software architectures using an adaptive multiobjective genetic algorithm that is designed to escape local minima. The use of multiobjective optimization allows a single system synthesis run to produce multiple designs which trade off different architectural features. Integrated circuit price, power consumption, and area are optimized under hard real-time constraints. MOCSYN differs from previous work by considering problems unique to single-chip systems. It solves the problem of providing clock signals to cores composing a system-on-a-chip. It produces a bus structure which balances ease of layout with the reduction of bus contention. In addition, it carries out floorplan block placement within its inner loop allowing accurate estimation of global communication delays and power consumption.

1   Introduction

The process of concurrently defining the hardware and software portions of an embedded system while considering dependencies between the two is called hardware–software codesign [1–4]. Time pressure makes it difficult for an engineer to explore the numerous alternative designs which have the potential to meet a given set of specifications. As a result, an engineer frequently selects a conservative architecture after little experimentation, resulting in a needlessly expensive system. The majority of past research in the area of hardware–software codesign has

Department of Electrical Engineering Princeton University Princeton, New Jersey 08544

This work was supported in part by an NSF Graduate Fellowship and in part by NSF under Grant No. MIP-9729441.

R. Lauwereins and J. Madsen (eds.), *Design, Automation, and Test in Europe–The Most Influential Papers of 10 Years DATE*, 291–311
© Springer 2008

attempted to ease the process of design exploration, allowing an engineer to examine a number of alternative implementations in the short time available. Co-synthesis is the automation of the generation of an embedded system architecture. Given an embedded system specification, a co-synthesis system selects hardware and software-processing elements, devices upon which tasks execute. In addition, the system assigns each task to a processing element and ensures that tasks which need to communicate with each other are capable of doing so. Finally, schedules are produced for tasks and communication, such that real-time constraints are met [5, 6]. Co-synthesis systems generate feasible, low-cost architectures without designer intervention.

It is possible to implement some embedded systems using a single integrated circuit (IC), thereby reducing cost and improving performance [7]. Economic and time pressures frequently make it impractical to do an in-house design for each component in a single-chip system. Fortunately, the number of intellectual property (IP) cores available from the industry has dramatically increased in the last year. Companies like Alta Group, VAutomatic Inc., Virtual Chips, and Intronic offer a wide range of IP cores, e.g. protocol processors, general-purpose processors, microcontrollers, digital signal processors, memory, and Data Encryption Standard engines.

Optimal co-synthesis is an intractable problem. Allocation/assignment and scheduling are each NP-complete for distributed systems [8]. As a result, it is not surprising that all co-synthesis systems which employ optimal mixed integer linear programming [9], [10] and exhaustive exploration [11] can only be applied to small instances of the co-synthesis problem. There are a number of co-synthesis algorithms which make the solution of large problem instances tractable. To achieve reasonable run-time, however, it is necessary to sacrifice the guarantee of solution optimality. Iterative improvement algorithms start with a complete solution and make local changes to it in an attempt to improve the solution's cost [5, 12, 13]. Constructive algorithms build a system by incrementally adding components to it [14, 15]. Simulated annealing algorithms have been successfully used to partition hardware-software systems [16]. Genetic algorithms have been applied to the hardware/software partitioning problem [17]. A multi-objective genetic algorithm was applied to the more general co-synthesis problem [18].

MOCSYN, which stands for multiobjective core-based single-chip system synthesis, differs from past work by considering a number of issues unique to core-based single-chip systems. MOCSYN determines the clock frequencies supplied to different cores. It generates priority-based bus structure of arbitrary topology, balancing ease of routing and bus contention minimization. In addition, it conducts floorplan block placement [19] within its inner loop, allowing estimates of global wiring delays and power consumption to be used during scheduling and cost calculation. Experimental results demonstrate that a global bus is, in general, inferior to the use of priority-based arbitrary bus topologies. Conducting block placement in the inner loop frequently results in an improvement in solution quality when compared with worst-case or best-case communication delay estimates.

2   Data Structures and Definitions

In this section, we describe the data structures used in MOCSYN and define basic co-synthesis terms.

Cost: A cost is a variable that a synthesis system attempts to minimize. Price, power dissipation, and IC area are examples of costs.

Task graph: As shown in Fig. 1, a task graph is a directed acyclic graph in which each node is associated with a task and each edge is associated with a scalar describing the amount of data that must be transferred between the two connected tasks. A task with an incoming edge, i.e. a task toward which an edge's arrow points, may execute only after receiving data from the other task to which the edge is connected. Thus, in Fig. 1, **DCT** is data-dependent on **NEG**. The *period* of a task graph is the amount of time between the earliest start times of its consecutive executions. A node without any outgoing edges is a *sink node*. A *deadline*, the time by which the task associated with the node must complete its execution, exists for every sink node. Other nodes may also have deadlines associated with them.

Multi-rate: A multi-rate system contains task graphs with different periods. It was shown in [20] that, in order to guarantee a valid schedule for a multi-rate system, each task graph must repeatedly execute until the least common multiple (LCM) of the periods, termed as the hyperperiod, of all the task graphs in the system has elapsed.

Core: A core executes one or more tasks. Multiple cores may be located on the same IC, upon which multiple tasks may execute simultaneously. The following information establishes the relationship between tasks and cores:

- A two-dimensional array indicating the relative worst-case execution time of each task on each core.

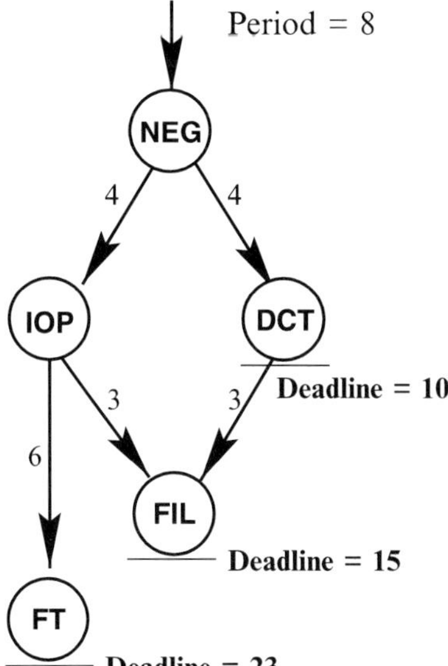

Fig. 1 Task graph

- A two-dimensional array indicating the relative average power dissipation of each task on each core.
- A two-dimensional array indicating the core types upon which each task type may be executed.

In addition, each core has a price which corresponds to royalties paid to the IP producer on a per-use basis. This price is zero for royalty-free IP cores. If IP has a one-time fee instead of, or in addition to, a per-use royalty, the price is equivalent to the one-time fee divided by the expected production volume. Each core has a width, a height, a maximum clock frequency, a variable indicating whether or not its communication is buffered, and an energy consumption per cycle dedicated to communication.

Core allocation: The information denoting the number of cores of each type present in an IC.

Link: A link is a potential point-to-point contact between a pair of cores. Any given link may be merged with other links during bus formation (see Section 3.7), thereby ceasing to exist.

Task assignment: The information denoting the core upon which a given task is executed.

Architecture: A set of allocation, assignment, and scheduling information which defines an embedded system.

3   The MOCSYN Algorithm

In this section, we give an overview of MOCSYN (see Section 3.1) and describe the algorithms of which it is composed. Section 3.2 describes the algorithm used to determine the clock frequency used for each type of core. Section 3.3 describes the initialization of MOCSYN's data structures. Section 3.4 describes how core allocation and task assignment are determined. Section 3.5 explains how links are prioritized. The block placement algorithm used by MOCSYN is described in Section 3.6. Bus topology generation is explained in Section 3.7. Scheduling is explained in Section 3.8. Finally, cost calculation is described in Section 3.9.

3.1   Algorithm Overview

In this subsection, we give a high-level description of the MOCSYN algorithm. MOCSYN carries out the following tasks:

1. Determine a *clock frequency* for each core type, subject to trade-off between execution time and power consumption.
2. Determine the *allocation* of cores to use.
3. Determine the tasks to *assign* to each core, subject to trade-off between ease of routing and minimization of bus contention.

4. Determine a *bus structure* to use on the IC.
5. Derive a *block placement* for the cores, allowing an estimation of wire delay, wire power consumption, and silicon area.
6. *Assign* each communication event to a bus.
7. *Schedule* the tasks on the cores and the communication events on the communication links.

MOCSYN uses a genetic algorithm to optimize embedded system architectures. In general, genetic algorithms have the ability to escape local minima. They allow solutions to cooperatively share information with each other. Genetic algorithms are especially useful for simultaneously optimizing more than one cost. Conventional iterative improvement and simulated annealing algorithms maintain only one solution at a time. Most single-solution optimization algorithms collapse all costs into a single value with a weighted sum [21, 22]. Genetic algorithms maintain a pool of solutions which evolve in parallel. This allows solutions to be ranked relative to each other. Genetic algorithms are capable of true multiobjective optimization, exploring the *Pareto-optimal set* of solutions, i.e. those solutions which are better than any other solution in at least one way. In a genetic algorithm, solutions are iteratively improved by *mutation*, randomized local changes to a solution, and *crossover*, during which information is shared between different solutions. In this paper, we concentrate on the unique problems encountered when designing single-chip core-based systems.

An overview of the MOCSYN algorithm is shown in Fig. 2. Initially an optimal, but potentially slow, algorithm determines the clock frequency to provide to

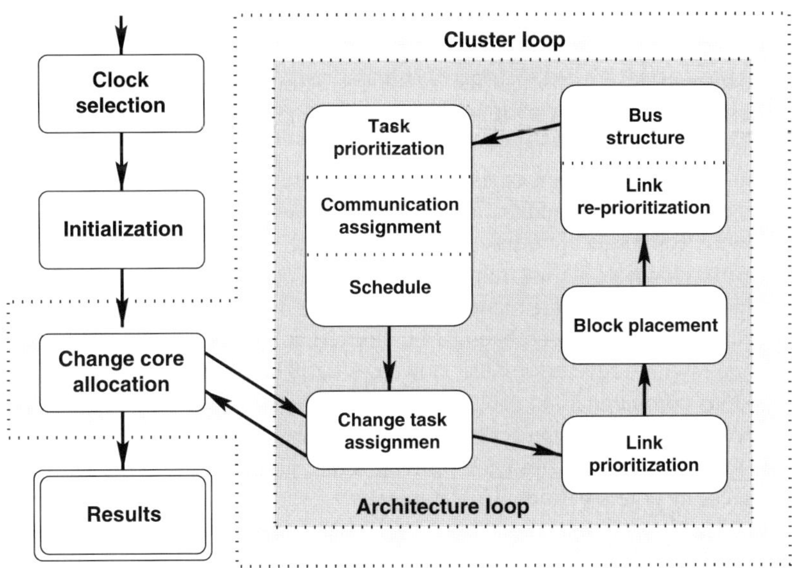

Fig. 2 MOCSYN overview

each core. Basic data structures are then initialized. MOCSYN is a hierarchical algorithm. It uses a genetic algorithm to improve the task assignments of individual architectures. More detailed information about this genetic optimization framework can be found in [23]. After this phase has been repeated an arbitrary (user-selectable) number of times, an attempt is made to improve the core allocation of a *cluster* of architectures, i.e. a collection of architectures which share the same core allocation but may have different task assignments. Within the architecture optimization loop, a number of deterministic algorithms are used to concurrently evaluate the core allocation and task assignment of each architecture. First, a priority is assigned to each link, i.e. the communication carried out between each pair of cores. These priorities are used to generate a block placement for the cores, ensuring that core pairs for which communication priority is high are located near each other. Links are reprioritized based on global wiring delay information which is extracted from the block placement. A bus structure which trades off potential bus contention for ease of routing is produced. At this point, tasks are prioritized and a schedule is generated for the tasks assigned to each core. Communication events are concurrently assigned to, and scheduled on, busses. At the completion of each architecture optimization loop, changes are made to the task assignments in an attempt to improve them. At the completion of each cluster optimization loop, changes are made to the core allocations in an attempt to improve them. Each of the sub-algorithms noted in Fig. 2 is described in the following sections.

3.2   Clock Selection

In this subsection, we discuss the problems associated with selecting a clock frequency for each core in an IC and describe the algorithm used in MOCSYN to solve these problems.

Communication between cores in an IC can be single-frequency synchronous, multi-frequency synchronous, or asynchronous [24, 25]. Single-frequency synchronous communication has the potential to keep communication overhead at a minimum. However, its use requires that all the cores which communicate with each other be clocked at the same frequency. When different cores have different maximum frequencies, all cores must be clocked at a frequency less than or equal to the maximum frequency of the slowest core. Thus, using a single-frequency synchronous communication protocol will generally force sacrifices in core speed. Multifrequency synchronous communication allows cores with different clock periods to communicate with each other at a rate proportional to the LCM of the communicating core's periods. Unfortunately, when cores have different minimum periods and efforts are made to allow each core to run near its maximum frequency, the LCM of the periods of communicating cores can be significantly higher than the period of any individual core, e.g. LCM(5, 7) = 35. This generally results in slow communication. A third option is asynchronous communication. Although it has a reputation for increasing communication overhead, we believe that it is the best available option for systems in which different cores are clocked at significantly different frequencies. Using asynchronous communication, communication speed is bounded only by the bus bandwidth and rate at which communicating cores can transmit and receive information. Using asynchronous communication has the additional advantage of making inter-core clock skew irrelevant. Past work provides a framework for automatically synthesizing asynchronous interface protocols [25].

Given that one is using asynchronous communication, the selection of clock frequencies for the cores comprising a single-chip system need not be constrained by communication considerations. However, there are a number of other problems which must be dealt with. Supplying each core with an arbitrary clock frequency would require a large number of frequency generators, e.g. analog timers based on RC delay or crystal oscillation. These components are difficult to integrate with conventional CMOS IC processes. Using discrete components is a poor option because each additional external component increases the price and area of an embedded system. Thus, a clocking approach which requires only one frequency source but allows nearly arbitrary frequencies to be delivered to each core would be advantageous.

We use an approach in which a single external oscillator is used to supply a base frequency. A cyclic counter or interpolating clock synthesizer associated with each core is used to divide this frequency by an integer, in the case of a cyclic counter, or multiply the frequency by a rational number, in the case of an interpolating clock synthesizer [26]. A description of the clock selection algorithm used in MOCSYN follows, which is capable of dealing with interpolating clock synthesizers. As the cyclic counter clock selection problem is a special case of the interpolating clock synthesizer clock selection problem, the algorithm used in MOCSYN is capable of solving either problem.

Given: A maximum external clock frequency ($Emax$) and a maximum frequency associated with each of the n cores ($Imax_1$, $Imax_2$,…, $Imax_n$).

Each core's clock frequency multiplier is a rational number, $M_i = N_i/D_i$, with a positive integer numerator N_i less than or equal to a user-supplied maximum, $Nmax$, and a positive integer denominator, D_i. A core's internal frequency, I_i, is equal to the external frequency (E) multiplied by its multiplier, M_i.

MOCSYN maximizes the average of the ratios of the core frequencies (I_i) to the core frequency maxima ($Imax_i$), i.e.

$$\left(\sum_{i=1}^{n} I_i / I\,max_i \right) / n$$

It is simple to determine an optimal external frequency (E) if the value of each multiplier (M_i) is known. Obviously, for an optimal E, $\exists i \in [1, n]$ such that $I_i = Imax_i$. Thus, one need only consider a small set of E's. $i \in [1, n]$, $E_i = Imax_i/M_i$. The $\max_{i=1}^{n} E_i$ for which $\nexists_{i=1}^{n} I_i > Imax_i$ is the optimal E for a given set of M's.

The only remaining problem is to determine an optimal set of M's. It is obvious that, for any given pair of $Imax$'s, $Imax_a$, and $Imax_b$, if $Imax_a \geq Imax_b$ then an optimal $M_a \geq M_b$. This observation allows the solution space of M's to be pruned.

Initially, all D's are equal to 1 and all N's are equal to $Nmax$. Therefore, all M's are equal to $Nmax$.

To maximize the average of core frequency to maximum frequency ratios, one need only repeatedly execute a simple algorithmic kernel, while keeping track of the best set of M's, until $E > Emax$. This kernel is shown in Fig. 3. Although, given a maximum highest internal clock frequency of $Imax_a$ and a minimum highest internal clock frequency of $Imax_b$, this algorithm takes $\&O; (n \cdot Nmax \cdot Imax_a / Imax_b)$ time, in practice it is fast (see Section 4).

Linear interpolating clock synthesizers are compatible with standard digital design tools and processes. Their use provides a significant advantage: one can distribute a base global clock frequency which is well below the maximum local clock frequencies, thereby reducing power consumption in the global clock distribution net. However, interpolating clock synthesizers are more complicated than cyclic counters. In addition, they are likely to require more area [26]. If one chooses to use cyclic clock division counters, instead of linear interpolating clock synthesizers, the same clock selection algorithm is used. However, $Nmax$ is set to 1.

3.3 Initialization

At the start of an optimization run, each cluster's core allocation is initialized. One of three initialization routines is randomly selected:

1. Add one core of a randomly selected type.

> For each i between 1 and n, inclusive,
> there is an array, A_i of size Nmax,
> which is contains integers.
>
> Each of these integers is the current denominator
> for the numerator equivalent to its index.
>
> Optimize E for the current M's.
>
> For all i's between 1 and n, inclusive if $I_i = Imax$:
> j ranges from 1 to $Nmax$, inclusive
>   find the j for which $j/(A_{ij} +1)$ is maximal
>   Increment A_{ij}
>   Set $D_i \leftarrow A_{ij}$
>   Set $N_i \leftarrow j$

Fig. 3 Clock selection kernel

2. Add one core of each type.
3. Repeatedly add cores of random types until a random number (ranging from one to twice the number of core types) has been added.

MOCSYN ensures that there is at least one core capable of executing each type of task in the input task graphs. It checks each task and adds an appropriate core to the allocation if none of the cores currently in the allocation are capable of executing the task. Each architecture's tasks are assigned using the task assignment algorithm described in Section 3.4. The global temperature of the genetic algorithm is set to one. This temperature decreases during the run of the algorithm and is used to control the probability of making a core allocation or task assignment change which decreases the quality of a solution [23]. At the beginning of an optimization run, random changes are made. As time progresses, MOCSYN becomes greedier, making only those changes which result in an improvement in core allocation or task-assignment quality. This property increases the probability that MOCSYN will escape local minima [27].

3.4 *Core Allocation and Task Assignment*

Core allocations are optimized by MOCSYN's genetic algorithm. The allocations change via crossover and mutation of the data structure which maintains the core allocations for each cluster of architectures. When mutation occurs, a core is added or removed from the core allocation. The probability of adding a core is equivalent to MOCSYN's global temperature, which gradually changes from one to zero during the run of the algorithm. This results in allocations being more likely to increase during the start of a run and more likely to be pruned near the end of a run. After removing a core, MOCSYN ensures that there is at least one core capable of executing each type of task present in the input task graphs using the method described in Section 3.3.

Core allocation crossover swaps portions of the core allocations of two clusters with each other. MOCSYN uses a novel method for selecting portions of the core allocations to be swapped. The probability of the allocations of two types of cores remaining together during a crossover, i.e. the probability that both are swapped or that both are not swapped, is proportional to the similarity between the data describing the core types, e.g. prices, execution time vectors, and power consumption vectors.

Task assignment mutation causes tasks to be reassigned to different cores. A task graph is randomly selected and a number of tasks within the graph are randomly selected for reassignment. The number of tasks to be reassigned is equivalent to the number of tasks in the task graph multiplied by the global temperature. For each task selected, a new core assignment is determined. A core's Pareto-rank is the number of other cores for which at least one property is inferior. First the properties of each core, when executing the given task, are used to determine the Pareto-ranks of all the cores capable of executing the task. Execution time, energy consumption, core area, and weight, a measure of the time required to execute the tasks already assigned to a core, are used in the Pareto-ranking process. An array of cores is sorted in order of increasing Pareto-rank. The core to which the selected task is re-assigned is determined by indexing into the sorted array by $\left|1(1 - \sqrt{flat - rand}).array - size\right|$ where flat_rand is a uniform random variable ranging from [0, 1) and array size is the size of the sorted array of cores.

Task assignment crossover causes the task assignments of one or more task graphs to be swapped between two architectures. The probability of the task assignments of two task graphs remaining together during a crossover is proportional to the similarity between the data describing the task graphs, e.g. periods and deadlines. A similar algorithm is used for both core allocation crossover and task assignment crossover.

3.5   Link Prioritization

This subsection describes the algorithm used by MOCSYN to prioritize links. Communication priority is composed of two values: slack and communication volume. *Slack* is the difference between the earliest finish time and latest finish time of a task. Thus, it is the amount of time by which a task's execution can be delayed, from its earliest possible execution time, without causing any other tasks to miss their deadlines. Earliest finish times are computed by considering task execution times during a topological search of the task graph, starting from the node with no incoming edges. Latest finish times are computed by conducting a backward topological search of the task graph, starting from the nodes which have deadlines.

Task graph edges, which signify communication, have a slack equivalent to the average of the slacks of the tasks they connect. Link priority determination is conducted before block placement and bus topology generation. Therefore, it is not possible to take communication time into account during communication priority determination. Hence, at this stage, slack is only estimated. This problem is corrected later, during link reprioritization. Communication volume is the quantity of communication which passes along a link. Link priority is a weighted sum of the reciprocals of the slacks of the task graph edges along it and its communication volume.

3.6   Floorplan Block Placement

This subsection describes the block placement algorithm used within MOCSYN's inner loop. This algorithm is based on previous work. Initially, a balanced binary tree of cores is formed, based on the priority of communication between core pairs. Accounting for the priority of communication between core pairs is an extension of the historical algorithm, which considered only the binary presence or absence of communication [28]. As a result, the time complexity of the partitioning algorithm is increased from $\mathcal{O}(n^2)$ to $\mathcal{O}(n^2 \cdot \log n)$ where n is the number of cores. Cores which are adjacent in the binary tree will be adjacent in the final block placement. After forming the binary tree, MOCSYN optimally determines the orientations of all of the cores such that the aspect ratio of the IC, i.e. the ratio between width and height, does not exceed a value specified by the user. Under this condition, IC area is minimized. This algorithm is based on past work and takes $\mathcal{O}(n \cdot \log n)$ time [29].

3.7 Bus Formation

This subsection describes the algorithm used by MOCSYN to produce an arbitrary bus topology.

MOCSYN recalculates link priorities using an algorithm similar to that described in Section 3.5. The global wiring delay information extracted from the block placement, however, is used to estimate communication time during this calculation.

In Fig. 4, the core graph represents four cores (**A, B, C,** and **D**). The edges connecting pairs of cores represent communication, i.e. no edges exist for core pairs between which there is no communication. The numeric labels on the edges denote the priority of the communication occurring between the connected cores. Thus, communication with a priority of seven occurs between core **A** and core **D**. The first step of MOCSYN's bus formation algorithm converts this core graph into a link graph, as shown in Fig. 4. For every pair of cores between which communication occurs, a node with the priority equivalent to that pair's communication priority is added to the link graph. Link graph nodes which share at least

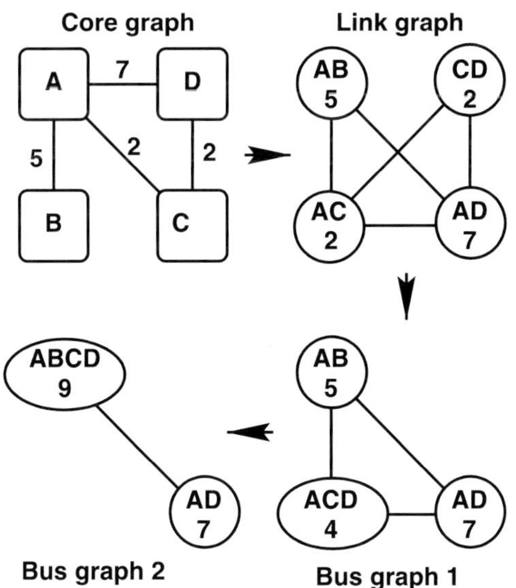

Fig. 4 Bus formation

one core are connected to each other with edges. Since core **A** communicates with core **B** with a priority of five, there is a node in the link graph, called **AB**, with a value of five. There is an edge between node **AB** and **AC** because they share core **A**. There is no edge between node **AB** and **CD** because they share no core.

The link graph is incrementally changed by merging the pair of nodes, between which there exists an edge and for which the sum of priorities is minimal, i.e. less than the sum of the priorities of any other pair of nodes between which an edge exists. The new node's name is the set union of the merged nodes' names. The new node's priority is the sum of the priorities of the nodes merged to form it. Thus, when node **AC** is merged with node **CD** in bus graph 1, the resulting node has the name **ACD** and a priority of four $(2 + 2)$. This algorithm is halted when the number of busses is less than or equal to a user-specified value.

After halting the bus formation algorithm, one is left with a bus graph in which each node represents a bus. Thus, in bus graph 2, there exists one global bus (**ABCD**) and one point-to-point link (**AD**). Note that the algorithm tends to form large common busses for multiple low-priority communications while producing smaller busses for high-priority communication. In this way, bus contention is reduced for high-priority communication while routing and multiplexing complexity is reduced for low-priority communication.

3.8   Scheduling

In this subsection, we describe the scheduling algorithm used in MOCSYN. Scheduling is NP-complete for distributed systems [8]. We, therefore, resort to a heuristic scheduling algorithm. MOCSYN uses a preemptive static critical path scheduling algorithm. The resulting schedule is static, i.e. the time at which each event is carried out is computed by MOCSYN to determine whether or not hard deadlines are met by the schedule. Such guarantees are not possible, in general, when task priorities are allowed to vary during the operation of the synthesized architecture.

We assume that uniform buffers are distributed throughout the communication networks in order to minimize communication delay. The use of regularly distributed buffers reduces the dependency of delay on wire length from $\mathcal{O}\ (len^2)$ to $\mathcal{O}\ (len)$ where len is wire length. Given the process parameters, and V_{DD}, optimal buffer spacing is calculated. This value is used to determine the RC delay between a pair of cores [30]. This value is divided by the bus width and multiplied by the number of digital voltage transitions to determine the delay for a communication event. Core execution time is equal to the number of execution cycles divided by the core's frequency.

MOCSYN targets multi-rate embedded systems. It ensures validity of schedules by scheduling copies of task graphs until the hyperperiod has been reached. It is, therefore, possible to have multiple task copies in the schedule for a single task specified in the input task graphs. When there are multiple copies of the same task graph, they are numbered in order of increasing start node earliest start time. Each copy's number is its *task graph copy number*. Task graphs may have periods which are less than the maximum deadline in the task graph. This makes it possible for the execution of multiple instances of the same task graph to overlap in time. MOCSYN handles this case correctly and is capable of interleaving tasks from different copies of the same task graph, as well as from different task graphs.

Before scheduling, MOCSYN assigns a priority to each task. This priority is the slack of the task. It differs from the slack computed in Section 3.5 because, at this point, a block placement has been generated. Thus, slack takes communication delay into account. Unfortunately, the effects of bus contention are unknown before scheduling is carried out.

When the scheduling algorithm begins, all tasks with no incoming edges are entered into a pending list which is sorted in order of decreasing slack. Ties are broken by ordering the tasks which share the same slack by increasing task graph copy number. Tasks are iteratively removed from the end of the pending list and scheduled. After a task is scheduled, tasks which are data-dependent upon it are checked to determine whether all of their dependencies have been satisfied. Tasks which pass this test are entered into the pending list and the pending list is sorted again before scheduling the next task.

Before scheduling an individual task, t, MOCSYN schedules all of its incoming edges, i.e. communication events. Each edge is scheduled on a bus connecting the core to which t is assigned and the core to which t's parent is assigned. When multiple busses are available, MOCSYN selects the bus upon which the communication event will complete at the earliest time. If either of the communicating cores does not have communication buffers, MOCSYN schedules the communication event to the unbuffered cores as well.

Every time a task is scheduled on a core, MOCSYN determines whether or not preemption is likely to result in an improved schedule. MOCSYN first tentatively schedules a task, t, to the earliest time slot on its core, which starts after its incoming edges have completed execution, and has a long enough duration to accommodate the task. MOCSYN then checks to see whether preempting the task, p, which is scheduled to the same processor as t, previous and adjacent to t, would result in a *net improvement*, where net improvement is defined as the $-$(increase in finish time for p) + (decrease in finish time for t) $-$ (t slack)+(p slack). If preemption results in a net improvement, there is enough time available on the core processor before the next scheduled task, and preempting p does not change the times at which it communicates with tasks on other cores, then the preemption is carried out.

3.9 Cost

In this subsection, we describe the manner in which an architecture's costs are calculated. As mentioned before, MOCSYN optimizes architecture price, area, and power consumption under hard real-time constraints. An architecture is invalid if any task with a deadline violates that deadline. Power consumption is the sum of the energy consumptions of all of an IC's tasks executed on all its cores, throughout the hyperperiod, in addition to the energy consumed in the global clock distribution and communication networks, divided by the hyperperiod. As discussed in Section 3.8, we assume regularly spaced buffers in the global communication network. In addition, the clock network is assumed to be constructed with buffered segments. Leakage current is assumed to be negligible. This allows delay and energy consumption to be estimated as linear functions of wire length and transition count, with constant factors derived from the process parameters and V_{DD}. Ultimately, three such constant factors are computed: communication wire delay factor, communication wire energy factor, and clock energy factor. The energy consumed by the global clock network is determined by estimating the total wire length of this network, multiplying this value by the number of transitions occurring during a hyperperiod, and also multiplying by the clock energy factor. The wire length estimate is derived from a minimal spanning tree of the core positions in the block placement. This provides a conservative estimate on wire length. A Steiner tree may be used in the final post-optimization routing operation. However, computation of minimal Steiner trees is time-consuming (NP-complete). Hence, it is not used in inner-loop routing estimates. Energy consumption in the global communication networks is similarly computed. A separate minimal spanning tree is computed for each bus. The transitions required for the communication events occurring on each bus are used to compute the bus energy consumptions.

An architecture's price is the sum of the prices of all the cores on the IC plus the area-dependent price of the IC. The area of the IC is equivalent to the total rectangular area required for its block placement.

4   Experimental Results

In this section, we present experimental results demonstrating the effects of a number of algorithms employed within MOCSYN. Section 4.1 shows the results produced by the clock selection algorithm when run on an example. In Section 4.2, we empirically determine the influence of a number of MOCSYN's specialized algorithms. Section 4.3 shows the result of running MOCSYN on a number of examples in multiobjective optimization mode.

Previous co-synthesis systems do not target the single-chip synthesis problem. As a result, there is not a body of examples with which MOCSYN's performance can be compared. It is, however, possible to experimentally determine the effects of the algorithms comprising MOCSYN. The examples discussed below attempt to determine how clock selection, block placement, and bus topology generation affect the solution of the single-chip synthesis problem. The data used in these examples are available via anonymous FTP at ftp://ftp.ee.princeton.edu/pub/dickrp /MOCSYN.

4.1 Clock Selection

MOCSYN automatically selects clock frequencies for each core using the algorithm described in Section 3.2. In this subsection, we examine the results produced by this algorithm when run on an example problem.

In the interest of decreasing the power consumed in the global clock distribution network, one may reduce the frequency of the base clock. There is a trade-off between power consumption and execution time. However, this relationship is not linear. Figure 5 shows the relationship between maximum reference clock frequency and the average proportion of maximum internal clock rates at which the cores are clocked for a set of eight cores, each of which has a random maximum internal frequency ranging from 2 to 100 MHz. Each sample point lies at the optimal reference clock frequency for a set of core multiplier values. The top solid line shows the average ratio of actual core frequencies to maximum core frequencies for linear interpolating clock synthesizers with a maximum numerator of eight. The bottom solid line corresponds to a cyclic counter clock divider. The dotted lines indicate the maximum ratio encountered before or at each frequency. The increase in power consumed by the clock reference frequency distribution network is approximately a linear function of frequency.

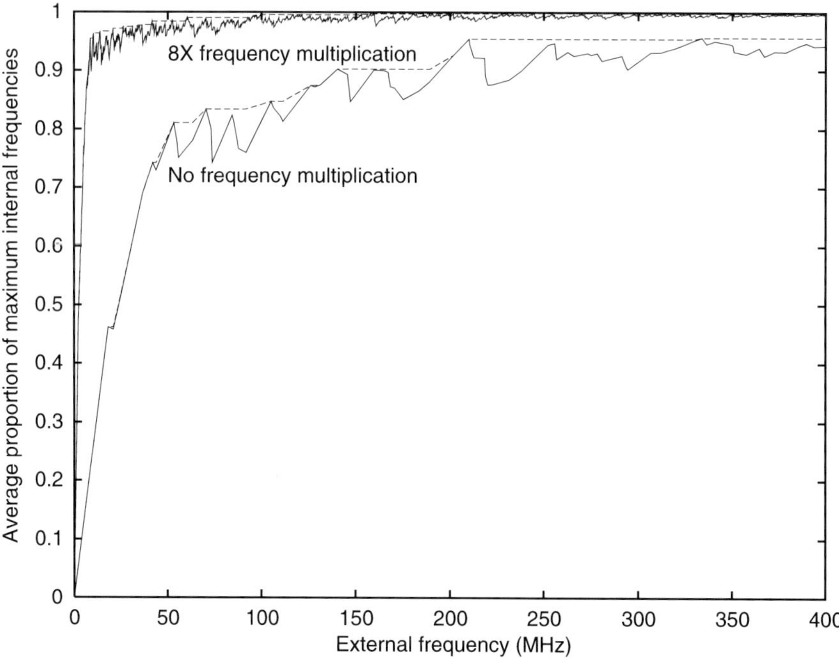

Fig. 5 Clock selection quality as a function of external frequency

As shown in Fig. 5, the quality of the internal clock frequencies is a sublinear function of reference clock frequency. If one were using an interpolating synthesizer with maximum numerator of eight for the cores in this example, choosing a maximum reference frequency greater than 100 MHz would not result in a significant increase in execution speed but may have a negative impact on system power consumption.

4.2   Feature Comparisons

This subsection empirically shows the influence of a number of the core-based synthesis algorithms used in MOCSYN.

Table 1 shows the results of synthesizing a number of ICs using MOCSYN with various sets of features enabled. For these examples, price was optimized under hard real-time constraints. If multiobjective optimization were used, it would be difficult to compare the solutions produced by different versions of MOCSYN because each problem might have more than one solution (see Section 4.3). Each example in Table 1 is produced with the aid of TGFF [31], a randomized task graph and core generator which allows correlation between different attributes. The examples are multi-rate. Each contains six task graphs with an average of eight tasks each and a variability of seven tasks. For each task with a deadline, the deadline is equal to $(depth + 1) \cdot 7{,}800\,\mu s$ where *depth* is the distance of a task, in nodes, from the start node of a task graph. Each communication event requires an average of 256 Kb, with a variability of 200 Kb, of data to be transferred. There are eight core types, each of which has an average price of 100 with a variability of 80. Each core has an average width and height of 6 mm and a width and height variability of 3 mm. The core maximum frequency average is 50 MHz with a variability of 25 MHz. Cores have buffered communication 92% of the time. Communication on cores requires an average of 10 nJ per cycle with a variability of 5 nJ per cycle. Tasks require an average of 16,000 cycles to execute with a variability of 15,000 cycles. Task preemption takes an average of 1,600 cycles with a variability of 1,500 cycles. Tasks dissipate an average of 20 nJ per cycle with a variability of 16 nJ per cycle. 57% of the core types are capable of executing any given task type. Communication wire delay factor, communication wire energy factor, and clock energy factor are calculated based on the 0.25 μm process parameters given in the literature [32], with a V_{DD} of 2.0 V. Communication networks are assumed to be 32 bits wide. Wire delay and power consumption per micron per transition are calculated based on the use of a buffer separation distance which optimizes delay per micron. The maximum clock reference frequency is 200 MHz and the maximum interpolating clock synthesizer numerator is eight. For each example, the same parameters are given to TGFF and MOCSYN. Only the random seed given to TGFF is varied, to produce different examples based on the same parameters.

The bottom two rows in Table 1 display the number of solutions which are superior and the number of solutions which are inferior to those produced by

Table 1 Feature comparisons

Example	MOCSYN price	Worst-case commun. price	Best-case commun. price	Single bus price
2	181	181	181	181
3	255	255		255
4	211	211	211	211
5	154	154	154	154
6	230			
7	174	174	174	174
8	219		219	
9	182	182		182
10	901			
11	166	166	166	166
12	405	405		
13	636			
14	166	368		166
15	151	151	151	151
16	389			
17	302			
18	315			
19	320	365	320	
20	176	176	176	176
21	212			
22	170	257	257	257
23	322	323	322	
24	421	421		
25	113	113	113	113
26	440			
27	173	173	173	173
28	858			
29	169	169	186	169
30	203	203		203
32	396	406		
33	317			
34	368			
35	694			
36	269			
37				132
38	460		460	
39	179	270	179	270
40	327	327		
41	273	273		273
42	426			
43	72			72
44	182			182
45	148	148	148	148
46	344	344		344
47	404	404		285
48	135			135
49	60	60	60	60
50	291	357		211
Better	0	0	0	3
Worse	0	26	31	24

MOCSYN with all its features turned on. The first column in Table 1 shows the random seed given to TGFF to generate the corresponding example. The second column shows the price of solutions produced by MOCSYN when carrying out block placement-based wire delay estimations and using up to eight busses organized to reduce bus contention. Empty price entries indicate examples for which no solution was found. Example 11 is omitted from Table 1 because no solution was ever found for them. Note that there is no guarantee that solutions exist for all of the problems produced by TGFF. Each of the examples in this section took less than two minutes to complete on a 200 MHz Pentium Pro running Linux.

The third column shows the price of solutions under the assumption that the distance in the block placement between each pair of cores is equal to the maximum distance between any pair of cores. Although this estimate may appear conservative, it is not possible to derive a tight bound on the maximum separation between any pair of cores without carrying out block placement in the inner loop. Thus, in practice, this estimate would probably be even more conservative if an inner-loop block placer were not available. Using worst-case communication delay assumptions never resulted in superior results to block placement-based communication delay assumptions. However, MOCSYN's performance with block placement based communication delay estimates was superior to its performance with worst-case estimates for 26 examples, many of which were unsolvable when the worst-case communication delay assumption was used.

The fourth column shows the price of solutions which result from carrying out optimization under the assumption that communication events take almost no time. After the optimization run is complete, solutions which are invalid due to unschedulability are eliminated. Best-case communication delay estimates never resulted in an improvement in solution quality over block placement-based communication delay estimates. However, the use of block placement-based communication delay estimates resulted in superior solutions for 31 of the examples.

The fifth column shows the price of solutions which result from allowing MOCSYN to carry out block placement in the inner loop to accurately estimate communication delay but allowing only a global bus to be used, instead of an arbitrary priority-based topology of up to eight busses. For three examples, this approach actually resulted in an improvement over the solutions produced when allowing up to eight busses. For these examples, there was a valid solution which contained only two or three cores, thereby reducing the amount of off-core communication. The availability of a large number of busses reduced bus contention, making it practical to use a larger number of cores. However, when only one bus was available, it became essential to minimize communication in order to minimize bus contention. Thus, with only one bus, MOCSYN was forced to concentrate its optimization efforts on solutions with allocations containing only a few cores. For a couple of examples, this focussed exploration of the solution space paid off. However, there is no guarantee that a solution with a small number of cores exists. Thus, for 24 examples, being forced to concentrate on solutions with a small number of cores resulted in inferior solutions to those produced when eight busses were available, or no solution at all. In general, the features employed by MOCSYN

Table 2 Multiobjective optimization

Example	Price	Area (mm^2)	Average power (mW)
	318	90	479.7
	358	96	443.3
1	543	196	424.1
	554	182	420.3
	612	216	384.6
	181	50	199.7
2	186	65	151.1
	250	91	128.1
3	166	72	47.9
	170	78	44.1
4	181	78	260.7
	211	66	363.3
	154	56	183.6
5	276	120	170.4
	483	232	164.4
6	405	176	172.7
	462	224	127.0
7	126	36	110.1
8	219	90	41.3
9	182	60	72.7
10	781	352	114.3

to synthesize core-based systems result in superior performance to simpler approaches.

4.3 *Multiobjective Optimization*

In this subsection, the results of using MOCSYN to conduct multiobjective optimization on a number of examples are presented. When MOCSYN is run in multiobjective optimization mode, it produces a set of solutions, each of which is superior, in some way, to at least one other solution. Table 2 shows the sets of solutions produced for ten examples. These are produced with the aid of TGFF. They use the same parameters as the examples in Section 4.2 with one exception: the average number of tasks in each task graph is related to the example number (ex) in the following manner: $1 + ex \cdot 2$. Thus, the six task graphs in Example 10 each has an average of 21 tasks. The variability in the number of tasks in each task graph is always one less than the average number of tasks in each task graph. MOCSYN took less than 7 min when run on each of these examples. For some of the examples, MOCSYN found numerous solutions which trade off price, area, and average power consumption.

5 Conclusions and Future Work

In this paper, we presented a method for the synthesis of core-based, single-chip, low-price, low-power, real-time, multi-rate, heterogeneous embedded systems. A multiobjective genetic algorithm, which allows exploration of the Pareto-optimal set of architectures instead of providing a designer with a single solution, was applied to a number of examples. MOCSYN's use of automatic clock selection, block placement-based communication delay estimation, and arbitrary bus topology generation allows it to solve the core-based synthesis problem.

Acknowledgments We would like to thank I-Jong Lin for his suggestions regarding an elegant clock frequency selection algorithm.

References

[1] W. H. Wolf, Hardware-software co-design of embedded systems, *Proc. IEEE*, vol. 82, 967–989, July 1994.

[2] M. Chiodo et al., Hardware/software co-design of embedded systems, *IEEE Micro*, vol. 14, 26–36, Oct. 1993.

[3] G. De Micheli, Computer-aided hardware-software codesign, *IEEE Micro*, 10–16, Aug. 1994.

[4] P. H. Chou, R. B. Ortega, and G. Borriello, The chinook hardware/software co-synthesis system, in *Proc. Int. Symp. System Synthesis*, 22–27, Sept. 1995.

[5] T.-Y. Yen, *Hardware-Software Co-Synthesis of Distributed Embedded Systems*, PhD thesis, Dept. of Electrical Engg., Princeton University, June 1996.

[6] D. D. Gajski, F. Vahid, S. Narayan, and J. Gong, *Specification and Design of Embedded Systems*, Prentice-Hall, Englewood Cliffs, NJ, 1994.

[7] R. Weiss, 32-bit cores drive systems-on-a-chip, *Computer Design*, pp. 82–89, Sept. 1996.

[8] M. R. Garey and D. S. Johnson, *Computers and Intractability: A Guide to the Theory of NP-Completeness*, W. H. Freeman and Company, NY, 1979.

[9] S. Prakash and A. Parker, SOS: Synthesis of application-specific heterogeneous multiprocessor systems, *J. Parallel Distributed Computers*, vol. 16, 338–351, Dec. 1992.

[10] M. Schwiegershausen and P. Pirsch, Formal approach for the optimization of heterogeneous multiprocessors for complex image processing schemes, in *Proc. European Design Automation Conf.*, pp. 8–13, Sept. 1995.

[11] J. D'Ambrosio and X. Hu, Configuration-level hardware/software partitioning for real-time systems, in *Proc. Int. Workshop Hardware/Software Codesign*, vol. 14, 34–41, Aug. 1994.

[12] T.-Y. Yen and W. H. Wolf, Communication synthesis for distributed embedded systems, in *Proc. Int. Conf. Computer-Aided Design*, pp. 288–294, Nov. 1995.

[13] J. Hou and W. Wolf, Process partitioning for distributed embedded systems, in *Proc. Int. Workshop Hardware/Software Codesign*, pp. 70–76, Mar. 1996.

[14] S. Srinivasan and N. K. Jha, Hardware-software co-synthesis of fault-tolerant real-time distributed embedded systems, in *Proc. European Design Automation Conf.*, pp. 334–339, Sept. 1995.

[15] B. Dave, G. Lakshminarayana, and N. K. Jha, COSYN: Hardware-software co-synthesis of embedded systems, in *Proc. Design Automation Conf.*, pp. 703–708, June 1997.

[16] D. Herrmann, J. Henkel, and R. Ernst, An approach to the adaptation of estimated cost parameters in the COSYMA system, in *Proc. Int. Workshop Hardware/Software Codesign*, pp. 100–107, Mar. 1994.

[17] D. Saha, R. Mitra, and A. Basu, Hardware software partitioning using genetic algorithm approach, in *Proc. Int. Conf. VLSI Design*, pp. 155–160, Jan. 1997.

[18] R. P. Dick and N. K. Jha, MOGAC: A multiobjective genetic algorithm for the co-synthesis of hardware-software embedded systems, in *Proc. Int. Conf. Computer-Aided Design*, pp. 522–529, Nov. 1997.

[19] W. Wolf, Floorplanning: The art of chip-level design, *Electron J*, 8–13, Oct. 1998.

[20] E. L. Lawler and C. U. Martel, Scheduling periodically occurring tasks on multiple processors, *Information Processing Lett*, vol. 7, 9–12, Feb. 1981.

[21] C. M. Fonseca and P. J. Fleming, Multiobjective genetic algorithms made easy: Selection, sharing and mating restrictions, in *Proc. Genetic Algorithms in Engineering Systems: Innovations and Applications*, pp. 45–52, Sept. 1995.

[22] J. L. Breeden, Optimizing stochastic and multiple fitness functions, in *Proc. Evolutionary Programming*, vol. 4, pp. 127–134, Mar. 1995.

[23] R. P. Dick and N. K. Jha, CORDS: Hardware-software co-synthesis of reconfigurable real-time distributed embedded systems, in *Proc. Int. Conf. Computer-Aided Design*, pp. 62–68, Nov. 1998.

[24] J. Smith and G. De Micheli, Automated composition of hardware components, in *Proc. Design Automation Conf.*, pp. 14–19, June 1998.

[25] M. Kishinevsky, J. Cortadella, and A. Kondratyev, Asynchronous interface specification, analysis and synthesis, in *Proc. Design Automation Conf.*, pp. 2–7, June 1998.

[26] M. Bazes, R. Ashuri, and E. Knoll, An interpolating clock synthesizer, *IEEE J Solid-State Circ*, vol. 31, 1295–1300, Sept. 1996.

[27] F. Romeo, *Simulated Annealing: Theory and Applications to Layout Problems*, PhD thesis, Dept. of Electrical Engg. & Computer Science, University of California, Berkeley, Mar. 1989.

[28] C. M. Fiduccia and R. M. Mattheyses, A linear-time heuristic for improving network partitions, in *Proc. Design Automation Conf.*, pp. 173–181, June 1982.

[29] L. Stockmeyer, Optimal orientations of cells in slicing floorplan designs, *Inform Control*, vol. 57, 91–101, May/June 1983.

[30] N. H. E. Weste and K. Eshraghian, *Principles of CMOS VLSI Design, A Systems Perspective*, Addison-Wesley, Reading, MA, 2 edn., 1993.

[31] R. P. Dick, D. L. Rhodes, and W. Wolf, TGFF: task graphs for free, in *Proc. Int. Workshop Hardware/Software Codesign*, pp. 97–101, Mar. 1998.

[32] J. Cong et al. Interconnect design for deep submicron ICs, in *Proc. Int. Conf. Computer-Aided Design*, pp. 478–485, Nov. 1997.

Minimum Energy Fixed-Priority Scheduling for Variable Voltage Processors*

Gang Quan and Xiaobo Sharon Hu

Abstract To fully exploit the benefit of variable voltage processors, voltage schedules must be designed in the context of work load requirement. In this paper, we present an approach to finding the least-energy voltage schedule for executing real-time jobs on such a processor according to a fixed priority, preemptive policy. The significance of our approach is that the theoretical limit in terms of energy saving for such systems is established, which can thus serve as the standard to evaluate the performance of various heuristic approaches. Two algorithms for deriving the optimal voltage schedule are provided. The first one explores fundamental properties of voltage schedules while the second one builds on the first one to further reduce the computational cost. Experimental results are shown to compare the results of this paper with previous ones.

1 Introduction

Low power design is an important issue for designing economical, complicated, and safe real-time systems. Since such systems usually have a time-varying computation load, how to appropriately modulate the system capability accordingly, by shutting down or dynamically varying the processor supply voltage, has been a major strategy to design low power systems. Over the past several years, many methods and techniques for minimizing power consumption for such systems has been published, e.g. [2, 9, 10]. Yet how to achieve the best energy efficiency for many of these systems remains unknown, and how close these approaches are to the optimal solutions is still a question.

---

*This work is supported in part by NSF under grant number MIP-9701416 and CCR-9988468, and by DARPA and Rome Laboratory, Air Force Material Command, USAF, under cooperative agreement FC 30602-00-2-0525 as part of the Power Aware Communication and Computation (PACC) program.

Department of Computer Science & Engineering University of Notre Dame Notre Dame, IN 46556, USA {gquan,shu@cse.nd.edu}

R. Lauwereins and J. Madsen (eds.), *Design, Automation, and Test in Europe–The Most Influential Papers of 10 Years DATE*, 313–323
© Springer 2008

Power-reduction techniques in general can be classified into two categories [14]: *dynamic* and *static*. Dynamic techniques are generally easy to implement and applied during run time. Examples of such techniques include [1, 4, 8, 11, 15–17]. Due to its inherent uncertainty and lack of complete knowledge about the timing constraints, no strong optimality results have been proven with these techniques. In [4], several dynamic voltage scheduling algorithms are proposed for real-time systems containing both periodic tasks and sporadic tasks. These approaches are based on the optimal voltage-scheduling algorithm in [19] and the optimal acceptance test in [18]. They are optimal in the senses that the schedule leads to the lowest power consumption for the periodic tasks and sporadic tasks which pass the acceptance test. However, these approaches cannot be easily extended to handle the cases where the tasks have predefined and fixed priorities, or where the timing information of the sporadic tasks are already known. In [1], a stochastic control approach is proposed. It describes the requests of real-time tasks and the state changes of the system components as discrete-time stationary Markov process. Under such formulation, power management becomes a stochastic optimization problem, and the result is optimal in the statistical sense. Unfortunately, this approach is not favorable for hard real-time systems such as embedded control applications with stringent timing requirements.

Static techniques are applied during design time, such as in the compilation and synthesis process. It takes advantage of the assumption that system specifications are known *a priori*. Several static power management policies have been investigated, e.g. [5, 6, 13, 19]. In [19], an static approach is proposed to find the optimal voltage schedule for real-time systems scheduled via the earliest deadline first (EDF) policy. Hong et al. [5] studied a more general processor model, where the voltage of the processor cannot change instantly. They proposed a static algorithm which can achieve the optimum in some very special scenarios. In [6], the low power non-preemptive scheduling problem is formulated as an integer linear programming problem. The system consists of a set of tasks with same arrival times and deadlines but different context switching penalties. None of the above approaches can be readily applied to address the optimal voltage scheduling problem for a fixed-priority real-time system.

In this paper, we develop an approach to schedule a fixed-priority (FP) real-time system on a variable speed processor which is guaranteed to consume the lowest possible energy. In our approach, we adjust the deadlines of the real-time jobs by carefully analyzing the preemptive effects among them. Then, we transform the low power fixed-priority scheduling problem into a set of low power EDF-based scheduling problems, and find the optimal voltage schedule based on the work in [19]. In regard to the high computation cost of this transformation, we propose a technique which reduces the computational cost significantly. Several experiments are conducted to compare the performances of some previous scheduling techniques with that in this paper. To our best knowledge, this is the first work ever that has provided the theoretical inherent limit to optimally schedule FP real-time jobs on a variable voltage processor in terms of energy saving.

This paper is organized as follows. Section 2 introduces the necessary background and provides several motivational examples. Section 3 explains our algorithm to find the optimal voltage schedule and section 4 discusses the techniques to improve the computational efficiency of our approach. Section 5 contains the experimental results which show the effectiveness and efficiency of our approach. Finally, section 6 concludes the paper.

2   Preliminaries and Motivational Examples

The real-time system that we are interested in consists of N independent jobs, $\mathcal{J} = \{J_1, J_2, \ldots, J_N\}$, arranged in the decreasing order of their statically assigned priorities. Each job, $J_i = (R_i, C_i, D_i)$, is characterized by its arrival time R_i, workload C_i (CPU cycles, for example), and deadlines D_i. Among these jobs, we assume that there is no such case as $R_p \leq R_q, D_p > D_q, p < q$, since a lower-priority job arriving at or after a higher priority one cannot possibly finish before the higher-priority job in a feasible FP system. The processor used in our system is similar to the one introduced in [3]. Note that in [3] the voltage can be varied nearly continuously as long as the number of bits of the speed control register is large enough. We assume that the processor speed is proportional to the supply voltage, and the power is a convex function of the voltage. For simplicity, we assume that the processor speed can be changed instantly. The problem we are interested in can be formulated as follows.

Definition 1 *Given a job set $\mathcal{J}$, find a set of intervals, $[t_s^k, t_f^k]$, and their corresponding speeds, $S = \{\overline{S}(t_s^k, t_f^k), k = 1, 2, \ldots K\}$, where $\overline{S} t(t_s^k, t_f^k)$ is a constant, such that if the processor operates accordingly, all the jobs can be completed by their deadlines and no other voltage schedules can consume less energy.*

An intuitive approach for this problem is to apply the EDF-based optimal voltage scheduling algorithm [19] (LPEDF). However, it has been shown in [13] that simply applying LPEDF to an FP job set may cause a job to miss its deadline. On the other hand, there do exist some cases that applying LPEDF will guarantee the schedulability of the jobs.

Consider the three systems shown in Fig. 1, each of which has three jobs. In this figure and the following figures in this paper, we use an up arrow to represent the arrival time of a job, and a down arrow to represent the deadline of the job. Note that, after the timing specifications of all the jobs are given, there is no fundamental

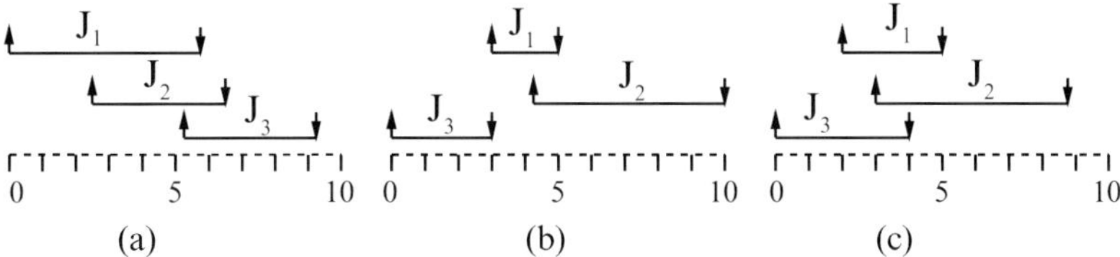

Fig. 1 Three real-time system examples

difference between the FP-based scheduling and the EDF-based scheduling, except that the jobs may have different "fixed" priorities. In Fig. 1a, the FP assignment is the same as that by EDF. The optimal voltage schedule for Fig. 1a found by applying LPEDF is certainly the optimal voltage schedule under the FP assignment. Furthermore, in some other cases, even though some real-time jobs have priorities different from the priority assignment by EDF, we can still use LPEDF to find the optimal voltage schedule. Figure 1b is such an example. In Fig. 1b, j_3 has lower priority but earlier deadline than both j_1 and j_2. However, note that, in a feasible schedule, the execution of j_3 never interferes with the execution of j_1 and j_2, since j_3 must finish before the arrival of j_1 and j_2. For this example, the optimal voltage schedule by EDF scheduling is also the optimal schedule by FP-based scheduling. Specifically, the job sets in Fig. 1a, b are called *primary job sets*, which are formally defined as follows.

Definition 2 *A job set $\mathcal{J}$ is called a primary job set if for any $J_p, J_q \in \mathcal{J}$, $p < q$, either $D_p \leq D_q$ or $R_p \geq D_q$.*

For a primary job set, we have the following lemma. (The proofs for the lemmas and theorems in the paper are omitted due to the page limit, interesting readers can refer to [12] for more details.)

Lemma 1 *The optimal voltage schedule for an FP primary job set $\mathcal{J}$ can be determined by applying LPEDF to $\mathcal{J}$.*

According to Lemma 1, we can find the optimal voltage schedule with LPEDF if the given real-time job set is a primary job set. Unfortunately, not all job sets are primary job sets. Figure 1c is such an example. For such a system, LPEDF no longer produces a valid voltage schedule. This is due to the fact that a lower priority job with earlier deadline might be preempted by other higher priority jobs with later deadlines, and thus make it difficult to meet the deadline with the processor speed computed by LPEDF. How can we find an optimal voltage schedule for this type of systems? In the next section, we introduce a technique to transform a given real-time job sets to a set of primary job sets and identify the optimal voltage schedule for the original job sets.

3   Optimal FP Voltage Schedule

In this section, we introduce our approach to search for the optimal voltage schedule for an FP real-time system, and provide the theoretical basis for our approach. The basic idea of our approach is to transform the complicated problem of determining the optimal voltage schedule for an FP job set to an easier problem: finding the lowest power consumption among the optimal voltage schedules for a number of primary job sets.

Two questions may arise for our approach: given a real-time job set, how to identify the necessary primary job sets; why the optimal voltage schedule for the original job set is among the optimal voltage schedules for the selected primary job sets. The following definitions and theorem will answer these two questions.

Definition 3 *The associative job sets of $\mathcal{J}$, denoted by $\mathcal{A}(\mathcal{J})$, are the job sets such that for any set $\mathcal{J}' \in \mathcal{A}(\mathcal{J})$, $C_i' = C_i$, $R_i' = R_i$, and $D_i' \leq D_i$ for $1 \leq i \leq N$.*

Definition 4 *Given two real-time job sets, $\mathcal{J}_1 = \{J_{11}, J_{12}, \ldots, J_{1N}\}$ and $\mathcal{J}_2 = \{J_{21}, J_{22}, \ldots, J_{2N}\}$, where $C_{1i} = C_{2i}$ and $R_{1i} = R_{2i}$, $1 \leq i \leq N$, job set $\mathcal{J}_2$ dominates $\mathcal{J}_1$ if $D_{1i} \leq D_{2i}$, $1 \leq i \leq N$, which is denoted by $\mathcal{J}_2 \geq \mathcal{J}_1$.*

Definition 5 *The nondominated associative primary (NAP) job sets of $\mathcal{J}$, denoted by $\mathcal{NAP}(\mathcal{J})$, are the job sets such that any $\mathcal{J}_i \in \mathcal{NAP}(\mathcal{J})$ is an associative primary job set of $\mathcal{J}$ and not dominated by any other associative primary job sets of $\mathcal{J}$.*

Theorem 1 *The optimal voltage schedule of $\mathcal{J}$ is the optimal voltage schedule of $\mathcal{J}_i \in \mathcal{NAP}(\mathcal{J})$ which consumes the lowest energy compared with that of any other job set in $\mathcal{NAP}(\mathcal{J})$.*

Since the optimal voltage schedule for any job set in $\mathcal{NAP}(\mathcal{J})$ can be computed with LPEDF (Lemma 1), according to Theorem 1, if we can find all the NAP job sets of $\mathcal{J}$, the voltage schedule with the lowest energy consumption must be the optimal schedule for $\mathcal{J}$. Now the problem becomes how to find all the NAP job sets. There are two challenges to achieve this goal: how to generate the NAP job sets and how to guarantee that all the NAP job sets are identified. The following lemma sheds some lights on constructing the NAP job sets.

Lemma 2 *Let $\mathcal{J}'$ be an associative primary job set of $\mathcal{J}$ and $D'_i = D_i$. Then the following must be true: (i)$D'_p \leq D_i$, if $p < i$ and $R_p < D_i$; (ii) $D'_p \leq R_i$, if $p > i$, and $D_p < D_i$; (iii)$D'_p \leq D_p$.*

With Lemma 2, we devise a procedure to construct the NAP job sets (see Algorithm 1). Specially, we fix, one by one, the deadline of each job, and modify the deadlines for the rest of the jobs to their largest possible values according to Lemma 2. Then, we remove this job and go through this procedure for the rest of the jobs. This procedure continues recursively until the rest of the jobs become a primary job set. Then we put back all the jobs whose deadlines have been fixed to the resultant job sets.

Algorithm 1 Construct the NAP job sets
  1:    **Search_Primary** $(\mathcal{J}, \mathcal{T})$
  2:    **Input**: $\mathcal{J} = (J_1 \ldots, J_N)$, where $J_i = (R_i, C_i, D_i)$
  3:    **Output**: A set $\mathcal{T}$ containing all the NAP job sets of $\mathcal{J}$.
  4:    **for** $J_k \in \mathcal{J}$, $k = 1 \ldots N$ **do**
  5:    Copy $\mathcal{J}$ to $\mathcal{J}_k$;
  6:    Adjust the deadlines of $J_i \in \mathcal{J}_k$, $i \neq k$ according to Lemma 2 with the deadline of $J_k \in \mathcal{J}_k$ fixed;
  7:    $\mathcal{J}_k = \mathcal{J}_k - J_k$;
  8:    **if** $\mathcal{J}_k$ is not a primary job set **then**
  9:    Search_Primary $(\mathcal{J}_k, \mathcal{T}')$;
10:    Add J_k to each job set in $\mathcal{T}'$;

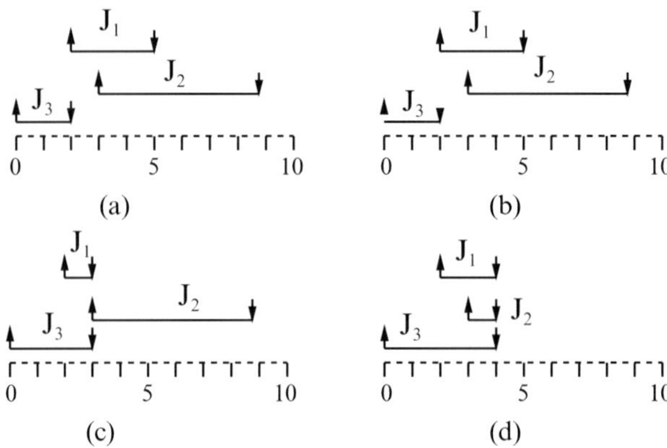

Fig. 2 NAP job sets output from Algorithm 1 for the real-time job sets shown in Fig. 1c. (a) is the result by first fixing J_1 to its deadline. (b) and (c) are the results by first fixing J_2 to its deadline. (d) is the result by first fixing J_3 to its deadline

11:     Add $\mathcal{T}_i' \in \mathcal{T}'$ to $\mathcal{T}$ if $\mathcal{T}_i'$ is a primary job set;
12:     **else**
13:     $\mathcal{J}_k = \mathcal{J}_k + J_k$;
14:     Add $\mathcal{J}_k$ to $\mathcal{T}$ if $\mathcal{J}_k$ is a primary job set;
15:     **end if**
16:     **end for**

The following lemma, together with Theorem 1, can demonstrate that our approach indeed produces the feasible optimal voltage schedule for an FP real-time system.

Lemma 3 *The job sets $\mathcal{T}$ output from Algorithm 1 cover all the NAP job sets of J.*

Figure 2 shows the NAP job sets constructed by applying Algorithm 1 to the system in Fig. 1c. The computation cost for finding the optimal schedule consists of two parts: the cost for constructing the NAP job sets (Algorithm 1), and the cost for searching the optimal schedule among these job sets (LPEDF). Note that the worst-case complexity of Algorithm 1 is $O(N!)$, where N is the number of jobs, and the complexity of LPEDF is $O(N^2)$ [19]. Therefore, the complexity to search for the optimal voltage schedule is $O(N! + MN^2)$, where M is the job sets output from Algorithm 1. When N increases, searching for the optimal schedule with this approach can be quite time consuming.

4   Improve the Computational Efficiency

In this section, we discuss our effort to improve the computational efficiency of Algorithm 1. After a careful study, we note that same primary job sets may be constructed more than once by Algorithm 1, and all these primary job sets are evaluated with LPEDF. For example, in Fig. 2, a and c are identical. This is because

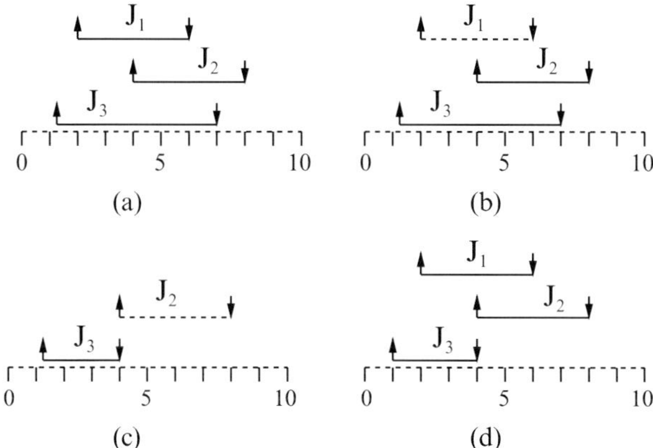

Fig. 3 Non-primary job sets may be generated from Algorithm 1. (a) original job sets. (b) fixing the deadline of J_1. (c) fixing the deadline of J_2. (d) putting back J_1 and J_2 and the result is not a primary job set

the same primary job set may be constructed in different recursive calls in Algorithm 1. This situation exasperates when the number of jobs becomes larger. Moreover, we note that not all the job sets constructed from Algorithm 1 are primary job sets. Figure 3 is such an example. In Algorithm 1, we simply discard these job sets. We refer the nonprimary job sets and extra copies of the same primary job sets as *redundant job sets*. Even though constructing such redundant job sets has no negative effect to the optimality of the results, they do make the algorithm quite inefficient.

One way to reduce the redundancy is to eliminate the identical copies of the same primary job sets once all the primary job sets have been constructed. However, a straightforward implementation of doing this will have a worst-case complexity of $O(M^2)$, where M is the total number of the job sets. When the number of jobs is large, M can be very high. On the other hand, constructing the redundant job sets is still an unnecessary effort, which can be extremely costly for large job sets. Therefore, we focus our effort on how to *avoid* generating these redundant job sets in the algorithm. The following theorem and corollary can help us greatly reduce the redundant job sets.

Theorem 2 *For job $J_q \in \mathcal{J}$, if there is any job J_p or $J_r \in \mathcal{J}$, $p < q < r$, such that (i) $R_q \leq R_r$, and $D_q \leq D_r$, or (ii) $R_p \geq D_q$, then the primary job sets constructed by first fixing the deadline of J_q are redundant.*

Corollary 1 *After removing the redundancy defined in Lemma 2, all the job sets output by Algorithm 1 are primary job sets.*

Theorem 2 identifies the recursive calls generating the job sets which can be generated from other recursive calls in Algorithm 1. Corollary 1 guarantees that if the recursive calls identified in Theorem 2 are removed, none of the non-primary job sets will be constructed. Based on Theorem 2 and Corollary 1, an improved algorithm is shown in Algorithm 2. Algorithm 2 is far more efficient than Algorithm 1 since the non-primary job sets are eliminated and the number of identical job sets are significantly reduced. The improvement achieved by Algorithm 2 will be demonstrated through experimental results in the next section.

5   Experimental Results

In this section, we use experiments to compare the voltage scheduling approach in this paper (**OPT**) with two other related approaches, the heuristic approach (**VSLP**) introduced in [13], and the approach (**LPFS**) presented in [16].

Algorithm 2 Improved algorithm for searching the NAP job sets

```
 1: Search_Primary2 (J, T )
 2: Input: J = {J_1,...,J_N}, where J_i = (R_i,C_i,D_i)
 3: Output: A set T containing all the NAP job sets of J
 4: for J_k ∈ J,k = 1...N do
 5:   if there is J_q, such that q > k,R_q > R_k, and D_q ≥ D_k, or q < k, and R_q ≥ D_k then
 6:     continue;
 7:   end if
 8:   Copy J to J_k;
 9:   Adjust the deadlines of J_i ∈ J_k, i ≠ k according to Lemma 2 with the deadline
       of J_k ∈ J_k fixed;
10:   J_k = J_k − J_k;
11:   if J_k is not a primary job set then
12:     Search_Primary2(J_k, T ');
13:     Add J_k to each job set in T';
14:     Add each job set in T' to T
15:   else
16:     J_k = J_k + J_k;
17:     Add J_k to T ;
18:   end if
19: end for
```

We also use experimental results to show the significant improvement of the computational efficiency achieved by Algorithm 2. All the experiments are conducted using ACER Travelmate 602, with a 650 MHz Pentium III processor, 128 M memory, and running Windows 98. In our experiments, we assume that the power is a quadratic function of the processor speed.

Our first experiment consists of 10 groups of randomly generated real-time systems with the number of jobs being 2,4,...,20. The arrival times and deadlines of these jobs are chosen to be uniformly distributed within [0, 50], [20, 100], respectively. These data are arbitrarily chosen without any special consideration. The execution time of each job is randomly generated from 1 to half of its dead-line to make the job sets easier to be schedulable under the maximum processor speed.

Only the job sets which are schedulable by our simulation test are used in our experiment. Each group contains at least 50 randomly generated schedulable job sets. To reduce statistical errors, the average power consumption and CPU time for each group using three approaches, i.e. **OPT, VSLP**, and **LPFS**, are used to compare their performance. Note, while **VSLP** and **LPFS** do not guarantee to always find the optimal voltage schedule for a given FP real-time system, they do find the optimal ones sometimes. Therefore, we also compare these two approaches in terms of the percentage of optimal solutions they find ($\mathcal{N}_{OPT}/N$) which are shown in the last two columns in Table 1.

Table 1 shows **OPT** has a much higher computational cost than **VSLP** and **LPFS**, and its CPU times increases rapidly as the number of jobs increase. This agrees with our theoretical analysis since the worst-case complexity is $O(N^3)$ for **VSLP**, $O(N^2 log N)$ for **LPFS**, and $O(N! + M log^2 M)$ for **OPT** (where N is the number of jobs, and M is the number of primary job sets constructed in **OPT**). Table 1 also shows that **VSLP** represents an excellent trade-off choice in searching for an optimal voltage schedule. First, from Table 1, the difference between the average power consumption of **VSLP** and that of **OPT** is within 2%, which is much better than that of **LPFS**. Secondly, **VSLP** can find the optimal voltage schedules in most cases, especially when the number of jobs is small. For example, for four jobs, 92% of the optimal schedules can be found by **VSLP**, but **LPFS** finds none. Finally, **VSLP** takes much less CPU time than **OPT**.

Our second experiment quantifies the improvement in terms of computation efficiency made by Algorithm 2. The same randomly generated systems are used. The average CPU times to find the optimal voltage schedules by applying Algorithm 1 or Algorithm 2 to construct the NAP job sets are collected and shown in Table 2. Table 2 shows the dramatic reduction of computational cost by applying Algorithm 2, especially for systems with a large number of jobs. Applying Algorithm 2 reduces not only the effort to construct the redundant job sets, but also the effort to search for the optimal voltage schedule among these sets. Since the number of

Table 1 Experimental results for comparing the three approaches: OPT, VSLP, and LPFS

No. Jobs	Energy			CPU(s)			$\mathcal{N}_{OPT}/\mathcal{N}$	
	OPT	VSLP	LPFS	OPT	VSLP	LPFS	VSLP	LPFS
2	4.35	4.35	5.74	0.00	0.00	0.00	96%	34%
4	12.8	12.9	19.6	0.00	0.00	0.00	92%	0%
6	26.2	26.9	36.1	0.01	0.00	0.00	82%	0%
8	33.9	34.1	45.7	0.02	0.00	0.00	75%	0%
10	45.5	46.6	57.0	0.12	0.00	0.00	70%	0%
12	49.2	50.1	57.6	0.31	0.00	0.00	75%	0%
14	50.0	50.7	59.8	0.58	0.01	0.00	75%	0%
16	57.0	58.8	68.7	2.78	0.01	0.00	75%	0%
18	57.4	59.1	65.8	6.41	0.02	0.00	75%	0%
20	63.2	65.4	72.4	30.30	0.01	0.00	52%	0%

Table 2 Computation efficiency improvement by
reducing the redundancy

No. Jobs	CPU Time(s)	
	Algorithm 2	Algorithm 1
2	0.001	0.001
4	0.002	0.v006
6	0.010	0.032
8	0.021	0.402
10	0.121	46.681
12	0.304	3147.070

redundant job sets increases drastically as the number of jobs increases, this explains the dramatic improvement by using Algorithm 2.

6   Summary

In this paper, we present an approach to finding an optimal voltage schedule in terms of energy saving for a variable voltage processor executing fixed-priority, real-time jobs. We introduce the concept of *nondominated, associative, primary* job sets and prove that an optimal voltage schedule of a given job set must be the same as that of one of NAP job sets. Two algorithms are developed to construct NAP job sets for a given job set with one improves on the other one. Experimental results are shown to compare our results with relevant previous ones.

The type of systems studied in this paper contains real-time jobs to be scheduled based on the fixed priority, preemptive scheme. Such a scheduling scheme is used widely in many real-world real-time systems due to its simplicity and predictability [7]. The static voltage scheduling approach adopted here can be readily used during the design process to fully exploit the timing information known *a priori*. Furthermore, the static approach can be supplemented by a dynamic voltage scheduling such as the one proposed in [16] to achieve the best overall result. The significance of the results presented here is that the theoretical limit in terms of energy saving for the systems of interest is established. Such results can be used as the standard to measure the quality of various heuristic approaches.

References

[1] L. Benini, A. Bogliolo, G. Paleologo, and G. Micheli. Policy optimization for dynamic power management. *IEEE Trans. CAD and Sys.*, 18(6):813–833, June 1999.

[2] L. Benini, A. Bogliolo, and G. Micheli. A survey of design techniques for system-level dynamic power management. *IEEE Trans. VLSI Sys.*, 8(3):299–316, June 2000.

[3] T. D. Burd and R. W. Brodersen. Design issues for dynamic voltage scaling. *ISLPED*, pp. 9–14, 2000.

[4] I. Hong, M. Potkonjak, and M. B. Srivastava. On-line scheduling of hard real-time tasks on variable voltage processor. *Proceedings of ICCAD*, pp. 653–656, 1998.

[5] I. Hong, G. Qu, M. Potkonjak, and M. B. Srivastava. Synthesis techniques for low-power hard real-time systems on variable voltage processors. *Proceedings of RTSS*, pp. 178–187, 1998.

[6] T. Ishihara and H. Yasuura. Voltage scheduling problem for dynamically variable voltage processors. *ISLPED*, pp. 197–202, August 1998.

[7] J. Liu. *Real-Time Systems*. Prentice Hall, NJ, 2000.

[8] Y. Lu, E. Chung, *T. Šimunić*, L. Benini, and G.D. Micheli. Quantitative comparison of power management algorithms. *DATE*, pp. 20–26, 2000.

[9] G. Micheli and L. Benini. System-level low power optimization: Techniques and tools. *Trans. Design Auto. of Electr. Sys.*, 5(2), April 2000.

[10] M.Pedram. Power minimization in ic design: principles and applications. *ACM Trans. Design Auto. of Electr. Sys.*, 1(1):3–56, January 1996.

[11] Q. Qiu, Q. Wu, and M.Pedram. Dynamic power management of complex system using generalized stochastic petrinets. *DAC*, pp. 352–356, 2000.

[12] G. Quan. *System level design techniques for real-time embedded systems*. PhD thesis, Department of CSE, University of Notre Dame, Notre Dame, IN, 2001.

[13] G. Quan and X. S. Hu. Energy efficient fixed-priority scheduling for real-time systems on voltage variable processors. *DAC*, pp. 828–833, 2001.

[14] J. Rabaey and M. Pedram. *Low Power Design Methodologies*. Kluwer, 1996.

[15] D. Ramanathan and R. Gupta. System level online power management algorithms. *DATE*, pp. 606–611, 2000.

[16] Y. Shin and K. Choi. Power conscious fixed priority scheduling for hard real-time systems. *DAC*, pp. 134–139, 1999.

[17] M. Srivastava, A. Chandrakasan, and R. Brodersen. Predictive system shutdown and other architectural techniques for energy efficient programmable computation. *IEEE Trans. VLSI Sys.*, 4:42–55, March 1996.

[18] T. Tia, J. Liu, J. Sun, and R. Ha. *A linear-time optimal acceptance test for scheduling of hard real-time tasks*. Technical report, Department of Computer Science, University of Illinois at Urbana-Champaign, Urbana-Champaign, IL, 1994.

[19] F. Yao, A. Demers, and S. Shenker. A scheduling model for reduced cpu energy. *IEEE Annual Foundations of Comp. Sci.*, pp. 374–382, 1995.

Synthesis and Optimization of Threshold Logic Networks with Application to Nanotechnologies

Rui Zhang, Pallav Gupta, Lin Zhong, and Niraj K. Jha

Abstract We propose an algorithm for efficient threshold network synthesis of arbitrary multi-output Boolean functions. The main purpose of this work is to bridge the wide gap that currently exists between research on the development of nanoscale devices and research on the development of synthesis methodologies to generate optimized networks utilizing these devices. Many nanotechnologies, such as resonant tunneling diodes (RTD) and quantum cellular automata (QCA), are capable of implementing threshold logic. While functionally correct threshold gates have been successfully demonstrated, there exists no methodology or design automation tool for general multi-level threshold network synthesis. We have built the first such tool, ThrEshold Logic Synthesizer (TELS), on top of an existing Boolean logic synthesis tool. Experiments with about 60 multi-output benchmarks were performed, though the results of only 10 of them are reported in this paper because of space restrictions. They indicate that up to 77% reduction in gate count is possible when utilizing threshold logic, with an average reduction being 52%, compared to traditional logic synthesis. Furthermore, the synthesized networks are well-balanced, and hence delay-optimized.

1 Introduction

The Semiconductor Industries Association (SIA) roadmap [1] predicts that complementary metal-oxide semiconductor (CMOS) chips will continue to fuel the need for high-performance systems for another 10–15 years. However, advancements in the material science and device community have enabled the creation of nanoscale devices (RTDs, QCA, to name a few) that have novel structures and properties. While CMOS is used to implement Boolean logic, many nanoscale devices implement threshold logic.

As progress is made in the material and physical understanding of nanoscale devices, research must be done at the logic level to fully harness the potential

Department of Electrical Engineering Princeton University Princeton, NJ 08544 {rzhang|pgupta| lzhong|jha}@ee.princeton.edu

R. Lauwereins and J. Madsen (eds.), *Design, Automation, and Test in Europe–The Most Influential Papers of 10 Years DATE*, 325–343
© Springer 2008

offered by these devices. Today, nanotechnologies are in their infancy and the development of computer-aided design methodologies for these nanotechnologies is crucial if any of them is to replace or augment CMOS. Among existing nanoscale devices [2], RTDs and QCA are two promising nanotechnologies that are of particular interest to us because they implement threshold logic.

In this paper, we present the first comprehensive methodology for multilevel threshold logic synthesis and optimization. Once a threshold network has been synthesized, it can be mapped onto a specific target nanotechnology. The algorithm in our methodology takes into account defect tolerances in the input weights, and the fanin restriction on a threshold gate. Taking these parameters into account improves the robustness of the synthesized network. Node sharing (i.e. a fanout node) is also preserved and thus, any advantage that is gained by preprocessing the network through a Boolean logic synthesis tool remains. The synthesized network is area and delay optimized. The novel contributions of this paper are as follows:

- This is the first *comprehensive* methodology for multilevel multi-output threshold network synthesis.
- Based on our methodology, we have built a threshold network synthesis tool on top of an existing Boolean logic synthesis tool.
- We formulate new theorems that describe properties of threshold logic and use them to our advantage in our methodology.

The remainder of this paper is organized as follows. Section 2 presents background material and previous work. Section 3 presents an example to motivate the need for our threshold network synthesis methodology. Section 4 presents some theorems and their proofs on the properties of threshold logic. Section 5 describes our synthesis methodology and its implementation in a tool in detail. We present our experimental results in Section 6 and conclude the paper in Section 7.

2　Background and Previous Work

In this section, we describe some preliminary concepts. Specifically, we describe what a threshold function is and its relationship to unateness. We also review algebraically factored and Boolean-factored expressions, and linear programming. Previous work on threshold network synthesis is also presented.

2.1　Threshold Logic

A linear threshold function f is a multi-input function in which each input, $x_i \in \{0,1\}$, $i \in \{1,2,\ldots,l\}$, is assigned a weight w_i such that f assumes the value 1 when the weighted sum of its inputs equals or exceeds the value of the function's threshold, T [3]. That is,

$$f(x_1, x_2, \ldots, x_n) = \begin{cases} 1 \text{ if } \sum_{i=1}^{l} w_i x_i \geq T + \delta_{on} \\ 0 \text{ if } \sum_{i=1}^{l} w_i x_i < T - \delta_{off}, \end{cases} \tag{1}$$

Parameters δ_{on} and δ_{off} represent defect tolerances that must be considered since variations (due to manufacturing defects, temperature changes, etc.) in the weights can lead to network malfunction. In this paper, we assume δ_{on} is zero and δ_{off} is one. In many past works, δ_{on} and δ_{off} are both assumed to be zero, thus providing no defect tolerance.

A linear threshold gate (LTG) is a multiterminal device that implements a threshold function. An LTG can be considered a generalization of a conventional logic gate. An l-input NAND and an l-input NOR gate can both be realized by a single LTG. Because both gates are functionally complete, any Boolean logic function can be realized by LTGs. However, not all functions can be realized by a single LTG. A function that can be realized by a single LTG is called a *threshold function*. A network of threshold gates is called a *threshold network*.

An LTG based on RTDs and heterostructure field-effect transistors (HFET) is shown in Fig. 1a. It outputs a logic 1 when $aw_a + bw_b - cw_c \geq T$, else a logic 0. It is called a monostable-bistable logic element [4] and its equivalent LTG is shown in Fig. 1b.

2.1   *Unateness*

A logic function, $f(x_1, x_2, \ldots, x_l)$, is said to be positive (negative) in variable x_i if there exists a disjunctive or conjunctive expression of f in which x_i appears in uncomplemented (complemented) form only. If f is either positive or negative in x_i, it is said to be *unate* in x_i. Otherwise, it is *binate* in x_i. Unateness is an important property for threshold logic because every threshold function is unate [5] (the converse is not true, however).

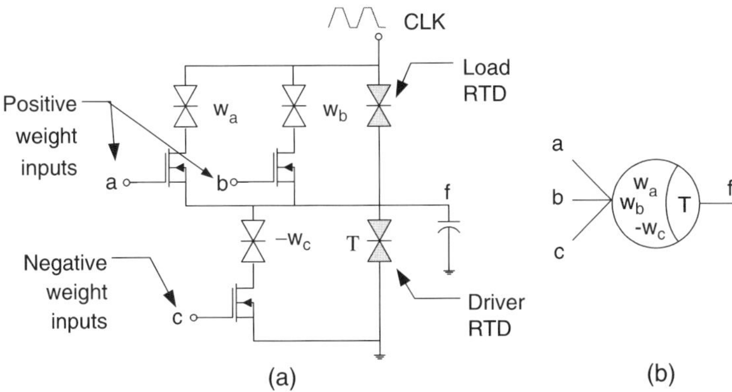

Fig. 1 A monostable–bistable logic element and its equivalent LTG

2.3   *Algebraically factored and Boolean-factored Networks*

A sum-of-products (SOP) expression, $f = \sum_{i=1}^{m} C_i$, is *algebraic* if no cube, C_i, is contained within another cube. That is, $\forall i, j, i \neq j, C_i \nsubseteq C_j$. An expression that is not algebraic is *Boolean* [6]. A factored form F is said to be algebraically factored if the SOP expression obtained by multiplying F out directly, without using the identities $xx- = 0$ and $xx = x$, and single-cube containment, is algebraic [6]. Otherwise, F is Boolean-factored.

2.4   *Linear Programming*

In linear programming, we have a $p \times q$ matrix $\mathbb{A}$, a $p \times 1$ vector $\mathbb{B}$, and a $1 \times q$ vector $\mathbb{C}$. We want to find a vector $\mathbb{X}$ of q elements such that the objective function $\mathbb{CX}$ is minimized subject to the p constraints given by $\mathbb{AX} \leq \mathbb{B}$. Integer linear programming (ILP) [3] is a special case of linear programming that requires all of the elements in $\mathbb{X}$ to assume integer values.

2.5   *Previous Work*

Research in threshold logic synthesis was done mostly in the 1950s and 1960s. In [7, 8], approximation methods are used to determine the input weights and threshold of a threshold function. In [5], admissible patterns on a Karnaugh map are used to determine whether a function is threshold or not. Unfortunately, because of their computational complexity, these methods are restricted to 10 or fewer variables. Linear programming and tabulation methods have been used in [3] to determine if a function is threshold or not. A CMOS implementation of threshold gates can be found in [9]. A survey of VLSI implementations of threshold logic can be found in [10]. In [11], a branch-and-bound algorithm is used to synthesize two-level threshold networks. Multilevel threshold network synthesis did not receive much attention in the 1950s and 1960s because efficient algorithms to factorize a multilevel network were unknown at that time. Most methods performed one-to-one mapping by replacing each Boolean gate in the network with a threshold gate. However, various algorithms exist today to compute the kernels and co-kernels of a network which can be used to perform algebraic or Boolean factorization [6]. In addition, methods have also been developed for Boolean network simplification. Finally, tools such as SIS [12] are available for network factorization and optimization.

3   Motivational Example

A small example is presented in this section to motivate the need for our threshold network synthesis methodology. Consider the Boolean network shown in Fig. 2a which has seven gates and five levels (including the inverter). If we simply

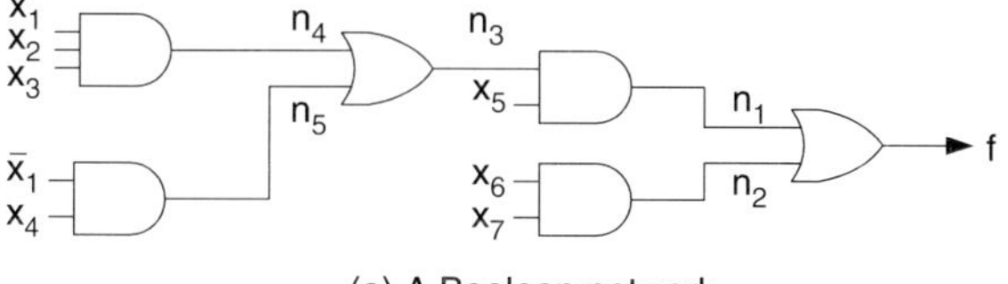

(a) A Boolean network.

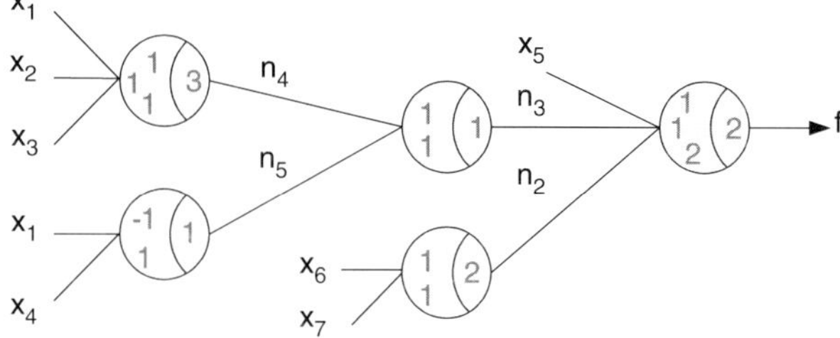

(b) The equivalent synthesized threshold network

Fig. 2   An example to motivate the need for our threshold network synthesis methodology

replace each gate with a threshold gate, the resulting network will contain seven threshold gates and five levels. This is a suboptimal network because some nodes in Fig. 2a can be collapsed into a single threshold node. However, choosing which node to collapse presents a problem. If we set the fanin restriction of a node to four, $f = n_1 \vee n_2$ can be collapsed to get $f = n_3 x_5 \vee x_6 x_7$. Now, we must determine if f is a threshold function or not. In this case, it turns out that f is not a threshold function. Consequently, we must split f into smaller nodes using efficient heuristics. We choose to split f as $f = n_3 x_5 \vee n_2$ where $n_2 = x_6 x_7$. We synthesize n_3 next. After collapsing, n_3 is expressed as $n_3 = x_1 x_2 x_3 \vee \bar{x}_1 x_4$. This is not a threshold function. Therefore, we split n_3 into two nodes to get $n_3 = n_4 \vee n_5$, where $n_4 = x_1 x_2 x_3$ and $n_5 = \bar{x}_1 x_4$. These three nodes are threshold functions. The synthesized threshold network is shown in Fig. 2b. It can be seen that the number of gates and levels has been reduced by 28.6% (seven to five) and 40% (five to three), respectively.

The above example demonstrates that a threshold network synthesis methodology should address the following key issues:

- It must be able to collapse the function of a node.
- It must be able to determine if a function is threshold or not.
- If a function is not threshold, it must be able to split the function into smaller functions using efficient heuristics.
- If there exist fanout nodes in the original Boolean network, it must preserve these nodes in the synthesized threshold network.

How a non-threshold function is split dictates the quality of the synthesized network. Furthermore, node sharing helps to prevent sub-network duplication during threshold network synthesis. It also helps maintain the original network structure somewhat.

4   Theorems on Threshold Logic

We present two new theorems that describe properties of threshold logic in this section. We utilize these theorems in our threshold logic network synthesis methodology. Along with the proof of each theorem, we demonstrate its application with an example.

Theorem 1: Given an expression for a unate function, $f(x_1, x_2,\ldots, x_l)$, replace literal x_i by literal $\overline{x}_j$, $i, j \in \{1, 2,\ldots, l\}$ and $i \neq j$, resulting in $g(x_1, x_2,\ldots,x_k)$, $k \in \{1, 2,\ldots,l\}$ and $k \neq i$. If g is not a threshold function, then f is not a threshold function.

Proof: We prove the contrapositive of the claim. That is, if f is a threshold function, then g is a threshold function. For simplicity, we assume $\delta_{on} = \delta_{off} = 0$. Assuming f is a threshold function, we have,

$$\sum_{k=1}^{l} w_k x_k \geq T \Rightarrow f=1, \tag{2}$$

$$\sum_{k=1}^{l} w_k x_k < T \Rightarrow f=0. \tag{3}$$

Note that Eqs. (2) and (3) represent 2^l inequalities for all combinations of variables x_1, $x_2,\ldots,x_l$. By replacing x_i with $\overline{x}_j$, we obtain the following 2^{l-1} inequalities:

$$\sum_{k=1, \, k \neq i}^{l} w_k x_k + w_i \overline{x}_j \geq T \Rightarrow g=1, \tag{4}$$

$$\sum_{k=1, \, k \neq i}^{l} w_k x_k + w_i \overline{x}_j < T \Rightarrow g=0. \tag{5}$$

Since $\overline{x}_j = 1 - x_j$, we obtain,

$$\sum_{k=1, \, k \neq i,j}^{l} w_k x_k + (w_j - w_i)x_j \geq T - w_i \Rightarrow g \Rightarrow 1, \tag{6}$$

$$\sum_{k=1, \, k \neq i,j}^{l} w_k x_k + (w_j - w_i)x_j < T - w_i \Rightarrow g \Rightarrow 0. \tag{7}$$

If assignments for w_i and T exist such that the inequalities in Eqs. (2) and (3) are satisfied, then the inequalities in Eqs. (6) and (7) can also be satisfied with the weight-threshold vector, $\langle w_1, w_2,\ldots,w_{i-1}, w_{i+1},\ldots,w_{j-1}, w_j - w_i, w_{j+1},\ldots, w_l; T - w_i \rangle$, the weights corresponding to the variable sequence $\langle x_1, x_2,\ldots,x_{i-1}, x_{i+1},\ldots, x_{j-1}, x_j, x_{j+1},\ldots, x_l \rangle$. Thus, g is also a threshold function. By proving the contrapositive, the proof is concluded.

As an application of Theorem 1, consider $f = x_1 x_2 \vee x_3 x_4$. To determine if f is threshold or not, we replace x_3 by $\overline{x}_1$. This results in $g = x_1 x_2 \vee \overline{x}_1 x_4$. Since g is binate in x_1, it is not a threshold function and, therefore, f is not a threshold function.

The relationship of the weights and threshold between functions containing x_i in positive and negative phase is given in [3]. That is, given a positive unate function, $f(x_1, x_2,\ldots, x_l)$, with weight-threshold vector $\langle w_1, w_2,\ldots, w_l; T\rangle$, if x_i is replaced by $\overline{x}_i$ to get $g(x_1, x_2,\ldots, x_{i-1}, \overline{x}_i, x_{i+1},\ldots, x_l)$, then the weight of x_i in g is simply $-w_i$. Furthermore, the threshold of g is $T - w_i$. We negate the original weight of each variable that appears in negative phase in g to get its new weight. To obtain the new threshold, we subtract the sum of the negated weights from the original threshold.

Theorem 2: If $f(x_1, x_2,\ldots, x_l)$ is a threshold function, then $h(x_1, x_2,\ldots, x_{l+k}) = f(x_1, x_2,\ldots,x_l) \vee x_{l+1} \vee x_{l+2} \vee \ldots \vee x_{l+k}$ is also a threshold function.

Proof: A threshold function can always be represented in positive unate form by substituting negative variables with positive variables. For simplicity, we assume that f is already expressed in this form. There exists a weight-threshold vector, $\langle w_1, w_2,\ldots,w_l; T\rangle$, for f since it is a threshold function. If any of the x_{l+j}, $j \in \{1,2,\ldots,k\}$, equals 1, h equals 1. Otherwise, h is equal to f. If we set the weight w_{l+j} of x_{l+j} to a value no less than $T + \delta_{on}$, for example, $w_{l+j} = T + \delta_{on}$, then the output is 1 when x_{l+j} is 1. When x_{l+j} and some other x_i, $i \in \{1,2,\ldots,l\}$, equal 1, the output is also 1. This is because we have represented the function in positive unate form, thereby guaranteeing that all the weights and threshold of f are positive. When all of x_{l+j} equal 0, h equals f. Thus, a weight-threshold vector for h always exists. Therefore, h is a threshold function.

To illustrate Theorem 2, let $f(x_1, x_2) = x_1\overline{x}_2$. First, we represent f in positive unate form as $g(x_1, y_2) = x_1 y_2$ where $y_2 = \overline{x}_2$. Since g is a threshold function with weight-threshold vector $\langle 1, 1; 2\rangle$, $h(x_1, y_2, x_3) = g(x_1, y_2) \vee x_3 = x_1 y_2 \vee x_3$ is also a threshold function with weight-threshold vector $\langle 1, 1, 2; 2\rangle$. Since $y_2 = 1-x_2$, $x_1\overline{x}_2 \vee x_3$ is also a threshold function with weight-threshold vector $\langle 1,-1, 2; 1\rangle$.

5   Methodology and Implementation

We present our multilevel threshold network synthesis methodology and its implementation in this section. Figure 3 gives a high-level overview of the main steps that comprise our methodology. The input to our methodology is an algebraically factored multi-output combinational network, G, and its output is a functionally equivalent threshold network, G_T. An algebraically factored circuit is used as an input because its nodes are more likely to be unate and hence possibly threshold functions. The weights and threshold are specified for each node in the synthesized network. The user can specify the fanin restriction and the defect tolerances in the weights in a threshold gate that are used during threshold network synthesis.

The synthesis algorithm begins by processing each primary output of network G. First, the node representing a primary output is collapsed. If the node represents a binate function, it is split into multiple smaller nodes which are then processed recursively. If the unate node is a threshold function, it is saved in the threshold network and the fanins of the node are processed recursively. Otherwise, the unate node is first split into two nodes. If either of the split nodes is a threshold function,

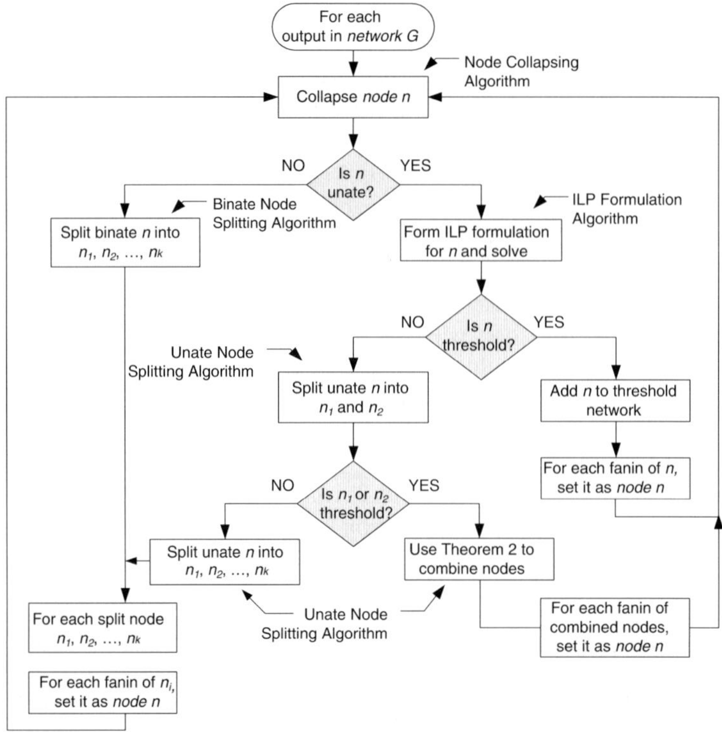

Fig. 3 Flow diagram providing a high-level overview of our threshold network synthesis methodology

Theorem 2 is used as a simplification step. If neither of the split nodes is a threshold function, the original node is split into multiple smaller nodes which are then processed recursively. The synthesis algorithm terminates when all the nodes in network G are mapped into threshold nodes. We describe each step in detail in the following subsections.

The variables that are used in our algorithms are defined as follows:

P	Set of primary inputs in network G.
S	Set of fanout nodes in network G.
n	A node in network G.
F_n	Set of fanins of node n.
$x_{(j)}$	The (j^{th}) fanin of a node.
ψ	Fanin restriction on a threshold gate.
l	Cardinality of F_n.
Z	Set of threshold functions.
K_n	Set of cubes of node n.
C_{ni}	The ith cube of node n.
$\vec{W}$	he weight-threshold vector, $\langle w_1, w_2, \ldots, w_1; T \rangle$.

5.1 Node Collapsing

The node collapsing algorithm is shown in Fig. 4. Given a node n, we keep collapsing it until one of the following conditions is met:

- All fanins of n are primary inputs and/or fanout nodes.
- The fanin of n exceeds ψ.

Note that the algorithm guarantees that the fanin of a node never exceeds ψ by choosing to undo the effects of node collapsing that was done in line 8 of the node collapsing algorithm.

To demonstrate node collapsing, let us consider the network with output node f shown in Fig. 5. Here, ψ is set to four, $l = 2$, $F_f = \{n_1, n_2\}$, $P = \{x_1, x_2, x_3, x_4\}$, $S = \{n_3\}$, and $f = n_1 \vee n_2$. Since the inputs to f are not primary inputs and l is less than ψ, we first collapse n_1 to get $f = x_1 n_3 \vee n_2$. Now, $l = 3$ and $F_f = \{x_1, n_2, n_3\}$. Since l is still less than ψ, we continue by collapsing n_2 to get $f = x_1 n_3 \vee n_3 x_4$. Now, we cannot collapse x_1 and x_4 since they are primary inputs. Furthermore, observing that n_3 is a fanout node, we do not collapse it either. Thus, the final result after collapsing is $f = x_1 n_3 \vee n_3 x_4$.

As demonstrated in the example, node sharing is preserved during node collapsing because the process stops once a fanout node is encountered. This implicitly helps to maintain some of the original network structure and provides guidance for better network decomposition. The benefit is profound when the network contains many fanout nodes.

Require: node n, $\psi > 0$

    n$'$ ← n

    l ← |F$_{n'}$|

3:   **if** $\forall$x in F$_{n'}$, x ∉ P **then**

      while l ≤ ψ **do**

        for all x in F$_{n'}$ **do**

6:          // **if** neither primary input nor fanout node

          if f ∉ P ∧ x ∉ S **then**

            substitute the function of x into n$'$

9:          l → |F$_{n'}$| // update l

          if l > ψ **then**

            n$'$ ← undo the substitution of x into n$'$

12:          break

        // **if** all fanins are primary inputs or fanout nodes

        if x in Fn$'$, x ∈ (P [S) **then**

15:          break

      return n$'$ // collapsed node

Fig. 4 The node collapsing algorithm

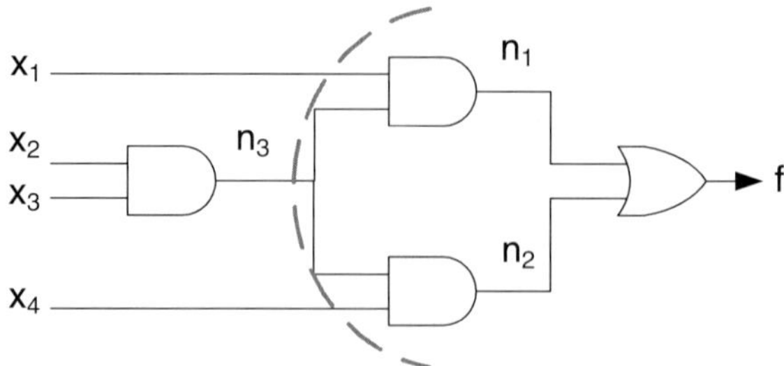

Fig. 5 Example network to demonstrate node collapsing on output *f*

5.2 *Formulating Synthesis as an ILP Problem*

Once a node has been collapsed into a unate function, it is necessary to determine whether it is threshold or not. We solve this problem by casting it in an ILP formulation. There are at most 2^l distinct cubes for a logic function of l variables and this leads to 2^l inequalities which represent the constraints. However, many of these constraints are redundant. We have devised a simple method to eliminate redundant constraints which makes the ILP formulation smaller and possibly faster to solve. The algorithm for formulating the ILP problem and determining the weight-threshold vector for a threshold function is shown in Fig. 6.

The algorithm is best demonstrated by an example. Given a unate function, $f(x_1, x_2,\ldots,x_l)$, if it contains variables in negative phase, we first transform these variables into other variables in positive phase using variable substitution. Consider $f = x_1\bar{x}_2 \vee x_1\bar{x}_3$, where x_2 and x_3 are in negative phase. By replacing $\bar{x}_2$ with y_2 and $\bar{x}_3$ with y_3, we get the positive unate function $g = x_1 y_2 \vee x_1 y_3$. The ILP formulation for g is as follows:

$$minimize \;:\; w_1 + w_2 + w_3 + T \tag{8}$$

$$subject\;to:\; w_1 + w_2 \geq T + \delta_{on} \tag{9}$$

$$w_1 + w_3 \geq T + \delta_{on} \tag{10}$$

$$w_2 + w_3 < T - \delta_{off} \tag{11}$$

$$w_1 < T - \delta_{off} \tag{12}$$

$$w_i \geq 0,\; integer,\; i = 1,2,3. \tag{13}$$

The objective function for this ILP problem is defined as the summation of the weights and threshold, in order to reduce threshold gate area. Since g has been transformed into a positive unate form, only 1 and don't care (−) will appear in its ON-set cubes (set of cubes for which the function is 1). The ON-set cubes of g are (1 1 −) and (1 − 1), where x_1, y_2, and y_3 is the variable sequence. To transform the

---

Require: positive unate node n

    $n_p \leftarrow$ n

    $n_n \leftarrow$ invert n

3:   **for** i = 1 to $|F_n|$ **do** // objective function

      print "w_i+"

    print "T"

6:   **for** i = 1 to $|K_{n_p}|$ **do** // ON-set inequalities

      for j = 1 to $|F_p|$ **do**

        if x_j 2 Cnpi **then**

9:      print "w_j+"

        print "_δ_{on} _ T"

      for i = 1 to $|K_{n_n}|$ **do** // OFF-set inequalities

12:    **for** j = 1 in $|F_n|$ **do**

        if xj $\notin$ C$_{nn_i}$ **then**

          print "w_j+"

15:    **print** "$\delta_{off} < T$"

    $\overrightarrow{W} \leftarrow$ solve ILP problem

    if $\overrightarrow{W} \neq \phi$ then

18:    **for** j = 1 in $|F_n|$ **do**

        if x$_j$ in negative phase in n **then**

          $W[T] \leftarrow W[T] - W[w_j]$ // subtract weight from original threshold

21:        $W[w_j] \leftarrow$?$W[w_j]$ // negate the weight

      return $\overrightarrow{W}$ // weight-threshold vector

    else

24:    return ϕ // no solution exists

Fig. 6 The ILP formulation algorithm

ON-set cubes into inequalities, the ith 1 value corresponds to w_i. We need not consider don't cares in the inequalities, because they represent redundancies in g. For example, the ON-set cube (1 1 −) corresponds to two inequalities, namely, $w_1 + w_2 \geq T + \delta_{on}$ and $w_1 + w_2 + w_3 \geq T + \delta_{on}$. The second inequality is redundant because once the first inequality is satisfied, the second inequality is automatically satisfied as well.

To compute the OFF-set cubes (set of cubes for which the function is 0) of g, we simply invert g to get $\bar{g}$ The ON-set cubes of $\bar{g}$ correspond to the OFF-set cubes of g. Because $\bar{g}$ is always in negative unate form, only 0 and − appear in its ON-set cubes. Continuing with the earlier example, we invert g to get $\bar{g} = \bar{x}_1 \vee \bar{y}_2\bar{y}_3$ after simplification. The ON-set cubes for $\bar{g}$ are (0 − −) and (− 0 0). For the OFF-set inequalities, the ith don't care corresponds to weight w_i. Therefore, the OFF-set inequalities for g are $w_2 + w_3 < T - \delta_{off}$ and $w_1 < T - \delta_{off}$, as given in Equations (11) and (12).

By requiring the variables to be integer-valued (i.e. constraint (13)), this ILP problem has an optimal solution. The weight-threshold vector for g is $\langle 2,1,1;3 \rangle$.

Using the relationship stated in Section IV, the final weight-threshold vector for f is $\langle 2, -1, -1; 1 \rangle$.

5.3 Unate Node Splitting and Combining

If the ILP problem for node n does not have a solution, the node must be split into smaller nodes to increase the likelihood of the split nodes being threshold functions. Figure 7 outlines the splitting process of a unate node. How a unate node is split is contingent upon one of the following conditions:

1. All of the variables appear exactly once.
2. Some of the variables appear in all the cubes.
3. The most frequent variable(s) does not appear in all the cubes.
4. A tie between the most frequent variables is broken randomly.

If all the variables appear only once, we simply split the node into two, with each node containing roughly equal number of cubes. For example, $n = x_1 x_2 \vee x_3 x_4 \vee x_5 x_6$ is split as $n_1 = x_1 x_2 \vee x_3 x_4$, $n_2 = x_5 x_6$, with $n = n_1 \vee n_2$. When a variable appears in all the cubes, we split the node into two by factoring this variable out of the node. For example, for $n = x_1 x_2 \vee x_1 x_3 x_4 \vee x_1 x_5 x_6$, $n_1 = x_1$ and $n_2 = x_2 \vee x_3 x_4 \vee x_5 x_6$, with $n = n_1 n_2$. If the above two conditions are not met, we split the node using the most frequently

Require: unate node n, $\Psi > 0$

    if all variables appear once **then** // condition 1

        $n = n_1 \vee n_2$ s.t. $|K_{n_1}| = |K_{n_2}|$

3:    **else if** $\forall c, c \in K_n$, variable $x_i \in c$ **then** // conditions 2 and 4

        $n_1 \leftarrow x_i$ // assuming correct phase

        $n_2 \leftarrow$ factor n w.r.t. n_1

6:    **else** // condition 3

        $x_i \leftarrow$ most frequently appearing variable

        $n_1 \leftarrow \Sigma c, x_i \in c \wedge c \in K_n$

9:      $n_2 \leftarrow \Sigma c_l, c_l \notin n_1 \wedge c_l \in K_n$

     if $n_1 \in Z$ **then** // assuming $|K_{n_1}| \geq |K_{n_2}|$

        combine nodes according to Theorem 2

12:    $\vec{v} \leftarrow n_1 \vee n_2$

    else if $n_2 \in Z$ **then**

        combine nodes according to Theorem 2

15:    $\vec{v} \leftarrow n_1 \vee n_2$

    else

        $k \leftarrow (|Kn| \leq \Psi)$? $|Kn|$: Ψ // pick the smaller one

18:$\vec{v} \leftarrow \sum_{i=1}^{k} n_i$ // split n into k smaller nodes

      return $\vec{v}$ // array of split nodes

Fig. 7 The unate node splitting algorithm

appearing variable. For example, given $n = x_1x_2 \vee x_1x_3 \vee x_4x_5$, we split on x_1 to get $n_1 = x_1x_2 \vee x_1x_3$ and $n_2 = x_4x_5$, with $n = n_1 \vee n_2$. This last condition reduces the likelihood of a function being non-threshold because there are fewer candidate variables to choose from in the split nodes to prevent the condition in Theorem 1 from being satisfied.

Once a node has been split, we choose the larger node (i.e. the one with more cubes) and check that node to see if it is a threshold function. If it is, we apply Theorem 2. Looking at the last example, since $n_1 = x_1x_2 \vee x_1x_3$ is a threshold function with weight-threshold vector $\langle 2, 1, 1; 3 \rangle$, function $n = x_1x_2 \vee x_1x_3 \vee n_2$ is also a threshold function with weight-threshold vector $\langle 2, 1, 1, 3; 3 \rangle$. Now, n_2 is processed by further collapsing, threshold checking, or splitting. If neither of the split nodes is a threshold function, the original node is split into k smaller nodes, where k is the smaller of ψ and $|K_n|$. After splitting, $n = \sum_{i=1}^{k} n_i$ which is a threshold function with weight-threshold vector $\langle 1,1,\ldots,1; 1 \rangle$. The split nodes, n_i, $i \in \{1,2,\ldots,k\}$, are then processed recursively.

5.4 Binate Node Splitting

If a node n is binate, we split it into k smaller nodes where k is the smaller of ψ and $|K_n|$. The algorithm for binate node splitting is shown in Fig. 8. We first split the binate node on the most frequently appearing binate variable. If the split nodes are binate, we repeat the process. Otherwise, we use the unate node splitting algorithm, as detailed in the previous subsection.

Require: binate node n, $\Psi > 0$
    $k \leftarrow (\,|K_n| \leq \Psi)$? K_n : Ψ // pick the smaller one
    $\vec{v} \leftarrow n$ // split on binate variables when needed
3:   **while** $|\vec{v}| \neq |k \wedge \exists p \in \vec{v}$, s.t. p has binate variable x_i **do**
    $n_1 \leftarrow p(x_i, \ldots)$
    $n_2 \leftarrow p(\overline{x}_i, \ldots)$ // assign the resulting split nodes to vector
6:   $\vec{v} \leftarrow \{\vec{v}-p\} \vee \{n_1,n_2\}$
    // split on unate variables when needed
    while $|\vec{v}| \neq k \wedge \exists p \in \vec{v}$, s.t. p is unate **do**
    $n_1 \leftarrow$ split unate node p
9:   //assign the resulting split nodes to vector
    $\vec{v} \leftarrow \{\vec{v} - p\} \vee n_1$
    return $\vec{v}$ // array of split nodes

Fig. 8 The binate node splitting algorithm

To demonstrate binate splitting, consider $n = \bar{x}_1 x_4 \vee x_2 x_3 \vee \bar{x}_2 x_4 x_5$, where ψ is five and $|K_n|$ is three. This node is split into three nodes. First it is split on the binate variable, x_2, to get $n_1 = \bar{x}_1 x_4 \vee x_2 x_3$ and $n_2 = \bar{x}_2 x_4 x_5$. Now, n_1 is further split into $n_1 = \bar{x}_1 x_4$ and $n_3 = x_2 x_3$. Thus, n is represented as $n = n_1 \vee n_2 \vee n_3$, which is a threshold function with weight-threshold vector $\langle 1, 1, 1; 1 \rangle$. Threshold network synthesis proceeds recursively by processing each of the split nodes.

5.5 *Complexity Analysis*

In this subsection, we perform the complexity analysis of the algorithms in our threshold network synthesis methodology. The complexity of the node collapsing algorithm is $O(1)$, because the total number of operations performed by the algorithm is proportional to the fanin restriction, which is constant. The complexity of both the unate and binate node splitting algorithms is $O(|F_n| \cdot |K_n|)$. Since ILP is NP-complete, our synthesis problem is NP-complete in theory. However, in practice, the ILP solver is implemented such that if the optimal solution cannot be found in a reasonable amount of time, it declares the problem as infeasible. If that happens, the splitting algorithms in our methodology create smaller problems for the ILP solver to solve. In this way, the threshold network synthesis problem can be solved efficiently in practice with our methodology.

5.6 *Implementation*

We implemented the proposed methodology in a tool called ThrEshold Logic Synthesizer (TELS) which has been integrated within SIS. This is the first multi-output multi-level threshold network synthesis tool to the best of our knowledge. The package currently consists of approximately 3,500 lines of C code. We integrated a linear programming tool called *LP_SOLVE* [13] in SIS to solve the ILP problems. The framework of TELS is shown in Fig. 9. Currently, it supports five commands, which perform one-to-one mapping, threshold synthesis and simulation, and displaying of network information.

6 Experimental Results

We present our experimental results in this section. The experiments were conducted on a 2.4 GHz Pentium IV machine with 768 MB RAM running Redhat Linux 8.0. We ran all the benchmarks in the MCNC benchmark suite through TELS. All the synthesized networks were simulated for functional correctness to validate our methodology.

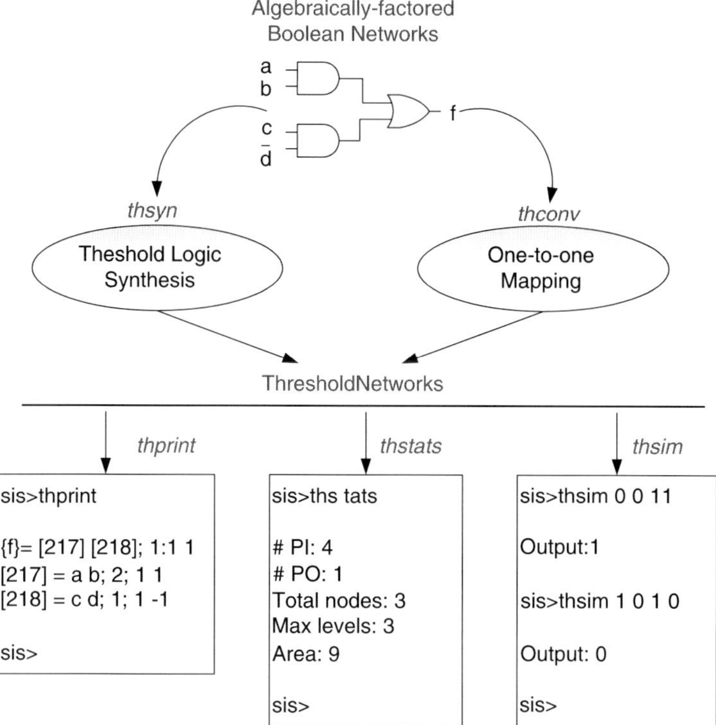

Fig. 9 The framework of the threshold network synthesis tool: TELS

6.1   *Threshold Gate Count and Area*

Due to space limitations, Table I lists the results for only 10 of the 60 benchmarks. In this table, one-to-one mapping refers to replacing each gate in the optimized Boolean network with a threshold gate. The optimized Boolean network was obtained by running the *script.boolean* script in SIS. The threshold network synthesis results were obtained by running the *script.algebraic* script and then synthesizing the threshold network from the resulting algebraically-factored network. The area of the network, A, was calculated using the following equation:

$$A = \sum_{g \in G_T} \sum_{i=1}^{l} (|w_i| + |T|) A_u, \tag{14}$$

where g is a threshold gate in network G_T, l is the number of inputs in gate g, w_i is the weight of input i, A_u is the unit area of an RTD with $w = 1$, and T is the threshold of the gate. In our case, we let A_u equal one. The HFET area is ignored because it is typically much smaller than the RTD area.

The total time required to algebraically factor and synthesize the benchmarks was less than one second in each case. On an average, 42% of the total execution time was

Table 1 Threshold Synthesis Results with Fanin Restriction Set to 3

Benchmark	One-to-one mapping			Threshold network synthesis		
	Gates	Levels	Area	Gates	Levels	Area
cm152a	28	4	99	13	4	69
cordic	92	9	307	39	8	219
cm85a	70	8	254	16	6	158
comp	181	12	625	70	9	435
cmb	41	7	142	16	7	103
term1	397	12	1,459	144	16	787
pm1	49	5	176	22	3	119
x1	428	10	1,589	144	10	968
i10	2,874	49	10,934	1,276	47	7,261
tcon	24	2	80	32	2	96

spent on threshold network synthesis while the remaining time was spent on factoring the network. Comparing the results, we see that 52% average reduction is possible in gate count. In some cases, such as *tcon*, we do worse than a one-to-one mapping because there exist Boolean functions that require more threshold gates than Boolean gates. However, this is not a significant problem because we can always choose the better of the two networks, thereby, guaranteeing that TELS will never output a network requiring more gates than that required for one-to-one mapping.

6.2 Trend of Threshold Gate Count with Change in Fanin Restriction

Figure 10 demonstrates the trend in the total number of threshold gates required for the *comp* benchmark as the maximum fanin restriction is relaxed from three to eight. There is a significant difference in gate count in the one-to-one mapping case as the fanin restriction is relaxed. This is because with larger allowed fanin, it is possible to decompose a factorized network better. However, there is no significant reduction in the number of threshold gates for TELS. This is understandable because as the allowed fanin increases, the likelihood of a function being threshold decreases. As reported in [3], all positive unate functions of three or fewer variables are threshold functions. However, 17 out of 20 and only 92 out of 168 positive unate functions of four and five variables, respectively, are threshold functions, not considering variable permutations. Thus, we can see that with increasing allowed fanin, the percentage of functions that are threshold decreases drastically. The overall conclusion is that a fanin restriction of three to five gives good results.

6.3 Parametric Variations in Weights

We performed experiments to gauge the impact of parametric variations in the input weights on the circuit functionality. We varied δ_{on} from zero to three and δ_{off} was

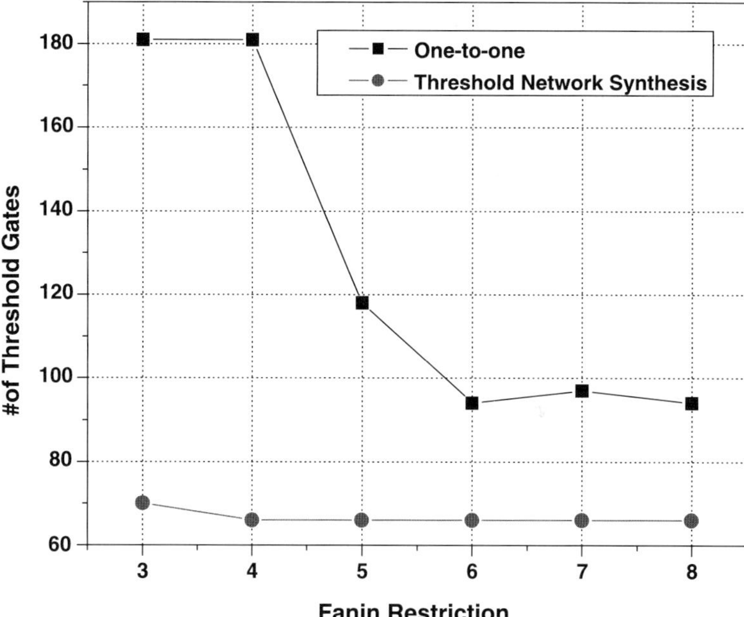

Fig. 10 Gate-count vs. fanin restriction for *comp*

fixed at one. The disturbed value w' was computed as $w' = w + v \times U(-0.5, 0.5)$, where w is the original value, v is the variation multiplier and $U(-0.5, 0.5)$ is a random variable uniformly distributed between -0.5 and 0.5. The circuit fails if there exists any input vector with which TELS generates a wrong output value under the disturbed weights during simulation. The failure rate is defined as the percentage of benchmarks that failed to pass simulation. The results shown in Fig. 11 demonstrate that as δ_{on} increases, the failure rate decreases. This is because the network is more robust. The tradeoff is that the network area increases as shown in Fig. 12 for the case $v = 0.8$.

7   Conclusions

In this paper, we introduced the first comprehensive threshold network synthesis methodology for multi-output multilevel networks. The algorithm in our methodology is recursive in nature and is based upon efficient heuristics that partition a logic function if it is determined to be non-threshold using an ILP formulation. Any logic sharing that occurs in the algebraically factored network is reflected in the threshold network. We have implemented the methodology on top of an existing logic synthesis tool to produce the first threshold network synthesis tool, and validated it by running the tool on a large number of benchmarks and

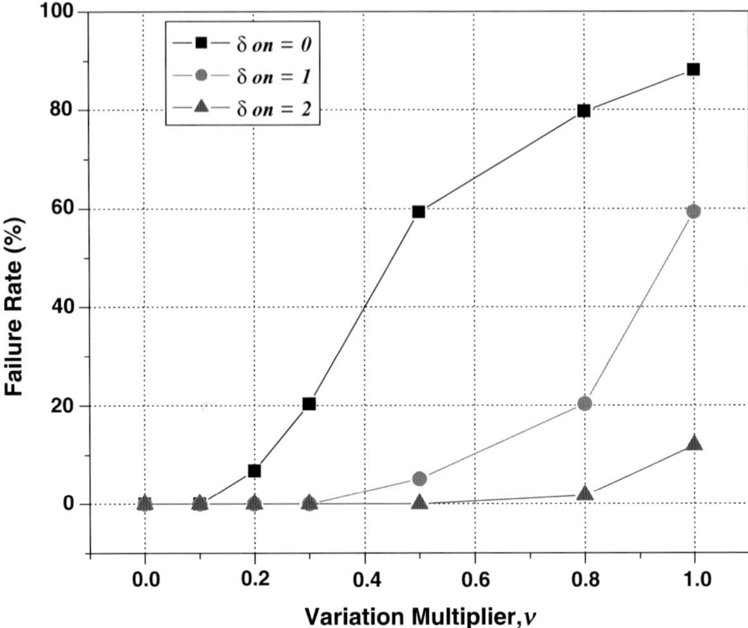

Fig. 11 The failure rate due to variations in the input weights

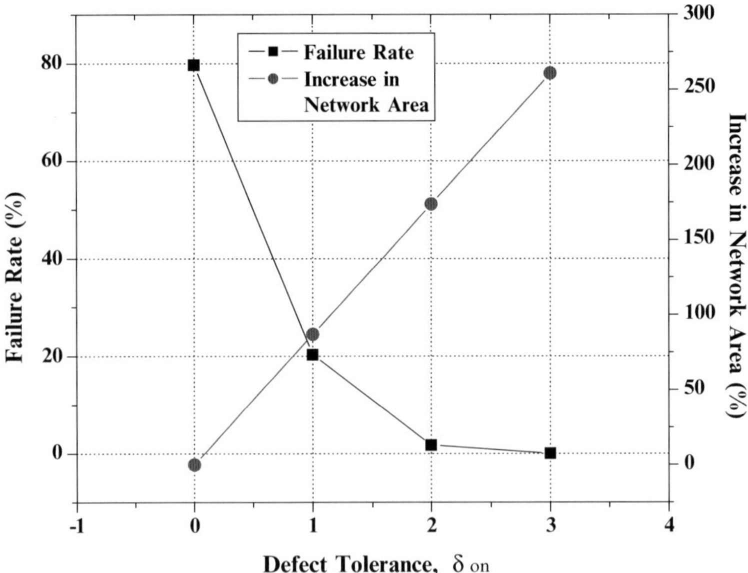

Fig. 12 The relationship between failure rate, network area, and defect tolerance due to variations in the input weights with variation multiplier $v = 0.8$

simulating the synthesized networks. Experimental results for the benchmarks show that the quality of the generated networks, in terms of total gate count and the number of levels, is very good.

Because this is the first tool for threshold network synthesis, there is room for improvement. For example, our method performs a backward traversal of the network from the outputs to the inputs. This makes the synthesized network somewhat dependent on the original network structure. Perhaps other approaches, such as divide and conquer, could also be used in threshold network synthesis. There may also exist better partitioning heuristics that might generate better results. Furthermore, perhaps different heuristics are required depending upon the optimization criteria. We hope that others will join in our efforts to improve upon this work. We also hope that integrating our methodology in commercial design automation tools will help pave the way for a smoother transition towards logic design using nanotechnologies.

Acknowledgments This work was supported in part by NSF under grant No. CCR-0303789.

References

[1] Semiconductor Industries Association Roadmap. http://public.itrs.net

[2] D. Goldhaber-Gordon et al. Overview of nanoelectronic devices, *Proc. IEEE*, vol. 85, no. 4, 521–540, Apr. 1997.

[3] S. Muroga, *Threshold Logic and its Applications*. New York, NY: John Wiley, 1971.

[4] K. J. Chen, K. Maezawa, and M. Yammoto, InP-based high-performance monostable-bistable transition logic elements (MOBILE's) using integrated multiple-input resonant-tunneling devices, *IEEE Electron Device Lett.*, vol. 17, no. 3, 127–129, Mar. 1996.

[5] Z. Kohavi, *Switching and Finite Automata Theory*. New York, NY: McGraw-Hill, 1978.

[6] G. D. Hachtel and F. Somenzi, *Logic Synthesis and Verification Algorithms*. Norwell, MA: Kluwer Academic, 1998.

[7] R. O. Winder, Threshold logic, Ph.D. dissertation, Princeton University, 1962.

[8] M. L. Dertouzos, *Threshold Logic: A Synthesis Approach*. Cambridge, MA: The M.I.T. Press, 1965.

[9] G. E. Sobelman and K. Fant, CMOS circuit design of threshold gates with hysteresis, in *Proc. Int. Conf. Circuits Systems*, vol. 2, May 1998, pp. 61–64.

[10] V. Beiu, J. M. Quintana, and M. J. Avedillo, VLSI implementations of threshold logic - A comprehensive survey, *Tutorial at Int. Joint Conf. Neural Networks*, July 2003.

[11] A. L. Oliveira and A. Sangiovanni-Vincentelli, LSAT - An algorithm for the synthesis of two level threshold gate networks, in *Proc. Int. Conf. Computer-Aided Design*, Nov. 1991, pp. 130–133.

[12] E. M. Sentovich et al. Sequential circuit design using synthesis and optimization, in *Proc. Int. Conf. Computer Design*, Oct. 1992, pp. 328–333.

[13] M. Berkelaar, Linear programming solver. http://www.cs.sunysb.edu/ algorith/implement/ lpsolve/implement.shtml

Part V
Physical Design and Validation

Physical Design and Validation

Jochen A. G. Jess

Abstract This paper reviews aspects of VLSI fabrication technology and the associated design and design automation techniques to set the stage for the presentation of those papers in physical design and validation that were selected as the most influential DATE publications in the last 10 years.

1 Introduction

THIS paper appears as the consequence of an invitation extended to the above author to set the stage for four papers that have been carefully selected as the most influential DATE publications in the area of physical design and validation in the last 10 years. To do this properly it seems fair to review the evolution of chip technology first, because it sheds a light on the challenges designers had to deal with in this period. Such a review is attempted in Section 2. Chapter 3 then provides a brief and admittedly incomplete exposure of the concepts created by designers and design automation engineers to meet those challenges. Section 4 briefly comments the selected papers. Eventually Section 5 winds up the paper with some disclaimers.

2 Evolution of Semiconductor Technology from 1997 Towards 2007 and Further

2.1 State of Affairs in 1997

The initiation of integrated circuits technology is accounted to Jack Kilby (TI) in 1957. Shortly after that event Moore's law entered the scene around 1965. Gordon Moore saw no essential obstacles to pursue the line of increasing density of functions on the real estate of a chip for at least 10 years. Essentially the law is now valid for almost 40 years. There was a fast but nevertheless smooth evolution. The density of the population of transistors on a single processor chip rose exponentially, doubling approximately every 18 months. From the mid-1970s

R. Lauwereins and J. Madsen (eds.), *Design, Automation, and Test
in Europe–The Most Influential Papers of 10 Years DATE*, 347–358
© Springer 2008

onwards CMOS became the essential standard for microprocessors (MPUs) and signal processors (DSPs). Also memories were dominantly based on MOS technology. Technology evolution concentrated on continuous improvement of the p- and n-MOS transistor. By 1980 some CMOS technology would comprise local oxidation to establish the field oxide that blocked current leaks between individual transistors, self-aligned polysilicon gate contacts and an n-well for the p-devices provided by Ion implantation. The oxide insulation between the Aluminum conductors was established by Si-oxide layers deposited by means of low-pressure chemical vapor deposition (LPCVD). The feature sizes at that time were around 3 to 1.5 µm. A chip would contain in the range of some ten thousands of transistors [1].

In the sequel, along with gradual improvements of the lithography and the process quality a number of fundamental breakthroughs removed serious obstacles on the road of increasing transistor density. One of them was anisotropic etching allowing for etching deep trenches with steep walls enabling effective electrical isolation between devices at close distance. Another important achievement was the chemical mechanical polishing (CMP). By this technique wiring layers could be kept reasonably planar. Consequently the number of wiring layers could be increased from, say, two layers up to five, eventually reaching about eight to ten. An important decision, taken about 1994, was to scale-down the power supply voltage and the logic swing, V_{DD}, from 5 V to lower values, beginning with 3 V. Reducing the logic swing degraded the noise margins of logic blocks and increased the noise sensitivity of logic signals. But, while the industry kept increasing the functional density on the chip, it allowed for downscaling of the power consumption of chips.

From 1992 onwards the industry felt that coordination of the evolution of chip technology needed more attention in view of the business risks involved. Various professional bodies from the US, ASIA, and Europe finally cooperated to establish the "International Technology Roadmap for Semiconductors" (ITRS). This roadmap documents the targets for the entire field of semiconductor technology, including design technology, over a period of about 15 years in the future. It is yearly rewritten or updated, respectively.

In 1997, just before DATE entered the conference scene, according to the ITRS documents, microprocessor chips would carry some 11 million transistors on about $300 \, mm^2$ Silicon real estate. With the global clock running at 375 MHz, local clocks running up to 750 MHz and the minimum logic swing downscaled to 2.1 V the processor would dissipate 70 W. The transistor channel length was 250 nm, the pitch of short wires was 750 nm. The total wire length amounted to 800 m distributed over six wiring layers. This is to say, the industry was way into the submicron era.

2.2 *Technology Challenges*

The sequence of ITRS documents of the last 10 years shows an impressive record of controlled progress, albeit that the industry outperformed the predictions of the roadmap consistently and continues to do so [2].

In 2007 the so-called half-pitch of poly lines of MOS transistors in an MPU will be down to 65 nm, which represents downscaling by a factor of 3.85 and up scaling of the transistor density by a factor of almost 16 as compared to 1997. While this pitch and its further downscaling present significant challenges to the lithographic techniques, an even bigger challenge is the gate oxide film. While the actual gate length is in the order of 28 nm the oxide thickness is much smaller, in the range of a few nano-meters, or equivalently, just some multiple of 10 atom layers, going down to just a few atom layers as scaling proceeds. This implies a whole series of parasitic effects with increased gate leakage current being the most prominent one. Increasing the carrier mobility in the MOS channel by "strained Silicon" and the use of gate isolation materials with a dielectric constant higher than that of Silicon oxide (so-called high-κ materials like Hafnium oxide) in combination with metal gate contacts relieves the leakage current issue to some extent. High-κ materials and strained Silicon improve the drive current of the transistor such that thicker gates with reduced leakage current are feasible. Metal contacting prevents the migration of carriers from the poly-Silicon gate dope material through the gate isolation film, thus reducing leakage current even further.

Other improvements of the transistor concern the reduction of the capacitive embedding of the transistor structure, for instance by "Silicon on insulator" technology. The channel region is shielded from the bulk region by a fully depleted layer buried in the substrate. In addition it was attempted to achieve more symmetry between p- and n-transistors by implanted dual-well structures.

Lately Copper started to replace Aluminum as the primary conducting material for metal interconnect, reducing the resistivity of metal lines by some 30%. For insulation so-called low-κ materials like SiLK and Xerogel were investigated that could reduce the dielectric constant by a factor of about two.

The progress of miniaturization involved a permanent flow of innovation in materials research. In addition lithography was challenged. Currently optical lithography with light of 193 nm wavelength is in use. This is five times larger than the pitches of the circuit structures. Clear prints of lines could only be achieved with techniques like "phase shift masks", "optical proximity correction" (OPC), and "resolution enhancement" (RET).

2.3   From 2007 Onwards

The 2006 update of the ITRS suggests continuity of the progress of miniaturization, be it that progress slows down gradually. Although the basic motor of success was the exponentially decreasing cost of an individual transistor there are other economic aspects impairing the continuation of the increase of transistor density in the line of Moore's Law. It is increasingly difficult to define products at the right scale of sales volume, which offer enough perspective to recover the cost of investment incurred by design, verification, mask- and prototype fabrication and test, that is, the non-recurring engineering cost (NRE cost). This holds in particular for the MPU-area.

Nevertheless there is already a new generation of lithography machines under development on the basis of 13.4 nm light sources ("Extreme UV"). Those machines would not have lenses — lenses absorb this kind of light. Rather the optical system is entirely based on mirrors.

Another promising approach includes stacking wafers on top of each other. To achieve this wafers have to be grinded and polished to become sufficiently thin before they are stacked. Vertical connections could be implemented using short Copper bonds of, say, 20 µm length. Less advanced approaches aim at vertical connection with ball bonding [3].

3   Challenges for Physical Design

3.1   Automation of Physical Design at 1997 and Before

3.1.1   Research and Basics

Research in design automation for electronic circuitry started in the early 1960s of the previous century. The first (SHARE) Design Automation Workshop was held in 1964 and was a predecessor of today's Design Automation Conference. The period till the early 1990s is characterized by a wealth of research and development of computing methods and prototype design software systems, both from academia and industry. Yet the design system implementations mostly disappeared in the archives of history. The reason for this was twofold: firstly the programming environments were not stable. Just when UNIX workstations and later PC's were powerful enough to handle the masses of data and the computational load involved they became a lasting foundation for design environments. Secondly the exponential growth of data associated with Moore's Law rendered algorithmic solutions outdated before they could be applied on a large scale in practice.

However in this period almost the entire spectrum of fundamental concepts governing today's Electronic Design Automation (EDA) design environments have been incepted. Below there is a small list of examples with specific relevance for physical design and validation:

* Circuit simulation, based on the sparse companion model, sparse L/U decomposition and implicit integration is still performed according to the principles developed at IBM Yorktown and UC Berkeley in the 1970s [4].
* Many modern floor-planning and placement algorithms are based on the min-cut principle [5], the eigenvalue placement [6] or the "force - directed" placement method [7].
* Global routing makes frequent use of the concept of "Steiner trees" [8].
* The "left-edge" algorithm was one of the first of a whole series of methods to deal with the problem of "channel-routing" [9].

- Maze routing and line search are at the roots of most modern routing principles and are special cases of breadth-first and depth-first graph-search algorithms [10–12].
- The corner stitching method to structure layout data is probably still in use with lots of software devoted to editing, displaying and manipulating layouts [13].
- The "scan line" algorithm is at the hart of many design rule check and extraction procedures [14].
- The principle of "simulated annealing" is applied in a whole range of optimization problems in physical design, beginning with constrained placement problems, for instance with gate arrays [15].
- Today's tuning algorithms owe to the posynomial programming approach [16].

The area of EDA also gained lots of profit from the general advancement of applied mathematics, in particular from mathematical programming.

3.1.2 Changing the Structure of the Semiconductor Industry from Vertical to Horizontal Organization

Before, say, 1985 the Semiconductor industry was largely vertically integrated. Large companies from the field of electronics maintained chip design teams and extensive fabrication lines. As the level of integration approached the margin of some 10,000 transistors per chip the need for more efficient computer-aided design environments became evident. Those companies developed captive design environments based on mainframe implementations of point tools like gate array and standard cell place and route, layout, and netlist editing tools and simulators.

This scenario changed quite drastically around 1985 by the advent of UNIX workstations. As mentioned before those workstations were powerful enough to serve most interactive and semiautomatic design, verification, and simulation tools. At the same time research institutions realized that the standardization of the hardware and software platform enabled large international research groups to pool their efforts. The concept of an "open" design environment was born, as opposed to the closed proprietary turnkey systems of the previous era. This trend was further enhanced by the availability of both local and global networking. Eventually personal computers became so powerful that the bulk of the design activity could be entrusted to them.

As a consequence the industry started to reorganize itself quite drastically. Foundry services allowed electrical companies to become "fabless". They created a portal to their fabrication lines providing the technology information needed to design circuits and systems. Simultaneously companies marketing chips would reduce their internal CAD support to the maintenance of libraries of reusable components and the associated models to assess their performance by simulation and validation tools. In turn the design tools were purchased from companies specializing on developing them. From then on there was a viable economic base for independent EDA companies to flourish.

Currently this trend has almost come to completion. The latest stage involved large electronic companies to split off their fabrication lines as separate companies.

The current organization of the semiconductor industry provided the flexibility necessary to sustain the pressure of Moore's Law, at least for a while.

3.1.3 Architecture of Open Design Environments

From 1985 onwards "top-down design" became the leading paradigm in EDA. For a while part of the EDA community cherished the ambitious concept of the "Silicon Compiler". This was understood to be a universal tool converting high-level behavior specifications of chips into provably correct layouts. This dream did not sustain. Yet the design environments displayed a vertically structured set of system representations and provided semiautomatic translation tools for synthesis and verification tools for analysis and equivalence checking. So there was "high level synthesis" establishing an RTL description, "logic synthesis" to obtain a gate netlist and the sequence of floorplan design, gate placement and global and local routing to eventually obtain a layout description in terms of rectangles. Layout- and network-validation tools in connection with so-called extractors complemented these transitions. For a while all this was embedded into a "framework" hosting all the necessary files and providing elements of design management, like for instance "version control". The "open" character of this environment allowed for including "private" data types and tools.

3.1.4 Loosing the "top-down" Design Paradigm

At the beginning of the 1990s the growing density of chips under submicron design rules was felt in various ways. To begin with the files capturing the design data became bigger and bigger. Specifically in the area of physical design the layout files would reach the multi-gigabyte range. It became increasingly tedious to send those files around over computer bus links or over internal and external network connections.

Another concern was the length of global connections such as buses and clock lines that could become up to 10 mm in addition to becoming very thin. The parasitic resistive and capacitive properties of such metal connections started to induce significant wire delays. In the first place, for a given conducting line, the delay increases with the square of length. In the second place the capacitance load, rather than decreasing with downsizing of the wires, tends to saturate to some constant value because of fringe effects and interaction with neighboring wires, as was demonstrated in a modeling experiment in [17]. To compensate for the delays of long wires buffer insertion was proposed [18]. For the design of DRAM type memories the physical design of the long word- and bit lines had been an art already before and became even more involved as we headed into the deep submicron era.

In the case of standard cell synthesis for timing, for instance, decisions on the level of logic design had to be revised after place and rout, because capacitive and resistive parasitic effects could only correctly assessed after all the geometric details of the layout were known.

3.2 *New Challenges and Solutions in the last 10 Years*

3.2.1 Design for Timing Closure of Logic Circuits

The original abstraction of Boolean logic, in its evolution to support design for meeting given clock speed requirements, had been enhanced to include "wire load models" (WRL) for standard logic gates. For a given logic gate in some component library the WRL is a lookup table, which for a set of selected fan-out values supplies estimates for the respective capacitive and resistive load. Delay estimates based on those WRLs, when relied on during logic synthesis, proved increasingly inaccurate because the unpredictable parasitic effects of the wiring became more and more relevant with every new technology generation. After the layout design was completed extraction created accurate models of the parasitic elements that regularly disclosed violations of the timing requirements. Thus redesign was necessary without anybody knowing how to control the design iterations towards convergence. The method of "gain based" design proved to be essential to resolve this problem [19].

Subsequently the design flow has been changed to the extent that in the first step a schedule is established for the logic operations. Logic synthesis, in addition to creating an appropriate logic structure, assigns a delay requirement to each individual logic gate. The procedure then enters the layout design phase. The result exhibits the exact shapes of the wires. The modeling of the logic circuit is then extended to include all the parasitic elements of the wire structure by elaborate extraction. Subsequently timing closure is achieved by adjusting the "gains" of the logic gates. As a result gates are just as "strong" as the timing requires. As a side effect, upon completion of a design, almost all logic paths through a combinational circuit are close to critical.

Practical experience with this method proved that many redesign cycles could be avoided. However, the logic libraries had to be adjusted to provide a large number of copies of each gate type for a sufficiently large number of gain values.

Subsequently detailed knowledge is available to evaluate the waveforms on all wires. This computation reveals a number of other phenomena such as delay violations introduced by cross talk between neighboring wires. A corrective go over the wiring can eliminate those problems by gain adjustment, wire sizing, relocation of wires or by the insertion of shielding wires.

Clock networks are usually designed separately by means of special techniques involving both buffer insertion and the sizing of wires. Of course there is an intimate relationship between the delays and skews of the combinational logic circuits, the set up and hold times of the latches holding the circuit state and the arrival times, the rise- and the fall times of the clock pulses. Since the clock switches at least twice per cycle the clock circuit is the busiest on the entire chip.

3.2.2 Power Issues

As mentioned earlier the power consumption in chips scales down with subsequent generations of technology because the supply voltage was reduced gradually. Yet as the functional density on the chips grew, still more and more worries about the entailing power density troubled the designers. Power consumption in a chip consists of three components:

- Power necessary to charge the wires that represent the logic signals ("dynamic power")
- Power lost in the transitions between logic values when both n- and p-transistors of a CMOS gate conduct during a certain period of time ("short-circuit power")
- Power entailed by the leakage current of transistors in "off"-state ("leakage power")

The first and the second contribution are directly proportional to the switching activity of the circuit. The first contribution scales with the capacitances to be charged and the second power of the logic swing. The second contribution is strongly dependant on the signal slew rates. Power assessment therefore requires, in addition to extracted circuit data, statistics of the probabilities of states and transitions. Those can be assembled by statistical simulation, for instance in the course of test pattern generation.

Power loss induced by the leakage current depends on the logic state of the circuit but has a DC-character, which means it is always there even if there is no logic payload activity. Given a certain power supply voltage, V_{DD}, there is a trade-off between leakage current and switching speed by controlling the threshold voltage of the transistors, V_T. With decreasing V_T switching speed and leakage current both increase. The increase of the leakage current is even exponential with decreasing V_T (which is the reason for the major concern it causes for deep submicron Si Technology). In some design flows therefore V_T is controlled to reduce leakage. But controlling V_T also changes the slew rate of the logic signals involved, implying a change of short-circuit power. An optimizing design strategy requires, therefore, accurate modeling tools.

Two other concerns are "hot spots" and electro-migration. Hot spots occur if in some region a lot of logic activity is concentrated. It is a matter of revising the placement to resolve this problem. Also certain signal wires may carry a lot of current while charging large capacitances over long distances, such as for instance buses. An erosion of metal particles as a consequence of high-current density threatens the reliability of those wires. The lifetime of those wires decreases according to the second power of the current density. The appropriate countermeasure is either to increase the thickness of the wire or to redistribute the current.

Special attention has to be paid to the design of the power grid, which distributes V_{DD} and ground to every component on the entire chip. The finite resistance of the power and ground lines narrows the logic swing for gates remote from the supply sources thus impairing their switching speed. Power lines are notoriously threatened by electro-migration. So careful design of the wire sizes is mandatory. In addition switching activity at one point of the chip generates noise that is distributed over the power grid. The countermeasure is the insertion of decoupling capacitances between power and ground. The design objective is to achieve the required noise reduction with as little decoupling capacitance as possible to curb the area penalty involved. Preferably use the "white space" (that is unused space in the layout) to position the decoupling capacitances.

3.2.3 Variability

Much like in any other technology manufacturing tolerances became more relevant with shrinking geometry. In any process step there is a certain variation of the parameters involved. In the traditional design flow "best case" and "worst case" corners of the performance accounted for those variations. Analysis tools had ways to propagate those corners to the results produced by tools like "static timing analysis". But by about the millennium change the designer's community realized that variations became disproportionally more relevant than before [20]. At the same time experiments suggested that the best-case-worst-case analysis yielded misleading results, leaving design performance on the table [21]. This situation has worsened ever since. In the meantime various methods are available to include statistical analysis into tools like static timing analysis, while foundries make an effort to supply statistical process data.

3.3 *From 2007 Onwards*

Obviously in trying to predict the evolution of physical design and validation the ITRS establishes a firm guideline as to which problems to expect. Variability, power issues, and noise problems are likely to dominate along with the old problem of handling the complexity of the integrated systems on a chip. In the first place there is a never-ending demand to have efficient techniques modeling the complicated interactions between electrical noise, speed, power and floating technology properties for the validation of designs. In addition the quest for more efficient design will demand new ways to regain top-down design flows lost by the interaction of previously separate design phases.

Currently a growing set of chip products starts to include various forms of power management, like:

- Suppressing the clock of latches in temporarily inactive parts of the circuit ("clock-gating")

- Switching off the power supply in temporarily inactive portions of the circuit (curbs also the leakage current)
- Matching clock speed with the required computation effort either statically or dynamically
- Tuning (statically or dynamically) V_{DD} and V_T to match the required performance (controlling V_T is possible by "back-gate biasing")

Additional challenges may arise from the integration of new components like sensors (camera back planes), light sources, MEMS components or carbon nano-tubes into the same chip.

4 The Selected Papers

The DATE papers selected in the physical design area for this book are a nice illustration of the way the emphasis changed over the 10-year period since the commencement of the DATE conference. The 1998-paper about "Interconnect Tuning Strategies" reflects the growing concern at the end of the previous century about the physical properties of the interconnect structure in as much as design had to account for them. It demonstrates the extent to which it matters to make the right choices.

The 2001-paper on inductance extraction again focuses on interconnect. This time the inductive properties are addressed as they became increasingly relevant for the bonding structures and the long wires in view of rising clock frequencies. The paper deals with the computational stability of discrete inductance models. The way of reasoning, in a sort of a "physical" style about this essentially mathematical issue, is appreciated as particularly elegant by this author.

The "Soft Error Tolerance Analysis" paper is taken from the 2005 issue of the conference proceedings. It supplies a good demonstration of how to capture a complex phenomenon in an as simple as possible style and yet provide sufficient accuracy.

Lastly the 2006-paper concerning the integration of an extremely sensitive "single photon" sensor array in a more or less standard sub-micron technology process is a stunning demonstration of "physical design" producing surprising combinations of high performance components.

5 Conclusion

This paper is in many ways incomplete. If it comes to the relevance of things the emphasis is somewhat misleadingly placed on microprocessor and ASIC design, perhaps because it is felt that they present the hardest challenges to design automation. But memories like SDRAM, embedded SRAM, or Flash memories are at least as important as system components and business factors. The same goes for field programmable devices and gate arrays. In the future the integration of a large number of different technologies on one die may present even bigger challenges. Also items like analog functions and radio front ends have not received the attention they would possibly deserve in the eyes of many experts.

But this paper had to limit its scope. The important objective is to prepare the reader to study the fine papers that follow hereafter.

References

[1] Johnson, N. L., V. V. Marathe, Bulk CMOS Technology for the HP-41C, Hewlett-Packard Journal, March 1980, see also http://www.hpmuseum.org/journals/hp41/41bc.htm
[2] The International Technology Roadmap for Semiconductors (ITRS), see http://www.itrs.net/
[3] Riley, G. A., Nailing IC's Together, Flip Chips Dot Com Tutorial 71, February 2007, see http://www.flipchips.com/tutorial71.html, see also IMEC's website http://www.imec.be.
[4] Pederson, D., A historical review of circuit simulation, IEEE Transaction on Circuits and Systems, Vol. 31, no. 1, Jan. 1984, 103–111.
[5] Breuer, M. A. 1977. Min-cut placement. Journal of Design Automation and Fault Tolerant Computing, Vol. 1, no. 4, 1977, 343–362.
[6] Hall, K. M., An r-dimensional quadratic placement algorithm. Management Science, Vol. 17, no. 3, 1970, 219–229
[7] Hanan, M. On Steiner's problem with rectilinear distance. Journal SIAM Applied Mathematics, Vol. 14, no. 2, 1966, 255–265.
[8] Cheng, C. K. Kuh, E.S., Module placement based on resistive network optimization, IEEE Transactions on Computer-Aided Design of Integrated Circuits and Systems, Vol. 3, no. 3, July 1985, 218–225.
[9] Hashimoto, A., J. Stevens, Wire Routing by Optimizing Channel Assignment within Large Apertures, Proc. 8th Workshop on Design Automation, Atlantic City, US, June 28–30, 1971, pp. 155–169.
[10] Lee, C. Y., An algorithm for path connections and its applications, IRE Transaction on Electronic Computers, Vol. EC-10, 1961. 346–365.
[11] Hightower, D. W., A solution to line-routing problems on the continuous plane, Proc. 6th Annual Conference on Design Automation, 1969, pp. 1–24.
[12] Tarjan, R.E., Depth first search and linear graph algorithms, SIAM Journal on Computing 1, 1972, 146–160.
[13] Ousterhout, J. K., Corner Stitching: A Data Structuring Technique for VLSI Layout Tools, IEEE Transaction on Computer Aided Design, Vol. 3, no. 1, 1984, 87–100.
[14] Bentley, J.L., D. Haken and R. Hon, Fast geometric algorithms for VLSI tasks, IEEE CompCon Spring '80, 1980, 88–92
[15] Kirkpatrick, S., C. D. Gelatt Jr., and M. P. Vecchi, Optimization by simulated annealing, Science, Vol. 220, no. 4598, 1983, 671–680.
[16] Fishburn, J.P., A. E. Dunlop. Tilos: A posynomial programming approach to transistor sizing. In IEEE Intl. Conf. on CAD, pp. 326–328, 1985.
[17] Schaper, R. L., D. I. Amey, Improved Electrical Performance Required for Future MOS Packaging, IEEE Transaction on Components, Hybrids, and Manufacturing Technology, Vol. CHMT-6, no. 3, September 1983, 283–289.
[18] Van Ginneken, L. P. P. P., Buffer placement in distributed RC-tree network for minimal Elmore delay, in Proc. IEEE Int. Symp. Circuits Systems, New Orleans, LA, 1990, pp. 865–868.

[19] Sutherland, I. E., R. F. Sproull, Logical Effort: Designing for Speed on the Back of an Envelope, Proc. of the 1991 UC Santa Cruz Conf. on Advanced Research in VLSI, MIT-Press, Cambridge, MA, USA, 1991, pp. 1–16.

[20] Orshansky, M., L. Milor, P. Chen, K. Keutzer, C. Hu, Impact of Systematic Spatial Intra-Chip Length Variability on Performance of High Speed Circuits, IEEE/ACM Intern. Conf. on CAD (ICCAD), 2000, pp. 62–67.

[21] Visweswariah, C., Death, Taxes and Failing Chips, Proc. of the 40th Conf. on Design Automation, ACM Press, New York, 2003, pp. 343–347.

Interconnect Tuning Strategies for High-Performance ICs*

Andrew B. Kahng, Sudhakar Muddu, Egino Sarto, and Rahul Sharma

Abstract *Interconnect tuning* is an increasingly critical degree of freedom in the physical design of high-performance VLSI systems. By interconnect tuning, we refer to the selection of line thicknesses, widths and spacings in multilayer interconnect to simultaneously optimize signal distribution, signal performance, signal integrity, and interconnect manufacturability and reliability. This is a key activity in most leading-edge design projects, but has received little attention in the literature. Our work provides the first technology-specific studies of interconnect tuning in the literature. We center on global wiring layers and interconnect tuning issues related to bus routing, repeater insertion, and choice of shielding/spacing rules for signal integrity and performance. We address four basic questions. (1) How should width and spacing be allocated to maximize performance for a given line pitch? (2) For a given line pitch, what criteria affect the optimal interval at which repeaters should be inserted into global interconnects? (3) Under what circumstances are shield wires the optimum technique for improving interconnect performance? (4) In global interconnect with repeaters, what other interconnect tuning is possible? Our study of question (4) demonstrates a new approach of offsetting repeater placements that can reduce worst-case cross-chip delays by over 30% in current technologies.

1   Introduction

With technology scaling, on-chip interconnect becomes an increasingly critical determinant of performance, manufacturability and reliability in high-end VLSI designs. Current and future designs are generally interconnect-limited, and the avail-

Silicon Graphics, Inc., Mountain View, CA 94039 {muddu,sarto,ashu}@mti.sgi.com, abk@cs.ucla.edu

*Principal author, to whom correspondence should be addressed: Dr. Sudhakar Muddu, Silicon Graphics, Inc., muddu@mti.Sgi.com. Andrew B. Kahng is associate professor fo Computer Science at VCLA

R. Lauwereins and J. Madsen (eds.), *Design, Automation, and Test in Europe–The Most Influential Papers of 10 Years DATE*, 359–375
© Springer 2008

Table 1 Selected technology projections from the 1997 SIA NTRS

SIA National Technology Roadmap (1997)						
Year	1997	1999	2001	2003	2006	2009
Minimum feature size - dense lines (nm)	250	180	150	130	100	70
High-end on-chip clock frequency (MHz)	750	1250	1400	1600	2000	2500
# Wiring layers	6	6–7	7	7	7–8	8–9
Minimum contacted M1 pitch (µm)	0.64	0.46	0.40	0.34	0.26	0.19
Metal height/width aspect ratio	1.8:1	1.8:1	2.0:1	2.1:1	2.4:1	2.7:1

able routing resource must be carefully balanced among signal distribution, power/ ground distribution, and clock distribution. Table 1 reproduces several technology projections from the 1997 SIA National Technology Roadmap for Semiconductors [1]. A notable deviation from the original 1994 Roadmap is that maximum on-chip clock frequencies will reach the gigahertz range even in the 180 nm process generation. The implications of technology scaling — particularly for system interconnect — are very complicated. Example considerations for a 7-layer metal (7LM) process might include:

- Local interconnect layers (e.g. M1–M3) should generally remain at near-minimum dimensions and pitch in order to achieve routing density (for an example analysis of interconnect density in 0.25 µm processes, see [10]). For short lines (e.g. several hundred microns or less), thinner metal offers less lateral coupling capacitance and driver loading, and thus locally improves circuit performance. At the same time, maximum wire width is limited by the aspect ratio upper bound. The resulting thin and narrow wires are highly resistive and also subject to reliability concerns; they are hence unsuitable for global interconnects, power distribution, etc. We also note that layers M2–M3 (and maybe M4) will support a mix of local and "near-global" wiring, e.g. long wires within a single block. The distribution of lengths and performance goals for these signals can vary considerably between designs; since shorter wires are better routed on thinner metal, these design-specific considerations will affect the interconnect.
- Power distribution layers (e.g. M6–M7, maybe M5), which typically also support the top-level clock distribution (mesh or balanced-tree), should be as thick as possible for reliability. IR drop and clock skew — as well as robustness under process variations — also suggest the use of thick wire on these layers. Thick wire additionally conserves area, but can suffer from increased lateral capacitive coupling.
- Global interconnect layers (e.g. M4—M6) support inter-block signal runs with length on the order of 3000–15000 µm. To satisfy delay and signal integrity constraints, at least three degrees of freedom are available: line width and spacing, repeater insertion, and shield wiring. Repeater insertion shields downstream capacitance and is the canonical means of converting "quadratic" *RC* delay into "near-linear" delay; this technique also improves edge rates and hence noise immunity. When lateral coupling capacitances are large, worst-case "Miller coupling" begins to dominate noise and delay calculations; this is alleviated by increasing the line spacing and/or adding shield wiring (i.e. wires connected to

ground), with future techniques possibly including dedicated ground and power planes interleaved with signal layers [5].[1] Another technique to reduce the lateral coupling capacitance is to interleave signal lines which do not switch at the same signal transistion period. The bus-dominated nature of global interconnects in building-block and high-performance designs only worsens the effects of coupling, since it results in longer parallel runs.

- All layers are subject to mutual pitch-matching, via sizing, etc. considerations. Hence, available widths and spacings on one layer are not independent of the widths and spacings on a second layer.

The above are only a few of the applicable design considerations; the net effect is that balancing interconnect resources is now extremely difficult as designs move into and beyond the quarter-micron regime.

1.1 Interconnect Strategies

Interconnect tuning is the selection *by a design team* of line thicknesses, widths and spacings in multilayer interconnect to simultaneously achieve: (i) distribution (available wiring density) for local signals, global signals, clock, power and ground; (ii) performance (signal propagation delay), particularly on global interconnects; (iii) noise immunity (signal integrity), again particularly on global interconnects; and (iv) manufacturability and reliability (e.g. required margins for AC self-heat or DC electromigration on interconnects, short-circuit power in attached devices, etc.). Today, interconnect tuning is a key activity in most leading-edge microprocessor projects. It is clearly an option whenever the design and fabrication are owned by a single entity; however, for high-volume projects even fabless design houses are exercising increasing influence on vendors' processes [10]. Nevertheless, this topic has received very little attention in the literature, with only a small handful of high-level treatments available.[2]

Our work is the first in the literature to attempt a wide-ranging study of interconnect tuning. We center on global wiring layers (e.g. M4 and M5 in a 6LM process),

---

[1] When two parallel neighboring lines $L1$ and $L2$ switch simultaneously in opposite directions, the driver of $L1$ sees the grounded line capacitance plus *twice* the coupling capacitance of $L1$ to $L2$. If $L2$ is quiet when $L1$ switches, then the driver of $L1$ sees the grounded line capacitance plus the coupling capacitance to $L2$. And if $L2$ switches simultaneously in the opposite direction, the driver of $L1$ sees only the grounded line capacitance. (In leading-edge processes, *each* neighbor coupling is of the same (and possibly greater) magnitude as the area coupling to ground.) The "coupling factor" or "switching factor" is often given in the range [0,2], and since most lines have two neighbors, the total coupling factor is in the range [0,4].

[2] For example, oh et al. [12] describes a characterization and analysis methodology and the need to break ideal scaling in deep submicron interconnect. [8] is another work that centers on analysis of a given multilayer interconnect process, as opposed to the underlying interconnect tuning. [3] and [6] are examples of system-level treatments based on Rent's rule for interconnect length distribution.

Table 2 Summary of M3 coupling capacitances extracted using Quick-Cap. Bottom M2 is a ground plane; top M4 is populated by crossover lines

Width, Space (μm)	Coupling capacitance per μm (aF)				
	Left neighbor	Right neighbor	Top plane	Bottom plane (ground)	Total
1.0,2.2	25.20	25.61	54.79	46.84	152.66
1.2,2.0	29.00	29.26	56.74	48.22	163.53
1.4,1.8	33.33	33.11	57.76	51.53	177.32
1.6,1.6	38.71	38.60	59.09	51.90	188.41
1.8,1.4	44.75	44.12	60.22	51.52	200.92

and interconnect tuning issues related to bus routing, repeater insertion, and choice of shielding/spacing rules for signal integrity and performance[3]. (Of necessity, our studies are for now independent of several other issues, e.g. wire tapering and choice of wire thickness.)

We address four basic questions.

1. How should width and spacing be allocated to maximize performance for a given line pitch?
2. For a given line pitch, what criteria affect the optimal interval at which repeaters should be inserted into global interconnects?
3. Under what circumstances are shield wires the optimum technique for improving interconnect performance?
4. In global interconnect with repeaters, what other interconnect tuning is possible?

We answer these questions using technology parameters from a representative 0.25 um CMOS process; this matches the process technology context for many current- and next-generation microprocessors. Coupling capacitance studies are performed with the commercial QuickCap 3-D field solver, and interconnect delay and noise coupling studies are performed with the commercial HSPICE simulator. Of particular interest is our study of question (4): we demonstrate that a new methodology for offsetting repeater placements can reduce worst-case cross-chip delays by over 30% in current technologies, versus traditional repeater insertion methodology.

2 Allocation of Width and Spacing for Given Pitch

Our first study seeks to determine how width and spacing should be optimally allocated for a given line pitch. In practice, the actual line width used is considerably greater than the minimum line width achievable in lithography. Thus, there is

---

[3] Even though the results presented in this paper are for aluminum interconnects with SiO2 dielectric, similar techniques can be applied for copper interconnects and low-K dielectrics.

Table 3 Delay estimates for various M3 line configurations. Driver and receiver buffer sizes: (wp = 100 μm, wn = 50 μm). Delay is computed from input of driver to input of receiver

| | 50% Threshold rise delay (ps) | | | | | | | | |
| | 4,000 μm M3 length | | | 5,000 μm M3 length | | | 6,000 μm M3 length | | |
Width space (μm)	Driver load delay	Int. delay	Total delay	Driver load delay	Int. delay	Total delay	Driver load delay	Int. delay	Total delay
1.0,2.2	106.19	113.99	220.17	132.74	168.36	301.10	159.28	233.09	392.37
1.2,2.0	115.00	100.72	215.73	143.76	149.26	293.02	172.51	207.14	379.65
1.4,1.8	126.61	92.80	219.41	158.27	138.04	296.31	189.92	192.10	382.02
1.6,1.6	138.77	87.12	225.89	173.46	130.04	303.04	208.15	181.41	389.56
1.8,1.4	151.24	82.84	234.08	189.04	124.03	313.08	226.85	173.41	400.26

freedom to tune the width and spacing once assumptions are in place for line thickness and target line length. We note that because very long inter-block lines will have repeaters inserted regularly (see Section 3 below), the maximum line length of interest is equal to the optimum interval between repeaters; this length ranges between 2,500 μm and 5,000 μm for global interconnect layers in leading-edge technologies.

We have performed detailed studies of "fast" M3 interconnect with 3.2 μm pitch, assuming that M2 crossunders are dense (i.e., can be approximated as a ground plane) [9] and explicitly modeling M4 crossovers. Dielectric modeling is based on actual layer data for a representative 0.25 μm CMOS process. QuickCap was used to extract coupling and area capacitances, summarized in Table 1. As is typical in such analyses, we assume worst-case coupling, i.e. a total coupling factor of 4.0 (worst-case coupling factor of 2.0 to each of the left and right neighbors of the (victim) line under analysis).

Table 3 shows HSPICE-computed line delays for M3 line lengths ranging from 4,000 μm to 6,000 μm. Again, dense M2 is assumed to be a ground plane, and M4 crossovers are modeled explicitly. The Table shows that (width,spacing) = (1.2,2.0) μm gives the best performance for the given line pitch.

3   Bounding the Interval Between Repeaters

A very basic study (in some sense a prerequisite to all other interconnect tuning) asks how often repeaters should be inserted into global interconnects. This is of course a chicken–egg problem, in that the optimum repeater interval depends on the interconnect tuning, and the interconnect tuning depends on the maximum run ever made without an intervening repeater. However, the following can be noted.

- A body of study shows that repeaters should be inserted at uniform intervals. In other words, there should be a constant inter-connect length (or interconnect delay) between each pair of adjacent repeaters; the first and last segments of the path are exceptions because in practice the driver and receiver sizes may not be the same as the repeater size. Actually, such theoretical results deviate from real-life practice. On any source-destination path the repeater sizes need not be the same. It may also be better to add repeaters in parallel in order to drive larger wire lengths. (This is not just for performance: repeaters locally affect device area and routing constraints. However, our studies have not yet addressed such layout issues. Using the same principle (and with certain types of methodology and chip planning constraints), it can be better to increase the size of the drivers inside the block as much as possible, which would increase the first segment length.

Assuming that the driver size and the receiver size are the same as the size of the repeaters inserted along the path, we calculate the total delay, optimal number of repeaters and optimal distance between the repeaters.

The total delay for a path with K repeaters is

$$T^k_{tot} = T_{first\_stage} + (K\text{-}1) * T_{Rep\_stage} + T_{Final\_stage}$$

The delay of the first stage is the total delay from the output of driver to the input of the first repeater, i.e., $T_{first\_stage} = T_{gd} + T_{int}$, where gate load delay is $T_{gd} = R_{rep} (C_{int}^{eff} + C_{rep})$, interconnect delay is $T_{int} = R_{int} (C_{int}/2 + C_{rep})$, and R_{rep}, C_{rep} are repeater output resistance and input gate capacitance. The effective capacitance at the gate output can be approximated as $C_{int}^{eff} = \alpha C_{int}$ where α is a constant between 1/6 and 1 [11]. Let L_p be the interconnect path length between driver and receiver. Then for optimal placement of repeaters the interconnect length between repeaters is $\dfrac{Lp}{K+1}$. Therefore, the total delay for the path is

$$T^K_{tot} = (K+1) * (T_{gd} + T_{int})$$

$$= (K+1) * R_{rep} \left(a * c * \frac{L_p}{K+1} + C_{rep} \right) \tag{1}$$

$$+ r * L_p \left(c * \frac{L_p}{2(K+1)} + C_{rep} \right)$$

where r, c are resistance and capacitance per unit length of the interconnect line. We compute the optimal number of repeaters that minimizes total delay by setting, $\dfrac{\partial T_{toi}}{\partial K}$ and obtain

$$K = \sqrt{\frac{rcL_p^2}{2R_{rep}C_{rep}}} - 1 \tag{2}$$

To minimize total delay, gate load delay and interconnect delay should be equal. If effective capacitance is not considered in the gate load delay computation, and with current technology trends, gate load delay will always be greater than interconnect delay. Under these conditions, to minimize total delay one can increase the time of flight (or wire length) between repeaters until slew time constraints become tight. In the current range of 0.35 μm and 0.25 μm process generations, global interconnects have repeaters inserted with periods ranging from 2,500 μm to 10,000 μm.

- Repeater insertion is also driven by pure interconnect delay, since larger time of flight implies larger slew time on the transition seen at the receiver. Edges with large slew times cause much larger gate delays, are more susceptible to noise, are more susceptible to process-distribution influenced delay variations, and also increase the short-circuit power dissipation. Even in today's designs, slew times above 600–700 ps cannot be tolerated. Thus, even without the delay minimization objective, edge rate control will force insertion of repeaters. In fact, some of the functionality of "post-layout optimization" tools for gate sizing and repeater insertion is driven by edge rate checks as opposed to signal delay reduction.
- In practice, repeaters will be implemented using inverters whenever possible, due to performance and area efficiency.

Table 4 summarizes M3 interconnect slew times for line width 1.0 μm and line spacing 1.2 μm (corresponding to a "dense" M3 routing pitch), and input slew time of 400 ps. All capacitance extractions were performed with QuickCap, and correspond to M4 and M1 as the top and bottom ground planes, respectively. Switching factors range from 4 (both neighbors switching in the opposite direction from the victim) to 2 (both neighbors quiet, or one neighbor switching in the opposite direction and one neighbor switching in the same direction with respect to the victim). We see that the M3 distance between repeaters has an upper bound of 5,000 μm due to edge rate considerations alone. Separate studies show that this upper bound on distance between repeaters is essentially unaffected by changes to the driver/receiver sizing or the input slew time.

4   Benefits of Shield Wiring

Our third study addresses the question of whether shield wiring is an effective means of improving delay and signal integrity performance of long global interconnects. We consider various width-spacing rules for M3 interconnect, in order to evaluate the utility of spacing vs. shielding techniques. Our evaluations are with respect to delay only; for all of the configurations, the assumed slew time upper bounds of approximately 600 ps imply that noise coupling will not be problematic. Figure 1 contrasts five pitch-matched width-spacing rules:

- **Rule 1**:   1.2 μm width, 1.0 μm spacing
- Single-V_{ss}:   1.2 μm width, 1.0 μm spacing, with every third line grounded (i.e. every signal line has one grounded neighbor to shield it)

Table 4 Summary of M3 interconnect slew times. M4 is top layer; M1 is bottom layer. Two combinations of width/spacing are shown, along with three different coupling factor assumptions. The input slew time is 400 ps and the output slew times are computed as 10–90% for rise time and 90–10% for fall time

Driver/receiver (wp,wn)(μm)	Width (μm)	Space (μm)	Length (μm)	SF	Delay (ps)	Rise time (ps)	Fall time (ps)
(130,65)/(130,65)	1	1.1	10,000	4	589	1679	1510
(130,65)/(130,65)	1	1.1	9,000	4	486	1421	1265
(130,65)/(130,65)	1	1.1	8,000	4	393	1187	1044
(130,65)/(130,65)	1	1.1	7,000	4	310	975	847
(130,65)/(130,65)	1	1.1	5,000	4	172	623	525
(130,65)/(130,65)	1	1.1	10,000	3	488	1405	1267
(130,65)/(130,65)	1	1.1	9,000	3	404	1193	1066
(130,65)/(130,65)	1	1.1	8,000	3	327	1001	885
(130,65)/(130,65)	1	1.1	7,000	3	259	828	723
(130,65)/(130,65)	1	1.1	5,000	3	147	538	458
(130,65)/(130,65)	1	1.1	10,000	2	388	1131	1026
(130,65)/(130,65)	1	1.1	9,000	2	323	966	869
(130,65)/(130,65)	1	1.1	8,000	2	263	817	728
(130,65)/(130,65)	1	1.1	7,000	2	209	682	601
(130,65)/(130,65)	1	1.1	5,000	2	120	456	393
(130,65)/(130,65)	1.4	1.6	10,000	4	366	1123	980
(130,65)/(130,65)	1.4	1.6	9,000	4	303	963	832
(130,65)/(130,65)	1.4	1.6	8,000	4	246	818	698
(130,65)/(130,65)	1.4	1.6	7,000	4	195	686	578
(130,65)/(130,65)	1.4	1.6	5,000	4	111	465	384
(130,65)/(130,65)	1.4	1.6	10,000	3	320	992	869
(130,65)/(130,65)	1.4	1.6	9,000	3	266	854	740
(130,65)/(130,65)	1.4	1.6	8,000	3	217	729	625
(130,65)/(130,65)	1.4	1.6	7,000	3	172	615	522
(130,65)/(130,65)	1.4	1.6	5,000	3	99	422	352
(130,65)/(130,65)	1.4	1.6	10,000	2	275	862	759
(130,65)/(130,65)	1.4	1.6	9,000	2	229	746	650
(130,65)/(130,65)	1.4	1.6	8,000	2	188	640	553
(130,65)/(130,65)	1.4	1.6	7,000	2	150	543	465
(130,65)/(130,65)	1.4	1.6	5,000	2	87	382	322

- **Rule 2**: 1.2 μm width, 2.1 μm spacing
- **Rule 3**: 2.2 μm width, 2.2 μm spacing
- Double-V_{ss}: 1.2 μm width, 2.1 μm spacing, with every other line grounded (i.e. every signal line has two grounded neighbors to shield it)

Again, QuickCap was used to extract capacitive couplings of a given victim line to its neighbor lines and the neighboring top/bottom layers; these results are shown in Table 5. Notice that the Rule 1, Rule 2, and Rule 3 have worst-case coupling factors = 4. On the other hand, the Single-V_{ss} rule has worst-case coupling factor = 3, and the Double-V_{ss} rule has worst-case coupling factor = 2. Table 6 shows the delay

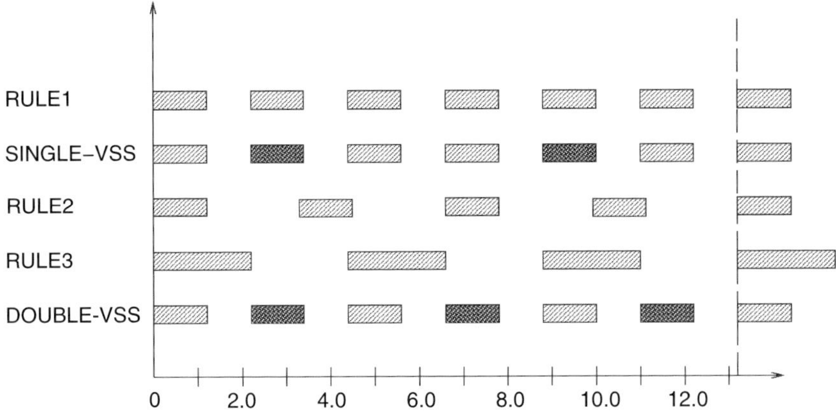

Fig. 1 Pitch-matched width-spacing rules. Rule 1 allows six lines per 13.2 μm; Rule 2 and the Single-V_{SS} rule (Rule 1 width/spacing, but every third line grounded) both allow four signal lines per 13.2 μm; and Rule 3 and the Double-V_{SS} rule (Rule 1 width/spacing, but every other line grounded) both allow three signal lines per 13.2 μm

Table 5 M3 coupling capacitances extracted using QuickCap for various interconnect tuning rules and combinations of bottom and top planes

			Coupling capacitance per μm (aF)				
M3 Rules	Width, space (μm)	Ground, top planes	Left neighbor	Right neighbor	Top plane	Bottom plane (ground)	Total
Rule 1	1.2,1.0	Substrate,M4 Line	68.23	68.15	43.68	14.79	195.03
Rule 1	1.2,1.0	M2,M4 Line	60.30	60.92	43.96	34.88	202.37
Rule 1	1.2,1.0	M2,–	74.67	74.23	–	42.99	192.44
Rule 2	1.2,2.1	Substrate,M4 Line	36.87	34.37	58.58	18.07	148.29
Rule 2	1.2,2.1	M2,M4 Line	26.96	27.10	58.51	48.72	160.41
Rule 2	1.2,2.1	M2,–	42.17	42.43	–	59.15	143.96
Rule 3	2.2,2.2	Substrate,M4 Line	35.09	36.50	77.61	22.14	171.52
Rule 3	2.2,2.2	M2,M4 Line	26.18	25.61	77.51	67.92	198.82
Rule 3	2.2,2.2	M2,–	44.33	43.86	–	73.23	162.14

performance for a 4,000 μm M3 line, under various bottom ground and top plane configurations. We observe:

- The Rule 3 provides 37% decrease in total delay, but since C_{eff} was not used in the gate load delay computation, actual delay reductions could be even greater.
- The Single-V_{SS} rule is *less effective* than the Rule 2; note that the two rules are equivalent in terms of effective routing density. Our studies have not yet addressed the routing interactions that can potentially affect this analysis. In particular, shield lines may be added to bring power and ground connections to repeater blocks.

Table 6 Delay estimates for a 4,000 μm M3 line, under various interconnect tuning configurations. Driver and receiver buffer sizes: (wp = 100 μm, wn = 50 μm). Delay is computed from input of driver to input of receiver

M3 Rules	Width, space (μm)	Ground, top planes	Driver load delay	Interconnect delay	Total delay	% Gain w.r.t. Rule 1
			50% threshold rise delay (ps)			
Rule 1	1.2,1.0	Substrate,M4 Line	173.04	116.88	289.92	–
Rule 1	1.2,1.0	M2,M4 Line	167.84	114.03	281.87	–
Rule 1	1.2,1.0	M2,–	178.03	119.62	297.65	–
Rule 2	1.2,2.1	Substrate,M4 Line	114.47	84.75	199.22	29
Rule 2	1.2,2.1	M2,M4 Line	112.50	83.66	196.16	30
Rule 1 with Single VSS	1.2,1.0	Substrate,M4 Line	137.41	97.34	234.75	17
Rule 1 with Single VSS	1.2,1.0	M2,M4 Line	136.17	96.66	232.83	17
Rule 1 with Single VSS	1.2,1.0	M2,–	139.14	98.28	237.42	16
Rule 2	1.2,2.1	M2,–	119.29	87.39	206.68	27
Rule 3	2.2,2.2	Substrate,M4 Line	126.91	49.95	176.85	37
Rule 3	2.2,2.2	M2,M4 Line	130.08	50.90	180.98	36
Rule 3	2.2,2.2	M2,–	130.40	50.99	181.39	36
Rule 1 with Double VSS	1.2,1.0	Substrate,M4 Line	99.74	78.11	177.85	37
Rule 1 with Double VSS	1.2,1.0	M2,M4 Line	104.34	80.83	185.17	34
Rule 1 with Double VSS	1.2,1.0	M2,–	121.14	78.53	199.67	29

- The Double-V_{SS} rule gives improved total delays compared with Rule 3, with the rules being equivalent in terms of effective routing density. However, the Rule 3 yields smaller interconnect delays, so that driver size reductions have greater potential for delay improvement. Thus, Rule 3 seems preferable. When two buses have activity patterns such that each is quiet when the other is active, then their lines can be interleaved such that they effectively follow the Double-V_{SS} rule. In such a case, interleaving is clearly superior to Rule 3, since the effective routing density is doubled.

- Gate load delays are larger than interconnect delays, suggesting that it is preferable to decrease line widths and increase line spacings. We also note that a dense M4 top layer decreases total delay, and a dense M2 bottom (ground plane) layer decreases total delay for smaller line widths only.

5   New Repeater Offset Methodology for Global Buses

Finally, we study another form of tuning that is possible for global interconnects. Our motivations are three-fold: (i) global interconnect is increasingly dominated by wide buses; (ii) present methodology designs global interconnects for *worst-case* Miller coupling; and (iii) present methodology routes long global buses using repeater *blocks*, i.e. blocks of co-located inverters spaced every, say, 4,000 μm.

We have proposed a simple method to improve global interconnect performance. The idea is to reduce the worst-case Miller coupling by offsetting the inverters on adjacent lines (see Fig. 2). In the previous methodology (Fig. 2a), the worst-case switching of a neighbor line (i.e. simultaneously and in the opposite direction to the switching of the victim line) persists through the entire chain of inverters. However, with offset inverter locations (Fig. 2b), any worst-case simultaneous switching on a neighbor line persists only for half of each period between consecutive inverters, *and furthermore becomes best-case simultaneous switching for the other half of the period!*.

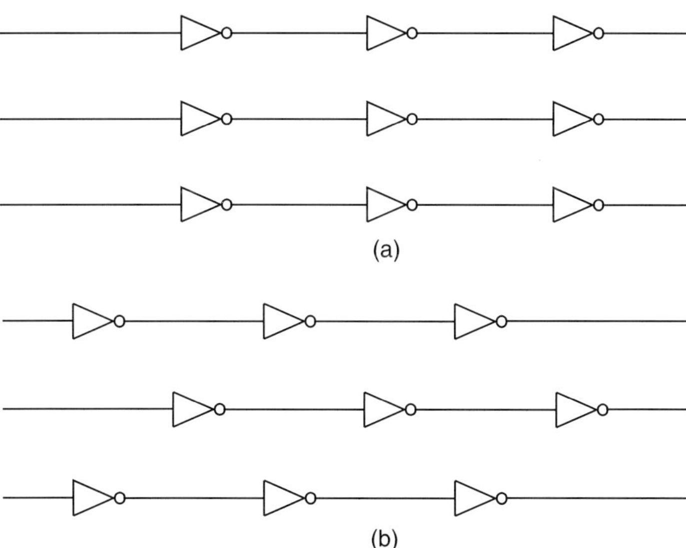

Fig. 2 Reduction of worst-case Miller coupling by offsetting inverters. In (a), inverters on the left and right neighbor lines are at phase = 0 with respect to the inverters on the middle line. In (b), inverters on the left and right neighbors are at phase = 0.5

To confirm the advantages of this method, the following experimental methodology was used.

- We study systems of three parallel interconnect lines, with lengths either 10,000 μm or 14,000 μm. These lines are stimulated by a waveform with risetime = falltime = 200 ps. The middle line is considered the "victim" for analysis purposes.
- We model two "technologies" representative of M3 and M4 in an 0.25 μm CMOS process. In each technology, line resistance is 50Ω per 1,000 μm. In Technology I, capacitive couplings to left neighbor, ground and right neighbor per 1,000μm are respectively 60fF, 80fF and 60fF. In Technology II, capacitive couplings to left neighbor, ground and right neighbor per 1,000 μm are respectively 80fF, 160fF, and 80fF.
- We assume a *period* between inverters (repeaters) of 4,000 μm. So that HSPICE cannot introduce any error in its RC analysis, we manually distributed the line and coupling parasitics into 40 μm segments, i.e. repeaters occurred every 100 segments, and line lengths were 250 or 350 segments. Each segment is modeled as a double-pi model.[4]
- We always place the inverters on the middle line with "phase = 0", i.e. at positions 4,000, 8,000, …, μm along the line. Inverters on the left and right neighbors are placed according to all combinations of phase = 0, 0.1, 0.2, …, 0.9 (again with respect to the period of 4,000 μm). There are 100 different phase combinations. Figure 2 shows the three-line configurations with left/right neighbor phase combinations of (0,0) and (0.5,0.5).
- We stimulate the three lines with the periodic waveform, with the first transition either rising (R) or falling (F). There are eight combinations of directions for the first transisions, i.e. RRR, RRF, …, FFF.
- Finally, we may offset the input waveforms of the left and right neighbors by −100 ps, 0ps or +100 ps with respect to the input waveform of the middle line. There are nine combinations of these input offsets.

Table 7 shows HSPICE delays for systems of three lines of length 10,000 μm, using Technology I, for all combinations of rising (R) and falling (F) initial transition on the input waveform. The Table shows delays for inverter phases (0,0) and (0.5,0.5) on the left and right neighbors of the middle line (phase 0). The effect of Miller coupling is clearly shown.

Table 8 shows the worst-case delays (with respect to all eight possible combinations of rising and falling inputs) for the middle line, for each combination of phases for the inverter locations on the left and right neighbor lines. Input offsets are all 0, i.e. the waveforms start at the same time. All four combinations of Technology and line length are shown. In every case, the optimum phase combination is (0.5,0.5), while the traditional phase combination of (0.0,0.0) is actually the

---

[4]This segmenting is chosen such that any finer-grain representation does not change the HSPICE-computed delays.

Table 7 HSPICE delays (ns) for three lines of length 10,000 μm, using Technology I, for all combinations of rising (R) and falling (F) initial transition on the input waveform. We show delays for inverter phases (0,0) and (0.5,0.5) on the left and right neighbors of the middle line (phase 0)

	Interconnect delay (ns)					
Input waveforms (left neighbor, victim, right neighbor)	Left, right neighbor buffer phases: 0,0			Left, right neighbor buffer phases:0.5,0.5		
	Left neighbor delay	Victim delay	Right neighbor delay	Left neighbor delay	Victim delay	Right neighbor delay
R, R, R	0.361	0.361	0.361	0.510	0.630	0.510
R, R, F	0.428	0.584	0.676	0.533	0.697	0.499
R, F, R	0.546	0.994	0.546	0.483	0.689	0.483
R, F, F	0.676	0.584	0.428	0.499	0.697	0.533
F, R, R	0.676	0.584	0.428	0.499	0.697	0.533
F, R, F	0.546	0.994	0.546	0.483	0.689	0.483
F, F, R	0.428	0.584	0.676	0.533	0.697	0.499
F, F, F	0.361	0.361	0.361	0.510	0.630	0.510

worst possible. The worst-case delay is reduced by anywhere from 25% to 30% when the repeaters are placed with optimum phase. Finally, Table 9 shows the same worst-case delays for the middle line, this time taken over all eight rise/fall combinations and all nine combinations of input waveform offsets. Again, even when the inputs do not switch perfectly simultaneously, the best phase combination is (0.5,0.5) and the worst phase combination is the traditional (0.0,0.0) methodology.

6 Conclusions

To our knowledge, this work has provided the first technology-specific studies of interconnect tuning in the literature. We have described experimental approaches to interconnect tuning issues related to bus routing, repeater insertion, and choice of shielding/spacing rules for signal integrity and performance. In particular, four questions have been addressed: allocation of width and spacing to maximize performance for a given pitch, finding the optimal interval for repeater insertion, assessing the potential benefits of shield wiring, and optimizing the insertion of repeaters in global buses. Our answers to these questions are at times surprising: in answering (3), we demonstrate that current shielding methodologies may be suboptimal when compared with alternate width/spacing rules, and in answering (4), we propose a new repeater offset technique that can reduce worst-case cross-chip delays by over 30% in current technologies. Ongoing efforts extend our interconnect tuning research to encompass layer thicknesses, more detailed analyses of noise coupling and tuning to meet noise margins, and the delay/noise behavior in emerging technology regimes (Cu interconnect and low-K dielectrics). Finally, we

Table 8 Worst-case middle line delays over all input rise/fall combinations, for each phase combination on left and right neighbors. Input offsets are all 0ps

		A. Line length 10,000 µm, Technology I									
		Right neighbor phase									
		0	0.1	0.2	0.3	0.4	0.5	0.6	0.7	0.8	0.9
	0	**0.994**	0.988	0.971	0.954	0.929	0.910	0.874	0.900	0.930	0.962
	0.1	0.988	0.974	0.960	0.938	0.911	0.885	0.854	0.881	0.917	0.952
	0.2	0.971	0.960	0.941	0.917	0.887	0.848	0.829	0.863	0.897	0.932
	0.3	0.954	0.938	0.917	0.890	0.855	0.806	0.801	0.834	0.872	0.912
Left neighbor	0.4	0.929	0.911	0.887	0.855	0.818	0.753	0.766	0.805	0.841	0.885
phase	0.5	0.910	0.885	0.848	0.806	0.753	**0.697**	0.735	0.778	0.822	0.867
	0.6	0.874	0.854	0.829	0.801	0.766	0.735	0.739	0.768	0.799	0.832
	0.7	0.900	0.881	0.863	0.834	0.805	0.778	0.768	0.796	0.827	0.859
	0.8	0.930	0.917	0.897	0.872	0.841	0.822	0.799	0.827	0.860	0.894
	0.9	0.962	0.952	0.932	0.912	0.885	0.867	0.832	0.859	0.894	0.924
		B. Line length 10,000 µm, Technology II									
		Right neighbor phase									
		0	0.1	0.2	0.3	0.4	0.5	0.6	0.7	0.8	0.9
	0	**1.437**	1.422	1.400	1.370	1.332	1.299	1.259	1.300	1.343	1.388
	0.1	1.422	1.405	1.379	1.347	1.306	1.258	1.234	1.278	1.324	1.372
	0.2	1.400	1.379	1.352	1.315	1.270	1.206	1.199	1.247	1.296	1.347
	0.3	1.370	1.347	1.315	1.274	1.223	1.144	1.158	1.208	1.261	1.314
Left neighbor	0.4	1.332	1.306	1.270	1.223	1.167	1.075	1.109	1.161	1.216	1.273
phase	0.5	1.299	1.258	1.206	1.144	1.075	**1.015**	1.069	1.124	1.180	1.239
	0.6	1.259	1.234	1.199	1.158	1.109	1.069	1.079	1.120	1.163	1.209
	0.7	1.300	1.278	1.247	1.208	1.161	1.124	1.120	1.160	1.203	1.250
	0.8	1.343	1.324	1.296	1.261	1.216	1.180	1.163	1.203	1.246	1.293
	0.9	1.388	1.372	1.347	1.314	1.273	1.239	1.209	1.250	1.293	1.339
		C. Line length 14,000 µm, Technology I									
		Right neighbor phase									
		0	0.1	0.2	0.3	0.4	0.5	0.6	0.7	0.8	0.9
	0	**1.474**	1.467	1.448	1.429	1.401	1.383	1.341	1.340	1.382	1.427
	0.1	1.467	1.454	1.439	1.414	1.385	1.356	1.308	1.324	1.370	1.417
	0.2	1.448	1.439	1.418	1.393	1.359	1.320	1.267	1.299	1.345	1.395
	0.3	1.429	1.414	1.393	1.362	1.328	1.276	1.217	1.267	1.319	1.375
Left neighbor	0.4	1.401	1.385	1.359	1.328	1.287	1.223	1.174	1.229	1.285	1.342
phase	0.5	1.383	1.356	1.320	1.276	1.223	**1.105**	1.146	1.203	1.263	1.323
	0.6	1.341	1.308	1.267	1.217	1.174	1.146	1.110	1.162	1.220	1.281
	0.7	1.340	1.324	1.299	1.267	1.229	1.203	1.162	1.192	1.240	1.287
	0.8	1.382	1.370	1.345	1.319	1.285	1.263	1.220	1.240	1.283	1.330
	0.9	1.427	1.417	1.395	1.375	1.342	1.323	1.281	1.287	1.330	1.377

(continued)

Table 8 (continued)

| | | D. Line length 14,000 μm, Technology II | | | | | | | | | |
| | | Right neighbor phase | | | | | | | | | |
		0	0.1	0.2	0.3	0.4	0.5	0.6	0.7	0.8	0.9
	0	**2.123**	2.108	2.085	2.052	2.011	1.983	1.925	1.938	1.995	2.056
	0.1	2.108	2.092	2.064	2.029	1.985	1.943	1.876	1.913	1.974	2.039
	0.2	2.085	2.064	2.036	1.996	1.947	1.889	1.816	1.878	1.944	2.012
	0.3	2.052	2.029	1.996	1.952	1.898	1.823	1.765	1.833	1.903	1.977
Left neighbor	0.4	2.011	1.985	1.947	1.898	1.837	1.743	1.703	1.778	1.854	1.932
phase	0.5	1.983	1.943	1.889	1.823	1.743	**1.590**	1.664	1.744	1.823	1.903
	0.6	1.925	1.876	1.816	1.765	1.703	1.664	1.630	1.686	1.763	1.843
	0.7	1.938	1.913	1.878	1.833	1.778	1.744	1.686	1.741	1.801	1.867
	0.8	1.995	1.974	1.944	1.903	1.854	1.823	1.763	1.801	1.860	1.925
	0.9	2.056	2.039	2.012	1.977	1.932	1.903	1.843	1.867	1.925	1.989

Table 9 Worst-case delays with all combinations of input offsets

| | | A. Line length 10,000 μm, Technology I | | | | | | | | | |
| | | Right neighbor phase | | | | | | | | | |
		0	0.1	0.2	0.3	0.4	0.5	0.6	0.7	0.8	0.9
	0	**1.090**	1.071	1.051	1.021	0.988	0.942	0.948	0.984	1.018	1.051
	0.1	1.071	1.054	1.026	0.995	0.957	0.905	0.920	0.958	0.997	1.035
	0.2	1.051	1.026	0.998	0.964	0.921	0.865	0.890	0.930	0.970	1.008
	0.3	1.021	0.995	0.964	0.924	0.876	0.825	0.854	0.894	0.936	0.980
Left neighbor	0.4	0.988	0.957	0.921	0.876	0.825	0.782	0.813	0.856	0.900	0.944
phase	0.5	0.942	0.905	0.865	0.825	0.782	**0.760**	0.791	0.824	0.860	0.900
	0.6	0.948	0.920	0.890	0.854	0.813	0.791	0.816	0.849	0.879	0.911
	0.7	0.984	0.958	0.930	0.894	0.856	0.824	0.849	0.880	0.911	0.945
	0.8	1.018	0.997	0.970	0.936	0.900	0.860	0.879	0.911	0.944	0.982
	0.9	1.051	1.035	1.008	0.980	0.944	0.900	0.911	0.945	0.982	1.016

| | | B. Line length 10,000 μm, Technology II | | | | | | | | | |
| | | Right neighbor phase | | | | | | | | | |
		0	0.1	0.2	0.3	0.4	0.5	0.6	0.7	0.8	0.9
	0	**1.526**	1.502	1.471	1.430	1.382	1.335	1.329	1.379	1.427	1.476
	0.1	1.502	1.475	1.440	1.396	1.343	1.284	1.292	1.345	1.398	1.449
	0.2	1.471	1.440	1.400	1.350	1.291	1.229	1.249	1.305	1.361	1.416
	0.3	1.430	1.396	1.350	1.295	1.231	1.171	1.200	1.258	1.315	1.373
Left neighbor	0.4	1.382	1.343	1.291	1.231	1.167	1.114	1.148	1.205	1.262	1.321
phase	0.5	1.335	1.284	1.229	1.171	1.114	**1.074**	1.124	1.175	1.226	1.279
	0.6	1.329	1.292	1.249	1.200	1.148	1.124	1.148	1.190	1.234	1.281
	0.7	1.379	1.345	1.305	1.258	1.205	1.175	1.190	1.234	1.280	1.328
	0.8	1.427	1.398	1.361	1.315	1.262	1.226	1.234	1.280	1.327	1.376
	0.9	1.476	1.449	1.416	1.373	1.321	1.279	1.281	1.328	1.376	1.425

(continued)

Table 9 (continued)

		C. Line length 14,000 µm, Technology I								
		Right neighbor phase								
		0	0.1	0.2	0.3	0.4	0.5	0.6	0.7	0.8	0.9
	0	**1.572**	1.551	1.530	1.502	1.465	1.419	1.391	1.429	1.474	1.521
	0.1	1.551	1.534	1.507	1.472	1.438	1.388	1.362	1.406	1.451	1.499
	0.2	1.530	1.507	1.474	1.442	1.400	1.345	1.323	1.373	1.423	1.475
	0.3	1.502	1.472	1.442	1.401	1.353	1.293	1.279	1.334	1.388	1.443
Left neighbor phase	0.4	1.465	1.438	1.400	1.353	1.297	1.241	1.231	1.288	1.348	1.406
	0.5	1.419	1.388	1.345	1.293	1.241	**1.171**	1.203	1.256	1.310	1.365
	0.6	1.391	1.362	1.323	1.279	1.231	1.203	1.206	1.247	1.291	1.339
	0.7	1.429	1.406	1.373	1.334	1.288	1.256	1.247	1.288	1.332	1.377
	0.8	1.474	1.451	1.423	1.388	1.348	1.310	1.291	1.332	1.374	1.424
	0.9	1.521	1.499	1.475	1.443	1.406	1.365	1.339	1.377	1.424	1.471

		D. Line length 14,000 µm, Technology II								
		Right neighbor phase								
		0	0.1	0.2	0.3	0.4	0.5	0.6	0.7	0.8	0.9
	0	**2.213**	2.190	2.157	2.116	2.069	2.031	1.974	2.027	2.084	2.147
	0.1	2.190	2.161	2.125	2.081	2.029	1.982	1.930	1.991	2.053	2.119
	0.2	2.157	2.125	2.085	2.035	1.977	1.920	1.879	1.946	2.013	2.084
	0.3	2.116	2.081	2.035	1.980	1.913	1.846	1.818	1.893	1.965	2.041
Left neighbor phase	0.4	2.069	2.029	1.977	1.913	1.837	1.775	1.750	1.831	1.909	1.989
	0.5	2.031	1.982	1.920	1.846	1.775	**1.666**	1.730	1.804	1.879	1.955
	0.6	1.974	1.930	1.879	1.818	1.750	1.730	1.713	1.773	1.835	1.901
	0.7	2.027	1.991	1.946	1.893	1.831	1.804	1.773	1.830	1.892	1.957
	0.8	2.084	2.053	2.013	1.965	1.909	1.879	1.835	1.892	1.951	2.015
	0.9	2.147	2.119	2.084	2.041	1.989	1.955	1.901	1.957	2.015	2.079

seek to develop more complete full-chip interconnect tuning approaches based on analyses of the interconnect structure, speed target, and power dissipation target for a given design.

References

[1] Semiconductor Industry Association, *National Technology Roadmap for Semiconductors*, December 1997.

[2] A. Deutsch, G. V. Kopcsay, C. W. Surovic, B. J. Rubin, L. M. Terman, R. P. Dunne and T. Gallo, Modeling and Characterization of Long On-chip Interconnections for High-Performance Microprocessors, *final report*, ARPA HSCD Contract C-556003, September 1995. Also appeared in *IBM Journal of Research and Development* 39(5), September 1995, 547–567.

[3] P. D. Fisher, Clock Cycle Estimations for Future Microprocessor Generations, *Proc. IEEE Innovative Systems in Silicon*, Austin, October 1997.

[4] L. Gwennap, IC Makers Confront RC Limitations, *Microdesign Resources Microprocessor Report*, August 4, 1997, pp. 14–18.

[5] D. P. LaPotin, U. Ghoshal, E. Chiprout and S. R. Nassif, Physical Design Challenges for Performance, *International Symposium on Physical Design*, April 1997, pp. 225–226.

[6] J. Meindl, GigaScale Integration: 'Is the Sky the Limit'?, *Keynote presentation slides*, Hot Chips IX, Stanford, CA, August 25–26, 1997.

[7] L. Scheffer, A Roadmap of CAD Tool Changes for Sub-micron Interconnect Problems, *International Symposium on Physical Design*, April 1997, pp. 104–109.

[8] R. F. Sechler, Interconnect design with VLSI CMOS, *IBM Journal of Research and Development*, January–March 1995, 23–31.

[9] J. Cong, L. He, A. B. Kahng, D. Noice, N. Shirali and S.H.-C. Yen, Analysis and Justification of a Simple, Practical 2 1/2-D Capacitance Extraction Methodology, *Proc. Design Automation Conference*, June 1997.

[10] L. Gwennap, IC Vendors Prepare for 0.25-Micron Leap, *Microprocessor Report*, September 16, 1996, pp. 11–15.

[11] A. B. Kahng and S. Muddu, Efficient gate Delay Modeling for Large Interconnect Loads, *Proc. IEEE Multi-Chip Module Conf.*, February 1996.

[12] S.-Y. Oh, K.-J. Chang, N. Chang and K. Lee, Interconnect modeling and design in high-speed VLSI/ULSI systems, *Proc. International Conference on Computer Design: VLSI in Computers and Processors*, October 1992, pp. 184–189.

Efficient Inductance Extraction via Windowing[†]

Michael Beattie and Lawrence Pileggi

Abstract We propose a new, efficient and accurate localized inductance modeling technique via windowing in a manner that is analogous to localized capacitance extraction. The stability and accuracy of this process is made possible by twice inverting the localized inductance models, and in the process exploit properties of the magnetostatic interactions as modeled via the susceptance (inverse inductance). Application of these localized double-inverse inductance models to actual IC bus examples demonstrates the significant improvement in simulation efficiency and overall accuracy as compared to alternative methods of approximation and simplification.

1   Introduction

For modern digital ICs the logic path delays can be dominated by the influence of parasitic capacitive and inductive coupling among the metal interconnect wiring. As technologies push the performance to its limits, it is necessary to find increasingly more detailed interconnect models to predict the signal delay more accurately. The growing complexity of today's integrated systems, however, makes this computationally very expensive. Increasing system size makes efficient analyses of parasitics and performance imperative. Reconciling these two contradicting requirements is an extremely difficult task. Full three-dimensional interconnect models are generally of unmanageable size and density such that they are not useful for analysis and simulation purposes without additional approximations and simplifications.

Of particular focus is the modeling of on-chip inductance and its interactions with on-chip capacitance. While operating frequencies are making on-chip inductance

[†]This work was supported by the MARCO/DARPA Gigascale Silicon Research Center (GSRC), the National Science Foundation, and a grant from Intel Corporation.

Carnegie Mellon University Department of Electrical and Computer Engineering Pittsburgh, PA 15232, USA beattie@ece.cmu.edu, pileggi@ece.cmu.edu

R. Lauwereins and J. Madsen (eds.), *Design, Automation, and Test in Europe–The Most Influential Papers of 10 Years DATE*, 377–388
© Springer 2008

evident, localizing the magnetic couplings for efficient extraction and analysis is challenging. Localized extraction techniques have been used with some success in the past for reducing the size of interconnect models. It has been demonstrated, however, that simple truncation—merely discarding long range couplings—can destroy the stability of the electromagnetic (EM) model. Shell models [4] have been applied successfully for stable localized extraction, but finding the correct shell sizes for a particular target accuracy is not straightforward.

In this paper we propose a novel localized extraction technique for inductance modeling via windowing—capture local interactions within small windows, then combine these localized models into a complete inductance matrix. This approach does not localize couplings in the inductance matrix directly, but begins with partial inductance models for small localized windows. These small, localized inductance matrices are inverted individually to generate the corresponding localized *susceptance matrices S*. As we will show in this paper, S has properties similar to the capacitance matrix, such as shielding, which can be exploited to combine the localized window models into a complete, *sparse*, susceptance matrix. We then invert this sparse susceptance matrix and produce a complete inductance matrix that we can further sparsify in an accurate and stable manner.

In [7] the direct use of the inverse of inductance for simulation has been suggested, but widespread commercial support for these models is not yet available.

2 Windowing for Inductance Extraction

The inductance extraction flow we are proposing in this paper consists of the following steps:

- Generate a partial inductance matrix $L^{(j)}$ for active conductor j and all conductors within a window around it.
- Find the current flowing through these conductors for which the total flux for active conductor j through its loop with infinity is unity and for each other conductors in the window the total flux is zero. This produces the susceptive couplings $S^{(j)}_{ij}$ between active conductor j and the group of conductors within a window around it.
- Repeat these two previous steps for all conductors in turn as active conductors.
- Merge all $S^{(j)}_{ij}$ submatrices into one complete, sparse susceptance matrix that approximates the magnetic interactions for the entire interconnect system.
- The combined susceptance matrix will be asymmetric, since $S^{(j)}_{ij}$ is typically not equal to $S^{(j)}_{ij}$. By choosing the value with the *smaller magnitude* for both entries, we can prove that the complete susceptance model can be rendered symmetric and stable.
- Invert the sparse, symmetric susceptance matrix to generate an inductance matrix. The truncation in the susceptance domain makes this matrix much easier to sparsify than the original inductance model.

This double-inverse inductance matrix can then be used for timing analysis or simulation without loss of generality. We present examples of the efficacy of these models in Section 5. But we will first begin by describing some important properties of capacitance and susceptance matrices which facilitates this extraction methodology.

3   Properties of EM Interaction Matrices

We assume a boundary element approach where the source (charge/current) density is constant for each section (panel/filament).

$$\psi_i^\alpha = (\phi_i; A_{i,x}; A_{i,y}; A_{i,z})$$

(1)

is the vector of average potentials for section i. The field type α is 0 for the electrostatic and 1, 2, or 3 for the xyz magnetostatic cases. The *discrete* source vector is

$$\gamma_j^\alpha = (q_j/\varepsilon \; ; \mu I_{j,x} l_{j,x} \; ; \mu I_{j,y} l_{j,y} \; ; \mu I_{j,z} l_{j,z})$$

(2)

where q_j and $I_{j,\{xyz\}}$ are the charge and the three current components for this section. The $l_{j,\{xyz\}}$ are the dimensions of section j. We can write the discretized electromagnetic interconnect interactions as

$$\vec{\psi}^\alpha \approx K^\alpha \vec{\gamma}^\alpha$$

$$K_{ij}^\alpha \equiv \frac{1}{W_i^\alpha} \frac{1}{W_j^\alpha} \frac{1}{4\pi} \int_{W_i^\alpha} \int_{W_j^\alpha} \frac{1}{\| \vec{r}_i - \vec{r}_j \|} dW_i^\alpha \, dW_j^\alpha$$

(3)

by defining the matrices K^α. W_i^α is the content (panel area or filament volume) of section i for field type α. These in turn form the four diagonal blocks of the *electromagnetic interaction matrix K* in

$$\psi \approx K\gamma$$

(4)

where $\vec{\psi}$ and $\vec{\gamma}$ are formed by concatenating $\vec{\psi}^\alpha$ or $\vec{\gamma}^\alpha$
The *capacitance matrix, $C = S^0$*, can be found by inverting the *potential matrix $P = K^0$*. The inverses of the *partial inductance matrices $L_x = K^1$, $L_y = K^2$ and $L_z = K^3$* are the *susceptance matrices S^1, S^2 and S^3* of the system. Together these four blocks form the *inverse interaction matrix S*.

Positive Definiteness of K: The term $K\vec{\gamma}$ represents (within the accuracy of the discretization) the electric scalar and magnetic vector potential. The combined energy stored in the electromagnetic field is (&$$$;, and must be non-negative (see [2][pg. 237]). In our matrix notation we then have

$$\vec{\gamma}^T K \vec{\gamma} \geq 0$$

(5)

which is zero if and only if no electric and magnetic sources are present. It follows that K, as well as S, must be positive definite.

Diagonal Elements of S are positive: K is positive definite, so we know

$$\forall \, \vec{\gamma} \neq 0 \; : \vec{\gamma}^{\,T} K \, \vec{\gamma} > 0 \tag{6}$$

If we choose $\vec{\gamma}$ to be the solution of $K\vec{\gamma} = e_i$ (which has to exist and be unique), where e_i is the unit vector with 1 as i^{th} element, then we have $\vec{\gamma} = Se_i$ and with (6) find

$$\forall \, e_i \; : e_i^T S^T KSe_i > 0 \tag{7}$$

With $SK = I$ we get

$$\forall \, e_i \; : e_i^T S^T e_i \equiv S_{ii} > 0 \tag{8}$$

This equates to all diagonal elements of S being positive, since S is a square matrix.

- **Off-Diagonals of S are negative or zero**: Next we show that all off-diagonal elements of S must be either negative or zero. For the electric case ($\alpha = 0$) it is shown in [8][pg. 223]. If we have unit potential at conductor i and zero potential everywhere else, then (8) tells us that we will have a positive charge on conductor i. This positive charge will create a positive potential at the locations of all other conductors and negative charge needs to be added to all those grounded conductors to ensure zero potential. Thus, for $\alpha = 0$ (electric fields), we find that $\forall j \neq: i: S_{ij}^0 < 0$.

For $\alpha = 1,2,3$, we have defined the currents to be directed in the positive x,y,z directions. So, similar to the electric case above, if we, for instance, require a unit vector potential along conductor i in x-direction, there must be, because of (8), a current flowing through conductor i in *positive* x-direction (see Fig. 1). This creates magnetic vector potential within the loops of the other segments j. To get a zero vector potential for all other segments j, currents in *negative* x-directions must flow through those segments. Similar for y- and z-directions. So the off-diagonal elements of S must be negative here as well (for the given current direction convention). So for the entire inverse interaction matrix we find

$$\forall j \neq i: S_{ij} < 0 \tag{9}$$

- **Diagonal Dominance of S**: Since sources $\vec{\gamma}$ are only found on or within the conductors in the system, we find that the electrostatic and magnetostatic potentials in the insulator satisfy Laplace' equation with the surfaces of the conductors and infinity being the boundaries of the domain:

$$\nabla^2 \vec{\Psi} = \vec{0} \tag{10}$$

 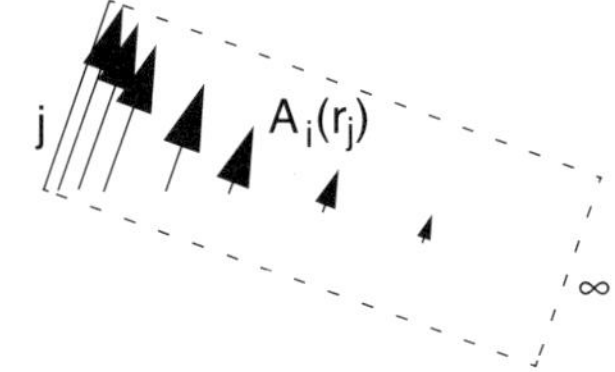

Fig. 1 Current in segment i in positive x-direction creates vector potential in j in positive x-direction as well. Compensating current in j must flow in negative x-direction

If we require the (electro- or magnetostatic) potential Ψ to be unity for all conductors, then each element of Ψ must be maximal on each conductor. This is a fundamental property of harmonic functions—solutions of Laplace' equation (see [9]). It follows that the gradient of each of the four elements of Ψ on the surface of each conductor must be pointing into the conductors. For the electric case this means all surfaces are positively charged, because the electric field points into the dielectric material everywhere.

For the magnetic case this means that the vector potential components A_x, A_y, and A_z are positive everywhere, since solutions of Laplace' equation are maximal and minimal on the boundary of their definition domain. Here the maximum (by construction unity) are the combined conductor surfaces, the minimum is infinity where the vector potential is zero by convention. Finally, since the vector potential components are positive everywhere, currents must flow in the positive xyz directions in all the conductors as well, since if not, there would be locations with negative vector potential components, creating a contradiction.

So combining the electrostatic and magnetostatic case, we find

$$\gamma = S\Psi = S\begin{bmatrix} 1 \\ \cdots \\ 1 \end{bmatrix} > \bar{0} \quad \text{so for each row}: \sum_j S_{ij} > 0 \tag{11}$$

The 'greater as' in the first part of (11) acts element for element. With (8) and (9), which give the signs of the S_{ij} (diagonals positive, off-diagonals negative!), we find diagonal dominance for S:

$$|S_{ii}| > \sum_{j \neq i} |S_{ij}| \tag{12}$$

- **Positive Definiteness of Truncated S**: S was shown to be positive definite. Due to its diagonal dominance, the positive definiteness of S is preserved when off-diagonal elements are set to zero—a property not available for K. For this we use the following theorem from linear algebra [12][pg. 349] for a matrix A:

$$\text{If all } A_{ii} > 0 \text{ and } A, A^T \text{ diagonal dominant } then \ A \text{ is positive definite} \tag{13}$$

Diagonal dominance and positivity of the diagonal elements of S have been shown previously. With (13) the positive definiteness of the sparsified S follows.

- **Shielding Effect in S**: To assist in understanding the physical interpretation of susceptance, we ask: *What* is the significance of jth column of S? The jth column of S is the amount of source (charge or current) necessary *on* or *flowing through* the conductors to force conductor j to unit (electric or magnetic) potential and all other conductors to zero potential. An individual term S_{ij}, when i and j are far removed, must include shielding effects for the electrostatic *as well as* for the magnetostatic fields. The 'source necessary' in some conductor i to force it to zero potential already takes into account that some of the original field of the reference (unit potential) conductor j has been compensated by charges / currents on zero potential conductors closer to j. This in turn means that the magnitude of the elements in S_{ij} drops off much faster with distance between i and j. This enhances sparsification, since elements which are 'large enough' in S are easier to distinguish from those that are 'too small', thereby forming a much smaller set than for the interaction matrix K.
- **Stability of S under Window Inversion**: Since the shielding effect renders all but a few short-distance couplings negligible, the idea to exploit this to make the inversion from K to S more efficient by only including those few neighbors of the current unit potential conductor in the inversion. That is, the inversion is restricted to *extraction windows* around each conductor, thereby replacing the inversion of a huge, dense $N \times N$ matrix—N being the total number of conductors in the system—by N times an inversion of much smaller $n_j \times n_j$ matrices. This results in N individual, small matrices $S(j)$, where n_j is the number of conductors to which segment j has significant couplings.

The $S(j)$ are all diagonally dominant, since the proof above applies to each of the small conductor subsets individually. Clearly, for this approach the coupling $S_{ij}(j)$ need not be equal to $S_{ji}(i)$, since the set of 'significant neighbors' is usually different for different segments i and j. However, when we assemble our sparse S' matrix for the entire system from the individual $S(j)$, we need to ensure symmetry of S'. To guarantee positive definiteness of S', we choose

$$S'_{ij} = S'_{ij} = \max\{S_{ij}(j),\, S_{ji}(i)\} \tag{14}$$

Since we know that all off-diagonal elements of any inverse interaction matrix are non-positive, this means we select the element with the *smallest magnitude*. This ensures the diagonal dominance of S' when assembling it from elements of the $S(j)$, while preserving the largest degree of accuracy. With (13) we then find that the sparse approximation S' is positive definite.

4 Double-inverse Inductance Models

Two major drawbacks using partial inductance to model magnetic interactions for on-chip interconnect are: (1) the slow, logarithmic decay of the couplings with the distance between the filaments; and (2) the absence of any shielding effect.

This makes localizing partial inductance difficult. In the previous section we have shown, however, that using susceptance rather than inductance for modeling the magnetic field interactions is much more beneficial. Using S directly in simulators and timing analysis tools as proposed in [7], however, is not readily supported by most programs available today.

We have shown in the last section that it is possible to ensure the stability of sparse susceptance matrices. We apply a windowing approach, solving many local inductance systems rather than one large problem. When assembling all partial results into a sparse global susceptance matrix, we have shown that the stability of the symmetric result is ensured by choosing the off-diagonal with the smaller magnitude.

The window size can be chosen to include only long-range susceptive couplings above a given magnitude threshold relative to the self terms. One percent cutoff means, for instance, that the window size was chosen such that increasing the window only added new susceptive couplings less than 1% of the self term for a given active conductor. The window sizes necessary are much smaller for susceptance than for the initial inductive model, due to the shielding effect for S.

The global, sparse susceptance matrix is then positive definite, as shown in Section 3, so inverting this sparse S matrix back into an inductance representation, using sparse matrix solving techniques [10], yields again a positive definite matrix.

The resulting double-inverted partial inductance matrix generally contains far fewer significant elements than the original inductance matrix, making further sparsification of L much easier (see Fig. 2). To preserve positive definiteness of the doubly inverted inductance matrix, we add the magnitude of off-diagonals which

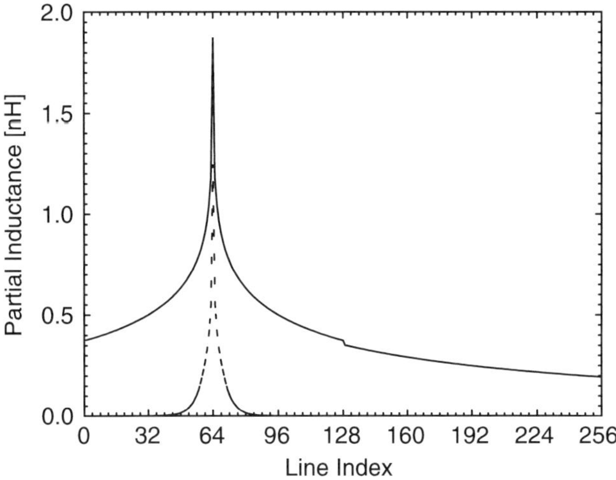

Fig. 2 Comparison of original inductive couplings with double-inverse L. Shown are couplings of all bus lines of example in Fig. 3 to line 64 (middle of first block). Solid line for original partial inductance, dotted line for double-inverse L (all elements of susceptance matrix smaller than 1% of maximum were discarded)

we cancel during sparsification to the corresponding diagonal elements. That this procedure preserves positive definiteness is easy to show. We define a symmetric matrix $M_{ij}(p,q)$ which is +1 for $i=j=p$, and $i=j=q$, either +1 or −1 for both $(i=p,j=q)$ and $(i=q,j=p)$, and 0 everywhere else.

$x^T M(p,q)x$ is always greater or equal to zero, which can be shown by explicitly calculating the expression. If a symmetric matrix A is positive definite, then the matrix $B=A+|A_{pq}|*M(p,q)$ must be positive definite as well. If we choose the sign of the off–diagonals in $M(p,q)$ opposite to the sign of A_{pq}, then $B_{pq}=B_{qp}=0$.

Since most of the off-diagonal terms of the double–inverse inductance matrix are very small (see Fig. 2), this procedure ensures the positive definiteness of the resulting sparse double–inverse inductance matrix while not significantly changing the diagonal elements. We use the same cutoff percentage threshold for the double–inverse inductance matrix as for the susceptance matrix.

This sparse inductance model can now efficiently and accurately represent magnetic interactions within interconnect without compromising stability. Examples will be presented in the following to demonstrate the efficacy of this approach.

5   Examples

5.1   Single Layer 2 × 128 Bit Bus

We demonstrate our inductance modeling method on a bus consisting of two blocks of 128 lines (W 1 μm, H 2 ∝m, Sp 1 μm, L 1,000 ∝m) with 16 μm additional gap between the two blocks (see Fig. 3). The driver resistance R_{Dr} is 70 Ω and the load capacitance C_{Ld} is 2 fF. Every 16th line is a return line (no driver resistance).

In Fig. 4 the far end node voltage responses for two lines of the structure are shown for 1 V step and 10 ps ramp inputs. As expected, the far end response reduces in magnitude with increasing distance from the active line. The reference results (white circles) are obtained by including all individual couplings, leading to high runtimes and memory consumption (see Table 1). Double-inverse inductance models are sparsified as described in Section 4 dependent on cutoff percentage (the smaller, the more accurate). For simple truncation we merely set to zero all off–diagonals which are zero for the double–inverse with the given cutoff threshold to ensure fair comparison of the efficiency and accuracy of the different methods.

The waveforms for our double–inverse inductance localization method are very close to the corresponding reference results, and especially the time interval containing the first few minima and maxima is captured very well by the double–inverse approximation leading to very high accuracy for interconnect timing analysis. The runtime and memory requirements, however, are significantly smaller than for the reference case. Speedup factors are shown in Table 1.

It should be noted that the double-inverse inductance model is effective at higher frequencies, which is where inductive effects have the most impact and

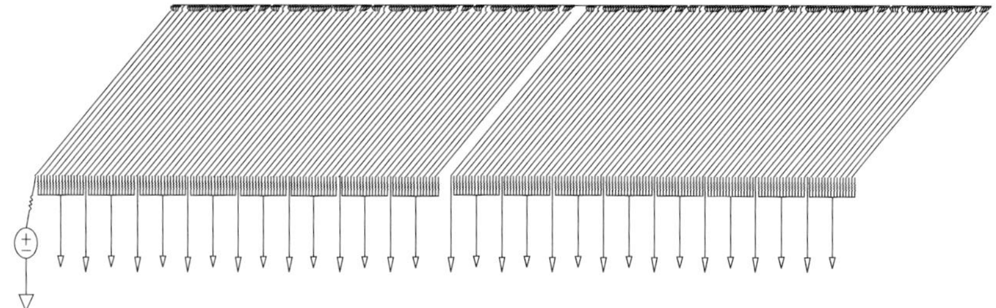

Fig. 3 2×128 bit bus. Leftmost line is active. Line numbering from left to right

Table 1 Simulation cost comparison for 2×128 bit bus example in Fig. 3

Symbols see Fig. 4:	○	●	◆	■	▲
Runtime (step) [s]	17550	38	87	680	32
Speedup Factor (step)	1.0	460	200	26	550
Runtime (ramp) [s]	20050	41	87	530	30
Speedup Factor (ramp)	1.0	490	230	38	670
Memory [MByte]	112800	5760	10040	14510	5760
Capacitance El.	32481	3280	4823	3280	3280
Inductance El.	32896	4201	6392	4201	4201
Sparsity [%]	0.0	88.5	82.8	88.5	88.5
HSpice internal El.	65890	7994	11728	19190	7994

predominantly determine the ringing and overshoot for timing analysis. For lower frequencies the damping of double inverse is less, therefore a low-amplitude, low-frequency oscillation around steady state remains while ringing is damped much quicker for the reference result. Overall, our double-inverse inductance model shows excellent agreement with the exact result.

For comparison we also tried to generate a corresponding waveform using simple truncation on the original partial inductance. However, this creates an ill-conditioned inductance model which leads to diverging waveforms. A more extensive study of this is available in [4]. This is observed in the results in Fig. 4 (squares) which coincide with the reference solution for about the first picosecond of simulation time, then diverge towards positive or negative infinity.

It is possible to avoid this stability problem by simply adding the truncated mutual partial inductances to the self term, much as described for our double-inverse inductance model in Section IV. However, since the off-diagonal terms are *significant in magnitude* for the partial L matrix during simple truncation, this will grossly overestimate the magnetic self coupling leading to much lower response waveform frequencies and very low accuracy (triangles in Fig. 4).

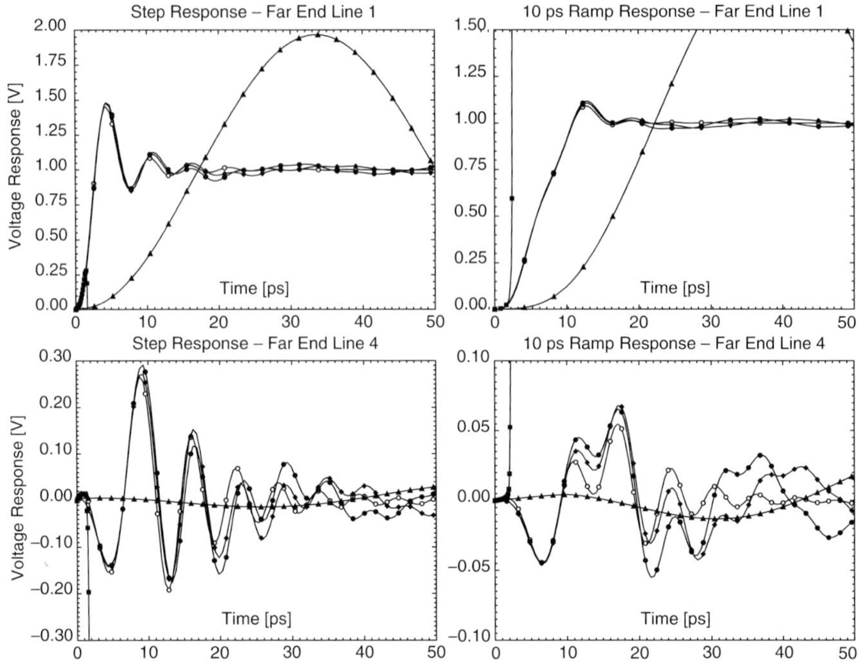

Fig. 4 Voltage responses at far ends of lines 1 and 4 of 2×128 bit bus. Curves with ○ are reference results for full L and C matrices. Approximations: Double–inverse inductance with cutoff 1% (●); double–inverse L with cutoff 0.5% (◆); simple truncation cutoff 1% (■); stabilized simple truncation cutoff 1% (▲)

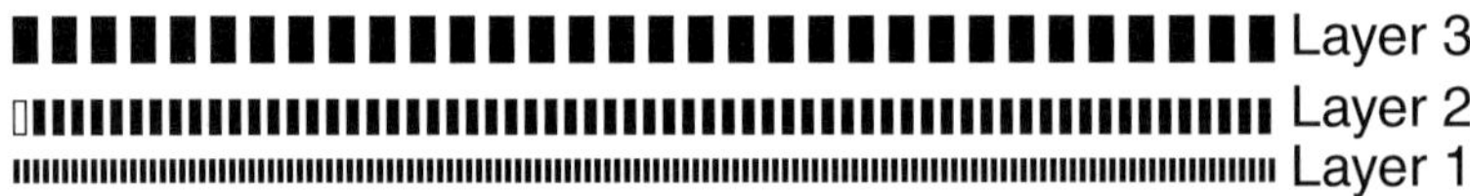

Fig. 5 Three Layer Bus Example. 128 bit on layer 1, 64 on layer 2 and 32 bit on layer 3. Active line shown white

5.2 Three Layer Bus Structure

Our second example consists of three parallel bus structures with the leftmost wire of the middle layer being active. All wires are $1,000\,\mu m$ long (cross-section in Fig. 5). Layer 1: W $1\,\mu m$, H $2\,\mu m$, Sp $1\,\mu m$, R_{Dr} 70 Ω, C_{Ld} 2 fF. Layer 2: W $3\,\mu m$, H $2\,\mu m$, Sp $1\,\mu m$, R_{Dr} 50 Ω, C_{Ld} 2 fF. Layer 3: W $6\,\mu m$, H $3\,\mu m$, Sp $2\,\mu m$, R_{Dr} 25 Ω, C_{Ld} 2 fF. Every eighth line is a return line.

The clear advantage of our double-inverse model over the simple truncation approach is also evident in the results for this example, shown here in Fig. 6. Our observations from the previous example apply in this case as well.

The presence of interconnect above and below the layer in which the active line is placed increases the number of wires to which coupling is significant for accurate simulation. Therefore, for the same cutoff threshold values, the sparsity of the approximate models is lower as for the single layer example. However, since the total number of conductors is smaller than for the 2×128 bit bus example, the runtimes and memory consumption are smaller as well.

Table 2 Simulation cost comparison for three layer bus example in Fig. 5

Symbols see Fig. 6:	○	●	◆	■	▲
Runtime (step) [s]	9390	72	310	1700	43
Speedup Factor (step)	1.0	130	30	5.5	220
Runtime (ramp) [s]	8950	75	310	1790	45
Speedup Factor (ramp)	1.0	120	29	5.0	200
Memory [MByte]	87230	7650	16070	29150	7650
Capacitance El.	25424	3125	4662	3125	3125
Inductance El.	25200	6962	12459	6962	6962
Sparsity [%]	0.0	80.0	66.1	80.0	80.0
HSpice internal El.	51073	10536	17570	27620	10536

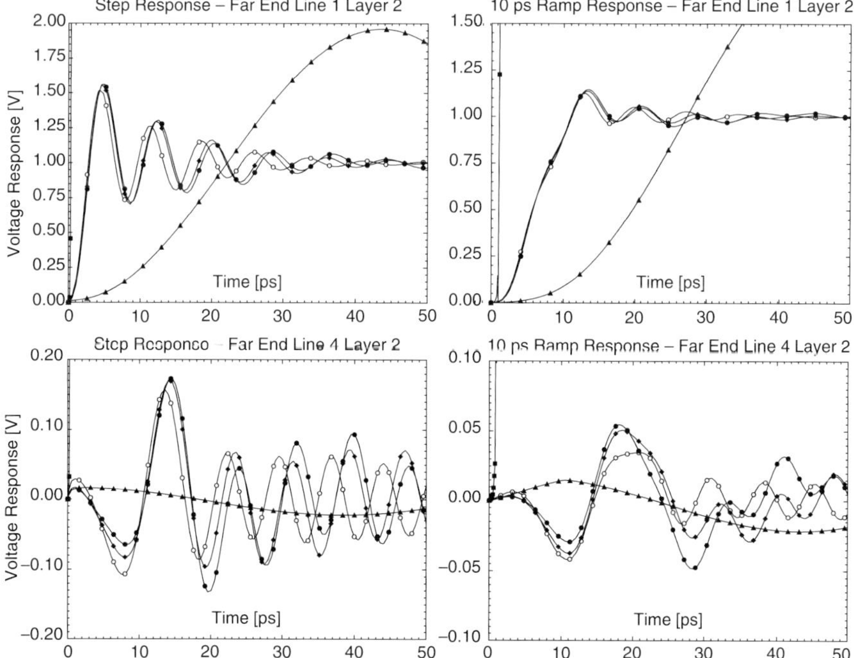

Fig. 6 Voltage responses at far ends of lines 1 and 4 from the left in the middle layer of Fig. 5. Curves with ○ are reference results for full L and C matrices. Approximations: Double-inverse inductance with cutoff 1% (●); double-inverse L with cutoff 0.5% (◆); simple truncation cutoff 1% (■); stabilized simple truncation cutoff 1% (▲)

Due to the higher density of surrounding conductors for each wire and the resulting lower sparsity of the inductance approximations, the speedup across the board is lower than for the previous example, but still quite significant.

6   Conclusions and Future Directions

We have presented a novel and accurate windowing technique for inductance extraction. This approach provides for sparsification of the inductance matrix in a manner that is more robust and accurate as compared with other simplification schemes. It is also shown to preserve model stability while avoiding high memory consumption and runtimes. This localization process requires efficient inversion of the small L and S submatrices that correspond to the applied windows.

Although this inductance localization approach provides for a substantial improvement over existing methods, neglecting long distance couplings or merging them into short distance couplings will always limit the modeling accuracy. To efficiently create accuracy beyond the level which can be provided by localized extraction requires *hierarchical models*. Efficient extraction of a hierarchical susceptance and double-inverse inductance model appears to be viable using existing hierarchical extraction approaches [5]. To integrate these hierarchical models into circuit netlists will be the focus of future research.

References

[1] A. Ruehli, *Inductance Calculations in a Complex Integrated Circuit Environment*, IBM J. Res. Dev., **16**, No.5 (September 1972).

[2] J. Jackson, *Classical Electrodynamics*, 2nd Ed., Wiley, New York (1975).

[3] E. Rosa, *The Self and Mutual Inductance of Linear Conductors*, Bul. Nat. Bureau of Standards, **4** (1908).

[4] M. Beattie, L. Pileggi, *IC Analyses Including Extracted Inductance Models*, 36th DAC (June 1999).

[5] M. Beattie, L. Pileggi, *Electromagnetic Parasitic Extraction via a Multipole Method with Hierarchical Refinement*, Proc. ICCAD 1999 (November 1999).

[6] L. Nagel, R. Rohrer, *Computer Analysis of Nonlinear circuits, Excluding Radiation (CANCER)*, IEEE J. Solid State Circ., **SC–6**, 162–182 (August 1971).

[7] A. Devgan, H. Ji, W. Dai, *How to Efficiently Capture On–Chip Inductance Effect: Introducing a New Circuit Element K*, Proc. ICCAD 2000 (November 2000).

[8] D. Ling, A. Ruehli, *Interconnection Modeling*, in: *Circuit Analysis, Simulation and Design*, Elsevier Science B.V., North–Holland, (1987).

[9] I. Bronshtein, K. Semendyayev, *Handbook of Mathematics*, 3 Ed., Van Nostrand Reinhold.

[10] G. Golub, C. Van Loan, *Matrix Computations*, 3rd Ed., Johns Hopkins University Press, Baltimore (1996).

[11] W. Press, S. Teukolsky, W. Vetterling, B. Flannery, *Numerical Recipes in C*, 2nd Ed., Cambridge University Press, New York (1992).

[12] R. Horn, C. Johnson, *Matrix Analysis*, Cambridge University Press, Cambridge (1985).

Soft-Error Tolerance Analysis and Optimization of Nanometer Circuits

Yuvraj Singh Dhillon, Abdulkadir Utku Diril, and Abhijit Chatterjee

Abstract Nanometer circuits are becoming increasingly susceptible to soft-errors due to alpha-particle and atmospheric neutron strikes as device scaling reduces node capacitances and supply/threshold voltage scaling reduces noise margins. It is becoming crucial to add soft-error tolerance estimation and optimization to the design flow to handle the increasing susceptibility. The first part of this paper presents a tool for accurate soft-error tolerance analysis of nanometer circuits (ASERTA) that can be used to estimate the soft-error tolerance of nanometer circuits consisting of millions of gates. The tolerance estimates generated by the tool match SPICE generated estimates closely while taking orders of magnitude less computation time. The second part of the paper presents a tool for soft-error tolerance optimization of nanometer circuits (SERTOPT) using the tolerance estimates generated by ASERTA. The tool finds optimal sizes, channel lengths, supply voltages and threshold voltages to be assigned to gates in a combinational circuit such that the soft-error tolerance is increased while meeting the timing constraint. Experiments on ISCAS'85 benchmark circuits showed that soft-error rate of the optimized circuit decreased by as much as 47% with marginal increase in circuit delay.

1 Introduction

Technology scaling has been the major factor behind the increasing computing power of microprocessors. Technology scaling roughly leads to a doubling of clock frequencies every generation, a 30% decrease in node capacitances every generation and a 30% reduction in supply voltages to reduce power consumption. All these factors are leading to a drastic increase in soft-error susceptibility of

Georgia Institute of Technology, Atlanta, GA 30332, USA {yuvrajsd, utku, chat} @ece.gatech.edu

This research was supported by NSF Information Technology Research Contract CCR 022-0259.

R. Lauwereins and J. Madsen (eds.), *Design, Automation, and Test in Europe–The Most Influential Papers of 10 Years DATE*, 389–400
© Springer 2008

combinational and memory circuits to alpha-particle and neutron strikes. Because of the reduced node capacitances, a smaller injected charge is needed to induce a glitch at a circuit node. Thus, low-energy particle strikes that earlier had no effect on a circuit can now cause soft-errors. Because of the reduced supply voltages, noise margins are reduced, which also increases the susceptibility to particle strikes. Increasing clock frequencies increase the probability of a soft-error getting latched. Furthermore, due to super-pipelining, the number of gates in pipeline stages have been reducing, which in turn reduces the electrical attenuation of glitches as they propagate to the latches.

Although these factors affect both memory and combinational elements, the overall soft-error rate of memories is not increased as much as combinational logic because memories are protected by techniques such as error-correcting codes (ECC). There has not been a need to protect combinational circuits because combinational circuits have a natural tendency to mask glitches due to three phenomena [1]. First, due to *logical masking*, a glitch might not propagate to a latch because of a gate on the path not being sensitized to facilitate glitch propagation. Second, due to *electrical masking*, a generated glitch might get attenuated because of the delays of the gates on the path to the output. Third, due to *latching-window masking*, a glitch that reaches the primary output might not still cause an error because of the latch not being open. The factors mentioned in the previous paragraph adversely affect all the above three factors in terms of soft-error tolerance. Due to decreasing number of gates in a pipeline stage, logical masking as well as electrical masking has been decreasing for new technology generations. Electrical masking has also been decreasing due to the reduction in node capacitances and supply voltages every generation. Furthermore, increasing clock frequencies have reduced the time window in which latches are not accepting data, thereby reducing latching-window masking also. Because of these factors, the soft-error rate (SER) of combinational logic is expected to rise 9 orders of magnitude from 1992 to 2011, when it will equal the SER of unprotected memory elements [2].

Generally, in mission-critical space applications combinational circuits are protected by using duplication/triplication and concurrent-error detection (CED) [3]. However, these methods have too high delay, area and power overheads to be used in commercial applications. Recently, low-cost methods for increasing soft-error tolerance of commodity applications using time-redundancy [4] and partial duplication [5] have been proposed. However, these methods still add additional delay overhead to the original circuit due to the presence of the checker circuit. Also, these methods have system level overheads (such as pipeline flushes) when an error is detected, either to correct the error or to do the computation again.

This paper proposes a novel, zero delay-overhead method for increasing the soft-error tolerance of nanometer CMOS combinational logic circuits. Using an optimal assignment of supply voltages, threshold voltages, sizes and channel lengths to gates in ultra-deep sub-micron circuits, the *electrical attenuation characteristics* of the gates in the circuits are enhanced without incurring any delay overhead. Multi-supply voltage and multi-threshold voltage designs are becoming increasingly common for low-power applications, however if these are infeasible, the method can still be used to just find optimal gate sizings for increased soft-error tolerance. This method can be used along with any of the traditional methods described above to greatly decrease the overhead of error detection and correction.

The paper is organized as follows. Section 2 describes characteristics of gates that affect the strike-induced glitches. Section 3 describes ASERTA, a tool for fast and accurate analysis of the soft-error tolerance of a circuit. Section 4 describes SERTOPT, a circuit optimization tool for enhancing the soft-error tolerance of circuits while meeting timing constraints. Section 5 gives experimental results. Section 6 concludes.

2 Glitch Tolerance Characteristics of Individual Gates

There are two characteristics of interest for a single gate in terms of soft error tolerance: *glitch generation* and *glitch propagation*. The glitch generation characteristics of a logic gate determine the shape and magnitude of the voltage glitch generated at the output of the gate due to a particle strike on the gate. The glitch propagation characteristics of a logic gate determine how the gate attenuates a glitch that is generated at some prior circuit node as it passes through the logic gate.

When a particle strikes a circuit node, the voltage magnitude of the corresponding glitch is dependent on the total capacitance of the node. The duration of the generated glitch is dependent on the delay of the gate that is driving the node. If the gate driving the node is fast, it will quickly discharge (or charge) the node back to its original value. Therefore, faster gates have better glitch generation characteristics in terms of the generated glitch width.

However, the behaviour is *opposite* for glitch propagation. Assuming a linear ramp at the output of the gate, for a gate propagation delay of d and glitch duration of w_i at the gate input, glitch duration at the output of the gate, w_o, can be approximated as follows:

$$
\begin{aligned}
w_o &= 0 \ if \ w_i < d \\
w_o &= 2 \cdot (w_i - d) \ if \ d < w_i < 2 \cdot d \\
w_o &= w_i \ if \ w_i > 2 \cdot d
\end{aligned}
\tag{1}
$$

This model is similar to the glitch amplitude attenuation model used in [6]. As seen from Eq. 1, a slow gate will attenuate a glitch at its output more compared to a fast gate. Therefore, slow gates have better glitch attenuation characteristics.

Figures 1 and 2 show SPICE simulation results for generated glitch width and propagated glitch width, respectively, for an inverter for different values of gate size, gate channel length, gate supply voltage (V_{DD}) and gate threshold voltage (V_{th}). The SPICE models are for 70 nm technology node [7]. The minimum and maximum values of the variables are indicated on the x-axis. Size of 1 means a gate width of 100 nm. It is clear that factors that slow down a gate (decrease in size, increase in channel length, reduction in V_{DD}, and increase in V_{th}) increase generated glitch width but also increase the attenuation of propagating glitches.

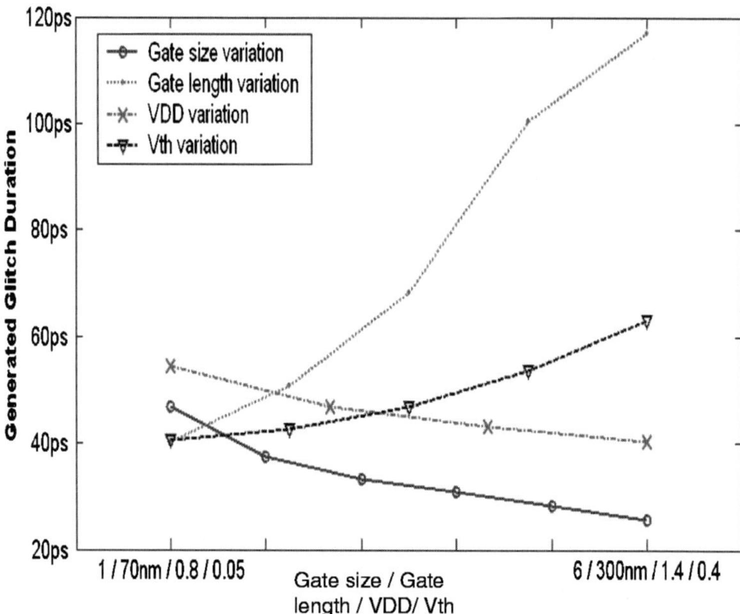

Fig. 1 Glitch generation characteristics for an inverter for an injected charge of 16fC

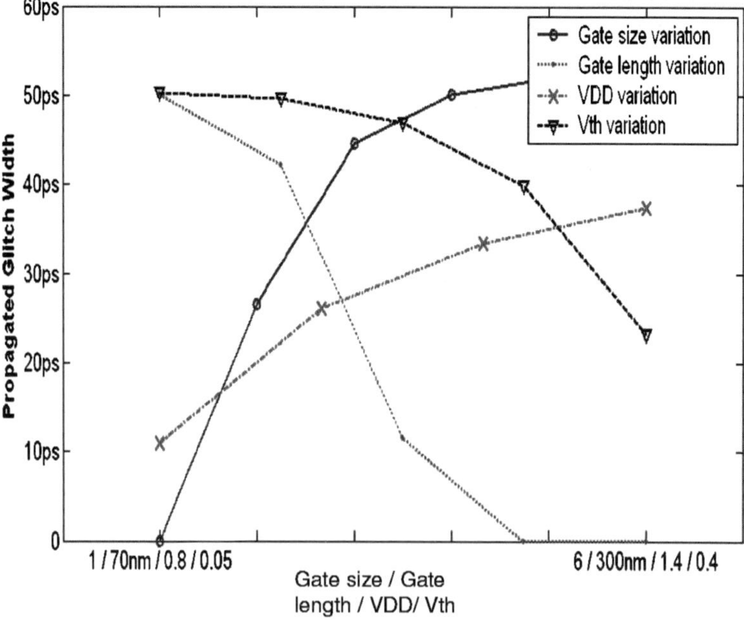

Fig. 2 Glitch propagation characteristics of an inverter for an input glitch of duration 50 ps

The insight gained from the SPICE simulation data is that only generated glitch width or propagated glitch width are not enough to characterize the "softness" of a gate as this might lead to erroneous conclusions. If only glitch propagation characteristics are considered as a measure of the "softness" of a gate (as in [8]), slowing down a gate would apparently always reduce the softness of the circuit; however, a slower gate will produce a bigger glitch at its output when it is subjected to a particle strike. Such a glitch can easily propagate to the output and cause an error. Slowing down all the gates at the primary outputs (POs) to attenuate all previous glitches (and hence to increase the soft-error tolerance of a circuit) is also not a viable solution as: (i) it is too expensive in terms of delay overhead, and (ii) it leads to very wide glitches being generated right at the latch inputs in the event of strikes at POs. Similarly, just speeding up all gates to "kill" the glitches generated at their outputs is also not viable as: (i) it would be too expensive in terms of area and power overheads, and (ii) a wide glitch generated due to a high energy strike would definitely propagate to the output because of little attenuation offered by the fast gates.

The conclusion drawn from the above discussion is that it is not possible to increase the soft-error tolerance of a circuit by just focussing on a few "soft" gates and trying to make them "hard". Gates hardened to resist glitch propagation cause generation of big glitches at their outputs and gates hardened to reduce generated glitch widths propagate glitches very easily. It is necessary to estimate the change in soft-error tolerance of the whole circuit after any optimization, as a "local" improvement of the softness of a gate might not lead to a "global" improvement in soft-error tolerance. The next section describes ASERTA, a tool for accurate estimation of the soft-error tolerance of a circuit.

3   Circuit Soft-error Tolerance Analysis

ASERTA models a particle strike at a node as a current source injecting (or removing) a fixed amount of charge into (or from) that node. If the node is at low voltage, charge is injected into the node and if the node is at high voltage, charge is removed by the current source. The opposites of these two cases do can not cause a voltage glitch to be generated and are neglected. A SPICE lookup table is constructed for generated glitch width (due to charge injected at gate output) for different types of gates, fan-ins, sizes, channel lengths, V_{DD}s, V_{th}s, and load capacitances. Although in reality the amount of charge injected (or removed) depends on the energy of the strike, for simplicity ASERTA assumes a fixed amount of injected charge. Future versions of ASERTA will have look-up tables for different amounts of injected charge.

SPICE lookup tables are also constructed for delays, static energies, dynamic energies, output ramp and gate input capacitances for different types of gates, fan-ins, sizes, channel lengths, V_{DD}s, V_{th}s, input ramps and load capacitances.

ASERTA uses linear-interpolation inside the lookup tables to compute output values for arbitrary values of input parameters. Using lookup tables allows ASERTA to have better accuracy than analytical models while still being much faster than SPICE. To estimate the soft-error tolerance of a circuit, ASERTA injects charge into every gate output, looks up the generated glitch width from the table and then propagates the generated glitch to the primary outputs (POs) taking into account the effects of logical and electrical masking. The sum total of the widths of the glitches reaching the POs is taken as a measure of the "Unreliability" of the circuit. The following subsections describe how ASERTA models logical, electrical and latching-window masking.

3.1   Logical Masking

Since actual signal values are not known, for every node ASERTA calculates the probability that there is at least one sensitized path from that node to a primary output. Calculation of the sensitization probability values from the input signal statistics is easy for circuits which do not have reconvergent fan-out. Sensitization probabilities for such circuits can be calculated as in [8]. However, finding the values for circuits with reconvergent fan-out is an NP-complete problem [9]. ASERTA uses zero delay simulation of the circuit with 10000 random inputs applied (as in [5]) to compute the probability, P_{ij}, that there is at least one path sensitized from output of gate i to primary output j. For primary output j, P_{jj} is 1. The static probability, p_i, of a node i being at logic 1 is obtained for all nodes using a commercially available tool, Synopsys Design Compiler, given a static probability of 0.5 at the primary inputs.

For all successor gates s of gate i, the probability that a glitch at i will be able to propagate through gate s to primary output j is calculated as follows:

$$\pi_{isj} = \frac{S_{is} \cdot P_{ij}}{\sum_{k \in \Psi} S_{ik} \cdot P_{kj}} \tag{2}$$

where Ψ is the set of successors of gate i and S_{is} is the probability that gate s is sensitized to gate i (i.e. all other inputs of gate s have non-controlling values). S_{is} can be obtained by multiplying together the static probabilities of the other inputs being 1/0 for a AND/OR gate. Note that π_{isj} is not taken to be just $S_{is}P_{sj}$ since π_{isj} should have the property that $\sum_{K \in \Psi} \pi_{ikj} \bullet P_{kj} = P_{ij}$. Also note that π_{isj} is an approximation to the actual probability value since in circuits with reconvergent fan-out, the

probability that gate s is sensitized to gate i conditions the probability of gate s having a path sensitized to a primary output.

The next subsection describes the procedure used in ASERTA for computing the glitch widths at POs for charge injected at every gate output.

3.2   *Electrical Masking*

As mentioned before, ASERTA computes the expected output glitch width, W_{ij}, at primary output j for generated glitch width, w_i, at gate i. To do this efficiently in one pass over the circuit, for every gate, the expected output glitch widths, WS_{ijk}, for 10 sample glitch widths, ws_k (k between 1 and 10) are computed.

   The output glitch widths are computed for all gates in reverse topological order (i.e. from POs to PIs) as follows:

1. Let current gate be i.
2. If gate i is a primary output, set $WS_{ijk} = ws_k$ for all k.
   Set $WS_{ijk} = 0$ for all other primary outputs j.
   Also, since gate is primary output, it will propagate generate glitch width, w_i, directly. Hence, set $W_{ij} = w_i$ and $W_{ij} = 0$ for all other primary outputs j.
3. If gate *i* is not a primary output, for all sample glitch widths, ws_k:
   For all successors s of gate i:
   Let d_s be the delay of gate s looked up from the SPICE tables.
   Calculate the glitch width, wo_{sk}, propagated to the output of gate s for input width of ws_k using Eq. 1.
   For each primary output j, look up the expected output glitch width, WE_{sjk}, for generated glitch width of wo_{sk} from the table of expected output glitch widths for gate s, linearly interpolating if necessary.

   Finally, Let $WS_{ijk} = \sum_{s \in \Psi} \pi_{isj} \bullet WE_{sjk}$

4. Compute W_{ij} by looking up the table of expected output glitch widths, WS_{ijk}, computed in step (iii), for a generated glitch width of w_i, again linearly interpolating if necessary. Now process the next gate.

At the end of this procedure, expected output glitch widths, W_{ij}, at primary output j for generated glitch width, w_i, for every gate i are known. The complexity of the procedure is $O(V+E)$, where V is the number of gates and E is the number of circuit edges.

Lemma 1: For a very wide glitch ww_i generated at output of gate i, the above procedure correctly computes the expected output glitch width at primary output j as $WW_{ij} = ww_i \bullet P_{ij}$, if it is assumed that ww_i is one of the sample glitch widths used above (say sample 1).

Proof: Since the generated glitch is very wide, it will pass through all gates on any path from i to j without attenuation. WS_{jj1} is correctly computed as ww_i at primary output j. Assume that WS_{rj1} is correctly computed for all successor gates r of a gate p between i and j as $ww_i \cdot P_{rj}$. Then, the expected width WS_{pj1} will be computed as:

$$WS_{pj1} = \sum_{r \in \Psi} \pi_{prj} \cdot WS_{rj1} = \sum_{r \in \Psi} \pi_{prj} \cdot ww_i \cdot P_{rj}$$

$$= ww_i \cdot \sum_{r \in \Psi} \pi_{prj} \cdot P_{rj} = ww_i \cdot P_{pj}$$

where WS_{rjl} can be used instead of WE_{rjl} because ww_i is wide enough to propagate through gate r without attenuation. By induction, WS_{ijl} is also computed as $ww_i \cdot P_{ij}$. Since ww_i is the first sample glitch width, WS_{ijl} is WW_{ij}.

3.3 *Latching-window Masking*

A glitch must arrive within the setup and hold times of the latch at the primary output to be captured. Since the exact time of the particle strike is unknown, it can be assumed to be uniformly distributed within the clock cycle. The probability of a glitch being captured by a latch is directly proportional to its duration. Hence, by summing up the expected output glitch widths, W_{ij}, for all primary outputs j, the total contribution of gate i to the circuit unreliability is obtained. However, this ignores the fact that the size, Z_i of a gate plays an important role in determining the particle flux incident on the gate. Hence, the actual contribution of gate i to circuit unreliability is:

$$U_i = Z_i \cdot \sum_j W_{ij} \tag{3}$$

The total unreliability of the circuit is:

$$U = \sum_i U_i \tag{4}$$

Figure 3 shows the unreliability numbers, U_i, for the gates in ISCAS'85 benchmark circuit "c432" calculated by ASERTA plotted along with values calculated by SPICE for 70nm technology node. In SPICE, the unreliability was computed by applying 50 random input vectors, injecting charge at every gate output i and using the width of the glitch at primary output j as W_{ij} in Eq. 3. Only the nodes that were at most five levels deep from the POs are plotted. It is seen that there is close matching. The correlation between the two series was computed to be 0.96. For the ISCAS'85 benchmark circuits, an average correlation of 0.9 was obtained.

The next section describes SERTOPT, a tool that uses the unreliability estimates generated by ASERTA to optimize nanometer circuits for increased soft-error tolerance by enhancing the electrical masking characteristics of gates in the circuits.

4   Circuit Soft-error Tolerance Optimization

SERTOPT uses a delay assignment variation method to minimize a cost function that is a weighted sum of circuit unreliability, circuit power consumption and circuit size. A designer can easily change the optimization constraints by changing the ratio of the weights.

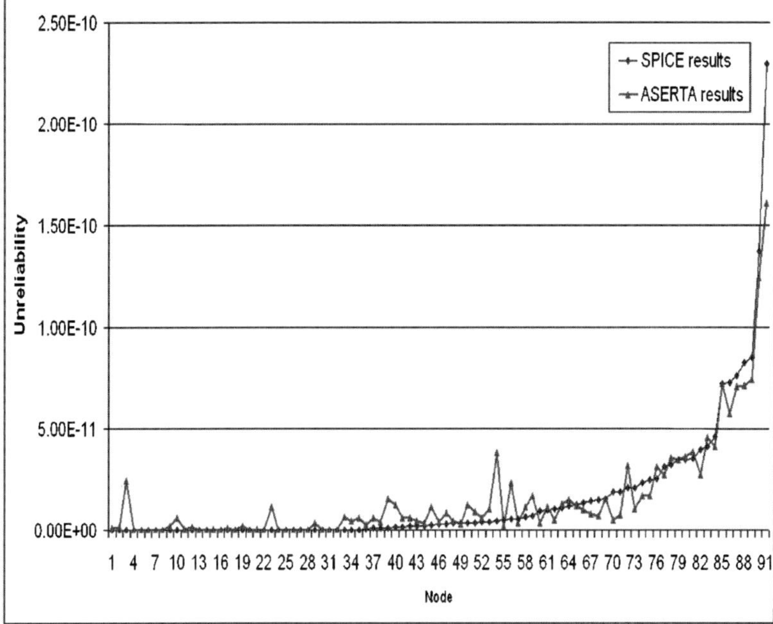

Fig. 3 Unreliability values obtained by SPICE and ASERTA for nodes in c432

The delay assignment variation method is based on the technique in [10, 11]. The circuit topology for a circuit with N gates and P paths from PIs to POs is represented with a binary topology matrix, **T**, defined as follows:

$$Tij = 1 \text{ if gate i lies on path j}$$

$$- 0 \text{ otherwise}$$

If the delays of the gates are represented by $\overline{d} = [d_1 \; d_2 \; \dots \; d_N]^T$, then $\overline{D} = T{\cdot}\overline{d}$ is the vector of path delays. The delay assignments of the gates are varied in an optimization iteration by adding $\overline{\Delta}$ such that $T \cdot \overline{\Delta} = 0$. This choice of $\overline{\Delta}$ lets the delays be varied without varying the path delay vector as shown below:

$$T \cdot (\overline{d} + \overline{\Delta}) = T \cdot \overline{d} + T \cdot \overline{\Delta} = \overline{D}$$

In other words, $\overline{\Delta}$ has to lie in the nullspace of **T** to guarantee that the timing constraints are met in every iteration.

To find the circuit parameters (gate sizes, lengths, V_{DD}s, V_{th}s) that are needed to match a delay assignment, SERTOPT traverses the circuit from POs to PIs in reverse topological order. The capacitive loads of the gates at the POs are known since these loads do not change. From these loads and the delay assignments for the

PO gates, the best matching sizes, lengths, V_{DD}s, V_{th}s available in the SPICE library that yield delays closest to the assigned delays are found and assigned to the PO gates. Once the parameters of these gates have been set, the capacitive loads offered by these gates to their predecessors can be found. The whole process is then repeated till the PIs are reached. The only constraint during the matching process is that only V_{DD} values greater than or equal to successor V_{DD} values are allowed to be used for a gate. This ensures that there is no low V_{DD} gate driving a high V_{DD} gate and eliminates the need for level-shifters.

Once the circuit parameters have been determined, SERTOPT calculates the circuit unreliability (U) using ASERTA and the circuit delay (T), total energy (E = dynamic energy + static energy) and area (A) using SPICE libraries. Note that although $\overline{\Delta}$ has been chosen from the nullspace of $\mathbf{T}$, the timing constraint might still be exceeded slightly because of the finite size library used for matching the delays. Hence, timing can also be included in the cost. The cost is computed as:

$$C = W_1 \cdot \frac{U}{U_{init}} + W_2 \cdot \frac{T}{T_{init}} + W_3 \cdot \frac{E}{E_{init}} + W_4 \cdot \frac{A}{A_{init}} \tag{5}$$

where U_{init}, T_{init}, E_{init} and A_{init} are the unreliability, delay, energy and area of the initial circuit.

The cost is minimized by using Sequential Quadratic Programming (SQP) to search for the optimal delay assignment giving lowest cost. However, simulated annealing, genetic algorithms or some other optimization algorithm can also be used.

5 Experimental Results

First, gate sizes were obtained for ISCAS'85 benchmark circuits by optimizing for speed using Synopsys Design Compiler. The gate sizes were then used with SPICE 70nm models [7] to compute the delays of the circuits for the 70nm technology. All the gates had a transistor channel length of 70nm, V_{DD} of 1 V and V_{th} of 0.2 V. The unreliability of the baseline circuits was estimated using ASERTA.

Then, SERTOPT was used to determine new gate sizes, channel lengths, V_{DD}s and V_{th}s for the circuits that would minimize unreliability while meeting the delay constraint of the baseline circuits. The results of the optimization are reported in Table 1. The second and third columns give the V_{DD} and V_{th} values used in the optimized circuits. Note that the values and numbers of V_{DD}s and V_{th}s to be used is a design variable. We report results for values that gave fair reduction in unreliability without compromising too much on power consumption. The maximum gate size used was the same as that for the baseline circuits. The gate channel lengths that SERTOPT was allowed to use for the optimization were 70, 100, 150, 250, and 300nm. The fourth, fifth, and sixth columns give the ratio of the area, energy consumption and delay of the optimized circuits to the corresponding values for

Table 1 Optimization results

Circuit	V_{DD}s used	V_{th}s used	Area	Energy	Delay	ASERTA	ASERTA/SPICE (50 random inputs)	
							Decrease in unreliability	
c432	0.8, 1	0.2, 0.3	×2	×2.2	×1.23	40%	44%	54%
c499	–	–	–	–	–	0%	0%	0%
c1908	0.8, 1, 1.2	0.1, 0.2, 0.3	×1.2	×1.8	×0.98	18%	6%	12%
c2670	0.8, 1, 1.2	0.1, 0.2, 0.3	×1.05	×1.3	×0.98	21%	42%	38%
c3540	0.8, 1	0.2, 0.3	×1.5	×1.6	×1.03	47%	35%	34%
c5315	0.8, 1, 1.2	0.1, 0.2, 0.3	×1.2	×1.9	×0.98	26%	–	–
c7552	0.8, 1	0.2, 0.3	×1.6	×1.6	×1.07	18%	–	–

the baseline circuits. As mentioned before, the delay constraint can sometimes be exceeded due to the finite sized library used. The seventh column gives the decrease in unreliability (as defined by Eq. 4) of the optimized circuits. The first sub-column gives the decrease calculated by ASERTA. The second and third sub-columns give the decrease in unreliability calculated by applying 50 random input vectors to the baseline and optimized circuits and measuring the average glitch width at the outputs using ASERTA and SPICE. They indicate the matching between ASERTA and SPICE for a small set of input vectors. Note that these numbers are different from the number in the previous column because 50 inputs do not capture the input statistics very well. The last 2 circuits were too big to be simulated by SPICE so we just report the unreliability reduction calculated by ASERTA.

The unreliability of "c499" could not be reduced. The reason is that "c499" is an error-correcting circuit for single-bit errors and ASERTA also models unreliability by injecting single-bit errors. A modelling scheme that takes into account simultaneous multiple-error injections could still be used with SERTOPT to reduce unreliability in the face of such errors.

ASERTA and SERTOPT have been implemented in MATLAB. They take 15 s and 20 min to run on "c432" and 200 s and 27 h to run on "c7552" respectively.[1]

6   Conclusion

This paper presented tools for the analysis and optimization of the soft-error tolerance of nanometer combinational circuits. The analysis tool, ASERTA, is able to accurately calculate (with average correlation of 0.9 with SPICE) the "unreliability" of circuits in orders of magnitude less computation time than SPICE. The optimization tool, SERTOPT, is able to reduce the unreliability of circuits by up to 47% by using a library with multiple V_{DD}s, V_{th}s, gate sizes and gate lengths.

---

[1] MATLAB is an interpreted language, hence slow. There is a ×10 speed-up expected by migrating to C.

References

[1] P. E. Dodd and L. W. Massengill, Basic mechanisms and modeling of single-event upset in digital microelectronics, *IEEE Trans. Nuclear Science*, vol. 50, 583–602, 2003.

[2] P. Shivakumar, M. Kistler, S. W. Keckler, D. Burger, L. Alvisi, Modeling the effect of technology trends on the soft error rate of combinational logic, *ICDSN*, 389–98, 2002.

[3] M. Nicolaidis and Y. Zorian, On-line testing for VLSI - a compendium of approaches, *JETTA*, vol. 12, 7–20, 1998.

[4] M. Nicolaidis, Time redundancy based soft-error tolerance to rescue nanometer technologies, *VTS*, 86–94, 1999.

[5] K. Mohanram and N. A. Touba, Cost-effective approach for reducing soft error failure rate in logic circuits, *ITC*, 893–901, 2003.

[6] M. Oman, G. Papasso, D. Rossi, C. Metra, A model for transient fault propagation in combinatorial logic, *IOLTS*, 111–15, 2003.

[7] Y. Cao, T. Sato, M. Orshansky, D. Sylvester, C. Hu, New paradigm of predictive MOSFET and interconnect modeling for early circuit simulation, *CICC*, 201–204, 2000.

[8] C. Zhao, X. Bai, S. Dey, A scalable soft spot analysis methodology for compound noise effects in nano-meter circuits, *DAC*, 894–899, 2004.

[9] F. N. Najm and I. N. Hajj, The complexity of fault detection in MOS VLSI circuits, *IEEE Trans. Computer-Aided Design of Integrated Circuits and Systems*, vol. 9, 995–1001, 1990.

[10] Y. S. Dhillon, A. U. Diril, A. Chatterjee, A. D. Singh, Sizing CMOS circuits for increased transient error tolerance, *IOLTS*, 11–16, 2004.

[11] Y. S. Dhillon, A. U. Diril, A. Chatterjee, H.H. S. Lee, Algorithm for achieving minimum energy consumption in cmos circuits using multiple supply and threshold voltages at the module level, *ICCAD*, 693–700, 2003.

A Single Photon Avalanche Diode Array Fabricated in Deep-Submicron CMOS Technology

Cristiano Niclass, Maximilian Sergio, and Edoardo Charbon

Abstract We report the first fully integrated single photon avalanche diode array fabricated in 0.35 μm CMOS technology. At 25 μm, the pixel pitch achieved by this design is the smallest ever reported. Thanks to the level of miniaturization enabled by this design, we were able to build the largest single photon streak camera ever built in any technology, thus proving the scalability of the technology. Applications requiring low noise, high dynamic range, and/or picosecond timing accuracies are the prime candidates of this technology. Examples include bio-imaging at cellular and molecular level, fast optical imaging, single photon telecommunications, 3D cameras, optical rangefinders, LIDAR, and low light level imagers.

1 Introduction

Imaging fast, time-correlated, molecular processes in physics and the life sciences is placing increasing pressure on camera technology. Over the last two decades, conventional imagers based on Charge-Coupled Devices (CCDs) and CMOS Active Pixel Sensor (APS) architectures have consistently achieved improved speed and sensitivity. However, very low photon counts are generally undetectable in these technologies or require deep cooling and highly optimized ultra-low-noise read-out circuitries.

Emerging imaging techniques have recently developed the need for single or low photon count sensitivity. There are numerous examples, where bio-luminescence, assay, and bio-scattering methods, among others, are currently the techniques of choice. In bioluminescence methods light is emitted as the result of a chemical reaction. The difficulty of detection is given by the extremely low emission intensity, typically in the microlux range [1]. Assay methods, based for example on fluorescence, may be used to determine the potential on a given cell *in vitro* or *in vivo* by means of Voltage-Sensitive Dyes (VSDs). A variety of techniques based on specific VSDs for example are utilized in neural activity analysis and research [2,3]. The challenge in these methods is usually in the detection of small variability

Ecole Polytechnique Fédérale de Lausanne (EPFL) CH-1015 Lausanne, Switzerland

R. Lauwereins and J. Madsen (eds.), *Design, Automation, and Test in Europe–The Most Influential Papers of 10 Years DATE*, 401–413
© Springer 2008

in the light intensity while maintaining high contrast in the presence of massive background illumination.

An increasing number of imagers also require timing accuracy. In time-correlated methods, for example, the arrival time of photons emitted by a synchronized source need be detected with picosecond accuracy. Examples of such need are in fluorescence decay measurements [4], Fluorescence Lifetime Imaging (FLIM) and Fluorescence Correlation Spectroscopy (FCS) [5,6], Förster Resonance Energy Transfer (FRET) [7], flow cytometry, etc. Single photon sensitivity can also be beneficial in reducing the complexity of fluorometer systems by replacing laser stimuli by LEDs [8] and by applying frequency domain [9] or space domain signal processing [10].

Other, nonbiological or medical imagers with high timing accuracy have also been sought over the years. Typical applications include time-of-flight based 3D cameras and fast cameras for natural or artificial gravity-driven flow analysis, fluid-dynamics modeling, and combustion optimization research, to name a few. However, the literature on the subject is very extensive [11,12].

To meet sensitivity and timing requirements, researchers have increasingly turned to nonsolid-state technologies, such as Photomultiplier Tubes (PMTs). In addition to their sensitivity to single photons, PMTs have several advantages in terms of noise, dynamic range, and timing accuracy. A major disadvantage is cost, size, and the fact that large arrays of PMTs are impractical. More compact multi-channel plate-based, multi-pixel devices have also been fabricated [13]. However, these devices generally require bulky vacuum chamber apparatuses.

Solid-state single photon detectors have been known for decades, however only recently researchers have succeeded in designing fully integrated single photon detectors in CMOS [14], thus triggering a renaissance in solid-state single photon detection. Recently, the emergence of multi-pixel arrays and new imaging techniques exploiting single photon detectors has accelerated the trend [15, 16]. These sensors are generally based on a device known as Single Photon Avalanche Diode (SPAD). A SPAD may be implemented as a reverse biased p-n junction where a lightly doped guard ring prevents premature discharge [14]. If the diode is biased above breakdown, the optical gain of the detector becomes virtually infinite, thereby ensuring Geiger mode of operation. In Geiger mode, when a photon triggers an avalanche in the multiplication region, a voltage pulse of appreciable amplitude is generated. By proper design, this pulse across the junction may be processed by conventional CMOS digital circuitry.

SPADs are useful in applications, where only a few photons must be detected and/or picosecond timing accuracy is required. Examples of SPAD-based designs in this context include FCS [17], high-speed imagers [18], time-of-flight 3D cameras [15, 16] and latchup/leakage test [19]. In these systems however, pitch and array size are still limited due to the technologies being used and the dynamic nature of SPAD signals which make it difficult to access large arrays of pixels simultaneously.

To the best of our knowledge, all published monolithically integrated SPAD arrays rely on a sequential read-out paradigm, where pixels are accessed one at a time, each for an arbitrary integration time. However, many applications require simultaneous column and row pixel access. Moreover, in general, significantly larger array sizes are necessary to provide the full advantages of SPAD technology. For example, a larger array of 1,000 or more Time-Correlated Single Photon Counting (TCSPC) pixels at high frame rate could significantly advance our understanding of neuro-transmitter biochemistry and pharmaceutical research [20]. In addition, to date CMOS SPAD imagers have been integrated only in near-micron processes. Deep submicron (DSM) technologies would enable larger arrays, more on-chip functionality, and overall better performance at significantly lower cost.

In this paper, we present the first SPAD ever integrated in a DSM technology. We demonstrate the lowest pitch ever achieved in a SPAD array *in any technology* and the largest SPAD streak camera ever designed. The array proposed in the paper has the lowest noise ever reported in a silicon single photon detector at room temperature.

Furthermore, the sequential read-out paradigm has been replaced with one that allows access to an arbitrary pixel in a column as soon as it detects a photon. In addition, all columns are independent from each other, thus enabling parallel column access. The novel readout scheme was implemented to carry the dynamic data generated by the pixels to the exterior of the array for processing. However, on-chip processing is possible with this scheme. The technique is scalable, thus the 4×112 pixel camera presented in this paper can be implemented in square or rectangular pixel arrays as well.

The paper is organized as follows: Section 2 describes the SPAD technology, while Section 3 describes the image sensor and camera prototype architecture. Finally, in Section 4, measurements are presented and summarized.

2   Single Photon Avalanche Diodes

SPADs are p-n junctions reverse-biased above breakdown voltage (V_{bd}) for single photon detection. When an avalanche photodiode is biased above V_{bd}, it remains in a zero current state for a relatively long period of time, usually in the millisecond range. During this time, a very high electric field exists within the p-n junction generating the avalanche multiplication region. Under these conditions, if a primary carrier enters the multiplication region and triggers an avalanche, several hundreds of thousands of secondary electron-hole pairs are generated by impact ionization, thus causing the diode's depletion capacitance to be rapidly discharged. As a result, a sharp current pulse is generated and can be easily measured. This mode of operation is commonly known as Geiger mode.

Conventional photodiodes, as those used in standard imagers, are not compatible with Geiger mode of operation since they suffer from premature breakdown when the bias voltage approaches V_{bd}. The main cause for premature breakdown is the fact that peak electric field is located only in the diode's periphery rather than in the planar region [14]. A SPAD is a photodiode specifically designed to avoid premature breakdown, thus allowing a planar multiplication region to be formed within the whole junction area.

Linear mode avalanche photodiodes are biased just below V_{bd}, thus they exhibit finite multiplication gain. Statistical variations of this finite gain produce an additional noise contribution known as excess noise. SPADs, on the contrary, are not concerned with these gain fluctuations since the optical gain is virtually infinite. Nevertheless, the statistical nature of the avalanche buildup is translated onto a detection probability. The probability of detecting a photon hitting the SPAD's surface, known as the *photon detection probability* (PDP), depends on the diode's quantum efficiency and the probability for an electron or for a hole to trigger an avalanche. Additionally, in Geiger mode, the signal amplitude does not provide intensity information since all the current pulses have the same amplitude. Intensity information is however obtained by counting the pulses during a certain period of time or by measuring the mean time interval between successive pulses.

Thermally or tunneling generated carriers within the p-n junction, which produce dark current in linear mode photodiodes, can trigger avalanche pulses. In Geiger mode, they are indistinguishable from regular photon-triggered pulses and they produce spurious pulses at a frequency known as *dark count rate* (DCR). DCR strongly depends on temperature and it is an important parameter for imagers since it defines the minimal detectable signal, thus limiting from below the dynamic range of the imager. Another source of spurious counts is represented by after-pulses. They are due to carriers temporarily trapped during a Geiger pulse in the multiplication region that are released after a short time interval, thus retriggering a Geiger event. After-pulses depend on the trap concentration as well as on the number of carriers generated during a Geiger pulse. The number of carriers depends in turn on the diode's parasitic capacitance and on the external circuit generally used to quench the avalanche. Typically, the quenching process is achieved by temporarily lowering the bias voltage below V_{bd}. Once the avalanche has been quenched, the SPAD needs to be recharged above V_{bd} so that it can detect subsequent photons. The time required to quench the avalanche and recharge the diode up to 90% of its nominal excess bias is defined as the *dead time* (DT). This parameter limits the maximal rate of detected photons, thus producing a saturation effect. The dead time consequently limits from above the dynamic range of the image sensor.

The statistical fluctuation of the time interval between the arrival of a photon at the sensor and the output pulse leading edge is defined as the *timing jitter* or *timing resolution*. Timing jitter mainly depends on the time a photo-generated carrier requires to be swept out of the absorption point into the multiplication region; in fully integrated SPADs it is generally a few tens of picoseconds [14].

During an avalanche, some photons can be emitted due to the electrolumines-cence effect [21]. These photons may be detected by neighboring pixels in an array of SPADs thus producing crosstalk. The probability of this effect is called optical crosstalk probability. This probability is much smaller in fully integrated arrays of SPADs in comparison to hybrid versions. This is due to the fact that the diode's parasitic capacitance in the integrated version is orders of magnitude smaller than the hybrid solutions, thus reducing the energy dissipated during a Geiger event. Electrical crosstalk on the other hand is produced by the fact that photons absorbed beyond the p-n junction, deep in the substrate, generate carriers that can diffuse to neighboring pixels. The probability of occurrence of this effect, whether it is optical or electrical, defines the *crosstalk probability*.

3 Sensor Architecture

3.1 Deep-Submicron CMOS SPAD

The SPAD, whose cross section is depicted in Fig. 1, consists of a circular dual junction structure: p+ anode/deep n-well/p-substrate. The p+ anode/deep n-well junction forms the avalanche multiplication region where the so-called Geiger breakdown occurs. The deep n-well/p-substrate junction allows the p+ anode to be biased independently from the p-substrate. It additionally prevents electrical crosstalk due to minority carriers diffusing in the substrate and improves the timing jitter of the SPAD [16].

A p-well guard-ring surrounds the p+ anode to prevent premature breakdown [14]. Breakdown voltage measurements have been done on the SPAD structure. In the 0.35 µm CMOS technology used, without the guard-ring, the breakdown voltage of a typical p+/deep n-well junction is about 11 V. The use of the guard-ring allowed the breakdown voltage to be increased by approximately 13 V. Conversely, the breakdown voltage of the deep n-well/p-substrate junction is typically 70 V, thus allowing the active junction to be biased above V_{bd}.

3.2 Digital Pixel and Event-driven Readout

The SPAD linear array consists of 4×112 pixels. Every pixel within a column can be accessed independently and simultaneously. The row readout is controlled

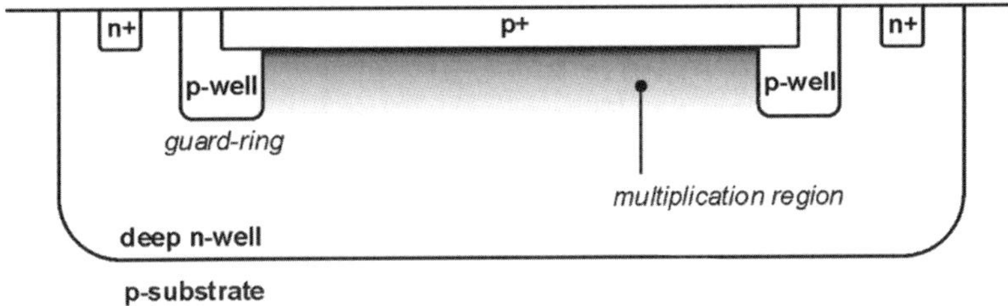

Fig. 1 SPAD cross section

directly by the pixels based upon an event, i.e. the detection of a photon. When a Geiger event occurs, a pixel generates a pulse synchronous with the photon arrival, thus enabling time-correlated measurements. A novel event-driven readout temporarily assigns the column readout bus to the pixel requesting it.

The pixel circuit consists of a SPAD, a quenching/recharge transistor (T_q), a CMOS inverter, and the pixel event-driven readout circuit as shown in Figure 2. The SPAD operates in passive quenching and passive recharge. The p+ anode is biased at a high negative voltage V_{op} equal to −23.4 V. This voltage is common to all the pixels in the array. The deep n-well cathode is connected to the power supply $V_{DD} = 3.3$ V through PMOS T_q, which eliminates the need for a quenching resistance [16]. The excess bias voltage (V_e), defined as the excess of bias voltage above V_{bd}, is thus equal to $|V_{op}| + V_{DD} - V_{bd} = 4.3$ V.

Upon photon arrival, the breakdown current discharges the SPAD's depletion region capacitance, reducing the voltage across the SPAD to close to V_{bd}. T_q is designed to provide a sufficiently resistive path to quench the avalanche process. After avalanche quenching, when the signal nRecharge is asserted, the SPAD recharges through T_q and progressively recovers its photon detection capability. The time required for quenching the avalanche and restoring the operating bias, i.e. DT, is typically less than 40 ns for the digital pixel. Signal nRecharge allows additionally the pixel to go to an inactive state when nRecharge is not asserted.

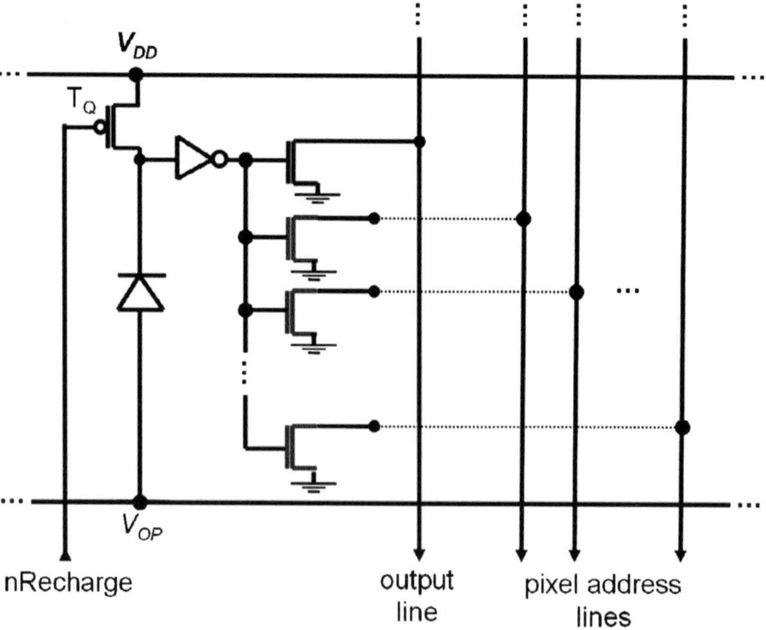

Fig. 2 Pixel circuit

At the cathode, an analog voltage pulse of amplitude approximately V_e [16] reflects the detection of a single photon. The inverter stage converts this analog voltage pulse into a robust digital pulse. The near-infinite internal gain inherent to Geiger mode operation leads to no further amplification.

The pixel event-driven readout circuit consists of an output NMOS transistor, a shared output line, N_{AT} address coding NMOS transistors, and N_{AT} shared address lines, where N_{AT} is given by $Ceiling(Log_2(N_{ROWS}))$, N_{ROWS} being the number of pixels within a column (Fig. 2). In this work, N_{ROWS} is 112, thus N_{AT} is 7. The source of the output transistor and the sources of all the address coding transistors are connected to GND. The drain of the output transistor is connected to the output line whereas the drains of the address coding transistors are either connected to the address lines or left unconnected so as to define a unique pixel address within the column.

Figure 3 shows the column readout circuitry based on the proposed event-driven topology, where only two pixels are sketched.

The output, address, and nRecharge lines forming the column readout bus are shared among all the pixels within a column and connected to V_{DD} via pull up resistors. Within the pixel, the output and address coding transistors are driven by the inverter stage. In idle state, nRecharge is asserted making all the SPADs recharged. The inverter output of all the pixels stay at GND until a photon is detected. During

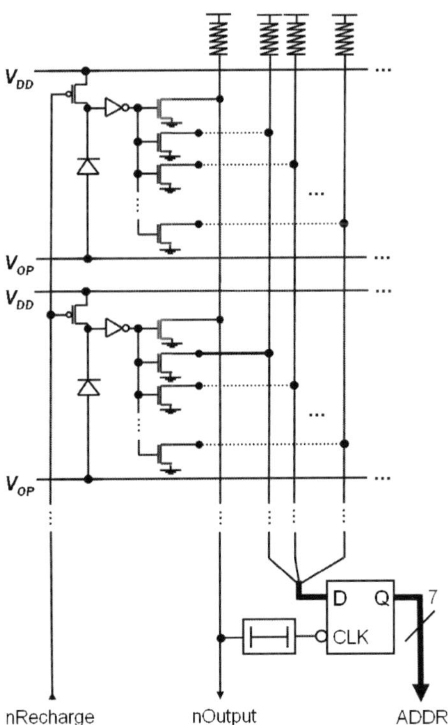

Fig. 3 Column readout circuit

this time, the output and address coding transistors of all pixels are blocked, thus leaving the lines of the column readout bus at V_{DD} through the pull up resistors. Upon photon detection within a pixel, the output transistor of that pixel pulls the output line down asserting nOutput in the end of the column. At the same time, the address coding transistors pulls the address lines that are connected for that particular pixel down, making the address of the pixel available on the input of a set of D-latches at the end of the column. Signal nOutput is additionally used to clock the D-latches through a delay line to hold the address of the pixel that detected the photon. The sensor readout circuitry, triggered by nOutput, reads out the address of the pixel (ADDR) that is stable until the next photon detection. In order to respect the setup time of the D-latches and to reduce the probability of address collisions when multiple photons are detected, the delay line is actively programmed to match precisely the setup time of the D-latches. When a second photon is detected by another pixel within the column, for instance few tens picoseconds later, nOutput is already asserted and the address of the previous pixel is already latched, therefore the second photon is simply missed. Due to the statistical properties of photons, this technique demonstrated to be very robust when light levels are moderate. When high light levels are required, ADDR can be coded using a conflict-free coding, though increasing the number of address lines and transistors per pixel.

Operating in time-correlated mode, e.g. using a time-to-digital converter, nOutput is used as trigger signal and it reflects the photon arrival with picosecond precision. Signal nRecharge can be continuously asserted in run mode to use the maximum of the column bandwidth or, on the other hand, it can be deasserted to leave the whole column in waiting state. Figure 4 plots a comparison between the signal counts of a typical pixel within a column and the corresponding signal count that the pixel would have with ideally parallel implementation. As can be seen, since the readout cycle time is limited by the SPAD dead time (i.e. 40 ns), a saturation effect appears when the signal count is increased.

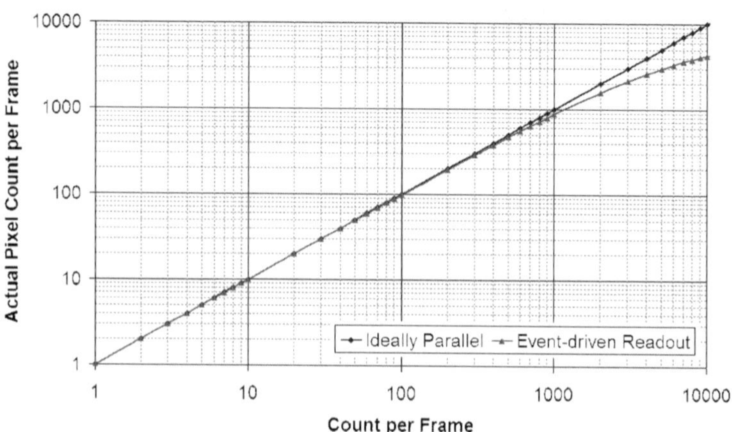

Fig. 4 Signal count for a typical pixel vs. an ideal parallel readout for a frame rate of 30fps

Fig. 5 Image sensor photomicrograph. Inset: pixel detail

A photomicrograph of the chip die is shown in Fig. 5, which also shows the layout of a pixel in the inset. The 11T digital pixel occupies a square area of $25{\times}25\,\mu$m. Each column is handled by a Column Control Unit (CCU). The CCU operates independently, thus an arbitrary number of columns N may be designed with a linear increase of complexity and power dissipation.

3.3 Camera Prototype

A camera prototype based on the proposed image sensor was built. The camera prototype was designed using an Altera Excalibur FPGA embedding an ARM processor core. Within the FPGA, banks of 16-bit counters were implemented and associated to each image sensor pixel. A Finite State Machine (FSM) generates and controls interface signals between the FPGA and the SPAD-based image sensor. Signal ADDR combined with a column address is used to access the corresponding counter to be incremented for each detected photon. In addition, the FSM handles the interface between the counters and the readout software running on the ARM processor.

Volatile and nonvolatile memories, communication links, power supply as well as optional measurement connectors were designed and implemented on the board level. Figure 6 shows the camera prototype. In order to use standard camera lenses, a c-mount thread was realized centered to the image sensor cavity.

4 Sensor Characterization

The suitability of DSM CMOS SPAD arrays is demonstrated through a variety of performance measures. DCR is log-plotted in Fig. 7, while a distribution of DCR across all the pixels is shown in the histogram of Fig. 8. Note that at room temperature 98.9% of the pixels exhibit a DCR below 10 Hz. To the best of our knowledge,

Fig. 6 Camera prototype

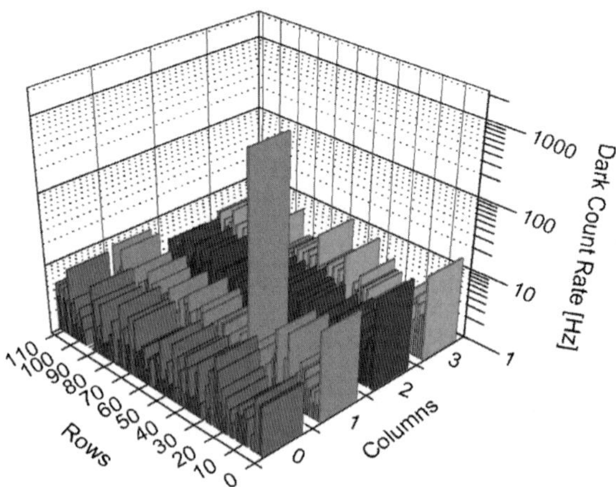

Fig. 7 Pixel map of dark count rate across the array

this is the best noise performance ever reported for SPADs implemented in any solid-state technology.

The sensitivity of the array is characterized in terms of its PDP as a function of photon wavelength and excess bias voltage V_e (plot of Fig. 9).

In this chip, the PDP varies from 0.1% to 5.3%. This performance measure was below expectations [16] due to reduced optical transparency in the passivation layers. The problem had been caused by an error in the mask generation. We estimate that the layers attenuate light by a factor of 2~5, thus reducing the expected PDP from a peak of 10~25% to the measured values. Nonetheless, note the wide range of wavelengths for which the SPAD is sensitive, covering NUV, visible light, and NIR.

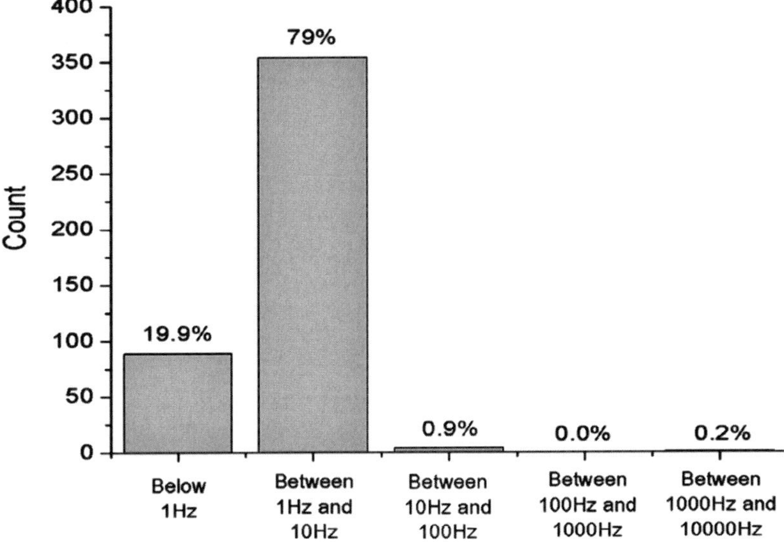

Fig. 8 Histogram of dark count rate

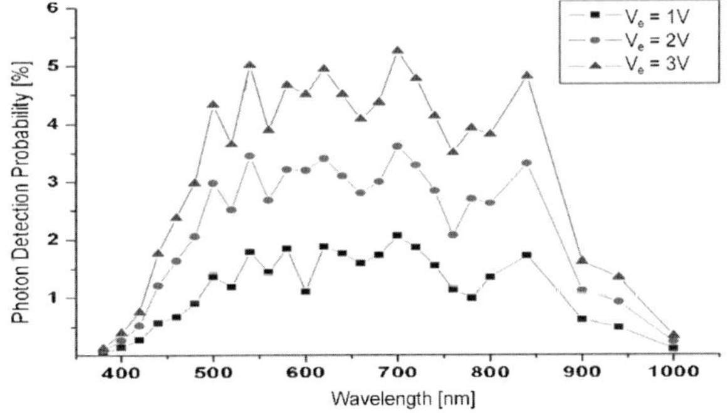

Fig. 9 Photon detection probability vs. photon wavelength for different values of excess bias voltage V_e

Other parameters completed the test, including pixel crosstalk, dead time, saturation, and image sensor power dissipation. Table 1 summarizes the overall performance of the imager.

5   Conclusion

We have reported on the first implementation of an integrated SPAD array in DSM technology. Due to a significant reduction in feature size, if compared to state-of-the-art technologies, it has been possible to design an array of 4×112 single photon

Table 1 Performance summary. All the measurements are reported at room temperature

Performance	Min.	Typ.	Max.	Unit
Pitch		25		μm
Array size		4×112		-
Avg. Dark Count Rate (DCR)		6		Hz
Dead Time (DT)			40	ns
Saturation			25	MHz
Wavelength range	350		1,000	nm
PDP	0.1		5.3	%
Power dissipation			7	mW
Overall crosstalk @ 30 μm		0.05		%

detectors. Thanks to a pitch of 25 μm, a lateral dimension of less than 3 mm was achieved. Another major innovation of this design is the readout scheme that allows simultaneous and independent pixel access upon photon arrival. The system is fully scalable, thus making a further expansion of SPAD-based multi-pixel sensors realistic. Preliminary measurements suggest that DSM technology is suitable to yield SPADs with equivalent or better performance than current implementations. Thus, applications, where sensitivity and timing accuracy are critical, will significantly benefit from this technology.

Acknowledgments This research was supported by a grant of the Swiss National Science Foundation. The authors are grateful to Zhen Xiao and Pierre-André Besse for helpful discussions and to Radivoje Popovic for continued encouragement.

References

[1] H. Eltoukhy, K. Salama, A. El Gamal, M. Ronaghi, R. Davis, A 0.18 μm CMOS 10^{-6} lux Bioluminescence Detection System-on-chip, *IEEE ISSCC*, 222–223, Feb. 2004.

[2] A. Grinvald et al. *In-Vivo* Optical Imaging of Cortical Architecture and Dynamics, Modern Techniques in Neuroscience Research, U. Windhorst and H. Johansson (Eds), Springer, 2001.

[3] S. Nagasawa, H. Arai, R. Kanzaki, and I. Shimoyama, Integrated Multi-Functional Probe for Active Measurements in a Single Neural Cell, *IEEE Intl. Conference on Solid-State Sensors, Actuators, and Microsystems*, Vol. 2, 1230–1233, Jun. 2005.

[4] J. C. Jackson et al. Characterization of Geiger Mode Avalanche Photodiodes for Fluorescence Decay Measurements, *Proc. SPIE*, Vol. 4650–07, Photonics West, San Jose, CA, Jan. 2002.

[5] A. V. Agronskaia, L. Tertoolen, H. C. Gerritsen, Fast Fluorescence Lifetime Imaging of Calcium in Living Cells, *Journal of Biomedical Optics*, Vol. 9, N. 6, 1230–1237, Nov./ Dec. 2004.

[6] P. Schwille, U. Haupts, S. Maiti, W. W. Webb, Molecular Dynamics in Living Cells Observed by Fluorescence Correlation Spectroscopy with One- and Two-Photon Excitation, *Biophysics Journal*, Vol. 77, 2251–2265, 1999.

[7] W. Becker, K. Benndorf, A. Bergmann, C. Biskup, K. König, U. Tirplapur, and T. Zimmer, FRET Measurements by TCSPC Laser Scanning Microscopy, *Proc. SPIE* 4431, ECBO, 2001.

[8] N. Chen, Q. Zhu, Time-resolved optical measurements with spread-spectrum excitation, *Optics Letters*, 1806–1808, Vol. 27, N. 20, Oct. 2002.

[9] P. Herman et al. Frequency-domain fluorescence microscopy with the LED as a light source, *Journal of Microscopy*, 176–181, Vol. 203, Pt. 2, Aug. 2001.

[10] J. Qu et al. Development of a Multispectral Multiphoton Fluorescence Lifetime Imaging Microscopy System Using a Streak Camera, *Proc. SPIE*, Vol. 5630, pp. 510–516, Jan. 2005.

[11] E. Charbon, Will CMOS Imagers Ever Need Ultra-High Speed? *IEEE Intl Conference on Solid-State and Integrated-Circuit Technology*, pp. 1975–1980, Oct. 2004.

[12] T. G. Etoh et al. An Image Sensor Which Captures 100 Consecutive Frames at 1,000,000 Frames/s, *IEEE Trans. Electron Devices*, Vol. 50, N. 1, 144–151, Jan. 2003.

[13] J. McPhate et al. Noiseless Kilohertz-frame-rate Imaging Detector based on Microchannel Plates Readout with Medipix2 CMOS Pixel Chip, *Proc. SPIE*, Vol. 5881, 88–97, 2004.

[14] A. Rochas, Single Photon Avalanche Diodes in CMOS Technology, Ph.D. Thesis, Lausanne, 2003.

[15] C. Niclass and E. Charbon, A Single Photon Detector Array with 64×64 Resolution and Millimetric Depth Accuracy for 3D Imaging, *IEEE Intl. Solid-State Circuit Conference*, 364–365. Feb. 2005.

[16] C. Niclass A. Rochas, P.A. Besse, and E. Charbon, Design and Characterization of a CMOS 3-D Image Sensor Based on Single Photon Avalanche Diodes, *IEEE Journal of Solid-State Circuits*, Vol. 40, N. 9, Sep. 2005.

[17] M. Gösch et al. Parallel Single Molecule Detection with Fully Integrated Single Photon 2×2 CMOS Detector Array, *Journal of Biomedical Optics*, Vol. 9, N. 5, 2004.

[18] C. Niclass et al. CMOS Imager Based on Single Photon Avalanche Diodes, *IEEE Transducers*, June 2005.

[19] F. Stellari et al. Testing and Diagnostics of CMOS Circuits Using Light Emission from Off-State Leakage Current, *IEEE Trans. Electron Devices*, Vol. 51, N. 9, 1455–1462, 2004.

[20] C. Petersen, *Personal communication*, Brain & Mind Institute (EPFL), 2005.

[21] R. H. Haitz, Studies on optical coupling between silicon p-n junctions, *Solid State Electronics*, Vol. 8, 417–425, 1965.

Part VI
Test and Verification

The Test and Verification Influential Papers in the 10 Years of DATE

T. W. Williams and R. Kapur

Abstract The history of test is looked at for IC Test to put in context the papers that are being highlighted at DATE. This paper focuses on the path that led to the adoption of structural test. ATE related historical aspects are not mentioned.

1 Introduction

Test technology has come a long way since the first few papers published on stuck-at faults, D algorithm, and Scan. Each paper has represented an incremental change to the technology that it makes it hard to recognize seminal work when it happens. Looking back we can identify papers that actually represent significant contributions to the technological mutation.

In reality the industry is still doing scan that was defined many years ago. The ATPG algorithms are fundamentally the same and the stuck-at faults are still the most popular failure mechanisms being modeled. Looking back scan has changed in many different ways. Some of the significant events in the change were:

- Scan was automated and made part of the synthesis process resulting in the adoption of DFT methods.
- Scan chains were used as part of a self-test solution.
- Scan chains are created based upon layout.
- Scan chains are aware of power aware methods used in the design.
- Scan compression technology was created.

Each of these concepts came about through papers published at the premier conferences in test.

Similarly, verification technology has changed over the years. There have been numerous papers on Binary Decision Diagrams, SAT solvers and improvements in simulation based verification methods. Verification represents the checking of the design before manufacturing as opposed to the test which verifies the integrity of the IC after manufacturing.

Synopsys Inc.

R. Lauwereins and J. Madsen (eds.), *Design, Automation, and Test in Europe–The Most Influential Papers of 10 Years DATE*, 417–422 417

2   Beginning Technologies

Right from IC number 1, test has played a critical role in the delivery of a successful chip. IC-manufacturing has never been at a stage where verification was not required.

Functional vectors were the initial source of tests. Very quickly the industry realized that some predictable alternative was required. This led into initial research on Scan [3] which was invented simultaneously at multiple locations across the globe. Scan gave access to internal locations of the design. This was not only used to test ICs, it was used to debug difficult problems with the non-working ICs.

Primarily two flavors of scan developed based upon the clocking schemes used. Edge triggered scan chains were the more popular style of the time. Level Sensitive Scan Designs were primarily used in IBM which was the only full scan house for a while. The adoption of scan happened over four decades with evangelists, conferences, and test problems providing the catalyst for adoption.

With scan design came along the use of ATPG-FS methods [2] that automated the test creation process. The D-Algorithm, PODEM, and Socretes represent key research outcomes that paved the way for successful scan based structural test. Test automation gained with the changes in compute power. Today, designs are handled flat with high stuck-at coverage assumed for any given ATPG-FS run.

3   Maturing Research

As scan developed, new areas were explored. The significant areas where work was done included:

1. Testability measures
2. Partial scan
3. Test economics

Testability measures provided the ability to classify faults as easy to detect and hard to detect. Using testability measure technology test points were determined, estimates of the detectability of faults were made, pattern counts were predicted or the overall efficiency of ATPG-FS was improved [1].

Partial scan approaches made scan attractive to the general designer by not adding scan in timing sensitive logic. As a result the negative impact of adding a multiplexer in the functional path was avoided.

A whole area of research covered topics such as Defect Level estimation [4, 10], or the area impact of scan. These topics are all unified under the umbrella of test economics. Research in this area allowed for us to answer fundamental questions on, how much testing is enough?

4 New Test Methods

Starting from stuck-at faults, other fault models were researched in the quest to cover all the defect mechanisms in the ICs. As a result a number of new test methods were published:

1. Stuck-open faults
2. Bridging faults
3. Pseudo-stuck-at faults (for Iddq)
4. Transition faults
5. Small delay faults
6. Gate exhaustive faults

A number of experiments have been performed to find the most effective recipe to detect the defects. Today the industry is relying on stuck-at and transition faults as the primary target for test generation.

While faults represent most of the test methods invented, there have been changes in the way tests are applied. Iddq and Logic-BIST represent such advances made in IC testing.

5 Scan Compression

In the early 2000s starting with Mentor Graphics – Test Kompress [7], a buzz was created around logic added in the scan path such that test data can be supplied from a small interface to a large number of scan chains. While the technology relied on concepts known before, such as logic BIST and reseeding, this was a dramatic change in the way IC-test is done which changes scan. Scan from becoming a method to apply the tests has become a backup method in place if test compression technology does not provide adequate results.

Following Test Kompress many more competitive technologies have been developed—some purely combinational [8, 9], and some sequential. Each has their own characteristics and trade-off points.

6 Target Fault Models: Back to the Future

The de facto standard fault model has been the Stuck at Fault model. However the industry drivers were looking for even better quality. The fact that so many designs are being synthesized results in paths having small slacks. With most paths being critical designs are more sensitive to the very more likely to occur, small delay defects [12]. Having scan clearly helped attain the high Stuck-at Fault coverage. Once high stuck-at coverage was achieved the drive for higher quality began. This brought transition fault testing in the forefront. However, it was shown in 1988 that transition fault testing was not very effective if one did not pay any attention to the path in which the transition fault was tested [10–12]. This notion gave rise to a new direction in how one looked at transition fault testing. These ideas lay dormant until 2003 when it was independently rediscovered. Today there is push by many in industry to test transition fault that are directed at the most likely delay defects, small delay defects.

7 Do Test Methods Die?

As improvements are made some things which were very useful at some point become less useful. Test technology has its own share of such technologies.

Iddq testing is one such technology that got outdated because of the underlying changes in silicon. As the leakage current of a good IC increased the delta increment of a defective IC's current has become insignificant enough for Iddq testing to become ineffective [5, 6].

The test industry has always struggled with tests generated for different fault models. While the Venn Diagram analysis of the effectiveness of different tests has yet to lead to concrete decisions on the utility of test methods, there is always the chance that going forward some test method is found redundant.

8 10 Years of Test and Verification

The last 10 years of test have seen leaps in core-based test technology and scan compression. In these years formal verification has seen significant adoption in the industry. As a result it should come to no surprise that the better papers at DATE address technology in these areas.

In 2000 a forward-looking paper by Lorena Angel and Michael Nicolaidis look at ways to deal with an efficient method of detecting temporal and soft errors which is seen to be the predominant failure mechanism in the future [13]. This is important to our industry since the number of soft errors is increasing with every technology.

In 2001 a paper by Ashish Giani, Shuo Shengy, Michael S. Hsiao, and Vishwani D. Agrawal take a new look at doing sequential Automatic Test Pattern Generation, ATPG [15]. Their approach is to look at the frequency spectrum of patterns which cause faults to get excited and tested. Using this information they go on to devolve an efficient fault dropping technique to speed up the process. Today there is more and more attention being given to this method of testing.

A paper in 2002 by Gonciari and Bashir M Al-Hashimi and Nicola Nicolici, addresses improvements in scan compression for system-on-chip designs. With a new variable length Huffman Coding method significant gains are achieved in Test Application Time and Test Data Volume [17]. Another paper by Larsson and Peng builds a framework for system-on-chip test to optimize the testing process by smart scheduling algorithms [14].

While the test industry struggled with scan compression and core based test, the seminal work for the future is being put in place by some researchers. Bernardi et. al. addressed issues in diagnosis which builds into the future of test playing a strong role in the yield improvement area [18].

Design Verification algorithms have used ATPG like algorithms as one of their engines to verify the integrity of the design. Goldberg and Novikov worked on BerkMin to introduce a new decision making procedure and methods for clause database management [16].

References

[1] M. J. Y. Williams and J. B. Angell. Enhancing Testability of Large Scale Integrated Circuits via Test Points and Additional Logic, IEEE Transaction on Computers, C-22, 46–60, 1973.

[2] S. Funatsu, N. Wakatsuki, and T. Arima, Test Generation System in Japan, *Proc. 12th Ann. Design Automation Conference*, pp. 23–25, 1975.

[3] E. B. Eichelberger and T. W. Williams A Logic Design Structure for LSI Testability, *Proc. 14th Design Automation Conference*, New Orleans, LA, pp. 462–468, June 1977.

[4] T. W. Williams and N. C. Brown, Defect Level as a Function of Fault Coverage, IEEE Transactions on Computers, Vol. C-30, No. 12, Dec. 1981, 987–988.

[5] T. W. Williams, R. H. Dennard, R. Kapur, M. R. Mercer, and W. Maley, Iddq Testing for High Performance CMOS - The Next Ten Years, *Proc. 1996 European Design and Test Conference*, Paris, France, pp. 578–583, March 1996.

[6] T. W. Williams, R. H. Dennard, R. Kapur, M. R. Mercer, and W. Maley, Iddq Test: Sensitivity Analysis to Scaling, in *Proc. 27th International Test Conference*, Washington DC, 1996 pp. 786–792.

[7] J. Rajski, J. Tyszer, M. Kassab, N. Mukherjee, R. Thompson, Kun-Han Tsai, A. Hertwig, N. Tamarapalli, G. Mrugalski, G. Eide, and Jun Qian, Embedded Deterministic Test for Low Cost Manufacturing test Test Conference, 2002. *Proc. International* 7–10 Oct. 2002 pp.(s):301–310.

[8] S. Samaranayake, E. Gizdarski, N. Sitchinava, F. Neuveux, R. Kapur, and T. W. Williams, "A Reconfigurable Shared Scan-In Architecture," *Proc. 21st IEEE VLSI Test Symposium*, 2003, pp. 9–14.

[9] N. Sitchinava, S. Samaranayake, R. Kapur, E. Gizdarski, F. Neuveux, and T. W. Williams' Changing the Scan Enable During Shift, *Proc. 22ed IEEE VLSI Test Symposium*, 2004, pp. 73–78.

[10] E. S. Park, M. R. Mercer and T. W. Williams, Statistical Delay Fault Coverage and Defect Level for Delay Faults, (with E. S. Park and M. R. Mercer), *Proc. 1988 International Test Conference*, Washington, DC, pp. 492–499, September 1988.

[11] T. W. Williams, Underwood, and M. R. Mercer, The Interdependence Between Delay-Optimization of Synthesized Networks and Testing, *Proc. 28th ACM/IEEE Design Automation Conference*, San Francisco, CA, pp. 87–92, June 1991.

[12] T. W. Williams, E. S. Park, B. Underwood, and M. R. Mercer "Delay Testing Quality in Timing-Optimized Designs,", *Proc. 1991 IEEE International Test Conference*, Nashville, TN, October 1991. pp. 897–905.

[13] L. Anghel and M. Nicolaidis, Cost Reduction and Evaluation of a Temporary Faults Detecting Technique, DATE 2000.

[14] E Larsson, Z Peng, An Integrated System-On-Chip Test Framework, DATE 2001.

[15] A Giani, S Sheng and M S Hsiao, Efficient Spectral Techniques for Sequential ATPG, DATE 2001.

[16] E. Goldberg and Y. Novikov, BerkMin: A Fast and Robust Sat-Solver, DATE 2002.

[17] P. Gonciari, B. Al-Hashimi, and N. Nicolici, Improving Compression Ratio, Area Overhead, and Test Application Time for System-on-a-Chip Test Data Compression/Decompression, DATE 2002.

[18] P. Bernardi, E. Sanchez, M. Schillaci, G. Squillero, and M. Sonza Reorda, An Effective Technique for Minimizing the Cost of Processor Software-Based Diagnosis in SoCs, DATE 2006.

Cost Reduction and Evaluation of a Temporary Faults-Detecting Technique

Lorena Anghel and Michael Nicolaidis

Abstract IC technologies are approaching the ultimate limits of silicon in terms of channel width, power supply and speed. By approaching these limits, circuits are becoming increasingly sensitive to noise, which will result on unacceptable rates of soft-errors. Furthermore, defect behavior is becoming increasingly complex resulting on increasing number of timing faults that can escape detection by fabrication testing. Thus, fault tolerant techniques will become necessary even for commodity applications. This work considers the implementation and improvements of a new soft error and timing error detecting technique based on time redundancy. Arithmetic circuits were used as test vehicle to validate the approach. Simulations and performance evaluations of the proposed detection technique were made using time and logic simulators. The obtained results show that detection of such temporal faults can be achieved by means of meaningful hardware and performance cost.

1 Introduction

Modern electronic circuits are highly complex systems and, as such, are prone to occasional errors or failures. Device size reduction, increased operating frequency and power supply reduction that accompany the process of very deep submicron scaling, reduce noise margins, and IC reliability, but also increase the impact and complicate the behavior of defects. This makes increasingly difficult to achieve acceptable reliability levels for future ICs and maintain acceptable cost and quality for IC testing. It seems that the most significant problem concerns the sensitivity of future IC generations face to various noise sources, and in particularly face to energetic particles. Such particles are for instance, cosmic neutrons produced by the sun activity. It is predicted that the energy and flow of these particles are sufficient for creating unacceptable soft-error rates in future IC generations, even at ground level. Alpha particles produced by the disintegration of radioactive isotopes contained in the material of electronic systems are becoming another cause of increasing soft-error rates in these technologies.

Tima, France

R. Lauwereins and J. Madsen (eds.), *Design, Automation, and Test in Europe–The Most Influential Papers of 10 Years DATE*, 423–438
© Springer 2008

One basic reason for increased sensitivity to noise is the reduction of V_{DD} voltage and geometry shrinking which implies the reduction of node capacitance. Thus, the charge stored on a circuit node ($Q = V_{DD} *$ Cnode) is drastically reduced. In fact, a significantly lower charge needs to be deposed by a particle strike in order to reverse the logic value of a node.

In memory cells, a particle striking a node of the cell produces a transient pulse that can be captured by the asynchronous loop forming the cell, resulting on a soft-error. Traditionally, only memories were protected against SEUs (single event upset) by means of error detecting and correcting codes. Deeper submicron scaling increases the sensitivity of logic networks too. A transient pulse created by a particle strike on a node of the network can be propagated to the network outputs. However, the transient pulse is often attenuated before reaching these outputs [BAZ 97]. In addition, reaching the network outputs does not necessary imply the occurrence of a soft-error. This will happen only if the pulse reaches an output at the time the latching edge of the clock is occurring. In this case, the erroneous value is stored in the latch connected to the output of the circuit.

Transient pulses wider than the logic transition time of a gate will propagate without attenuation and can affect the correct operation of the circuit. Transient pulses induced by particle strikes have a width of few hundreds of picoseconds (the exact value depends on circuit topology and particle energies). Thus, their duration is becoming higher than the transition time of gates in deep submicron technologies. Therefore, transient pulses are not attenuated even for relatively low-energy particles. In addition, as the clock frequencies are increasing significantly, the probability of latching a transient pulse is increased by the same factor. In fact the more frequent are the latching edges of the clock, the higher is the probability to have a transient pulse coinciding with a latching edge. Due to these trends, error rates in logic parts risk to become as high as error rates in memories.

Another significant problem concerns timing errors. Such errors are gaining importance in very deep submicron and nanometer technologies. Process parameter variation, various defect types (shorts, opens, etc.) often affect circuit speed. These faults are increasing the path delays and may result on timing errors unacceptable in the future high-frequency integrated circuits. In addition, these timing errors require very complex test conditions. The huge number of paths of modern circuits, together with other timing critical conditions such as cross talk, or ground bounce, make ATPG for timing faults computationally unfeasible, and test length unrealistic. It is therefore unavoidable that a significant number of circuits with timing faults will pass fabrication tests.

In this context, it will become mandatory to design the future ICs to be tolerant for soft errors and timing faults. Because this protection will be needed for any product, including commodity ones, traditional fault tolerant design such as TMR can not be used due to its high cost. The most economic solution is to use a concurrent checking scheme combined with a retry procedure. With such a scheme, soft-error tolerance can be achieved by a retry operation after each error-detection. The same principle will be used for timing error tolerance, but the clock cycle has to be reduced during the retry procedure. One possible concurrent checking scheme is based on self-checking design. In some situations this approach may offer low-cost concurrent error detection (e.g. self-checking multipliers using arithmetic codes [PET 58, PET 72] [AVI 73, ALZ 99]). In other circuit cases, self-checking design may require high hardware cost, for instance in the case of arbitrary logic functions.

Several other design solutions for achieving soft-error tolerance were proposed in [NIC 99]. The idea is to take advantage of the temporal nature of transient faults, and achieve tolerance of these faults by using time redundancy. This should lead on a significant reduction of hardware cost. Among these solutions, in the present work we have considered a soft-error detecting scheme that can be implemented without any performance penalty, and can also be used to detect timing errors. In order to evaluate the scheme we have used several types of multipliers and adders as test vehicles.

The scheme of soft-error detection is described in Section 2. Evaluations of speed degradation and area overhead are shown in Section 5. Simulations of transient faults have been performed in order to validate the fault tolerance merits of the design and are presented in Section 7.

2   Concurrent Checking Based on Time Redundancy

Since the transient faults are manifested for a limited duration of time, one can design circuits such that the correct values are present on the circuit outputs for a time duration greater than the duration of the transient fault. A time-domain majority voter can be used to determine the value of the circuit output for a majority of time. Various transient-fault tolerant implementations of this principle were proposed in [NIC 99]. However, these schemes affect the performance of the circuit, so they can be used without performance penalty only for blocks that are not in the critical path of an IC. A technique performing transient error detection by means of time redundancy is also proposed in the same paper. This technique is considered here because it avoids performance penalty.

Figure 1 shows the simplified scheme of the detection technique. The combinational circuit is monitored by a circuit able to detect the occurrence of soft-errors produced by a transient fault affecting the combinational circuit. The detection circuit of Fig. 1a is composed of two latches per circuit output and one comparator. The input of one latch is coming directly from the combinational circuit output, while the input of the second latch is coming from the same output, but is delayed by δ = Dtr + Dsetup (where Dtr is the maximum transient pulse duration that has to be tolerated, and Dsetup is the setup time of the latches). The latching edge of Ck should occur at t_0 + Dtr + Dsetup, where t_0 is the instance that the latching edge occurs in a standard circuit design. It is obvious that the delay δ introduces a reduction of the circuit speed.

An improved alternative is shown in Fig. 1b. At instant t_0 the clock Ck captures the output of the combinational circuit in Output-Latch.

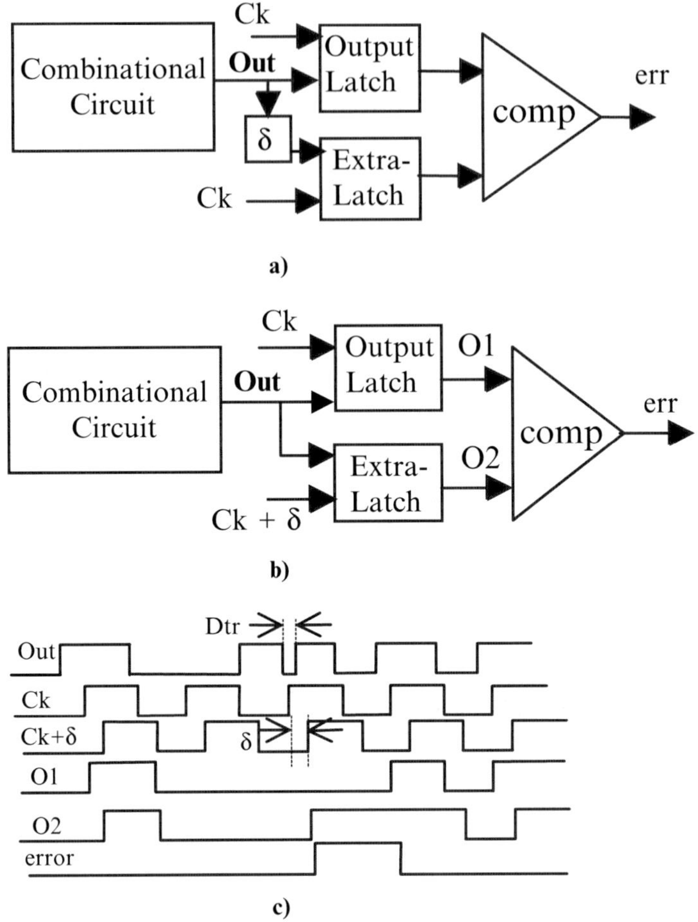

Fig. 1 (a, b) Transient fault detection implementation using a latch and a comparator; 1c Transient fault detection principle

A second clock having its latching edge shifted by δ latches this output in Extra-Latch at instant $t_0 + \delta$. A comparator indicates an error occurrence by a pass/fail signal. The designer determines the maximum transient pulse duration Dtr that must be tolerated in order to achieve an acceptable level of soft-error rates, and sets $\delta = $ Dtr + Dsetup.

From this construction, it is obvious that no transient pulse of a duration lower than Dtr can be captured by both Output-Latch and Extra-Latch (Fig 1c). Thus, the comparator detects any transient of duration lower than Dtr that could result on a latched erroneous value.

Due to the shifted clock of Extra-Latch, this scheme can also involve a performance penalty. However, a careful analysis allow us to avoid this penalty. In fact, the output results of the combinational circuit are latched in Output-Latch at t_0. These results can be sent to their destination without added delay. But, applying new inputs on the combinational circuit at the normal operation speed is not yet possible, since it may affect the value captures by Extra-Latch at $t_0+\delta$. This drawback can be eliminated if we exploit the delays of the combinational circuit. Let us consider the following situations:

1. The propagation delay Dmin, of the shortest path of the circuit respects the relationship Dmin $> \delta$. Since no circuit output is affected by the new input values before t_0 + Dmin, then the circuit of Fig. 1b can operate at the same clock frequency as the conventional design without affecting the values latched by Extra-Latch.
2. Dmin $< \delta$. In this case, one can add delay elements in the shortest paths (use gates with higher delays or add inverters and/or buffers) to meet condition Dmin $> \delta$.

This scheme requires a low hardware cost corresponding to some delay elements, an Extra-Latch per circuit output and a comparator (note that Output-Latch is also the functional latch of the circuit). In addition, no performance penalty occurs.

Finally, it is worth to note that the scheme of Fig. 1b can also be used to detect timing errors, as far as the extra delay do not exceed δ. However, when a retry procedure is activated to correct a possible soft-error, the timing error can be produced again. In order to cope with this situation, one can implement a clock control circuit (driven by the error detection signal). This circuit will increase the clock cycle by duration higher or equal to δ. After the clock-cycle increase, the retry procedure is activated again to correct the error.

The shifted clock (Ck+δ used in Fig. 1b) can be obtained from Ck by using delay elements. Another possibility is to fix the same δ for all the blocks of an IC and use a second clock distribution network to bring Ck+δ. To ensure a precise value for δ, careful implementation is needed for delay blocks, or for the clock distribution network, in order to keep the clock skews within acceptable values. A more convenient solution consists on using the rising edge of Ck as the latching edge of Output-Latch and the falling edge of Ck as the latching edge of the Extra-Latch. This solution requires the modification of the duty cycle of Ck, so that the high level of the clock must have duration $T_1=\delta$. However, if $T_1 > \delta$, it is also possible to leave unchanged the duty cycle of Ck, but we need to introduce a delay equal to $T_1-\delta$ on the input of Extra-Latch (see Fig. 2).

Thus, while the latching edge of Extra-Latch occurs at $t_0 + T_1$, the captured value is the value presented on Output at the instant $t_0 + \delta$. This way we can use a single-clock distribution network.

Another technique allowing concurrent checking of temporary faults is presented in [FRA 94]. This technique is proposed for timing faults, but it can detect transient faults as well. It uses a transition detector to monitor the signals to be checked. Transitions occurring during the steady state phase of the monitored signals due to timing faults or transient faults are thus detected. Due to this difference, the technique proposed in [FRA 94] can not detect soft errors produced by particles striking the nodes of the Output-Latch. In addition, the implementation of the transition detector can not be done by means of standard logic design techniques.

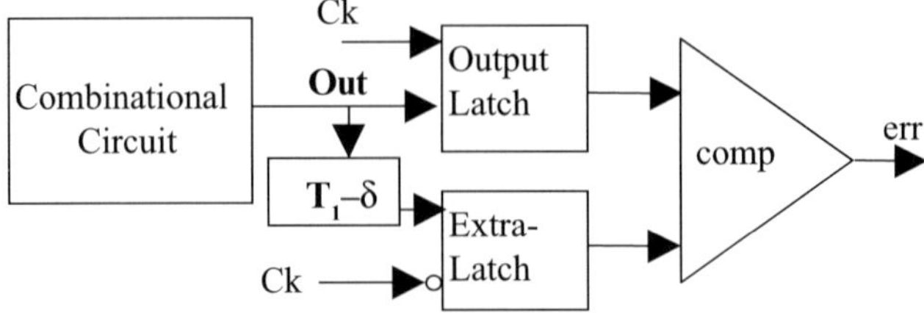

Fig. 2 Clock network simplification

The detection of signal transition is also exploited in [MET 98] for checking timing faults and soft errors. This paper uses an improved transition detector able to detect its own faults.

3   Fault Modeling

The fault modeling includes transients faults externally generated which affect a single node at a time. When a particle (e.g. heavy ion) strikes a semiconductor material, it creates electron-hole pairs in its track. If this particle goes into a depletion region, the electrons and holes created will drift according to the electric field in the depletion region. This movement of electrons and holes in opposite direction causes charges to be collected on the opposite sides of the depletion region. Typically the duration of such pulses is a few hundreds picoseconds.

Simulations considering this family of pulses are necessary if we know the exact shape of the transient pulse for a particular particle strike and we want to determine the exact shape of the circuit response. However, if we look for an average evaluation of a detection technique that has to work for various pulse shapes and duration, exact pulse shape is not of importance. In the present analysis, we considered square pulse shape. Duration of several hundreds of picoseconds was considered for the injected pulses. The evaluation of the present technique requires a good accuracy concerning the timing characteristics of the pulse reaching the circuit latches (width of the pulse and instance that the pulse reaches the latch). Electrical simulations (e.g. SPICE3) are very slow on large circuits and we can not use this kind of tool to perform several hundreds of thousands of simulations at it was done in the evaluation. Thus, we have used the Verilog-XL logic simulator that offers also the necessary timing accuracy. A more dedicated simulator for transient faults could also be used for these experiments [CHA 93]. Transient pulses of duration up to 1ns have been used to validate the principle. This duration is much higher than the practical transient pulses induced by particle strikes.

As concerning injection of timing faults, we added delays within the circuit paths. Delays up to 20% of the maximum delay of the circuit were considered. Obviously this model does not represents the complex mechanisms creating timing errors in ICs, but again, the efficiency of the detection scheme does not depend on the timing fault mechanism, but on its effects (extra delay on output value).

4 Circuit Samples

To perform our experiments and validate the approach, we considered various multiplier and fast adder structures. Brown multipliers are unbalanced structures having a wide dispersion of propagation delays. For large operand sizes, the signal delays make this kind of multiplier unacceptable in practical applications, so multiplier structures with logarithmic delays were also investigated (Wallace trees, Booth encoding structures). To further improve the speed, various fast adder structures were used for the carry propagate adder forming the last stage of the multiplier. Booth multipliers using Wallace trees and fast carry propagate adders are the faster and most balanced structures.

The experimented adder structures include carry lookahead adders using CLU units [HWA 79, WES 94], as well as Brent and Kung [BRE 82], Kogge and Stone [KOG 73], Sklanski [SKL 60], and Han and Carlson [HAN 87] adders using Brent and Kung cells in various cascading structures. These adders present optimal trade-offs on speed, area and fan-out.

Each multiplier and adder was configured as in Fig. 1b. The clock frequency was established according to the critical path of the functional circuit. That is, the circuit is operating at its maximum speed, as in the case of a design not using fault tolerant mechanisms.

These structures were considered for evaluating the hardware cost of the technique. However, concerning softerror detection evaluation, we have taken into account examples of adders and multipliers of 8×8 size only, due to the large number of simulations to be performed.

5 Experimental Results

To avoid performance lost, we have implemented only the technique described in Fig. 1b, that is, the circuit must be constrained to have Dmin > δ. This was done by performing constrained circuit synthesis using Synopsys optimization and timing analyzing tools. The multipliers and adders were synthesized by using $0.35\,\mu$m AMS technology ($3.3\,$V), which was available to us.

Table 1 shows the results of these experiments over various multiplier structures (see column 1). The experiments were performed for values of δ equal to 0.46, 0.6, 0.8, and 1. We have also considered the case $\delta = 20\%$Dmax, which can be used to detect timing errors of a delay up to 20% of the maximum delay of the circuit. Table 1 shows the critical path delay (CP) for each multiplier and the area overhead. The first component of the area overhead (PAO) corresponds to the cost required for achieving Dmin > δ. We remark that the area overhead is low, especially when considering the multipliers of practical interest such as Wallace and Booth-Wallace. The second component of the overhead columns (TAO) shows the total area overhead (overhead for Dmin > δ and overhead for the Extra-Latches and the comparator). For δ values up to $1\,$ns (which largely exceed practical width of transient pulses produced by particle strikes), the area overhead is within low

Table 1 Area overhead results for different sizes and structures of multiplier

	CP (ns)	$\delta = 0.46$ (ns)	$\delta = 0.6$ (ns)	$\delta = 0.8$ (ns)	$\delta = 1$ (ns)	20% Dmax	20% Dmax improved
				PAO(%)/TAO(%)			
Brown 16	32	0.2/4.3	1.2/5.3	3.2/7.3	4.5/8.5	44.2/46.8	16.3/19.5
Brown 32	66	0.4/2.4	0.5/2.5	1.5/3.5	2.6/4.6	18.2/19.8	6.9/8.68
Wallace 8	7	0.9/7.5	3.1/9.5	3.4/9.9	3.5/10	10.8/17.3	5.7/11.4
Wallace 16	11	0.1/3.5	0.2/3.6	0.4/3.8	0.5/4	15.3/18.3	6.2/9.2
Wallace 32	18	.05/1.3	.09/1.4	.15/1.5	0.2/1.5	3.5/4.8	0.06/1.3
Booth-Wall 16	11	0.03/4	0.7/4.7	1.1/5.2	1.2/5.3	17.9/21.3	4.36/8
Booth-Wall 32	16	.09/2.3	.12/2.3	.16/2.3	0.2/2.4	3.2/5.4	0.1/2.26
Booth 8	8	0.3/6.5	1/7.2	1.2/7.5	2.4/8.8	19.7/25.3	4.2/9.6
Booth 16	14	0.2/3.9	0.3/4	0.4/4.2	0.9/4.8	39.5/42	12.6/15.1
Booth 32	20	0.1/2.6	.16/2.7	.2/2.78	0.3/2.9	3/5.62	0.09/2.6

values and becomes insignificant as the multiplier size increase. Of course these results concern a 0.35 µm process but indicate that the technique is of low cost. For $\delta = 20\%$ Dmax, used for timing fault detection, the area overhead is again low, except for Brown multipliers. We will see later how we can reduce this cost to obtain the results of the last column.

Let us further analyze the technique by considering the two components of the area overhead. The first component includes an Extra-Latch per output, plus the cost of the omparator (one XOR gate and one OR gates per output). The percent overhead due to this component is determined by the (area per output) *APO* ratio = *(circuit area)/(# outputs)*. The higher is *APO* ratio, the lower is the overhead. The second component is the area required to implement the constraint Dmin > δ (PAO). For circuits with balanced delays, this cost will be low. However, for those circuits where a large number of paths have much lower delays than the critical path, the cost may become high. Although modern synthesis tools, used to reduce critical path delays, trend to generate circuits with balanced delays, there should be still some cases of unbalanced circuits.

Multipliers result on low hardware cost, because they have high *APO* ratio and are usually delay-balanced (especially the fast ones). For nonfavorable cases, we have experimented the technique over various adder structures. Adders have a low *APO* (area per output) ratio. In addition, there is a large amount of paths with low delay with respect to the critical path (e.g. the delay between an input Ai or Bi and the output Si is only two XOR gates for any i). Table 2 gives the results for various adder cases. We observe that the cost is much higher than in the case of multipliers, reducing the interest of using the for adders.

In the following section we propose some improvements of this technique in order to further reduce the hardware cost.

6   Cost Reduction

The proposed technique is detecting transient faults by comparing the values present on an output at two instances distant by δ. The same detection capability will be achieved if the second instance is $t_0-\delta$, as shown in Figure 3. In this figure, a part of the outputs uses the principle of figure 1b (clocks Ck and Ck+δ), and another part uses the modified principle (clocks Ck and Ck−δ). The second principle can be used only for outputs that are ready early enough (i.e. having their maximum delay lower than or equal to Dmax-δ). To reduce the number of clock signals, we can use the normal clock for the Extra-Latch 1, as shown in Fig. 2.

The normal clock Ck can also be used to latch the input of the Extra-Latch 2 at the same time as for the Output-Latch (1,2), but a delay δ must be added at the input of this latch. This way, the actual value latched in Extra-Latch 2 is the value present at Out 2 at time $t_0-\delta$.

Since the two latching instances of Out 2 differ by δ, the technique detects all transients with duration Dtr<δ-Dsetup. As concerning the detection of timing faults (delays lower than or equal to δ), we observe that the principle applied on Out 1 clearly detects them. As for Out 2, since its worst delay is less than Dmax-δ, by adding the faulty delay δ, we obtain a delay lower than Dmax. That is, the delay fault cannot affect the value used by the system (i.e. captured by Output-Latch 2).

The principle of Fig. 3 is interesting since it removes the constraint Dmin $> \delta$, from the outputs with delays not exceeding Dmax-δ (Out2), and thus, it can be used for cost reduction.

A second cost-reduction principle is shown in Fig. 4. This principle has to be used for circuits with low *APO* because in this case, the Extra-Latch may represent a significant overhead and the principle shown in Fig. 4 will result on significant cost reduction.

Figure 4 works as follows: consider the delay of the comparator to be Dcomp. The output value of the comparator is latched at time $t_0 + \delta$ + Dcomp. This value is the result of the comparison of the values present on the inputs of the comparator

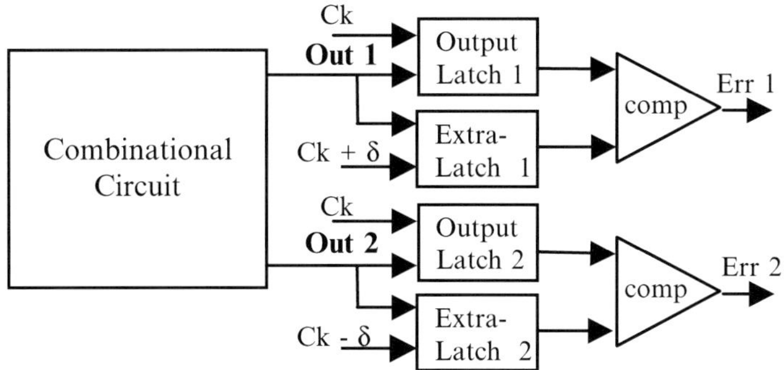

Fig. 3 Dmin < δ relaxation

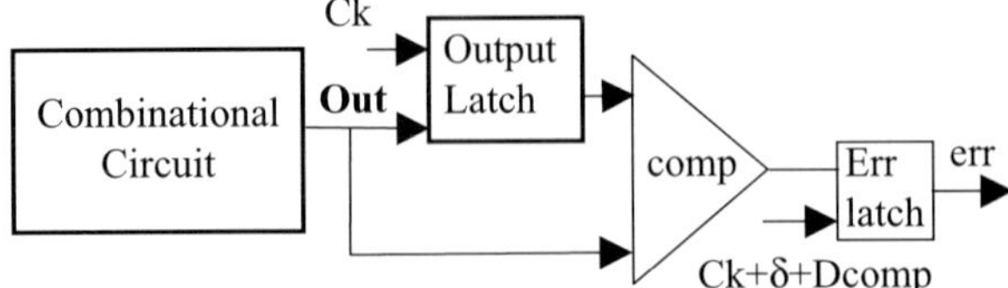

Fig. 4 Extra-Latch removal

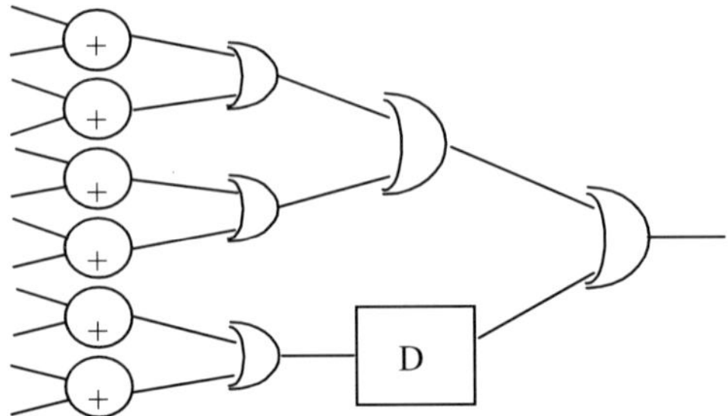

Fig. 5 Comparator with balanced delays

at time $t_0 + \delta$. These values are on the one hand, the content of Output-Latch, which is holding the value presented on Out at t_0, and on the other hand the values present on Out at time $t_0 + \delta$.

Thus, the scheme of Fig. 4 is equivalent to the scheme of Fig. 1b.

This scheme uses two clocks (Ck and Ck + δ + Dcomp). One can simplify it to use only one Ck for both Output-Latch and Error-Latch. In this case, the Error-Latch will be implemented to use the falling edge of the Ck as latching edge. In addition, a delay equal to T_1 - δ - Dcomp has to be added on the second input of the comparator, where T_1 is the duration of the high level of the Ck. One can also use the rising edge as latching edge of Error Latch. In this case, the delay added on the comparator input will be T- δ - Dcomp, with T the period of Ck.

Note that this scheme requires that all paths between the inputs and the output of the comparator have similar delays. This will happen when the number of its inputs is a power of 2. If this is not the case, delay elements have to be added. This is illustrated in Fig. 5, when the comparator receives 6 input pairs. An element of delay equal to an OR gate was added to balance the circuit.

The two cost reduction principles are then used to reduce the cost for the case of multipliers considering $\delta = 20\%$Dmax and for the case of adders. The last column of table 1 shows the obtained results for the multipliers applying the improved principle of Fig. 3. We remark that area cost is reduced for all multiplier cases. Tables 3 and 4 show respectively, the use of the principle of Figs. 3 and 4 in the case of adders.

It results a drastically reduction of the hardware overhead with respect to the results shown in Table 2. The best results are obtained in Table 4, excepting 14 adder cases (in gray in Table 3). The columns of practical interest are the ones

Table 2 Area overhead of different sizes and structures of adders according to the principle of Fig. 1b

		TAO(%)				
	CP (ns)	$\delta = 0.46$	$\delta = 0.6$	$\delta = 0.8$	$\delta = 1$	20% Dmax
Brent&Kung 8	3	39.25	48.54	59.54	62.42	48.54
Brent&Kung 16	4.2	34.99	45.46	52.43	58.10	53.4
Brent&Kung 32	5.1	32.55	42.54	50.03	55.05	55.05
Han Carlson 8	2.8	37.53	47.3	53.6	70.2	47.3
Han Carlson 16	3.6	31.71	40.60	48.92	51.68	45.8
Han Carlson 32	4.2	27.69	36.8	41.33	48.39	42.1
Kogge&Stone 8	2.2	31.72	33.75	38.47	46.84	31.72
Kogge&Stone 16	2.5	26.6	27.4	31.21	37.52	27.0
Kogge&Stone 32	3.1	22.06	22.82	26.53	31.14	22.82
Sklanski 8	2.3	34.97	38.0	44.66	54.39	34.97
Sklanski 16	2.5	29.35	32.5	37.66	44.1	30.6
Sklanski 32	4.9	26.7	29.29	33.48	38.65	37.1
CLA 8	3.1	37.62	43.21	53.66	70.21	43.21
CLA 16	3.6	37.74	44.1	50.10	67.49	50.10
CLA 32	5.2	35.97	42.65	46.98	65.48	68.8

Table 3 Area overhead for adders according to the implementation of Fig. 3

		TAO(%)				
	CP (ns)	$\delta = 0.46$	$\delta = 0.6$	$\delta = 0.8$	$\delta = 1$	20% Dmax
Brent&Kung 8	3	25.41	31.50	34.98	39.33	31.50
Brent&Kung 16	4.2	24.60	25.24	29.38	39.37	32.2
Brent&Kung 32	5.1	22.02	23.01	26.97	35.84	35.84
Han Carlson 8	2.8	24.29	28.31	34.65	45.87	26.16
Han Carlson 16	3.6	22.01	26	31.63	37.66	25.8
Han Carlson 32	4.2	18.64	20.02	23.50	33.71	25.2
Kogge&Stone 8	2.2	19.89	22.12	25.25	28.76	19.89
Kogge&Stone 16	2.5	19.16	20.1	23.66	27.84	19.8
Kogge&Stone 32	3.1	14.83	15.83	17.95	22.91	15.83
Sklanski 8	2.3	22.16	25.41	30.48	38.41	22.16
Sklanski 16	2.5	21.06	23.89	27.54	32.93	22
Sklanski 32	4.9	17.98	19.96	21.89	25.84	21.9
CLA 8	3.1	23.88	26.51	30.90	43.63	26.51
CLA 16	3.6	26.97	30.64	33.31	35.76	33.31
CLA 32	5.2	24.87	26.24	27.07	38.84	40.2

corresponding to $\delta = 0.6\,$ns and $\delta = 0.8\,$ns (they will allow high-detection efficiency for transient faults with a duration of several hundreds of ps), and the column corresponding to 20%Dmax, which is used for timing fault detection. We observe that the cost varies from 10.62% (Kogge&Stone 32 adder in column $\delta = $ 20%Dmax, Table 4) to 40.2% (CLA32 adder in column $\delta = $ 20%Dmax). The cost is quite low for the Kogge&Stone adder, varying from 10.62% to 12.7%

(column $\delta = 20\%$Dmax) and from 14.19% to 18.87% (column $\delta = 0.8$ns). This is interesting because this is the faster adder implementation (see column CP-Critical Path). In addition, the transient fault detection efficiency for this kind of adders is high as illustrated in the next section.

7 Transient Fault Injection Simulations

The error-detection efficiency of the scheme is verified by using several 8×8 adders and multipliers. Transient pulses of 0.65 ns width have been injected. We have considered also the case of delay fault with a duration equal to 20% Dmax (1.4 ns) for a Wallace 8×8 multiplier.

Since transient faults occur randomly, they can affect any node and occur at any instance during the clock cycle. In our simulations, the instance of the transient pulse injection is evenly distributed over the clock cycle. In fact the injection is repeated over several clock cycles by shifting within a clock cycle the instance of the injection at a 0.2 ns (or 0.1 ns in the case of adders). The injection locations were selected randomly, and simulations were performed for a large number of pseudorandom test patterns. On the total, 260,000 simulations were performed on each selected circuit.

Figure 6 presents several cases of the experiments done on a 8×8 Wallace multiplier. The rising edge of clock latches the inputs of the multiplier and also its outputs in Output-Latch (M signals in Fig. 6). The rising edge of ck + δ latches the multiplier outputs in Extra-Latch. A transient fault (SE signal on Fig. 6) of 0.65 ns width has been injected at the node computing the partial product 8. As explained above, the experiment is repeated several times with the transient pulse shifted of 0.2 ns in each clock cycle. This guarantees that the pulse will reach the circuit outputs at the moment of the latching edge of the clock for at least one clock cycle.

In the windows shown in Fig. 6A, we can a case of transient fault detection. In this case, a transient fault is latched at instance t_1 by Output-Latch on three outputs of the multiplier (output M12, M11, and M10). The correct values are latched later, at t_2, by the Extra-Latch. This case corresponds at a transient pulse injected early in the clock cycle and for this case the errors are latched only by Output Latch. Another possible case corresponds to transient pulses latched only by the Extra Latch. It corresponds at pulses injected at the end of the clock cycle. It is also possible that the transient pulse is captured by Output Latch on a group of outputs, and by the Extra Latch on another group of outputs. In all cases, an error message is delivered at the output of the comparator. In all these cases the delayed clock used for Extra-Latch ensures that no error affecting the same outputs are latched in both Output-Latch and Extra-Latch.

Note however that there is a possibility that an output error escapes detection. This will happen when the transient pulse width is amplified during its propagation through the circuit paths. The mechanisms creating this amplification are reconvergent fan-outs with different delays. They can propagate the original pulse through several paths, which reconverge and concatenate several pulses into a single one.

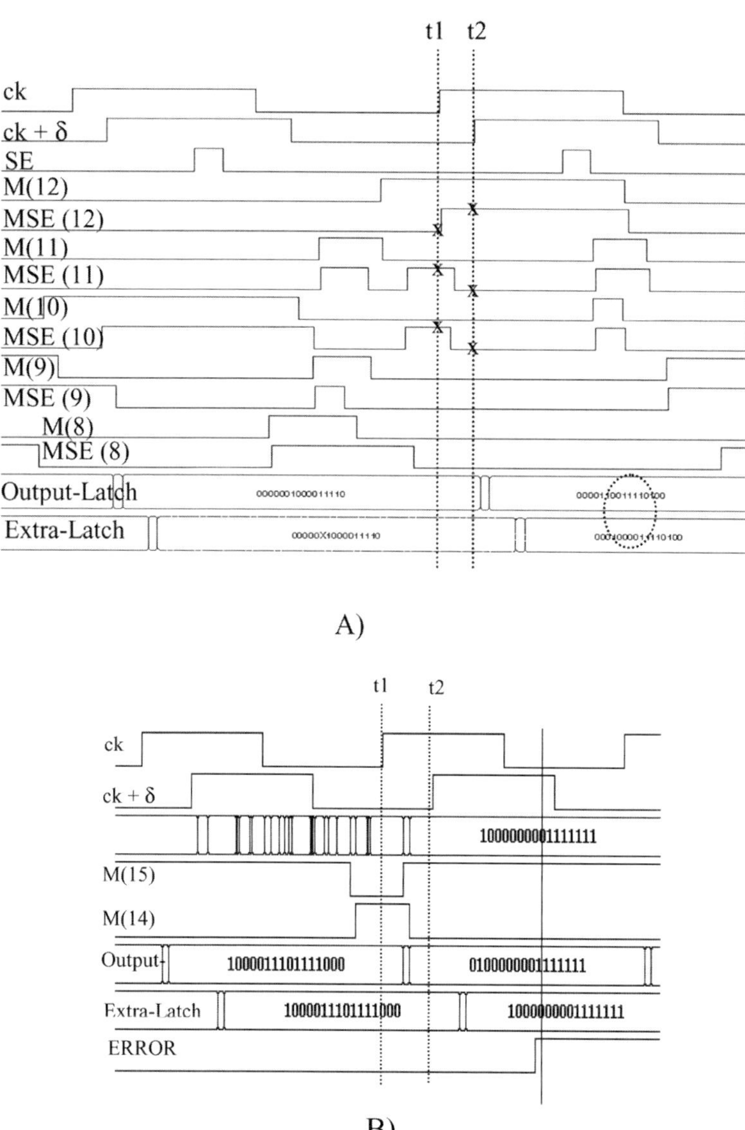

A)

B)

Fig. 6 (A) Transient fault injection logic simulation (M and MSE are the outputs of the fault-free and faulty multiplier, respectively); (B) timing faults injection on the Wallace 8×8 multiplier

This pulse can be larger than the original one due to the different delays of the propagating paths. Multiple pulses could also be created at the final node, but this will be a rare situation, since the delay difference between the paths should exceed the width of the original pulse. Due to this phenomenon, the scheme can not detect

Table 4 Area overhead for adders according to the implementation of Fig. 4

	CP (ns)	$\delta = 0.46$	$\delta = 0.6$	$\delta = 0.8$	$\delta = 1$	20% Dmax
				TAO(%)		
Brent&Kung 8	3	16.79	26.19	36.37	38.55	26.19
Brent&Kung 16	4.2	15.46	25.98	32.57	37.88	35.1
Brent&Kung 32	5.1	14.56	24.57	31.87	36.72	36.72
Han Carlson 8	2.8	15.7	25.64	33.47	37.91	21.19
Han Carlson 16	3.6	13.85	22.79	30.76	33.19	24.5
Han Carlson 32	4.2	12.31	21.44	25.81	32.72	27.1
Kogge&Stone 8	2.2	12.7	14.82	18.87	26.62	12.7
Kogge&Stone 16	2.5	11.84	12.73	16.18	22.21	12.2
Kogge&Stone 32	3.1	9.84	10.62	14.19	18.69	10.62
Sklanski 8	2.3	14.01	17.20	23.07	32.11	14.01
Sklanski 16	2.5	12.58	15.77	20.61	26.81	14.9
Sklanski 32	4.9	11.88	14.5	18.54	23.56	18.8
CLA 8	3.1	14.96	20.67	30.33	46.13	20.67
CLA 16	3.6	17.3	23.75	29.31	46.33	29.31
CLA 32	5.2	17	23.7	27.83	46.15	48.6

all output errors. We have computed the global efficiency of this scheme (ratio of errors occurring at the Output-Latch detected versus total number of the errors occurring in the Output-Latch).

For a transient pulse of a duration of 0.65 ns the experiments performed over 8×8 Wallace multiplier (260,000 simulations) show an efficiency of 99.4% for $\delta = 0.8$ ns. The efficiency is increased to 99.7% if δ is increased by 0.1 ns ($\delta = 0.9$ ns). We observe that the escapes are divided by two when δ is increased by 100 ps. This shows that the efficiency can be increased significantly by slightly increasing δ.

The efficiency calculated for 8-bit adder structures, for 0.65 ns transient pulse duration, is 99.8% for Sklanski and Brent&Kung adder, 99.1% for Han and Carlson adder, and 100% for the other structures (CLA, Kogge&Stone). The results show that Kogge&Stone adder is the most advantageous adder structure for the present scheme (highest transient detection efficiency and lowest cost, in addition to its highest speed). Sklanski adder is the next structure of interest in terms of speed, detection efficiency and hardware overhead.

Figure 6B presents the simulation of a timing fault in a 8×8 Wallace multiplier. The timing fault is injected by adding a delay of 1.4 ns in a path segment leading to the outputs M14 and M15. The fault is activated by using an input transition that sensitizes the longest path of the multiplier. The fault creates a timing error on outputs M14 and M15 that are captured by Output-Latch. Due to the shifted clock used in Extra-Latch the timing fault disappeared when M14, M15 are latched in Extra-Latch. Thus, the output of the comparator detects the error. This simulation is presented to illustrate the detection mechanism for timing faults. This scheme achieves 100% detection of timing faults as far as the total delay added in any path of the circuit will not exceed δ. For this reason, we do not need to use fault simulation to compute the efficiency of the scheme. As far as the total delay does not exceed δ, the detection is guaranteed, independently to the mechanism creating the delay fault and the faulty delay distribution along the circuit paths.

8 Conclusions

In the advent of very deep submicron and nanometer technologies, geometry shrinking, reduced power supply, and increasing operating speeds are making circuits increasingly sensitive to soft-errors and in particularly to single event upsets. At the same time, small delay deviation provoked by distributed or localized defects result on timing faults difficult to detect by fabrication testing. Due to these trends, it is becoming mandatory to design timing-error and soft-error tolerant ICs in order to maintain acceptable levels of reliability. Since this kind of solutions has to be used for any application, including commodity ones, traditional fault tolerant approaches are of reduced interest due to their high cost.

This paper considers a new scheme that exploits the temporal nature of timing errors and soft-errors in order to detect them by means of time redundancy. The scheme was experimented over various multiplier and adder structures. The experiments show that the new scheme requires low hardware cost and no performance penalty and achieves complete timing error and very high soft-error detection. Thus, the scheme is of high interest for achieving increasing robustness to timing errors and soft-errors and push aggressively the limits of technological scaling. The hardware cost is higher for adders, but it remains low for the more interesting (faster) adder cases. Further cost reduction could be obtained by obtained by combining the principles of Figs. 3 and 4 into a single circuit.

References

[**ALZ99**] I Alzaher-Noufal, M. Nicolaidis A Tool for Automatic Generation of Self-Checking Multipliers Based on Residue Arithmetic Codes, 1999 DATE Conference, March 1999, Munich, Germany.

[**AVI 73**] AVIZIENIS A., Arithmetic Algorithms for Error-Coded Operands IEEE Trans. Comput., Vol. C-22, No. 6, 567–572, June 1973.

[**BAZ97**] M. BAZE, S. BUCHNER, Attenuation of Single Event Induced Pulses in CMOS Combinational Logic IEEE Trans. Nuclear Science, Vol. 44, No. 6, December 1997.

[**BRE 82**] R. Brent and H. Kung, A Regular Layout for Parallel Adders, IEEE Trans. Computers, Vol. C-31, No. 3, 260–264, March 1982.

[**CHA 93**] H. CHA et al. A Fast and Accurate Gate-level Transient Fault Simulation Environment, Proceedings of FTCS, June 1993

[**FRA 94**] Franco P., McCluskey E. J., On-Line Delay Testing of Digital Circuits, 12th IEEE VLSI Test Symposium, Cherry Hill, NJ, April 1994.

[**HWA 79**] K. Hwang, Computer Arithmetic, Principles, Architectures and Design, Wiley, New York, 1979

[**HAN 87**] T. Han, D. Carlson, Fast area efficient VLSI Adders, 8th Symposium on Computer Arithmetic, pp. 49–56, May 1987.

[**KOG 73**] P.M. Kogge and H. Stone, A Parallel Algorithm for Efficient Solution of a General Class of Recurrence Equations, IEEE Trans. Computers, Vol. C-22, No. 8, 786–792, August 1973

[**MET 98**] C. Metra, M. Favalli, B. Ricco, On-Line Detection of Logic Errors due to Crosstalk, Delay, and Transient Faults, ITC October 18–23, 1998, Washington, DC.

[**NIC 99**] M. NICOLAIDIS, Time Redundancy Based Soft-Error Tolerance to Rescue Nanometer Technologies, In proceedings VTS 99.

[**PET 58**] PETERSON W.W. On Checking an Adder, IBM J. Res. Develop. 2, 166–168, April 1958

[**PET 72**] PETERSON W.W., WELDON E.J., Error-Correcting Codes, 2nd edn, MIT press, Cambridge, M, 1972

[**SKL 60**] J. Sklanski, Conditional-sum Addition Logic, IRE Trans. Electronic Computers, Vol. EC-9, No. 2, 226–231, June 1960.

[**WES 94**] N.H.E. Weste and al. Principles of CMOS VLSI Design- A System Perspective, 2nd edn, Addison-Wesley, 1994.

An Integrated System-on-Chip Test Framework

Erik Larsson and Zebo Peng

Abstract[1] In this paper we propose a framework for the testing of system-on-chip (SOC), which includes a set of design algorithms to deal with test scheduling, test access mechanism design, test sets selection, test parallelization, and test resource placement. The approach minimizes the test application time and the cost of the test access mechanism while considering constraints on tests, power consumption and test resources. The main feature of our approach is that it provides an integrated design environment to treat several different tasks at the same time, which were traditionally dealt with as separate problems. Experimental results shows the efficiency and the usefulness of the proposed technique.

1 Introduction

The increasing complexity of digital systems has led to the need of extensive testing and long test application times. It is therefore important to schedule the tests as concurrently as possible and to design an access mechanism for efficient transportation of test data in the system under test.

When developing the test schedule, conflicts and limitations must be carefully considered. For instance, the tests may be in conflict with each other due to the sharing of test resources; and power consumption must be controlled, otherwise the system may be damaged during test. Furthermore, test resources such as external testers support a limited number of scan-chains and have a limited test memory which also introduce constraints on test scheduling. For the test designer, it is also important to get an early impression on the systems overall test characteristics in order to develop an efficient test solution.

Research has been going on in developing techniques for test scheduling, test access mechanism design and testability analysis. For example, a technique to help

Embedded Systems Laboratory Separtment of Computer and Information Science, Linköpings Universitet, Sweden.

[1]This work has partially been supported by the Swedish National Board for Industrial and Technical Development (NUTEK).

R. Lauwereins and J. Madsen (eds.), *Design, Automation, and Test in Europe–The Most Influential Papers of 10 Years DATE*, 439–454
© Springer 2008

the designer determine the test schedule for SOC with Built-In Self-Test (BIST) is proposed by Benso et al. [1]. In this paper, we combine and generalize several approaches in order to create a framework for SOC testing where:

- Tests are scheduled to minimize the test time
- A test access mechanism is designed and minimized
- Test sets for each block with test resource are selected
- Test resources are floor-planned
- Tests are parallelized (i.e. long scan-chains are divided into several scan-chains of shorter length)

Furthermore, the above tasks are performed under test, power consumption and test resource constraints.

The rest of the paper is organised as follows. After an overview of related work in Section 2, a system modelling technique is introduced in Section 3. Factors affecting the test scheduling and an algorithm which takes them into account in test scheduling and test access mechanism design are then presented in Section 4 and 5, respectively. The paper is concluded with experimental results and conclusions in Section 6 and 7.

2 Related Work

Zorian proposes a test-scheduling technique for fully BISTed systems where test time is minimized while power constraints are considered [2]. In order to reduce the complexity of the test controller, tests are scheduled in sessions where no new tests are allowed to start until all tests in a session are completed. Furthermore, tests at blocks placed physically close to each other are grouped in the same test session in such a way that the same control line can be used for all tests in a group. The advantage is that the routing of control lines is minimized.

In a fully BISTed system, each block has its own dedicated test generator (test source) and its own test response evaluator (test sink); and there might not be any conflicts among tests, i.e. the tests can be scheduled concurrently. However, in the general case, conflicts among tests may occur. Garg et al. [3] propose a test scheduling technique where test time is minimized for systems with test conflicts and for core-based systems a test scheduling technique is proposed by Chakrabarty [4]. Chou et al. [5] propose an analytic test scheduling technique where test conflicts and power constraints are considered. Another test scheduling approach is proposed by Muresan et al. [6] where constraints among tests and power consumption are also considered. In the latter approach, favour is given to reduce the test time by allowing new tests to start even if all tests in a session are not completed. The drawback is the increasing complexity of the test controller. Note also that in the approaches by Chou et al. [5] and by Muresan et al. [6] the systems to be tested are not restricted to fully BISTed systems.

The conflicts among tests can be reduced by using a wrapper such as Boundary scan [7], TestShell [8], or P1500 [9]. These techniques are all developed to increase test isolation and to improve test data transportation.

Usually, several test sets can be used to test a block in the system under test. Sugihara et al. [10] propose a technique for selecting test sets where each block may be tested by one test set from an external tester and one test set from a dedicated test generator for the block.

The effect on test application time for systems tested by one test set per core using various design styles for test access with the TestShell wrapper is analysed by Aertes et al. [11]. Furthermore, the impact on test time using scan-chain parallelization is also analyzed by Aertes et al. [11].

The use of different test resources may entail constraints on test scheduling. For instance, external testers have limitations of bandwidth due to that a scan chain operates usually at a maximum frequency of 50 MHz [12]. External testers can usually only support a maximum of 8 scan chains [12], resulting in long test application time for large designs. Furthermore an external tester's memory is limited by its size.

3 System Modelling

An example of a system under test is given in Fig. 1 where each core is placed in a wrapper in order to achieve efficient test isolation and to ease test access. Each core consists of at least one block with added DFT technique and in this example all blocks are tested using the scan technique. The test access port (*tap*) is the connection to an external tester and the test resources, *test generator* 1, *test generator* 2, *test response evaluator* 1 and *test response evaluator* 2, are implemented on the chip.

The system is tested by applying several set of tests where each test set is created at some test generator (source) and the test response is analysed at some test response evaluator (sink).

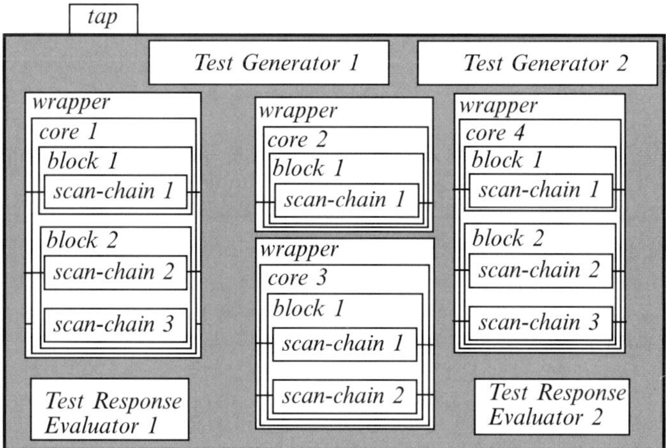

Fig. 1 An illustrative example

The system in Fig. 1 can be modelled as a *design with test, DT = (C, R_{source}, R_{sink}, p_{max}, T, source, sink, core, block, constraint, memory, bandwidth)*, where:

$C = \{c_1, c_2, ..., c_n\}$ is a finite set of cores; each core consists of a finite set of blocks, $c_i = \{b_{i1}, b_{i2}, ..., b_{nm}\}$. Each core consists of at least one block and each block $b_{ij} \in B$ is characterized by:

$p_{idle}(b_{ij})$: idle power,
$par_{min}(b_{ij})$: minimal parallelization degree, and
$par_{max}(b_{ij})$: maximal parallelization degree;
$R_{source} = \{r_1, r_2, ..., r_p\}$ is a finite set of test sources;
$R_{sink} = \{r_1, r_2, ..., r_q\}$ is a finite set of test sinks;
p_{max}: maximal allowed power at any time;
$T = \{t_1, t_2, ..., t_0\}$ is a finite set of tests, each consisting of a set of test vectors. Several tests form a *block tests (BT)*. And each block, b_{ij}, is associated with several block tests, BT_{ijk} $(k=1,2,...,l)$. Each test t_i is characterized by:

$t_{test}(t_i)$: test time at parallelization degree 1, $par(t_i)=1$,
$p_{test}(t_i)$: test power at parallelization degree 1, $par(t_i)=1$,
$t_{memory}(t_i)$: memory required for test pattern storage.
source: $T{\rightarrow}R_{source}$ defines the test sources for the tests;
sink: $T{\rightarrow}R_{sink}$ defines the test sinks for the tests;
core: $B{\rightarrow}C$ gives the core where a block is placed;
block: $T{\rightarrow}B$ gives the block where a test is applied;
constraint: $T{\rightarrow}2^B$ gives the set of blocks required for a test;
memory(r_i): memory available at test source $r_i \in R_{source}$;
bandwidth(r_i): bandwidth at test source $r_i \in R_{source}$.

In the above definitions, test time, t_{test}, test power consumption, p_{test}, idle power, p_{idle}, and memory requirement, t_{memory}, are given for each of the tests. The maximal and minimal degree of parallelization for a test is given by par_{max} and par_{min} which determine how much a scan-chain may be divided. For instance, if $par_{max}(b_{31})=2$ and $par_{min}(b_{31})=1$ for block 1 at core 3 in Fig. 1, then the scan flip-flops are connected into a single scan-chain $(par(b_{31})=1)$ or two scan-chains $(par(b_{31})=2)$.

4   The SOC Test Issues

In this section the different issues considered by our SOC test framework are discussed.

4.1   *Test Scheduling*

Scheduling the tests means that the start time and end time for each test is determined in order to satisfy all constraints. In our approach, the test bus used to transport the test data is also determined by the scheduling algorithm. The basic difference

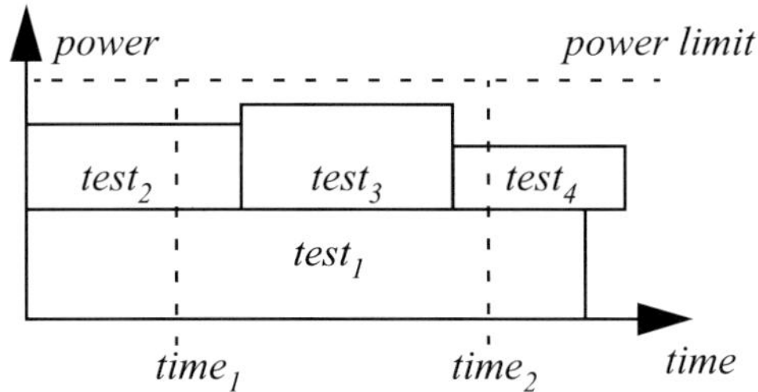

Fig. 2 Example of test scheduling

of our scheduling approach compared to previously proposed approaches is illustrated in Fig. 2. In the approaches by Zorian [2] and Chou et al. [5] no new tests are allowed to start until all tests in a session are completed. In their approaches $test_3$ and $test_4$ would not be allowed to be scheduled as in Fig. 2. However, in the approach proposed by Muresan et al. [6], $test_3$ is allowed to be scheduled as in Fig. 2 if it is completed no later than $test_1$. It means that $test_4$ is still not allowed to be started before *test*$_1$ finishes.

In our approach it is optional if tests may start before all tests in a session are completed or not. If it is allowed, $test_3$ and $test_4$ can be scheduled as in Fig. 2, which gives more flexibility, but entails usually a more complex test controller.

Let a schedule S be an ordered set of tests such that:

$$\{S(t_i) < S(t_j) \mid t_{start}(t_i) \leq t_{start}(t_j),\ i \neq j, \forall t_i \in S, \forall t_j \in S\},$$

where $S(t_i)$ defines the position of test t_i in S; $t_{start}(t_i)$ denotes the time when test t_i is scheduled to start, and $t_{end}(t_i)$ its completion time:

$$t_{end}(t_i) = t_{start}(t_i) + t_{test}(t_i).$$

For each test, t_i, the start time and the bus for test data transportation have to be determined before it is inserted into the schedule, S.

Let the Boolean function $scheduled(t_i, time_1, time_2)$ be true if test t_i is scheduled in such a way that the test time overlaps with the time interval $[time_1, time_2]$, i.e.,

$$\{t_i \in S \wedge \neg(t_{end}(t_i) < time_1 \vee t_{start}(t_i) > time_2)\}.$$

An example to illustrate the function *scheduled* for a set of scheduled tests is shown in Fig. 3.

The Boolean function $scheduled(r_i, time_1, time_2)$ is true if a source r_i is used by a test t_j between $time_1$ and $time_2$, i.e.:

$$\{\exists t_j \in S \mid r_i = Source(t_j) \wedge scheduled(t_j, time_1, time_2)\}.$$

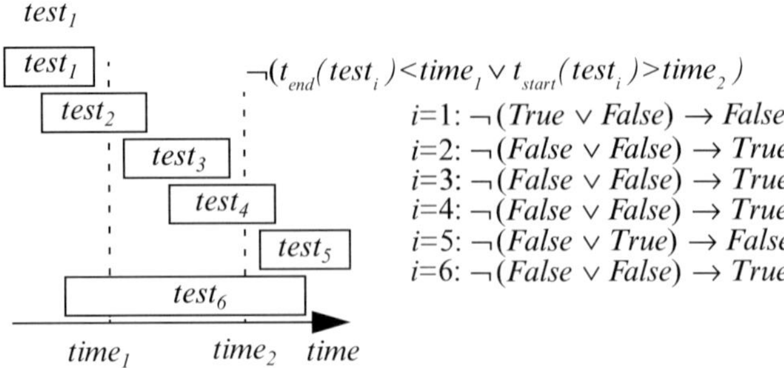

Fig. 3 The function *scheduled*.

A similar definition is used if a sink r_i is scheduled (used by any test) between $time_1$ and $time_2$.

The Boolean function *scheduled*(*constraint*(t_i), $time_1$, $time_2$) is true if:

$$\{\exists t_j \in S \mid block\,(t_j) \in constraint\,(t_i) \wedge$$

$$scheduled(t_j, time_1, time_2)\}.$$

The Boolean function *scheduled*(w_i, $time_1$, $time_2$) is true when a wire w_i is used between $time_1$ to $time_2$:

$$\{\exists t_j \in S \mid w_i \in bus(t_j) \wedge scheduled\,(t_j, time_1, time_2)\},$$

where $bus(t_j)$ is the set of wires allocated for test t_j.

4.2 Power Dissipation

In this paper, an additive model used by Zorian [2], Chou et al. [5] and Muresan et al. [6] for power consumption is assumed. Let $p_{sch}(time_1, time_2)$ denote the peak power between $time_1$ to $time_2$, i.e.:

$$max\left\{ \sum_{\forall t_j scheduled(t_j, time)} P_{test}(t_i) - P_{idle}(block(t_i)) + \right.$$

$$\left. \sum_{\forall b_{i,j} \in B} P_{idle}(b_{ij}),\ time \in [time_1, time_2] \right\},$$

where $scheduled(t_i, time) = scheduled(t_i, time, time)$.

As an example, applying the function $p_{sch}(time_1, time_2)$ on the schedule for a system with 4 tests as in Fig. 2, with $time_1$ and $time_2$ as indicated in the figure, returns $p_{test}(test_1) + p_{test}(test_3) + p_{idle}(block(test_2)) + p_{idle}(block(test_4))$ since it gives the peak power consumption between $time_1$ and $time_2$.

In our approach, the maximal power consumption should not exceed the power constraint, p_{max}, for a schedule to be accepted. That is, $p_{sch}(0, \infty) \leq p_{max}$.

4.3   Test Source Limitations

A test generator may use a memory for storing the test patterns. In particular, external test generators use such a memory with a limited size which may lead to additional constraints on test scheduling [12].

The function $memory_{alloc}(r_i, time_1, time_2)$ gives the peak allocated memory between $time_1$ and $time_2$ for a given source r_i, i.e.:

$$max\left\{ \sum_{\forall t_j scheduled(t_j, time) \wedge r_i = source(t_j)} t_{memory}(t_j), time \in [time_1, time_2] \right\}.$$

A test resource may have a limited bandwidth. For instance, external tester may only support a limited number of scan chains at a time or there could be a limit in the available pins for test. This information is given in the attribute bandwidth for each test resource.

The function $bandwidth_{alloc}(r_i, time_1, time_2)$ gives the maximal number of buses allocated between $time_1$ and $time_2$ for a given source r_i, i.e.:

$$max\left\{ \sum_{\forall t_j scheduled(t_j, time) \wedge r_i = source(t_j)} |t_{bus}(t_j)|, time \in [time_1, time_2] \right\}.$$

4.4   Test Floor-planning

In the general case it is not feasible to assume that all cores can be tested with only one BIST structure. A block may be tested by several test sets produced and analysed at different test resources. Furthermore, test resources may be shared among several blocks at different cores. It is therefore important to consider the routing of the test data access mechanism. And an efficient placement of test resources in the system under test must be created in order to minimize the routing cost associated with the test access mechanism.

4.5   *Test Set Selection*

Each test set is defined by a test source and a test sink. For a test set, its test power consumption, test memory requirement and test application time are defined as discussed in Section 3. We assume that an arbitrary number of test sets can be used to test a block.

Due to that the test resources are defined for each test set it is possible to make a comparison of different test sets not only in terms of the number of test vectors but also in respect to test resources and test memory requirement. This information should be taken into account in our algorithm.

4.6   *Test Parallelization*

The test time for a test may be reduced if it is parallelized. This is because dividing a scan-chain into several scan-chains of shorter length will shorten the test application time. Formulas for calculating the test time for scan-based designs are defined by Aertes et al. [11]. Similar to Aertes et al. [11] we assume that the scan-chain may be divided into equal portions. To simplify the problem, the degree of *parallelization* is assumed to be linear with respect to test time and test power consumption. The test time $t'_{test}(t_i)$ for a test t_i after parallelization is given by:

$$t'_{test}(t_i) = \left\lceil \frac{t_{test}(t_i)}{par(block(t_i))} \right\rceil,$$

where $t_{test}(t_i)$ is the test time when parallelization=1 and $par(block(t_i))$ is the degree of parallelization for the block where t_i is applied.

Assuming that the product *time×power* is constant, we let the test power $p'_{test}(t_i)$ for a test t_i after parallelization be given by:

$$p'_{test}(t_i) = p_{test}(t_i) \times par(block(t_i)),$$

where $p_{test}(t_i)$ is the test power when parallelization=1.

The parallelization at a block can not be different for different test sets; the original scan-chain can not be divided into n chains at one moment and to m chains at another moment where $m{\neq}n$. The function $par(b_{ij})$ denotes the common parallelization degree at block b_{ij}.

4.7   *Test Access Mechanism*

A test infrastructure transports, and controls the transportation of, test data in the system under test. It transports test patterns from test sources to the blocks and the test response from the blocks to the test sinks.

The test designer faces mainly two problems, namely:

- Designing and routing the test access mechanism
- Scheduling the test data transportation

The system can be modelled as a directed graph, $G=(V,A)$, where V consists of the set of blocks, B, the set of test sources, R_{source}, and the set of test sinks, R_{sink}, i.e. $V=B\cup R_{source}\cup R_{sink}$.

An arc $a_i\in A$ between two vertices v_i and v_j indicates a test access mechanism (a wire) where it is possible to transport test data from v_i to v_j. Initially no test access mechanism exists in the system, i.e. $A=\emptyset$. However, if the functional infrastructure may be used, it can be included in A initially.

When adding a test access mechanism between a test source and a core or between a core and a test sink, and the test data has to pass through another core, c_i, several routing options are possible:

1. Through the logic of core c_i using the transparent mode of the core
2. Through an optional bypass structure of core c_i;
3. Around core c_i where the access mechanism is not connected to the core

The advantage of alternatives 1 and 2 above is that the test access mechanism can be reused. However, a delay may be introduced when the core is in transparent mode or its by-pass structure is used. A test wrapper such as the TestShell has a clocked by-pass structure and the impact on the test time using it is analysed by Aertes et al. [11].

In the following, we assume that by-pass may be solved by a non-delay mechanism or that the delay due to clocked by-pass is negligible.

A test wire w_i is a path of edges $\{(v_0,v_1),..,(v_{n-1},v_n)\}$ where $v_0\in R_{source}$ and $v_n\in R_{sink}$.

Let Δy_{ij} be defined as $|y(v_i)-y(v_j)|$ and Δx_{ij} as $|x(v_i)-x(v_j)|$, where $x(v_i)$ and $y(v_i)$ are the *x-placement* respectively the *y-placement* for a vertex v_i.

Initially, the test resources may not be placed. In this case, their placement must be determined by our algorithm described in the next section.

The distance between vertex v_i and vertex v_j is given by:

$$dist(v_i, v_j) = \sqrt{(\Delta y_{ij})^2 + (\Delta x_{ij})^2}.$$

The information of the nearest core in four direction, *north*, *east*, *south*, and *west*, are stored for each vertex and the function $south(v_i)$ of vertex v_i gives the closest vertex south of v_i and it is defined as:

$$south(v_i) = \left\{\left(\frac{\Delta y_{ij}}{\Delta x_{ij}} > 1 \vee \frac{\Delta y_{ij}}{\Delta x_{ij}} < -1\right),\right.$$

$$y(v_j) < y(v_i), i \neq j, min\{dist(v_i, v_j)\}.$$

The functions $north(v_i)$, $east(v_i)$ and $west(v_i)$ are defined in similar ways. The function $insert(v_i, v_j)$ inserts a directed arc from vertex v_i to vertex v_j *if and only if* the following is true:

$$\{south(v_i, v_j) \vee north(v_i, v_j) \vee west(v_i, v_j) \vee east(v_i, v_j)\}.$$

The function $closest(v_i, v_j)$ gives a vertex, v_k, which is in the neighbourhood of v_i and has the shortest distance to v_j. The function $add(v_i, v_j)$ adds arcs from v_i to v_j in the following way: (1) find $v_k = closest(v_i, v_j)$; (2) add a wire from v_i to v_k; (3)if $v_k = v_j$, terminate otherwise let $v_i = v_k$ and go to (1).

5   The Algorithm

In this section the issues discussed above are combined into an algorithm. The algorithm assumes that the tests are initially sorted according to a key k which characterizes *power(p)*, *test time(t)* or *power×test time(p×t)*.

Let P be an ordered set with the tests ordered based on the key k. If new tests are allowed to be scheduled even if all tests in a session are not completed the function $nexttime(t_{old})$ gives the next time where it is possible to schedule a test:

$$\{t_{end}(t_i) \mid min(t_{end}(t_i)), t_{old} < t_{end}(t_i), \forall t_i \in S\},$$

otherwise function $nexttime(t_{old})$ is defined as:

$$\{t_{end}(t_i) \mid max(t_{end}(t_i)), t_{old} < t_{end}(t_i), \forall t_i \in S\}.$$

The algorithm is depicted in Fig. 4 and it can basically be divided into four parts for:

- Constraint checking
- Test resource placement
- Test access mechanism design and routing
- Test scheduling

A main loop is terminated when there exists a block test (BT) for all blocks where all tests within the BT are scheduled. In each iteration of the loop over the tests in P a test *cur* is checked.

If the *parallelization degree* is fixed for the block, *i.e.* some tests have been scheduled for the block, $par = par(b_{ij})$ otherwise it is computed:

$$par = min\{par_{max}(b_{ij}), \lfloor (p_{max} - p_{sch}(time, t_{end})) / p(cur) \rfloor,$$

$$bandwidth(v_a, time, t_{end}) - bandwidth_{alloc}(v_a, time, t_{end}),$$

which is the minimum among the available power and the available bandwidth of the test source.

> *Sort T according to the key (p, t or p X t) and store the result in P;*
> *S=Ø, time=0;*
> *until $\forall b_{pq} \exists BT_{pqr} \forall t_s \in S$ do*
>   *for all cur in P do*
>   *b_{ij}=block(cur); v_a=source(cur);*
>   *v_b=c_i; v_c=sink(cur);*
>   *par=determine parallelization degree;*
>   *t_{end}=time+$\lceil t_{test}(cur)/par \rceil$;*
>   *$p_{test}(cur)=p_{test}(cur)Xpar$;*
>   *if all constraints are satisfied then*
>   *¬scheduled(v_a, 0, t_{end}) floor-plan v_a at v_b;*
>   *¬scheduled(v_c, 0, t_{end}) floor-plan v_c at v_b;*
>   *for all required test resources*
>   *new=length of a new wire w_j;*
>   *u=number of wires connecting v_a, v_b and v_c, and are not*
>   *scheduled from time to t_{end};*
>   *v=number of wires connecting v_a, v_b and v_c;*
>   *for all min(v-u,par) w_j*
>   *extend=extend+length of an available wire(w_j);*
>   *if (par>u)*
>   *extend=extend+newX(par-u);*
>   *move=par(v_a) X min{dist(v_a, v_b),dist(v_b, v_c)};*
>   *if (move≤min{extend, new X par})*
>   *v_x, v_y=min{dist(va, vb), dist(vb, vc)}, dist(va, vb)>0,*
>   *dist(v_b,v_c)>0*
>   *add par(v_a) wires between v_x and v_y;*
>   *if (v_x=source(cur)) then floorplan v_a at v_b;*
>   *if (v_y = sink(cur)) then floorplan v_c at v_b;*
>   *set parallelization;*
>   *for r = 1 to par*
>   *if there exists a not scheduled wire during time to t_{end}*
>   *connecting v_a, v_b and v_c it is selected*
>   *else*
>   *if (length of a new wire < length of extending a wire w_j)*
>   *w_j=add(v_a, v_b) + add(v_b, v_c);*
>   *else extend wire;*
>   *schedule cur and remove cur from P;*
>   *time = nexttime(time).*

Fig. 4 The system test algorithm

A check is also made to determine if all constraints are fulfilled, i.e. it is possible to schedule test *cur* at *time*:

- $\neg \exists t_f (t_f \in BT_{ijk} \wedge t_f \in S \wedge cur \notin BT_{ijk})$ checks that another block test set for current block is not used,

- $par \geq par_{min}(b_{ij})$ checks that the current parallelization degree is larger than the minimal level,
- $\neg scheduled(v_a, time, t_{end})$ checks that the test source is not scheduled during *time* to t_{end},
- $\neg scheduled(v_c, time, t_{end})$ checks that the test sink is not scheduled during *time* to t_{end},
- $\neg scheduled(constraint(cur), time, t_{end})$ checks that all blocks required for *cur* are not scheduled during *time* to t_{end}, and
- the available memory test source v_a is checked to see if: $memory(v_a) > t_{memory}(cur) + memory_{alloc}(v_a, time, t_{end})$.

Then the placement of the test resources are checked. If the test resources are placed it is checked if they are to be moved.

When the placement of the test resources for the selected test is determined, the corresponding test access mechanism is designed and routed. The basic question is if some existing wires can be used or new wires must be added.

If no routed connection is available connecting all required blocks, the distance for adding a completely new connection is re-calculated due to a possible moving of test resources.

The *extend wire* step in the algorithm extends needed parts to connect the test resources and block with a given wire.

The computational complexity for the above algorithm, where the test access mechanism design is excluded in order to make it comparable with other approaches, comes mainly from sorting the tests and the two loops. The sorting can be performed using a sorting algorithm at $O(n \times log\ n)$. The worst case for the loops occurs when only one test is scheduled in each iteration resulting in a complexity given by:

$$\sum_{i=0}^{|P-1|} (P-i) = \frac{n^2}{2} + \frac{n}{2}$$

The total worst case execution time is $n \times log + n^2/2 + n/2$ which is of $O(n^2)$. For instance, the approach by Garg et al. [3] and by Chakrabarty [4] both have a worst case complexity of $O(n^3)$.

6 Experimental Results

We have performed experiments to show the efficiency of the proposed algorithm.

6.1 Benchmarks

We have used the System S presented by Chakrabarty [4], and ASIC Z design presented by Zorian [2] with added data made by Chou et al. [5] (see the floor-plan in Fig. 5). We have also used one design consisting of 10 test presented by Muresan

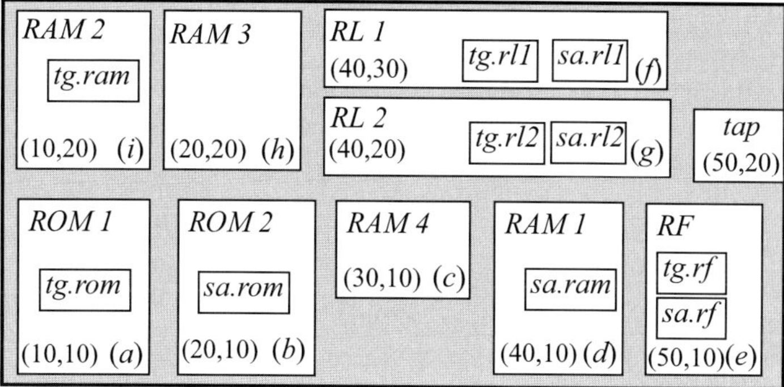

Fig. 5 ASIC Z floor-plan

Table 1 Characteristics of the industrial design

Test	Block	Test	Test time	Idle power	Test power	Test port
	A	Test A	515	1	379	scan
	B	Test B	160	1	205	testbus
	C	Test C	110	1	23	testbus
	E	Test E	61	1	57	testbus
Block-level tests	F	Test F	38	1	27	testbus
	I	Test I	29	1	120	testbus
	J	Test J	6	1	13	testbus
	K	Test K	3	1	9	testbus
	L	Test L	3	1	9	testbus
	M	Test M	218	1	5	testbus
	A	Test N	232	1	379	fp
Top-level tests	N	Test O	41	1	50	fp
	B	Test P	72	1	205	fp
	D	Test Q	104	1	39	fp

et al. [6] and an industrial design with characteristics given in Table 1. The power limitation for the industrial design example is 1,200 mW and only one test may use the test bus or the functional pins (fp) at a time. Furthermore block-level tests may not be scheduled concurrently with top-level tests.

6.2 Test Scheduling

We have compared our algorithm using initial sorting based on *power(p)*, *time(t)* and *power×time(p×t)* with the approaches proposed by Zorian [2] and Chou et al. [5]. We have used the same assumptions as Chou et al. [5] and the results are in Table 2. Our approaches results, in all cases, in a test schedule with three test sessions (*ts*) at a test time of 300 time units which is 23% better than Zorian's approach and 9% better than the approach by Chou et al. [5].

Table 2 ASIC Z test scheduling

	Zorian		Chou et al.		Our algorithm	
ts	Blocks	Time	Blocks	Time	Blocks	Time
1	RAM1, RAM4, RF	69	RAM1, RAM3, RAM4, RF	69	RL2, RL1, RAM2	160
2	RL1, RL2	160	RL1, RL2	160	RAM1, ROM1, ROM2	102
3	RAM2, RAM3	61	ROM1, ROM2, RAM2	102	RAM3, RAM4, RF	38
4	ROM1, ROM2	102				
Test time:		392		331		300

Table 3 Results on the designs by Chakrabarty [4] and Muresan [6] as well as the industrial design

Design	Approach	Test time	Improvement
Chakrabarty's design case [4]	Chakrabarty ours(t)	1204630	–
		1152810	4.3%
Muresan's design case [6]	Muresan	29	–
	ours(p)	28	3.4%
	ours(t)	28	3.4%
	ours(p×t)	26	10.3%
Industrial design	designer	1592	–
	ours(p)	1077	32.3%
	ours(t)	1077	32.3%
	ours(p×t)	1077	32.3%

In System S, no power constraints are given and therefore only test scheduling using initial sorting of tests based on time is performed. Our approach finds the optimal solution, see Table 3.

We have also compared our technique with the technique proposed by Muresan et al. [6]. In this case we use the same assumption as Muresan et al. [6] which assumes that new tests can start even if all tests are not fully completed in the current test session. In all cases our technique achieve better solutions, see Table 3.

Finally, the results on an industrial design are in Table 3 where the industrial designer's solution is 1592 time units while our test scheduling achieve a test time of 1077 time units in all sorting variations which is 32.3% better

All solutions using our technique were produced within a second on a Sun Ultra Sparc 10 with a 450 MHz processor and 256 Mbyte RAM.

6.3 Test Resource Placement

In the ASIC Z design all blocks have their own dedicated BIST structure. Let us assume that all ROM blocks share one BIST structure and all RAM memories share another

Table 4 Results on ASIC Z

Initial sorting	Test time	Test access mechanism
power	300	360
time	290	360
power×time	290	360

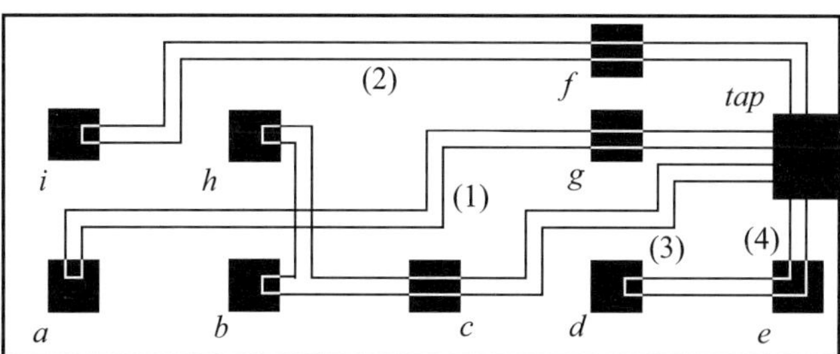

Fig. 6 ASIC Z with test data access mechanism

BIST structure; the rest of the blocks have their own dedicated BIST structure. Using our placement strategy the test resources in ASIC Z will be placed as in Fig. 5.

6.4 *Test Access Mechanism Design*

Assume the floor-planning of ASIC Z as in Fig. 5 where each block is placed according to its (x, y) coordinates. For instance, RAM2 is placed at (10,20), which means that the center of RAM2 has x-coordinate 10, and y-coordinate 20. Assume that all tests are scan-based tests applied with an external tester allowing a maximum of 8-scan chains to operate concurrently.

In this experiment we allow a new test to start even if all tests are not completed, see results in Table 4.

The test schedule and the test bus schedule achieved with initial sorting of tests according to power×time and considering idle power is in Fig. 7. The total test access mechanism length is 360 units and it is routed as in Figure 6. All solutions were produced within a second on a Sun Ultra Sparc 10 with a 450 MHz processor and 256 Mbyte RAM.

7 Conclusions

For complex systems such as SOCs, it is a difficult problem for the test designer to develop an efficient test solution due to the large number of factors involved. In this paper we propose a framework where several test-related factors are considered in

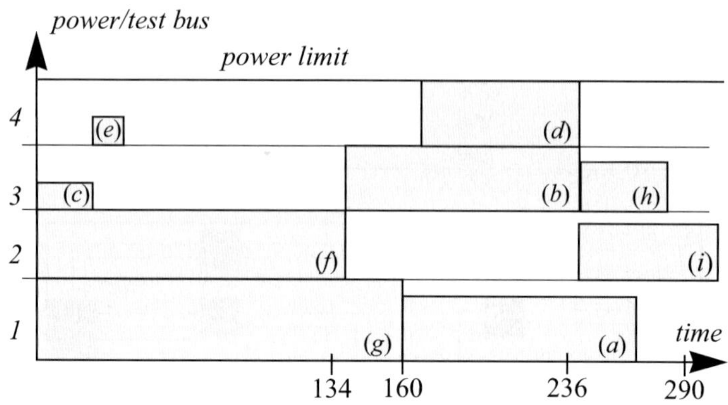

Fig. 7 Test schedule for ASIC Z

an integrated manner in order to support the test designer to develop an efficient test solution for a complex system. An algorithm has been defined and implemented, and experiments have been performed to show its efficiency.

References

[1] A. Benso, S. Cataldo, S. Chiusano, P. Prinetto, Y. Zorian, A High-Level EDA Environment for the Automatic Insertion of HD-BIST Structures, *JETTA*, Vol. 16.3, 179–184, June 2000.

[2] Y. Zorian, A Distributed BIST Control Scheme for Complex VLSI Devices, *Proc. VLSI Test Symp.*, 4–9, April 1993.

[3] M. Garg, A. Basu, T.C. Wilson, D.K. Banerji, J.C. Majithia, A New Test Scheduling Algorithm for VLSI Systems, *Proc. Symp. on VLSI Design*, pp. 148–153, November 1999.

[4] K. Chakrabarty, Test Scheduling for Core-Based Systems, *Proc. Int. Conf on CAD*, pp. 391–394, January 1991.

[5] R. Chou, K. Saluja, V. Agrawal, Scheduling Tests for VLSI Systems Under Power Constraints, *IEEE Trans. VLSI Systems*, Vol. 5, No. 2, 175–185, June 1997.

[6] V. Muresan *et al.*, A Comparison of Classical Scheduling Approaches in Power-Constrained Block-Test Scheduling, *Proc. Int. Test Conf.*, pp. 882–891, 3–5 October 2000.

[7] H. Bleeker et al. Boundary-Scan Test: A Practical Approach, *Kluwer Academic*. ISBN 0-7923-9296-5, 1993.

[8] E.J. Marinissen et al. A Structured and Scalable Mechanism for Test Access to Embedded Reusable Cores, *Proc. International Test Conf.*, pp. 284–293, October 18–23, 1998.

[9] IEEE P1500 Web site. http://grouper.ieee.org/groups/1500/.

[10] M. Sugihara, H. Date, H. Yasuura, A Test Methodology for Core-Based System LSIs, *IEICE Trans. Fund.* Vol. E81-A, No. 12, 2640–2645, December 1998.

[11] J. Aerts, E. J. Marinissen, Scan Chain Design for Test Time Reduction in Core-Based ICs, *Proc. International Test Conference*, pp. 448–457, 1998.

[12] G. Hetherington et al. Logic BIST for Large Industrial Designs: Real Issues and Case Studies, *Proc. International Test Conference*, pp. 358–367, 1999.

Efficient Spectral Techniques for Sequential ATPG

Ashish Giani[1], Shuo Sheng[1], Michael S. Hsiao[1], and Vishwani D. Agrawal[2]

Abstract We present a new test generation procedure for sequential circuits using spectral techniques. Iterative processes of filtering via compaction and spectral analysis of the filtered test set are performed for each primary input, extracting inherent spectral information embedded within the test sequence. This information, when viewed in the frequency domain, reveals the characteristics of the input spectrum. The filtered and analyzed set of vectors is then used to predict and generate future vectors. We also developed a fault-dropping technique to speed up the process. We show that very high fault coverages and small vector sets are consistently obtained in short execution times for sequential benchmark circuits.

1 Introduction

Simulation-based test generation began with random test generation, which used a pseudo-random pattern generator [1]. However, random testing generally results in large test sets [2] and they are useful for circuits without *random-pattern-resistant* faults [3]. Weighted random patterns have been found to yield better fault coverages in circuits that contain such random-pattern-resistant faults [4, 5]. In these approaches, the probability of obtaining a 0 or 1 at a particular input is biased towards detecting random resistant faults. However, the difficulty arises when no one set of weights may be suitable for all faults. In sequential circuits, faults may need a biased internal state in addition to biased input values, making it more difficult to obtain a good set of weights.

Recently, static compaction [6, 7] has been used to aid test generation. A specific feature of vector-restoration based compaction [6, 7] is that the resulting compacted test set guarantees to retain the original fault coverage. Various test generation methods based on compaction have been proposed [8–12] in which repeated calls

---

[1]Department of Electrical and Computer Engineering, Rutgers University, Piscataway, NJ

[2]Bell Labs, Lucent Technologies, Murray Hill, NJ

R. Lauwereins and J. Madsen (eds.), *Design, Automation, and Test in Europe–The Most Influential Papers of 10 Years DATE*, 455–464

© Springer 2008

to static compaction on modified test sets are performed. During each iteration, the test set is first modified by appending new vectors, then static compaction is called to remove any unwanted vectors. This process is repeated until a satisfactory coverage or a maximum number of iterations has been reached. In [8, 9], new vectors are appended to the test set by randomly choosing vectors from the compacted test set obtained in the previous iteration, while in [10–12], spatial and temporal correlations among test vectors are used to append new vectors to the test set.

In this work, we view the sequential circuit as a black-box system that is identifiable and predictable from its input–output signals, instead of viewing it as a netlist of primitives. In studying a signal, what we care most is the predictability of the signal. If the signal is predictable, we can use a portion of it (the past and the current) to represent and reconstruct its entirety. Testing of sequential systems, then, becomes the problem of constructing a set of waveforms, which when applied at the primary inputs of the circuit, can excite and propagate targeted faults in the circuit. These input waveforms (spectrums) have specific spectral characteristics, as exhibited in all signals. In order to capture the spectral characteristics of a given signal, a clean representation for that signal is desired (wider spectrums lead to more unpredictable/random signals). Thus, any embedded noise should be filtered. Static test set compaction reduces the size of the test set by removing any unnecessary vectors while retaining the useful ones. In other words, static compaction *filters* unwanted noise from the derived test sequence, leaving a cleaner signal (narrower spectrum) to analyze. Taking this idea to test generation, the spectral information obtained not only helps to identify embedded spectral information, it also offers a *new* way for testing sequential circuits by predicting intelligent vectors based on the vectors we have so far. Vectors generated from the narrow spectrum have better fault detection characteristics.

We also developed a technique to speed up the test generation process. Instead of using fault sampling during compaction to reduce the execution time, previously detected faults are periodically removed from the target fault list with the corresponding compacted vector sequence saved. In other words, compaction is performed using only the remaining faults. This significantly reduces the work during each iteration of compaction.

2 Overview and Motivation

Because what we care about most is the information embedded in the input signals (test set we already have), we want to employ signal-processing techniques to extract this information. Frequency decomposition is the most commonly used technique in signal processing. A signal can be projected to a set of independent waveforms that have different frequencies. This set of waveforms, each represented as a vector, forms a basis matrix. The projection operation (a postmultiply to the basis matrix) reveals the quantity each basis vector contributes to the original signal. This quantity is called decomposition coefficient. Subsequently, enhancing the important frequencies and suppressing the unimportant ones ("noise"), we expect that we can have a new and higher-quality signal that will help test generation.

In choosing the projection matrix, Hadamard transform is a well-known non-sinusoidal orthogonal transform in signal processing. It consists of only 1s and −1s, which makes it a good choice for the signals in VLSI testing (1 = logic 1, −1 = logic 0). Each basis in the Hadamard matrix is a distinct sequence that characterizes the switching frequency between 1s and −1s. For these reasons, we will use Hadamard transform for our analysis.

Figure 1 presents the overall framework of the spectral test generation procedure. Initially (iteration 0), the test set simply consists of random vectors. A call to static compaction will *filter* any unnecessary vectors. Using the Hadamard transform on the test set obtained, we *analyze* and *identify* the predominant pattern at each primary input. We generate test patterns based on this identified spectrum and filter out any unwanted random bits. At the same time, we can generate vectors spanning the *likely* vector space using only the basis vectors. This can potentially help drive the circuit into hard-to-reach states that require specific vectors at the primary inputs, making it easier to detect hard-to-detect faults faster. This process is repeated until either (1) desired fault coverage is obtained, or (2) maximum number of iterations is reached.

Hadamard matrices are square matrices containing only 1s and −1s, and can be generated using the following recurrence relation:

$$H_h(k) = \begin{bmatrix} H_h(k-1) & H_h(k-1) \\ H_h(k-1) & -H_h(k-1) \end{bmatrix}, \quad k = 1, 2, \ldots, n,$$

where $H_h(0) = 1$ and $n = log_2 N$. For example, with k = 1 and k=2, above equation yields

$$H_h(1) = \begin{bmatrix} 1 & 1 \\ 1 & -1 \end{bmatrix}, \quad H_h(2) = \begin{bmatrix} 1 & 1 & 1 & 1 \\ 1 & -1 & 1 & -1 \\ 1 & 1 & -1 & -1 \\ 1 & -1 & -1 & 1 \end{bmatrix}$$

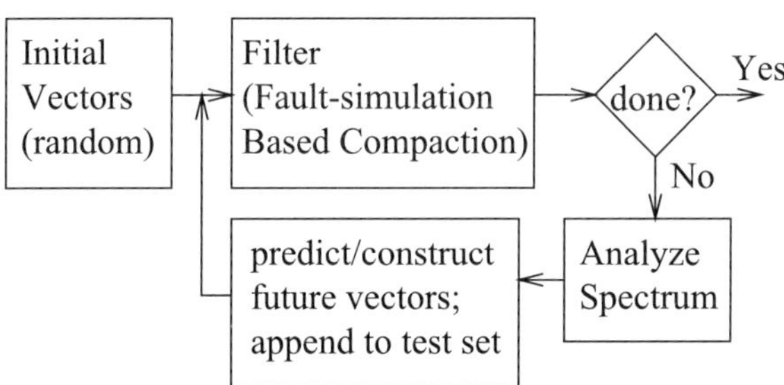

Fig. 1 Test generation framework

From this definition of the Hadamard matrix, we can observe that $H(n) \times H(n)^T = nI_n$, where I_n is the $n \times n$ Identity Matrix. Given only 1s and -1s in the matrices, multiplication can essentially be computed using additions and subtractions. Moreover, the inverse transform of a Hadamard matrix is the same as the forward transform, making reconstruction straightforward.

Each row / column in a Hadamard matrix is a basis vector, carrying a distinct frequency component. Taking $H_h(2)$ for illustration, the four basis vectors are [1 1 1 1], [1 0 1 0], [1 1 0 0], and [1 0 0 1]. Any bit sequence of length 4 can be represented as a linear combination of these basis vectors. For instance, the vector [1 0 0 0] can be written as $-1 \times [1, 1, 1, 1] + 1 \times [1, -1, 1, -1] + 1 \times [1, 1, -1, -1] + 1 \times [1, -1, -1, 1]$. Therefore, what we can do is project the test sequence to the Hadamard bases, filter out certain frequencies and do an inverse transform to get the *de-noised* sequence.

3   Test Generation Approach

The goal of the spectral technique is to generate / construct intelligent vectors from one iteration to the next. Based on the predominant spectra identified using the Hadamard transform, we construct new test patterns. Algorithm 1 illustrates the construction of such vectors.

Algorithm 1:

Let a_i be the input bit sequence for primary input i.

for (each primary input i in test set)
    coefficient vector $c_i = H \times a_i$
for (each value in the coefficient matrix $[c_0, ..., c_n]$)
    if (absolute value of coefficient < cutoff)
       Set the coefficient to 0.
    else
    Set the coefficient to 1 or -1, based on its abs value.
for (each primary input i)
    extension vector $e_i = modified\ c_i \times H$
if (weight > 0)
    Extend the vector set with value 1 to PI i.
else if (weight < 0)
    Extend the vector set with value -1 to PI i.
else if (weight == 0)
    Randomly extend the vector set with either 1 or -1

We will illustrate this algorithm with an example. Consider a subsequence of eight 4-input vectors. We first replace each '0' with with a -1:

$$
\begin{array}{cccc}
PI_0 & PI_1 & PI_2 & PI_3 \\
1 & 1 & 1 & 0 \\
1 & 0 & 0 & 1 \\
1 & 1 & 0 & 0 \\
1 & 1 & 0 & 1 \\
1 & 1 & 1 & 1 \\
1 & 0 & 0 & 0 \\
1 & 1 & 0 & 0 \\
1 & 0 & 0 & 1
\end{array}
\qquad \textit{replace } 0 \textit{ with} -1 \mapsto \qquad
\begin{array}{cccc}
PI_0 & PI_1 & PI_2 & PI_3 \\
1 & 1 & 1 & -1 \\
1 & -1 & -1 & 1 \\
1 & 1 & -1 & -1 \\
1 & 1 & -1 & 1 \\
1 & 1 & 1 & 1 \\
1 & -1 & -1 & -1 \\
1 & 1 & -1 & -1 \\
1 & -1 & -1 & 1
\end{array}
$$

Next, we perform spectral analysis. Consider the bit stream for primary input PI_1. Multiplying $H_h(3)$ with the bit stream for PI_1, we obtain its coefficient vector:

$$
\begin{bmatrix}
1 & 1 & 1 & 1 & 1 & 1 & 1 & 1 \\
1 & -1 & 1 & -1 & 1 & -1 & 1 & -1 \\
1 & 1 & -1 & -1 & 1 & 1 & -1 & -1 \\
1 & -1 & -1 & 1 & 1 & -1 & -1 & 1 \\
1 & 1 & 1 & 1 & -1 & -1 & -1 & -1 \\
1 & -1 & 1 & -1 & -1 & 1 & -1 & 1 \\
1 & 1 & -1 & -1 & -1 & -1 & 1 & 1 \\
1 & -1 & -1 & 1 & -1 & 1 & 1 & -1
\end{bmatrix}
\begin{bmatrix} 1 \\ -1 \\ 1 \\ 1 \\ 1 \\ -1 \\ 1 \\ -1 \end{bmatrix}
=
\begin{bmatrix} 2 \\ 6 \\ -2 \\ 2 \\ 2 \\ -2 \\ -2 \\ 2 \end{bmatrix}
$$

If the cutoff for coefficients is set to 4, then the coefficient vector c_i is modified by replacing every coefficient whose absolute value is less than 4 with 0. Thus, the new c_1 becomes $[0\ 1\ 0\ 0\ 0\ 0\ 0\ 0]^T$ Now, multiplying the new coefficient $[c_1]^T$ with $H_h(3)$ yields:

$$
\begin{bmatrix} 0 & 1 & 0 & 0 & 0 & 0 & 0 & 0 \end{bmatrix} \times H_h(3) = \begin{bmatrix} 1 & -11 & -11 & -11 & -1 \end{bmatrix}
$$

We will extend the test set for PI_1 with $(1, -1, 1, -1, 1, -1, 1, -1)$. Here, we have filtered out the random pattern that appeared in the input bit stream, since bit #4 has been changed from 1 to -1.

Now let us consider a different primary input PI_3. Multiplying H with PI_3 yields coefficients shown below:

$$
\begin{bmatrix}
1 & 1 & 1 & 1 & 1 & 1 & 1 & 1 \\
1 & -1 & 1 & -1 & 1 & -1 & 1 & -1 \\
1 & 1 & -1 & -1 & 1 & 1 & -1 & -1 \\
1 & -1 & -1 & 1 & 1 & -1 & -1 & 1 \\
1 & 1 & 1 & 1 & -1 & -1 & -1 & -1 \\
1 & -1 & 1 & -1 & -1 & 1 & -1 & 1 \\
1 & 1 & -1 & -1 & -1 & -1 & 1 & 1 \\
1 & -1 & -1 & 1 & -1 & 1 & 1 & -1
\end{bmatrix}
\begin{bmatrix} -1 \\ 1 \\ -1 \\ 1 \\ 1 \\ -1 \\ -1 \\ 1 \end{bmatrix}
=
\begin{bmatrix} 0 \\ -4 \\ 0 \\ 4 \\ 0 \\ -4 \\ 0 \\ -4 \end{bmatrix}
$$

If the cutoff for coefficients is again 4, then the coefficient vector is changed to
[0 −1 0 1 0 −1 0 −1]. Multiply the new coefficient with H_8 yields the following
extension vector:

$$\begin{bmatrix} 0 & -1 & 0 & 1 & 0 & -1 & 0 & -1 \end{bmatrix} \times H_h(3) = \begin{bmatrix} -2 & 2 & -2 & 2 & 2 & -2 & -2 & 2 \end{bmatrix}$$

Extending the compacted vector set for PI_3, we obtain scaled vector $(-1, 1, -1,
1, 1, -1, -1\ 1)$. This is the same as the input sequence, which means that there was
no random noise that needed to be filtered out.

Similarly, we obtain extended vector sets for PI_0 and PI_2 as $(1, 1, 1, 1, 1, 1, 1, 1)$
and $(1, -1, -1, -1, 1, -1, -1, -1)$. Thus, we append the following newly generated
vectors to the test set:

$$\begin{array}{cccc}
1 & 1 & 1 & -1 \\
1 & -1 & -1 & 1 \\
1 & 1 & -1 & -1 \\
1 & -1 & -1 & 1 \\
1 & 1 & 1 & 1 \\
1 & -1 & -1 & -1 \\
1 & 1 & -1 & -1 \\
1 & -1 & -1 & 1
\end{array}$$

Extending the compacted vector set in the above manner would lead to the extended
vector set having only twice as many vectors as the compacted vector set. In order
to append more vectors, we modify Algorithm 1 as follows: ($H_h(4)$ is used in
Algorithm 2 and the values of the coefficients range between +16 and −16)

Algorithm 2:

For each cutoff value of 6, 8, and 10 do
    Perform Algorithm 1.
    Randomly pick p vectors from the extended test set, and hold each vector an
arbitrary number of cycles.

Increasing the cutoff value is equivalent to fine-tuning the filter, leading to bit
patterns very similar to the original input stream. This technique leads to the generation of a sizeable number of "useful" vectors, which helps detect hard-to-detect
faults rapidly.

4 Speeding up Test Generation with Fault Dropping

One of the ways to reduce the computation costs of simulation-based test generation is by reducing the number of faults during simulation. Fault sampling is used
during the compaction process in [8, 9], in which a sample of 128 or 256 randomly

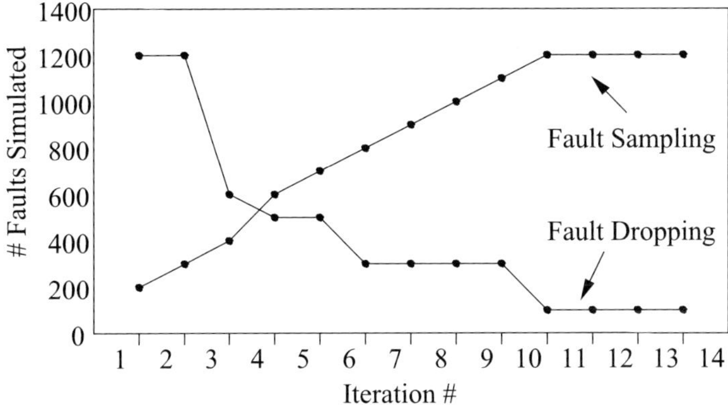

Fig. 2 Fault sampling vs. fault dropping

chosen faults were used as targets during fault simulation and compaction. As more faults become detected, the target fault list is replenished, until all the faults are included in the target fault list.

Instead of using fault sampling to reduce the test generation time, we developed a *fault-dropping* technique to speed up test generation. In fault dropping, detected faults are periodically removed from the target fault list, reducing the time taken for compaction. Figure 2 illustrates the difference in the two concepts. In the fault sampling approach, test generation starts with a smaller number of target faults, and gradually add new faults to the target list. This is illustrated by the rising curve. On the contrary, the fault-dropping technique starts with the entire fault list, and remove the detected faults periodically to reduce the target fault list. This is illustrated by the falling curve. When faults are removed, the test sequence that detected those faults must be saved. Finally, these saved sequences are combined and a final compaction is called to produce the final test set. In smaller circuits that required only a few iterations to obtain the desired fault coverages, faults are removed every 1 or 2 iterations. For larger circuits, we remove faults less frequently - 4 or 5 iterations between removal of faults.

5   Experimental Results

All experiments were conducted on an Ultra SPARC 10 with 256 MB of RAM for ISCAS89 [16] and ITC99 [17] benchmark circuits. The 16×16 Hadamard matrix $H_h(4)$ is used for all circuits. Results for four test generators are compared: HITEC [14], STRATEGATE [15], PROPTEST [9], and the proposed spectral method. HITEC is a deterministic test generator, STRATEGATE is a genetic-based test generator and PROPTEST is a compaction-based test generator.

Table 1 reports the results: the circuit name is first given, then, the number of faults detected, test set size, and execution time (in minutes) are reported for each

Table 1 Test generation results

Circuit	HITEC [14]			STRATEGATE [15]			PROPTEST [9]			Spectral-Based		
	Det	Vec	Time	Det	Vec	Time	Det	Vec	Time	Det	Vec	Time
s382	301	1463	90.0	364	1486	8.1	364	572	0.5	364	567	1.0
s400	341	1845	72.0	384	2424	15.5	382	677	0.7	382	588	1.5
s713	476	173	0.1	476	176	1.3	476	104	0.1	476	89	0.4
s1196	1239	435	0.1	1239	574	1.5	1239	224	0.4	1239	244	1.2
s1238	1283	475	0.1	1282	624	1.5	1283	235	0.4	1283	255	1.1
s1423	723	150	834.0	1414	3943	76.2	1416	1049	8.8	1416	927	16.3
s1488	1444	1170	16.5	1444	593	7.5	1444	426	1.9	1444	384	3.2
s1494	1453	1245	9.6	1453	540	7.6	1453	454	2.2	1453	388	2.9
s5378	3231	912	1104.0	3639	11571	2268.0	3643	672	35.6	3643	734	43.5
b01	–	–	–	133	86	0.1	–	–	–	133	50	0.1
b04	–	–	–	1168	2680	10.3	–	–	–	1168	158	0.8
b08	–	–	–	463	1567	0.7	–	–	–	463	387	1.5
b11	–	–	–	1003	4883	50.0	1004	419	1.6	1004	962	2.5
b12	–	–	–	1488	33113	9659.0	1470	3697	27.5	1645	4464	24.2

Time reported in minutes

Different platforms were used for different test generators

ATPG method. Note that the different platforms were used for different test generators:. HP 9000 J200 for HITEC, Sun UltraSPARC 1 for STRATEGATE, Pentium II for PROPTEST, and Sun UltraSPARC 10 for the spectral technique. PROPTEST [9] used fault samples of 256 initially, and gradually increase the sample size until all the undetected faults are targeted in later iterations. On the other hand, the spectral technique used the proposed fault-dropping technique. We fixed an upper bound of 125 iterations as the terminating condition.

From Table 1, we observe that in all cases, the spectral technique was able to obtain very high fault coverages very quickly, and the test sets were also very compact. For instance, in circuit b12, our spectral technique detected 1645 faults, 175 more faults than best reported coverages. For the rest of the circuits, the spectral technique was able to reach the maximum coverages as well. In terms of test set sizes, the spectral technique results in smaller test sets for many of the circuits. The execution times were less than HITEC and STRATEGATE for most cases, and they were slightly longer than PROPTEST due to different platform, programming style, and the extra effort needed to obtain highly compact test sets and to perform spectral analysis.

To see whether the spectral technique is truly effective in generating intelligent vectors, we compare the number of faults detected at each iteration of the test generation process. We implemented a compaction-based test generator similar to [8], and used that test generator as a comparison. Figure 3 shows the number of faults detected in circuit b12 over the first 100 iterations by our spectral technique and by our implementation of [8]. As shown in the figure, the spectral technique detects 1,600 faults by the fourth iteration, while [8] reached 1400 faults (200 fewer) after 60 iterations. Figure 4 shows comparisons of the number of iterations needed to reach maximum fault coverages by our spectral technique and by our implementation

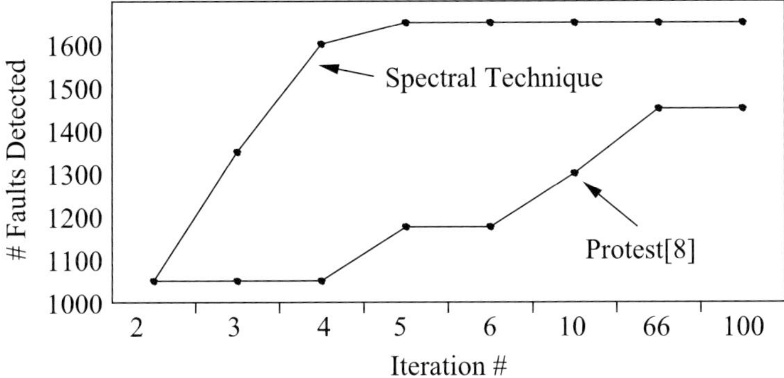

Fig. 3 Faults detected per iteration for b12

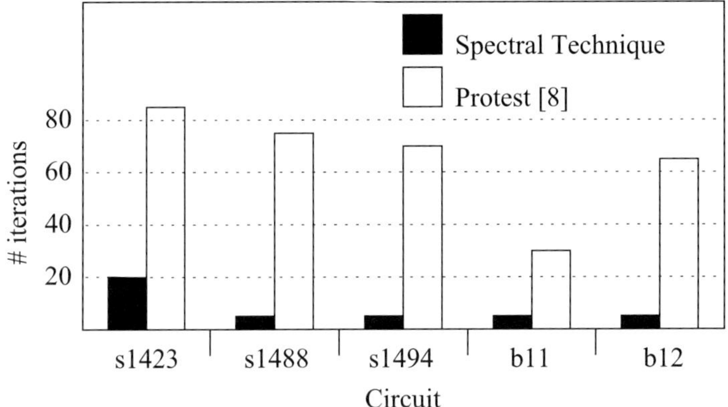

Fig. 4 Number of iterations needed

of [8]. As illustrated in this figure, the spectral technique consistently reached the final fault coverage in much fewer iterations than [8]. For example, in s1488, only 3 iterations were needed by the spectral technique, while the compaction-based technique [8] required 79 iterations. This demonstrates the effectiveness and potential of the spectral technique.

6 Conclusions

We have presented a novel spectral technique for test generation. Static compaction is first used to filter unwanted vectors. Then, Hadamard transform is used to analyze input spectrums. This technique identifies inherent periodicity of the input bits. Vector sequences, then, can be represented as a linear combination of the basis vectors (columns of the Hadamard matrix). Experiments conducted using this spectral technique showed that very high fault coverage with small test sets can be rapidly obtained in a few iterations. Together with the use of the proposed fault dropping method, a considerable computation cost reduction was achieved. In circuits such as b12, we achieved a noteworthy increase in the fault coverage when compared to any previously reported technique.

References

[1] M. A. Breuer, A random and an algorithmic technique for fault detection test generation for sequential circuits, *IEEE Trans. Computers*, vol C-20, no. 11, 1364–1370, Nov. 1971.

[2] M. Abramovici, M. A. Breuer and A. D. Friedman, *Digital System Testing and Testable Design*, New York, NY: Computer Science Press, 1990.

[3] V. D. Agrawal, When to use random testing, *IEEE Trans. Computers*, vol C-27, no. 11, 1054–1055, Nov. 1978.

[4] M. F. Alshaibi and C. R. Kime, Fixed-biased pseudorandom built in self test for random pattern resistant circuits, *Proc. Intl. Test Conf.*, 1994, pp. 929–938.

[5] F. Muradali, T. Nishada, and T. Shimizu, Structure and technique for pseudo random-based testing of sequential circuits, *J. Electron Testing: Theory Appl.*, 107–115, Feb. 1995.

[6] I. Pomeranz and S. M. Reddy, Vector restoration based static compaction of test sequences for synchronous sequential circuits, *Proc. Intl. Conf. Computer Design*, 1997, pp. 360–365.

[7] R. Guo, I. Pomeranz, and S. M. Reddy, Procedures for static compaction of test sequences for synchronous sequential circuits based on vector restoration, *Proc. Design, Aut., and Test in Europe*, 1998, pp. 583–587.

[8] R. Guo, I. Pomeranz, and S. M. Reddy, A fault simulation based test pattern generator for synchronous sequential circuits, *Proc. VLSI Test Symp.*, 1999, pp. 260–267.

[9] R. Guo, S. M. Reddy and I. Pomeranz, PROPTEST: a property based test pattern generator for sequential circuits using test compaction, *Proc. Design Aut. Conf.*, 1999, pp. 653–659.

[10] A. Jain, V. D. Agrawal, M. S. Hsiao, On generating tests for sequential circuits using static compaction, *Intl Test Synthesis Workshop*, March 1999.

[11] S. Sheng, A. Jain, M. S. Hsiao and V. D. Agrawal, Correlation analysis of compacted test vectors and the use of correlated vectors for test generation, *IEEE Intl. Test Synthesis Workshop*, March, 2000.

[12] A. Giani, S. Sheng, M. S. Hsiao and V. D. Agrawal, Correlation-based test generation for sequential circuits, *Proc. IEEE North Atlantic Test Workshop*, 2000, pp. 76–83.

[13] A. V. Oppenheim, R. W. Schafer and J. R. Buck, *Discrete-Time Signal Processing*. Englewood Cliffs, New Jersey: Prentice Hall, 1999.

[14] T. M. Niermann and J. H. Patel, HITEC: A test generation package for sequential circuits, *Proc. European Design Aut. Conf.*, 1991, pp. 214–218.

[15] M. S. Hsiao, E. M. Rudnick and J. H. Patel, Dynamic state traversal for sequential circuit test generation, *ACM Trans. Design Aut. Electronic Systems*, vol. 5, no. 3, 548–565, July, 2000.

[16] F. Brglez, D. Bryan, and K. Kozminski, Combinational profiles of sequential benchmark circuits, *Int. Symposium on Circuits and Systems*, 1989, pp. 1929–1934.

[17] S. Davidson and Panelists, ITC '99 benchmark circuits – preliminary results, *Proc. Int. Test Conf.*, 1999, pp. 1125. also at www.cerc.utexas.edu/itc99-benchmarks/bench.html

BerkMin: A Fast and Robust Sat-Solver

Evgueni Goldberg[1] and Yakov Novikov[2]

Abstract We describe a SAT-solver, BerkMin, that inherits such features of GRASP, SATO, and Chaff as clause recording, fast BCP, restarts, and conflict clause "aging". At the same time BerkMin introduces a new decision making procedure and a new method of clause database management. We experimentally compare BerkMin with Chaff, the leader among SAT-solvers used in the EDA domain. Experiments show that our solver is more robust than Chaff. BerkMin solved all the instances we used in experiments including very large CNFs from a microprocessor verification benchmark suite. On the other hand, Chaff was not able to complete some instances even with the timeout limit of 16h.

1   Introduction

Given a conjunctive normal form (CNF) F specified on a set of variables $\{x_1,\dots,x_n\}$, the satisfiability problem is to satisfy (set to 1) all the disjunctions of F by some assignment of values to variables from $\{x_1,\dots,x_n\}$. A disjunction of F is also called a clause of F. Many problems such as ATPG [16], logic synthesis [5], equivalence checking [6, 9], and model checking [4] reduce to the satisfiability problem.

In the last decade substantial progress has been made in the development of practical SAT algorithms [1, 2, 8, 11–13, 15, 18]. All of them are search algorithms that aim at finding a satisfying assignment by variable splitting. Search algorithms of that kind are descendants of the DPLL-algorithm [7].

DPLL-algorithm can be considered as a special case of general resolution which is called tree-like resolution. It was shown in [3] that there is exponential gap between the performance of tree-like resolution and that of general resolution.

Modern SAT-solvers have made at least two steps towards general resolution trying to eliminate the drawbacks and limitations of pure tree-like resolution. First,

[1] Cadence Berkeley Labs(USA), Email: egold@cadence.com

[2] Academy of Sciences (Belarus), Email: nov@newman.bas-net.by

R. Lauwereins and J. Madsen (eds.), *Design, Automation, and Test in Europe–The Most Influential Papers of 10 Years DATE*, 465–478
© Springer 2008

they record so called conflict clauses [15], which are implicates of the original CNF. Adding conflict clauses allows one to prune many of the branches of the search tree that are yet to be examined [1, 13, 15, 18]. The deduced implicates are just added to the current CNF which we will also refer to as (clause) database.

Second, some of the state-of-the-art SAT-solvers use the strategy of restarts when the SAT-solver abandons the current search tree (without completing it) and starts a new one. So instead of one complete search tree the SAT-solver constructs a set of incomplete (except the last one) trees. In [1, 10] the usefulness of restarts was proven experimentally. Restarts are effectively used in Chaff [13].

We introduce a new SAT solver called BerkMin (BerkMin stands for Berkeley–Minsk, the cities where the authors live). BerkMin can be considered as the next representative of the family of SAT-solvers that includes GRASP [15], SATO [18], Chaff [13]. BerkMin uses the procedures of conflict analysis and nonchronological backtracking introduced in GRASP, fast BCP suggested in SATO, and Chaff's idea of reducing the contribution of "aged" conflict clauses into decision making. Besides, BerkMin uses restarts.

At the same time BerkMin introduces many new features into decision-making and clause-database management. First, the set of conflict clauses is organized as a chronologically ordered stack (the top clause is the one deduced the last). If in the current node of the search tree there are unsatisfied conflict clauses, the next branching variable is chosen among the free variables, whose literals are in the top unsatisfied conflict clause. Second, we introduce a heuristic to pick which out of two possible assignments to the chosen branching variable should be examined first. In the case of using a single search tree, spending time on the selection of the branch to be examined first makes sense only for satisfiable CNFs. (For unsatisfiable CNFs both branches are "symmetric" i.e. the search tree size is not affected by whichever branch is examined first.) However, when using restarts the symmetry between the two alternative branches is broken even for unsatisfiable CNFs.

Third, our procedure of computing the activity of variables is different from that of Chaff. The activity of variables in conflict making is used by Chaff to single out good candidates for branching variables. For computing the activity of a variable x Chaff counts the number of occurrences of x in conflict clauses. This may lead to overlooking some variables that do not appear in conflict clauses while actively contributing to conflicts (e.g. if these variables are deduced). BerkMin solves this problem by taking into account a wider set of clauses involved in conflict making. Fourth, we use a new procedure of clause database management performed after the current search tree is abandoned. The novelty of the procedure is that the decision whether a database clause should be removed is based not only on its size (the number of literals). It is also based on the "activity" of this clause in conflict making and its "age".

We experimentally compare the performance of BerkMin with that of Chaff that is currently considered as the best SAT-solver used in the EDA domain. Experiments clearly show that BerkMin is more robust than Chaff. By greater robustness of BerkMin we mean that it is able to solve more instances than Chaff in a reasonable amount of time. Though Chaff is faster on some instances, BerkMin does not have problems solving them. On the other hand, we give examples of CNFs that are relatively easy for BerkMin and that cannot be solved by Chaff even with the timeout limit of 16 hours.

In particular, BerkMin solved all the instances of the benchmark suite fvp-unsat2.0 [17] in less than 2 h. This class consists of large unsatisfiable CNFs (except one instance) describing verification of pipelined microprocessors. The proof of unsatisfiability of the instance pipe6.cnf from this suite by the best SAT-solver takes 60 h of CPU time on a SUN computer with clock frequency 336 MHz [17]. BerkMin solved this instance in less than 20 min on a SUN computer with clock frequency 450 MHz. A more complex instance pipe7.cnf that no SAT-solver has been able to complete was solved by BerkMin in 1 h.

2 Basic Notions

Given a CNF F, a DPLL-algorithm-based SAT-solver looks for a satisfying assignment that is also called a solution to the satisfiability problem. Search is organized as a binary tree. If no solution is found after completing search tree examination, then F is unsatisfiable. At each node of the search tree the following three steps are performed: (a) choosing the next variable to split on and selecting the value, 0 or 1, to be first assigned to the chosen variable; (b) running the *Boolean Constraint Propagation* (BCP) procedure; (c) performing conflict analysis and backtracking if a conflict is encountered.

The choice of the next branching (splitting) variable is usually based on a heuristic. Different heuristics are used for different classes of CNFs (e.g. [1, 8, 11–13, 15, 18]). The heuristics used by BerkMin are described in Section 4.

After choosing a branching variable (say variable y) and its first value, the *BCP procedure* is initiated. It performs as follows. Suppose that in the first branch value 1 is assigned to variable y. Then all the clauses having the positive literal of y are satisfied and so removed from the current CNF. From all the clauses having the negative literal of y, this literal is removed. This may result in producing *unit clauses* i.e. clauses having only one literal left. Suppose that $\sim x$ is such a unit clause. This clause can be satisfied only by the assignment $x = 0$. This assignment is called *deduced* from clause $\sim x$. After making the deduced assignment (in our case $x = 0$) the procedure of discarding satisfied clauses and removing literals that are set to 0 (in our case positive literals of x) is performed again. This may lead to producing new unit clauses. The BCP procedure is performed until 1) a solution if found or 2) all clauses of the current CNF contain at least two literals or 3) an empty clause is produced (this situation is called a conflict). Before producing an empty clause two unit clauses having the opposite literals of the same variable must appear, for example z and $\sim z$. Deducing the value of z from either clause makes the other clause unsatisfiable. The efficiency of the BCP procedure can be improved by using the technique suggested in SATO [18] that was also incorporated into Chaff [13].

We omit the description of this technique. In BerkMin, our own version of SATO's BCP procedure is implemented.

Suppose that BCP has been completed and no solution is found in the current node. If no conflict is encountered then the algorithm moves away from the root to the next node. If there is a conflict, the algorithm backtracks. Suppose that in the last chosen assignment leading to a conflict, x was equal to 0. The simplest back-tracking is chronological: assignment $x = 0$ is undone and all deduced assignments obtained by the BCP procedure, initiated by $x = 0$, are undone as well. The aim of backtracking is to restore the CNF as it was before making the assignment $x = 0$. If assignment $x = 1$ has not been examined yet then it is made now and the BCP pro-cedure is initiated. Otherwise, the algorithm backtracks to the first node of the search tree where the alternative assignment has not been examined yet. However, conflict analysis sometimes allows to backtrack nonchronologically, skipping a set of nodes of the search tree. The technique of conflict analysis and nonchronological backtracking is given in detail in [15]. In Section 3 we give a short description of this technique.

## 3	Nonchronological Backtracking and Conflict Analysis

The notion of conflict analysis is crucial for understanding nonchronological back-tracking. Let F be a CNF consisting of the set of clauses of the initial CNF F^* and implicates of F^* deduced during search. Let C be a clause of F. Suppose that a con-flict is caused by clause $C = \sim a \lor x \lor \sim c$ being unsatisfiable. That is, variables a,x,c have been assigned $a = 1$, $x = 0$, $c = 1$ which resulted in removing all literals from clause C. A set R of value assignments to variables of F is called an assignment leading to a conflict on C (or a *conflict assignment* for short) if after making the assignments from R and running the BCP procedure, clause C becomes unsatisfiable.

A trivial assignment leading to a conflict on $C = \sim a \lor x \lor \sim c$ is the set $\{a = 1, x = 0, c = 1\}$ since after making these value assignments all literals of C are removed. However, this conflict assignment is of no interest because it does not provide any new information. Suppose, that we found out that $R=\{x = 0, y = 1, z = 1\}$ is also an assignment leading to a conflict on C. Then clause $x \lor \sim y \lor \sim z$ specified by set R is an implicate of the current CNF F and can be added to F. Clause $x \lor \sim y \lor \sim z$ record-ing conflict assignment $R=\{x = 0, y = 1, z = 1\}$ is called a **conflict clause**.

Suppose that assignment $x = 0$ of $R=\{x = 0, y = 1, z = 1\}$ is chosen (or deduced) in the current node. Let us assume that values $y = 1$ and $z = 1$ are deduced at nodes $v\sim$ and v' respectively. Suppose also that $v > v\sim > v'$ where $n_1 > n_2$ means that node n_2 is closer to the root than n_1. After adding conflict clause $x \lor \sim y \lor \sim z$ to the current CNF we can back-jump from node v right to node $v\sim$ (skipping the nodes that may lie between $v, v\sim$). The point is that current CNF $F_{v'}$ at node v' is different from the one obtained at node v' before, because clause $x \lor \sim y \lor \sim z$ has been added to the database. Since variables y and z are assigned value 1 at nodes v' and v', then at node

v' clause $x \vee \sim y \vee \sim z$ of CNF F_v is just equal to x and so assignment $x = 1$ can be deduced from it. Before adding clause $x \vee \sim y \vee \sim z$ this deduction was not possible. With this clause in the current CNF we can back-jump to node v' to make this deduction.

Conflict analysis and conflict clause derivation are described in [15]. We sketch here the main idea. Suppose that at the current node v clause $\sim a \vee x \vee \sim c$ becomes unsatisfiable. We need to find a conflict assignment R that includes *only one* assignment "initiated" at node v. (This is either an assignment made to the branching variable of node v or an assignment deduced in the BCP procedure performed at node v.) The rest of the assignments of R are made at nodes located above node v (i.e. closer to the root).

Such a conflict assignment can be constructed from the trivial conflict assignment $R_1 = \{a = 1, x = 0, c = 1\}$ by performing the BCP procedure "backwards". For example, suppose that all the value assignments of R_1 were made at node v and assignment $a=1$ was deduced from clause $a \vee x \vee \sim z$ (due to assignments $x=0$, $z=1$ made before). Now R_1 can be replaced with a non-trivial conflict assignment $R_2 = \{x = 0, c = 1, z = 1\}$. (This conflict assignment is obtained from R_1 by replacing assignment $a = 1$ with $x = 0, z = 1$.) Suppose that assignment $z = 1$ was deduced at node v' located above v. Then conflict assignment R_2 contains only two assignments ($x = 0$ and $c = 1$) that were made at node v. Proceeding with the "reverse" BCP procedure we will eventually produce a conflict assignment containing only one assignment made at node v. Suppose, for example, that assignment $c = 1$ was deduced from clause $c \vee \sim y \vee \sim z$ and assignment $y = 1$ was made at node v'' located above v. By replacing $c = 1$ with $y = 1, z = 1$ we obtain from R_2 conflict assignment $R_3 = \{x = 0, y = 1, z = 1\}$. R_3 satisfies our requirement because only one assignment of R_3, namely $x=0$, was made at node v. The corresponding conflict clause is $x \vee \sim y \vee \sim z$.

Formally, the deduction of clause $x \vee \sim y \vee \sim z$ can be described by the following chain of resolutions: $(\sim a \vee x \vee \sim c) \wedge (a \vee x \vee \sim z) \rightarrow x \vee \sim c \vee \sim z$, $(x \vee \sim c \vee \sim z) \wedge (c \vee \sim y \vee \sim z) \rightarrow x \vee \sim y \vee \sim z$. Only the final result of this chain (in our case clause $x \vee \sim y \vee \sim z$) is added to the current CNF. The intermediate conflict assignments (like clause $x \vee \sim c \vee \sim z$) are thrown away. The clauses of the current CNF that are used in the deduction of the final conflict clause will be called *clauses responsible for the conflict*. In our example, clauses $\sim a \vee x \vee \sim c$, $a \vee x \vee \sim z$, $c \vee \sim y \vee \sim z$ are responsible for the conflict. Clauses responsible for a conflict are identified during the reverse BCP procedure used to construct the conflict clause to be added to the database.

4 BerkMin's Decision Making

When designing the decision-making strategy of BerkMin we tried to take into account the following two facts. (1) Current SAT-solvers can quickly solve fairly large "real-life" CNFs. This suggests that in such CNFs short implicates can be deduced from a small subset of clauses by branching on a small subset of variables. (2) The set of variables responsible for conflicts may change very quickly. Suppose, for example, that a variable x of the CNF to be tested for satisfiability describes the output of a gate feeding many AND gates. Then in the branch $x = 0$ the variables corresponding to the rest of the inputs of these AND gates are not involved in conflict making. However, in the branch $x = 1$ all these variables immediately become "active" again.

The two facts above imply that the choice of branching variables for "real-life" CNFs should be very dynamic. It must quickly adapt to all the changes of the set of relevant variables caused by new value assignments. Chaff's idea of conflict clause "aging" is a significant contribution to creating such a dynamic decision making strategy.

Here is how this idea is implemented in BerkMin. Each variable is assigned a counter (denote it by $ac(z)$) that stores the number of clauses, responsible for at least one conflict, that have a literal of z. The value of $ac(z)$ is updated during the reverse BCP procedure. As soon as a new clause responsible for the current conflict is encountered, the counters of the variables, whose literals are in this clause, are incremented by 1. The values of all counters are periodically divided by a small constant assignment to a variable x is aimed at having uniform distribution of positive and negative literals in conflict clauses of the database. For this purpose, for each literal l cost function $lit\_ac(l)$ is computed. $lit\_ac(l)$ gives the number of conflict clauses generated so far that contain literal l. Initially $lit\_ac(l)$ is equal to 0. As soon as a conflict clause C is added to the database, for each literal l of C the value of $lit\_ac(l)$ is incremented by 1. The value of $lit\_ac(l)$ is not divided by a constant (as it is done for counters $ac(z)$). Besides, the value of $lit\_ac(l)$ is not recomputed after some conflict clauses are removed from the database according to the rules described in section 5.

If x is the next branching variable, the literal l, $l \in \{x, \sim x\}$ with the largest value of $lit\_ac(l)$ is selected. If $lit\_ac(x) = lit\_ac(\sim x)$, literal l, $l \in \{x, \sim x\}$ is selected at random. Then x is assigned the value setting the chosen literal l to 1. So all clauses having literal l become satisfied and no conflict clause can contain l. On the other hand, the opposite literal (i.e. $\sim l$) is set to 0 and so conflict clauses may include l. So, setting the literal with the largest value of $lit\_ac(l)$ to 1, we reduce the gap between the numbers of occurrences of x and $\sim x$ in conflict clauses of the database.

Now we describe how the branch to be examined first is selected when the current top clause is a clause of the original CNF. We will call a clause *binary* if it contains only two literals. For each literal l, a cost function $nb\_two(l)$ is computed that approximates the number of binary clauses in the "neighborhood" of literal l. Function $nb\_two(l)$ is computed as follows. First, the number of all binary clauses containing literal l is calculated. Then for each binary clause C containing literal l, the number of binary clauses containing literal $\sim v$ is computed where v is the other literal of C. The sum of all computed numbers gives the value of $nb\_two(l)$. This cost function can be considered as an estimate of the power of BCP performed after setting l to 0. The greater the value of $nb\_two(l)$ is the more assignments will be deduced from the binary clauses containing literal $\sim v$ after setting literal l to 0 in clause C. This is the reason why, given the next branching variable x, the literal l, $l \in \{x, \sim x\}$, with the greatest value of nb_two (l) is selected. If nb_two $(x) = nb\_two$ $(\sim x)$ then $l \in \{x, \sim x\}$ is chosen at random. Then x is assigned the value setting literal l to 0. To reduce the amount of time spent on computing nb_two (l) we use a threshold value (in our experiments it was equal to 100). As soon as the value of nb_two (l) exceeds the threshold its computation is stopped. It should be noticed that the idea of taking into account binary clauses in decision making is not new. For example, it is successfully used in SATZ [12] (though cost functions of SATZ are different from ours).

BerkMin uses a relatively complex decision-making procedure. However this complexity has been justified in numerous experiments. This fact refutes the claim made in [14] that for SAT-solvers using clause recording the quality of decision making is of no great importance.

5   Clause Database Management

Before starting the next iteration (i.e. building a new search tree), some clauses are "physically" removed from the database. This allows one to reduce memory used for database allocation. In the process of removing clauses BerkMin's data structures are partially or completely recomputed to fit them into smaller memory blocks.

A fraction of clauses is removed "automatically" due to retaining some value assignments deduced in the last iteration. Namely, all the value assignments, that were deduced from unit conflict clauses (if any) and in the BCP procedure triggered by the assignments deduced from unit conflict clauses, are retained in the new iteration. All the clauses that are satisfied by the retained assignments are removed from the current CNF.

The rest of the clauses to be removed are selected using heuristics described below. Our approach is based on the following hypothesis. Recently deduced clauses are more valuable because it took more time to deduce them from the original set of clauses.

From the view point of clause removal heuristics, the set of conflict clauses is a queue. New conflict clauses are added to the tail of this queue while clauses to be considered for removal are located in its head. In the experiments, from the head part of the queue whose size is 1/16th of the whole queue, all clauses with more than 8 literals were removed. From the rest 15/16 of the queue clauses containing more than 42 literals were removed. However, BerkMin always keeps in the queue the last conflict clause and a fraction of "active" clauses regardless of how many literals they have. The activity of a clause C is measured by the number of conflicts for which C has been responsible so far. In the head part of the queue only clauses with activity value greater than 60 are considered as active. In the rest of the queue active clauses are the ones whose activity is greater than 7. The threshold value of the activity for the clauses of the head part of the queue is increased every 1,024 nodes (the starting value is 60). So large clauses that are not used in conflicts any more will be soon removed from the database.

The current search tree is abandoned after generating 550 conflict clauses (that are added to the tail of the queue). Before starting a new search tree BerkMin starts removing clauses trying to get rid of at least 1/16 of the conflict clauses forming the queue. If after applying the rules described above, the number of removed clauses was less than 1/16 of the queue, the threshold on the size of clauses removed from the head part decreased by one literal. (However, after the threshold value reaches 4, it does not decrease any more.) In our experiments it was almost always possible to remove 1/16 of the queue, so greater than 1. (This constant is equal to 2 for Chaff and 4 for BerkMin.) This way the influence of "aged" clauses is decreased and preference is given to recently deduced clauses.

In Chaff, only literals of the conflict clauses added to the current CNF are taken into account when computing the "activity" of a variable z. (So if a literal of z occurs in a clause responsible for the current conflict but z is not in the conflict clause, Chaff does not change the value of the activity of z.) Taking into account a wider set of clauses responsible for conflicts allows BerkMin to get a more accurate estimate of the activity of variables. If a variable z is frequently assigned through deduction it may drive many conflicts without appearing in conflict clauses. So if, when computing the activity of z, one takes into account only conflict clauses, this activity will be underestimated.

Currently, BerkMin has two decision-making procedures. The first procedure is to pick the variable z with the maximum value of $ac(z)$ as the next branching variable. However, it is BerkMin's second choice. The main decision making procedure (that was used in all experiments) is based on chronological conflict clause ordering.

The reason for using such an ordering is that we believe that Chaff's strategy is not "dynamic" enough. In particular, it cannot adjust quickly to changes of the set of relevant variables mentioned in the beginning of the section. Consider the following example. Suppose that Chaff divides counters by 2, say, every 100 conflicts. Suppose that immediately after dividing the counters, a change of relevant variables occurred. Then for the next 20–30 conflicts the choice of branching variables will be dominated by an "obsolete" set of active variables.

Here is how the problem is solved in BerkMin. The clause database is organized as a stack. The clauses of the initial CNF are located at the bottom of the stack and each new conflict clause is added to the top of the stack. In the process of assigning values some clauses of the stack get satisfied. However there is always an unsatisfied clause that is the closest to the top of the stack. We will call this clause the *current top clause*. If the current top clause is a conflict clause then it is just the most recently deduced conflict clause that is not satisfied yet.

Let C be the current top clause. The idea is to choose the next branching variable among free (unassigned yet) variables of C, if it is a conflict clause. Among the variables of C, the variable z with the largest value of $ac(z)$ is selected. Note, that variable z may look very "passive" in the set of all the currently free variables.

However, the activity of these variables may be the result of a very different set of conflicts that happened before deducing C or would occur after the deduction of C. These conflicts may involve quite different sets of variables. Choosing z we take into account the fact that these active variables may be irrelevant to the conflict which had led to the deduction of C and to "similar" conflicts i.e. the ones involving sets of variables that are close to that of C.

Let ~z be the literal of variable z that is in the top clause C. Suppose that z is chosen as the next branching variable. BerkMin does not try to satisfy C immediately by assigning $z = 0$. It has a separate procedure for choosing the branch to be examined first. This procedure is further on. However, it is not hard to see that clause C will become satisfied after no more than $n-1$ assignments to branching variables where n is the number of literals of C. Indeed, if C is not satisfied by the chosen assignment to z and no conflict is produced during the following BCP procedure, C remains the current top clause. Then a new free variable is selected from the ones whose literals are in C. So eventually either C will be satisfied by an assignment to a branching variable (or by deduced assignment) or after removing from C all literals but 1 it will be satisfied during the BCP procedure.

If the current top clause is a clause of the original CNF then all the unsatisfied clauses in the stack are also clauses of the original CNF. In this case the most active free variable of the current CNF (i.e. free variable z with the greatest value of $ac(z)$) is selected.

Before describing how the branch to be examined first is selected, we would like to make a few comments. It is obvious that if the initial CNF is satisfiable, there may be asymmetry between the two alternative branches regardless of whether or not the used algorithm has restarts. However, if the initial CNF is unsatisfiable and the algorithm builds a single search tree, the two alternative branches are symmetric. That is, the search tree size is not affected by the choice of the branch to be examined first.

Indeed, suppose that x is the next branching variable and branch $x = 0$ is examined first. The size of the subtree built in the alternative branch $x = 1$ is, in general, affected by conflict clauses deduced in branch $x = 0$. However these conflict clauses cannot contain literal x because a clause containing literal x is satisfied by $x = 1$. So the only kind of conflict clauses that affect computations in branch $x = 1$ are the ones that do not contain a literal of x. But the conflicts described by such clauses are symmetric in x and can be deduced in either branch. So branches $x = 0$ and $x = 1$ "interact" only through conflict clauses that are symmetric in x. Then the size of subtrees constructed in branches $x = 0$ and $x = 1$ does not depend on which one is examined first. However, in case of using restarts, asymmetry between the alternative branches exists even for unsatisfiable CNFs. The reason is that, when branching on a variable x, the algorithm does not commit to examining the alternative branch if no solution was found in the branch examined first.

Let us describe now the procedure used for branch selection. First we consider the case when the current top clause is a conflict one. In this case the choice of the first the clause database was kept small. It should be noted that the described procedure of clause removal makes BerkMin incomplete because there is a possibility of looping. Indeed, it is possible that the algorithm removes and then deduces the same set of clauses. For example, if all the clauses deduced in the current iteration have more than 42 literals and their activity is less than 7, all of them will be removed from the database. Then after restarting the algorithm, the same set of clauses will be deduced and then removed again and so on.

A simple way of eliminating the possibility of looping is as follows. One conflict clause deduced since the last restart is marked and forever forbidden to be removed from the database (unless it is satisfied by an assignment retained from the previous iteration). Then we guarantee that the number of marked clauses in the database grows monotonically and so no looping is possible. The number of marked clauses to be kept can be reduced n times by marking a clause only after performing n iterations (restarts). This technique is not hard to implement. However, we have not done that because BerkMin never looped in our experiments.

6   Experimental Results

BerkMin was written from scratch in Microsoft's Visual C++ under Windows-95. Then it was ported to Solaris using GNU's C++ compiler gcc. In the experiments, we compared BerkMin with Chaff [13]. The binary of Chaff was downloaded from [19]. Both programs were run on the same SUNW, Ultra-80 system with clock frequency 450 MHz and 4 Gbytes of memory.

The results of the experiments are shown in Tables 1–3. Tables 1 and 2 are mostly self-explanatory. Table 1 contains instances for which BerkMin's and Chaff's performances are comparable. Table 1 gives results on two classes of instances from the Dimacs benchmark suite (hole and par 16) that can be downloaded from [20]. All the other Dimacs instances that were used in [13] to estimate Chaff's performance are too easy for both programs. Besides, Table 1 contains results on the class blocksworld, that is also available at [20], and the "easy" classes of CNFs from [17] encoding verification of pipelined microprocessors. The result of the program that had the best runtime on a class of instances is shown in bold. It

Table 1 Benchmarks on which Chaff's and BerkMin's performances are comparable

		Chaff		BerkMin	
Class of benchmarks	Number of instances	Time (s)	Number of aborted	Time (s)	Number of aborted
Blocksworld	7	52.5	0	**9.0**	0
Hole	5	**79.9**	0	339	0
Par 16	10	33.9	0	**13.6**	0
sss 1.0	48	41.6	0	**13.4**	0
sss 1.0a	8	**17.4**	0	17.9	0
sss-sat 1.0	100	383.9	0	**254.4**	0
fvp-unsat1.0	4	**589.1**	0	1637.4	0
vliw-sat 1.0	100	**2602.9**	0	7305.0	0

Table 2 Benchmarks on which BerkMin dominates

Class of benchmarks	Number of instances	Chaff		BerkMin	
		Time (s)	Number of aborted	Time (s)	Number of aborted
Beijing	16	468.1 (>120,468.1)	2	**494.0**	0
Miters	5	1515.9 (>121,515.9)	2	**3477.6**	0
Hanoi	3	1320.6 (> 61,320.6)	1	**1401.3**	0
fvp-unsat2.0	22	19679.4 (> 139,679.4)	2	**6869.7**	0

is worth mentioning that in class fvp-unsat1.0 Chaff is faster only on one instance (9vliw_bp_mc.cnf). All the 100 CNFs of the class vliw-sat1.0 are obtained by modification of 9vliw_bp_mc.cnf. So Chaff has better performance on classes fvp-unsat1.0 and vliw-sat1.0 due to one instance.

Table 2 contains results on more complex classes of CNFs. Class Beijing can be downloaded from [20]. Class Hanoi from the Dimacs benchmark suite consisting of two CNFs hanoi4.cnf and hanoi5.cnf was extended by a more complex example hanoi6.cnf (courtesy of Henry Kautz). Class fvp-unsat2.0, that can be downloaded from [17], consists of large CNFs encoding microprocessor verification. Finally, class Miters consists of CNFs obtained by encoding equivalence checking of artificial combinational circuits (We used artificial circuits because their complexity was easy to control).

Each class of Table 2 contains at least one CNF that Chaff was not able to solve without exceeding the timeout limit (60 000 s). In the column "time" describing Chaff's performance, the upper number gives the total time spent on the instances that Chaff finished. The lower number is equal to the upper number plus 60,000 times the number of aborted instances in the class.

In our opinion, the main conclusion that can be drawn from Table 2 is that BerkMin is more robust than Chaff. One more argument substantiating this point of view is the following. The web site [21] gives statistics on the performance of 23 SAT-solvers (including a version of Chaff) on a representative set of instances. This set includes the 16 CNFs of class Beijing which are all satisfiable except one CNF. Each of the 23 SAT-solvers had at least two CNFs of the Beijing class that it was not able to solve in the timeout limit (10 000 s) Interestingly, there are no "universally" hard CNFs in this class. That is a CNF that cannot be solved by one SAT-solver can be finished in a few seconds by another, which suggests that these SAT-solvers are not robust. At the same time, BerkMin was able to solve all the 16 CNFs of the Beijing class in about 8 minutes.

Table 3 gives some details of Chaff's and BerkMin's performance on a few instances from classes fvp-unsat1.0, Hanoi, and fvp-unsat2.0. Chaff was aborted on the 3 instances marked with the star symbol. The two columns (Database size)

Table 3 Details of Chaff's and BerkMin's performance on some instances

		Chaff			BerkMin			
Instance number	Satisfiable	(Database size) / (Initial CNF size)	Number of decisions	Time (s)	(Database size) / (Initial CNF size)	(Largest CNF size)/ (Initial CNF size)	Number of decisions	Time (s)
9vliw_bp_mc	no	2.42	1,124,797	**567.4**	1.88	1.04	2,384,485	1625.0
hanoi5	yes	31.49	504,463	1313.4	8.68	2.38	194,672	**71.2**
hanoi6*	yes	129.04	5,580,228	49021.1	19.58	4.19	1,948,717	**1328.7**
4pipe	no	3.01	412,358	354.6	1.49	1.08	144,036	**40.9**
5pipe	no	1.65	461,275	333.7	1.09	1.01	213,859	**71.8**
6pipe*	no	15.97	5,580,228	49131.2	1.71	1.05	1,371,445	**1015.6**
7pipe*	no	7.37	11,779,016	48053.1	1.95	1.05	3,357,821	**3673.2**

/ (Initial CNF size) give the ratio of the total number of generated conflict clauses and clauses of the initial CNF to the number of clauses of the initial CNF. Table 3 shows that BerkMin has better performance because it builds smaller search trees and its clause database is smaller. Besides, column (Largest CNF size)/(Initial CNF size) shows that the number of clauses BerkMin had to keep in memory at the same time was at most 4 times the number of clauses in the initial CNF. Unfortunately, Chaff does not report the size of the largest intermediate CNF. However judging by the data output by the Unix utility called "top", Chaff took much more memory than our program. BerkMin was developed on a PC with 128 Mbytes of memory. So it was designed to reduce the probability of exceeding that limit. In experiments BerkMin took more than 128 Mbytes of memory only on a few instances of very large CNFs. On the other hand, Chaff often used 1 Gbytes of memory and more.

7   Conclusions

This paper describes a new SAT solver, BerkMin, that is more robust than Chaff, currently the best SAT solver in the EDA domain. BerkMin has four new features distinguishing it from the predecessors. Three of them develop Chaff's idea that recently deduced clauses are the most important to satisfy. First, all clauses are chronologically sorted so that BerkMin always tries to satisfy the most recently deduced clause that is left unsatisfied. Second, BerkMin gets a better estimation of the activity of variables by taking into account a set of clauses which is wider than what Chaff takes. Third, BerkMin uses a new clause database management strategy that takes into account not only the size of a clause but its activity in conflict making and its "age". The fourth new feature of BerkMin is its branch selection strategy.

Acknowledgments   We would like to thank the anonymous reviewers for their insightful recommendations for improving this paper.

References

1. L. Baptista, J.P. Marques-Silva. The Interplay of Randomization and Learning on Real-world Instances of Satisfiability. In: Proceedings of AAAI Workshop on Leveraging Probability and Uncertainty in Computation. July 2000.
2. R.J.J. Bayardo, R.C. Schrag. Using CSP Look-Back Techniques to Solve Real-World SAT Instances. In: Proceeding of the 14th National Conference on Artificial Intelligence (AAAI'97), Providence, Rhode Island, 1997, pp. 203–208.
3. E. Ben-Sasson, R. Impagliazzo, A. Wigderson. Near Optimal Separation of Treelike and General Resolution. In: Proceedings of SAT-2000: 3rd Workshop on the Satisfiability Problem. May 2000. pp. 14–18.
4. A. Biere, A. Cimatti, E.M. Clarke, M. Fujita, Y. Zhu. Symbolic Model Checking Using SAT Procedures Instead of BDDs. In: Proceedings of Design Automation Conference, DAC'99. 1999.

5. R.K. Brayton et al. Logic Minimization Algorithms for VLSI Synthesis. Kluwer Academic, 1984.

6. J. Burch, V. Singhal. Tight Integration of Combinational Verification Methods. In: Proceedings of International. Conference on Computer-Aided Design. 1998.

7. M. Davis, G. Longemann, D. Loveland. A Machine Program for Theorem Proving. In: Communications of the ACM. 1962. V. 5. P.394–397.

8. O. Dubois, P. Andre, Y. Boufkhad, J. Carlier. SAT Versus UNSAT. In: Johnson and Trick, Second DIMACS Series in Discrete Mathematics and Theoretical Computer Science, American Mathematical Society, 1996, pp. 415–436.

9. E. Goldberg, M.Prasad. Using Sat for Combinational Equivalence Checking. In: Proceedings of Design, Automation and Test in Europe Conference. 2001. pp.114–121.

10. C.P. Gomes, B. Selman, H. Kautz. Boosting Combinational Search Through Randomization. In: Proceedings of International Conference on Principles and Practice of Constraint Programming. 1997.

11. J.W. Freeman. Improvements to propositional satisfiability search algorithms. Ph.D. thesis, Department of Computer and Information science, University of Pennsylvania, Philadelphia, 1995.

12. C.M. Li. A constrained-based approach to narrow search for Satisfiability. Information processing letters. 1999. V. 71. pp. 75–80.

13. M.W. Moskewicz, C.F. Madigan, Y. Zhao, L. Zhang, S. Malik. Chaff: Engineering an Efficient SAT Solver. In: Proceeding of the 38th Design Automation Conference (DAC'01), 2001.

14. J.P.M. Silva. The Impact of Branching Heuristics in Propositional Satisfiability Algorithms. In: Proceedings of the 9th Portuguese Conference on Artificial Intelligence (EPIA), September 1999.

15. J.P.M. Silva, K.A. Sakallah. GRASP: A Search Algorithm for Propositional Satisfiability. IEEE Transactions of Computers. 1999. V. 48. 506–521.

16. P. Stephan, R. Brayton, A. Sangiovanni-Vencentelli. Combinational Test Generation Using Satisfiability. IEEE Transaction on Computer-Aided Design of Integrated Circuits and Systems. 1996. Vol. 15. 1167–1176.

17. M. Velev. CMU benchmark suite. Available from http://www.ece.cmu.edu/~mvelev.

18. H. Zhang. SATO: An Efficient Propositional Prover. In: Proceedings of the International Conference on Automated Deduction. July 1997. pp. 272–275.

19. http://www.ee.princeton.edu/~chaff/spelt3/chaff2_20010323_spelt3-bin-solaris.tar

20. http://www.satlib.org/benchm.html

21. http://www.lri.fr/~simon/satex/satex.php3

Improving Compression Ratio, Area Overhead, and Test Application Time for System-on-Chip Test Data Compression/Decompression

Paul Theo Gonciari[1], Bashir M. Al-Hashimi[1], and Nicola Nicolici[2]

Abstract This paper proposes a new test data compression/decompression method for systems-on-a-chip. The method is based on analyzing the factors that influence test parameters: compression ratio, area overhead and test application time. To improve compression ratio, the new method is based on a Variable-length Input Huffman Coding (VIHC), which fully exploits the type and length of the patterns, as well as a novel mapping and reordering algorithm proposed in a preprocessing step. The new VIHC algorithm is combined with a novel parallel on-chip decoder that simultaneously leads to low test application time and low area overhead. It is shown that, unlike three previous approaches [2, 3, 10] which reduce some test parameters at the expense of the others, the proposed method is capable of improving all the three parameters simultaneously. For example, the proposed method leads to similar or better compression ratio when compared to frequency directed run-length coding [2], however with lower area overhead and test application time. Similarly, there is comparable or lower area overhead and test application time with respect to Golomb coding [3], with improvements in compression ratio. Finally, there is similar or improved test application time when compared to selective coding [10], with reductions in compression ratio and significantly lower area overhead. An experimental comparison on benchmark circuits validates the proposed method.

1 Introduction

Due to the increased complexity of System-on-a-Chip (SoC), testing is an important factor that drives time-to-market and hence, the cost of the design [14]. In addition to the standard design for test (DfT) issues for SoCs [14], a new emerging problem is the large volume of test data [16]. This is because the cost of Automatic Test Equipment (ATE) grows significantly with the increase in speed (i.e. operating

---

[1]Department of ECS, University of Southampton United Kingdom

[2]Department of ECE, McMaster University Canada {p.gonciari, bmah} @ ecs.soton.ac.uk nicola@ece.mcmaster.ca

R. Lauwereins and J. Madsen (eds.), *Design, Automation, and Test in Europe–The Most Influential Papers of 10 Years DATE*, 479–495
© Springer 2008

frequency), channel capacity, and memory size [16]. In order to support large volume of test data with limited channel capacity for future SoCs, ATE requires modifications and additional expenses. Besides, when using conventional test data transfer, the volume of test data has a direct impact on test application time, which adds to the manufacturing cost of SoCs. There are two main potential solutions for closing the ever-increasing gap between ATE speed, channel, and memory requirements vs. SoC complexity. As shown in the ITRS Roadmap [1], the first solution is the well-known built-in self-test (BIST). However, to make the existing embedded cores *BIST-ready*, a considerable redesign effort is required, as well as performance penalties are incurred. Moreover, depending on the business model, many embedded cores cannot be modified by the system integrator [16]. The second solution for reducing the speed, channel capacity and memory requirements of ATE is the use of test data compression/decompression methods. This solution does not introduce performance penalty and guarantees full reuse of the existing embedded cores, as well as the ATE infrastructure, with minor modifications required during system test preparation (compression) and test application (decompression). Hence, the second solution is an *efficient complementary alternative to BIST*, and requires further investigation.

Test data reduction received increased attention over the last several years. The methods for reducing test data range from test vector compaction [6, 15], test data compression methods which reduce the amount of data transmitted to the ATE [8] (e.g. with the test data stored on a workstation the amount of data transfered to the ATE is reduced), to methods which require an on-chip decompression architecture [2, 3, 5, 9–12]. To obtain not only test data reduction but also test application time reduction, when compared to already compacted test sets, an on-chip decompression architecture is required. A compression algorithm for test data should have two features: lossless and simple decompression. The first requirement is easily fulfilled by all known lossless compression strategies (i.e. Huffman coding, Lempel-Ziv compression algorithm, arithmetic coding [4]) however, to meet the second requirement the compression algorithm has to be carefully chosen. Iyengar et al. [9] proposed a new built-in scheme for testing sequential circuits based on statistical coding. Results indicate that the method is suitable for circuits with a relatively small number of primary inputs. To overcome this disadvantage a selective coding algorithm was proposed [10]. The method splits the test vectors into *fixed-length* block size patterns and applies Huffman coding to the newly created set. The method carefully selects the patterns which are coded using the Huffman codes while the rest of the patterns are prefixed. Although the method leads to fast on-chip decoding and hence, low test application time, by extending the block size the on-chip decoder increases in complexity, and consequently leads to very high area overhead. In [11] test vector compression is performed using run-length coding, where the method relies on the fact that successive test patterns in a test sequence often differ in only a small number of bits. Therefore, the initial test set T_D is transformed in T_{diff}, a difference vector sequence, in which each vector is computed by the difference between consecutive vectors in the initial sequence. Despite its performance, the method is based on variable-to-fixed-length codes and, as demonstrated in [3], these are less effective than variable-to-variable-length codes. Chandra et al. [3] introduced a compression method based on variable-to-variable-length Golomb codes. Both methods [11, 3] use the difference vector sequence. In order to regenerate the test set, inside the chip, a scan chain architecture called cyclical scan register (CSR) has to be used. Since Golomb codes are only dependent on the size of the run and not on its frequency, in [2] a new method based on frequency-directed run-length codes is proposed, and by means of experimental results, it is shown that the compression ratio can be improved. In addition to the above approaches which require an on-chip decoder, other methods were proposed which assume the existence of an embedded processor [5, 12]. The method in [12] is based on storing the differing bits between two test vectors, while the method in [5] uses regular geometric shapes formed only from 0s or 1s to compress the test data. Regardless of its benefits, in systems where embedded processors are not available or where embedded processors have access only to a small number of processing elements, this embedded processor method is not applicable. Therefore, only the approaches that are equally applicable to every SoC (i.e. they do not assume the existence of an embedded processor) are relevant to the proposed method.

The aim of this paper is provide a new compression method using an on-chip decoder where the three test parameters (compression ratio, area overhead and test application time) are simultaneously reduced. Unlike previous approaches [2, 3, 10] which reduce some test parameters at the expense of the others, simultaneous reduction is achieved by accounting for multiple factors that influence test parameters. The rest of the paper is organized as follows. Section 2 gives the preliminaries and the motivation for the proposed approach. A detailed analysis of the factors that influence test parameters is given and previous approaches and their characteristics are outlined. Section 3 describes the new compression method, the novel decompression architecture and analyzes test application time. Sections 4 and 5 provide the experimental results and conclusions respectively.

2   Preliminaries and Motivation

Testing in test data compression environments (TDCE) implies sending the compressed test data from the ATE to the on-chip decoder, decompressing the test data on chip and sending the decompressed test data to the circuit under test (CUT). There are two major components in TDCE: the **compression method**, used to compress the test set off-chip, and the associated **decompression method**, based on a **on-chip decoder**, used to restore the initial test set on-chip. The on-chip decoder is composed out of two units: a unit to identify a compressed code and a unit to decompress it. If the two units can work independently (i.e. decompressing the current code and identifying a new code can be done simultaneously), then the decoder is called **parallel**. Otherwise the decoder is referred to as **serial**. Testing in TDCE can be characterized by the following three parameters: **(a) compression**

ratio which identifies the performance of the compression method, the memory and channel capacity requirements on the ATE; **(b) area overhead** imposed by the on-chip decoder (dedicated hardware or on-chip processor); **(c) test application time** given by the time needed to transport and decode the compressed test set.

There are a number of factors which influence the above parameters: the *mapping and reordering algorithm*, which prepares the test set for compression by mapping the "don't cares" in the test set to '0' or '1' and reordering the test set, the *compression algorithm*, the *type of input patterns* (fixed or variable length), the *length of the pattern*, and the *type of the on-chip decoder* (serial or parallel). Each of the three test parameters is influenced by these factors as follows:

(a) **Compression ratio**: Compression algorithms that use various types of patterns and different lengths exploit different features of the test set. Using mapping and reordering of the initial test set, those features can be emphasized. Therefore, the compression ratio is influenced firstly by the the *mapping and reordering algorithm*, and then by the *type of input patterns* and the *length of the pattern*, and finally by the *compression algorithm*.

(b) **Area overhead**: Area overhead is influenced firstly by the *nature of the decoder*, and then by the *type of the input pattern* and the *length of the pattern*. If the decoder has a serial nature then synchronization between the two units (code identification and decompression) is already at hand. However, if the decoder has a parallel nature, the units have to synchronize one with each other, and hence increasing the area overhead. The *type of the input pattern* requires different type of logic to generate the pattern on-chip. For example, the use of fixed-length input patterns requires a shift register, while the use of variable-length input pattern (runs of 0s for example) requires a counter. On the other hand, the *length of the pattern* impacts the size of the decoding logic, and hence it influences area overhead.

(c) **Test application time (TAT)**: TAT is firstly influenced by the **compression ratio**, and then by the *type of the on-chip decoder*, the *length of the pattern*. To illustrate the factors that influence TAT, consider the ATE operating frequency as the reference frequency. Minimum TAT (min_{TAT}) is given by the size of the compressed test set in ATE clock cycles. However, this TAT can be obtained only when the on-chip decoder can process the currently compressed bit before the next one is sent by the ATE. In order to do so, the relation between the frequency at which the on-chip decoder works and the ATE operating frequency have to meet certain conditions. The *frequency ratio* is the ratio between the on-chip test frequency and the ATE operating frequency. Consider that the min_{TAT} is obtained when the frequency ratio is greater than an *optimum frequency ratio*. Since the, min_{TAT} is given by the size of the compressed test set, increasing the compression ratio would imply further reduction in TAT. However, this reduction happens *only* if the optimum frequency condition is met. Since real environments can not always satisfy the optimum frequency ratio condition, then a natural question is what happens if this condition is not met? TAT in this cases is dependent on the *type of on-chip decoder*. If the on-chip decoder has a serial nature [2, 3], then TAT is heavily influenced by changes in the frequency ratio. However, if the decoder has a parallel nature [10], the influences are rather minor. Since the *length of the pattern* determines the optimum frequency ratio, consequently it also influences TAT.

In the following it is shown that previous approaches [2, 3, 10] for test data compression/decompression have efficiently solved some of the test parameters at the expense of the others.

(i) **Selective coding (SC) [10]**: This method splits the test vectors into patterns of size b (b is the size of the fixed input pattern used for coding called also the block size) and applies Huffman coding to a carefully selected number of patterns while the rest of the patterns are prefixed. The SC decoder has a parallel nature. Although, because of its parallelism, the on-chip decoder yields good TAT. However, the choice of a *fixed-length* input pattern and the use of prefixed codes requires shift registers of length b which cause great area overhead. Also, making use of *fixed-length* input patterns, the method is restrained to exploiting the test set features for compression. Therefore, *due to a fixed-length input pattern* the main drawbacks of this approach are low compression ratios and large area overhead.

(ii) **Golomb coding [3]**: This method assigns a variable-length Golomb code, of group size m_g, to a run of 0s. Golomb codes are composed from two parts, a prefix and a tail, and the tail has the length given by log m_g. Using runs of 0s as input patterns, the method has two advantages: firstly the decoder uses counters of length log m_g instead of shift registers, thus leading to small area overhead; secondly it better exploits the features of the test set for greater compression. Since the method in [3] uses a serial on-chip decoder, TAT is heavily influenced by changes in the frequency ratio. Also as shown later in this paper, Golomb is a particular case of the proposed Variable-length Input Huffman Coding, and hence it constantly leads to inferior compression.

(iii) **Frequency-Directed Run-Length (FDR) coding [2]**: The FDR code is composed from two parts, a prefix and a tail, where both parts have the same length. In order to decompress a FDR code, the on-chip decoder has to identify the two parts. Because the code is not dependent on a group size as Golomb codes, the decoder has to detect the length of the prefix in order to decode the tail. In order to do so, the FDR code requires a more complicated decoder with fixed area overhead. Therefore, regardless of the good compression ratios the area overhead of FDR is a disadvantage.

As illustrated above, current approaches for test data compression, efficiently address some of the test parameters at the expense of the others. Therefore, the motivation of this paper is to seek to exploit the factors that influence the test parameters: *mapping and reordering algorithm, compression algorithm type of input patterns, length of the pattern* and the *type of on-chip decoder*. Consequently a new compression method, that leads to simultaneous improvement in compression ratio, area overhead and test application time, is proposed in the following section.

3   New Test Data Compression/Decompression Method Based on *Variable-length* Input Huffman Compression (VIHC)

This section introduces a new compression method based on Huffman coding. It is important to note that the previous approach [10] which has used Huffman coding in test data compression have employed only patterns of *fixed-length* as input to the Huffman coding algorithm. As outlined in the previous section *fixed-length* input patterns restrict exploitation of the test set features for compression. This problem is overcome by the approach proposed in this section which uses patterns of *variable-length* as input to the Huffman algorithm and can efficiently exploit a reordered test sequence (Section 3.1). This *fundamental distinction* does not influence only the compression, but also provides the justification for employing a *parallel decoder* using counters (Section 3.2.1) that will lead not only to significantly lower area overhead, but will also facilitate TAT reduction when the compression ratio and frequency ratio are increased (Sections 3.2.2 and 4).

To illustrate the use of the proposed method, consider the example given in Fig. 1. First, the maximum acceptable length of the runs of 0s is selected ($m_h = 4$) and referred to as the group size. Using the group size, the initial test vector (t_{init}) is divided into runs of 0s of length smaller or equal to 4, which are referred to as patterns (Fig. 1a). From these patterns and the number of occurrences a dictionary is build as shown in Fig. 1b. Using the dictionary the Huffman tree is built (Fig. 1c). To build the Huffman tree, the list of patterns are ordered in the descending order of their occurrences. The first two patterns, the ones with the smallest number of occurrences, are merged into a new pattern with the occurrence given by the sum

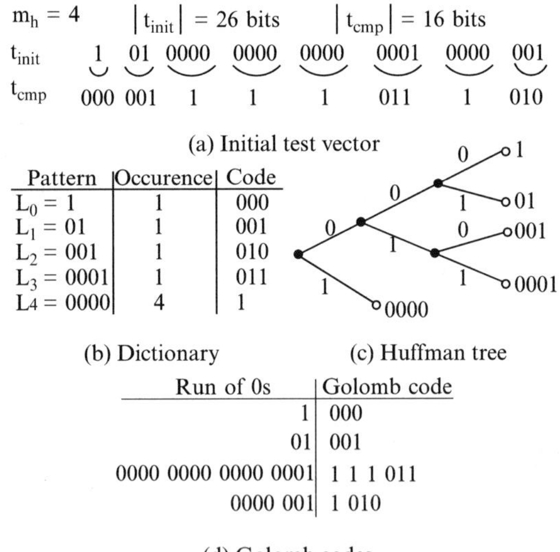

Fig. 1 VIHC, the particular case

of the two patterns' occurrences. Using the remaining patterns and the merged pattern, a new list is created, and the process is repeated until the list comprises only one merged pattern. Each merged pattern in the list represents a node in the Huffman tree, while the composing patterns are its descendents. The Huffman code[1] is constructed by assigning binary 0s and 1s to each segment starting from a node of the Huffman tree (Fig. 1c). Using the obtained Huffman code, the initial test vector (t_{init}) is compressed (t_{cmp}). The above can be formalized as follows.

Let $L = \{L_0,\ldots,L_{m_h}\}$ be a set of unique patterns obtained after the test set was divided into runs of 0s of maximum length m_h and $P = \{n_0,\ldots,n_{m_h}\}$ the number of occurrence of each pattern. $n = \Sigma_{i=0}^{mh} |L_i| * n_i$ is the size of the test set, where $|L_i|$ denotes the length of the pattern L_i. With L and P the Huffman codes are obtained. For a pattern L_i the Huffman code is denoted by c_i, and the length of the code is given by w_i. It should be noted that the terms "Huffman code" and "codeword" will be used interchangeably and the term "group size" is preferred to "block size". It is interesting to note that there are $m_h + 1$ patterns and thus $m_h + 1$ leafs in the Huffman tree. There are m_h patterns formed from runs of 0s ending in 1 ($L_0 \ldots L_{m_h} - 1$) and one pattern formed from runs of 0s only (L_{m_h}).

Before presenting the procedures used to compress the test sets, the following observation has to be made:

Observation 1 *For a given group size m, Golomb coding is a particular case of the proposed coding method (VIHC).*
The above can be seen in Fig. 1 where, if the Golomb codes for a group size of 4 is used (the codes where computed for the required runs of 0s in Fig. 1d), the t_{init} would be compressed to the same t_{cmp} as in the case of VIHC. This is true only for a particular pattern distribution, i.e. when (*a*) $n_m \geq n_{i_0} + n_{i_1} + \ldots + n_{i_{\frac{m}{2}-1}}$, $\forall$, *distinct* $i_0 \ldots i_{\frac{m}{2}-1}, < m$ and (*b*) $n_k < n_i + n_j$, $\forall i, j, k < m, i \neq j \neq k$, where m represents the group size and n_i is the number of occurrences of pattern L_i, as defined above.

In general, however, the codes differ. This is because the particular pattern distribution, when Golomb leads to same compression as VIHC, is generally not respected. This is illustrated in Fig. 2, where by a simple change in the t_{init} from Fig. 1a, the bit 24 is set from 0 to 1, the size of the compressed vector ($t_{cmp}^{\,G}$) obtained by Golomb is greater than the size of the compressed vector ($t_{cmp}^{\,V}$) obtained by VIHC. The following observation indicates the relation between the compression obtained by Golomb and VIHC.

Observation 2 *For a given group size m, the compression obtained by the Golomb coding is smaller or equal than the compression obtained by VIHC.*
The above can be easily shown by using the fact that the Huffman code is an optimal code ([4, Theorem 5.8.1]), i.e. any other code will have the expected length greater than the expected length of the Huffman code. Since, the Golomb code is a particular case of VIHC (the VIHC in this particular case is referred to as $VIHC_G$),

---

[1] For detailed description of the Huffman algorithm, the reader is referred to [4]

$$m_h = 4 \quad |t_{init}| = 26 \text{ bits} \quad |t^{V}_{cmp}| = 17 \text{ bits} \quad |t^{G}_{cmp}| = 19 \text{ bits}$$

t_{init}	1	01	0000	0000	0000	0001	0000	1 01
t^{V}_{cmp}	00	011	1	1	1	010	1	00 011
t^{G}_{cmp}	000 001	1	1	1	011	1	000 001	

(a) Test vectors

Run of 0s	Golomb code
1	000
01	001
0000 0000 0000 0001	1 1 1 011
0000 1	1 000

(b) Golomb codes

Pattern	Occ.	Code
$L_0 = 1$	2	00
$L_1 = 01$	2	011
$L_3 = 0001$	1	010
$L_4 = 0000$	4	1

(c) VIHC codes

Fig. 2 VIHC, the general case

thus it is optimal only for a particular pattern distribution. For other pattern distributions the $VIHC_G$ code is not optimal, thus the expected length of the $VIHC_G$ code is greater. This concludes the observation.

It is important to note that Observation 1 assures that the VIHC decoder proposed in this paper (Section 3.2.1) can also be used to decompress a Golomb compressed test set.

3.1 Compression Algorithm

This section presents the procedures used to implement the proposed method (VIHC). In this paper, without loss of generality, only the compression of the difference vector sequence (T_{diff}) is taken into account. The initial test set (T_D) is partially specified and the test vectors can be reordered. This is because ATPG tools generate a small number of care bits in every test vector [13]. This gives great flexibility to the *mapping and reordering algorithm,* which in a pre-processing step prepares the test set for compression by mapping the "don't cares" in the test set to '0' or '1' and reordering the test set such that VIHC can increase its compression. Thus, the compression algorithm has three main procedures: (1) prepare initial test set, (2) compute Huffman code and (3) generate decoder information. The first two procedures are used for compression only, while the last one determines the decoding architecture described in Section 3.2.

1. **Prepare initial test set**. The procedure consists of two steps. In the first step, the "don't cares" are mapped to the value of the previous vector on the same position. In the second step the test set is reordered such that the number of 1s in the difference between two test vectors is minimum and the length of the minimum run of 0s is maximum. This will be exploited by the *variable-length* input patterns used for Huffman code computation. The reordering algorithm

has a complexity of $O(|T_D|^2)$, where $|T_D|$ represents the number of test vectors in the test set.

2. **Huffman code computation**. Based on the chosen group size (m_h) the dictionary of *variable-length* input patterns L and the number of occurrences P are determined from the reordered test set. This adds an additional computational step when compared to Golomb coding which assigns a precomputed code for each run length. Using L and P the code is computed by the Huffman algorithm [4].

3. **Generate decoder information**. This procedure computes the necessary information to build the decoder. For each Huffman code c_i a binary code b_i is assigned. The binary code is composed from the *length* of the initial pattern of $\lceil log_2(m_h + 1) \rceil$ bits and a *special* bit which identifies the cases when the initial pattern is composed of runs of 0s only or runs of 0s ending in 1, thus $b_i = (|L_i|, 0)$ for $i < m_h$, and $b_{m_h} = (|L_{m_h}|, 1)$. The *Control and Generation Unit* (see Section 3.2.1) uses the binary code to generate the pattern.

3.2 *Decompression Architecture*

This section introduces the new decoding schemes (Section 3.2.1) for the VIHC compression method, assuming a generic decompression architecture (Fig. 3) and provides the TAT analysis (Section 3.2.2) of the VIHC decoder. The VIHC decoder, proposed in Section 3.2.1, uses a novel parallel approach, in contrast to the previous Golomb [3] serial decoder. It should be noted that for the decompression of T_{diff}, a CSR architecture [3,11] is used after the VIHC decoder in Fig. 3. This work assumes that the ATE is capable of external clock synchronization as shown in [7].

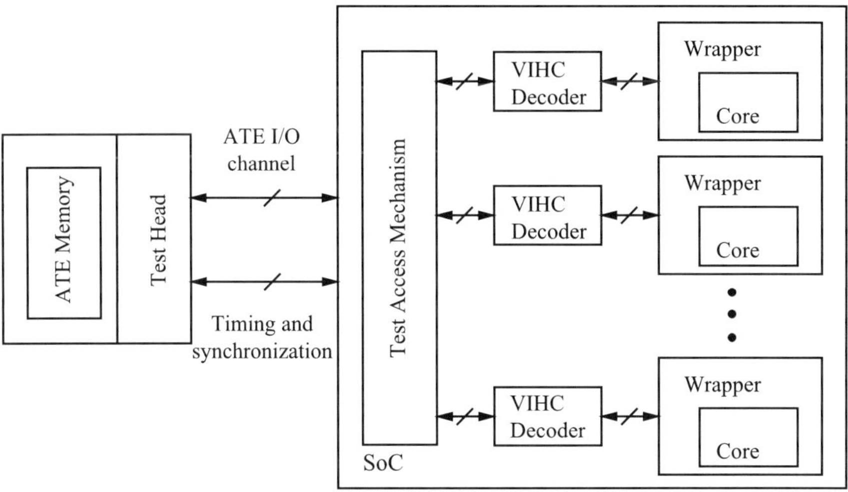

Fig. 3 Generic decompression architecture [3]

3.2.1 VIHC Decoder

A block diagram of the VIHC decoder is given in Fig. 4. The decoder is formed from a *Huffman decoder* [9] (Huff-decoder) and a *Control and Generator Unit (CGU)*. The Huff-decoder is a finite state machine (FSM) which detects a Huffman code and outputs the corresponding binary code. The CGU is responsible for controlling the data transfer between the ATE and the Huff-decoder, generate the initial pattern and the scan clock for the CUT. The *data in* line is the input from the ATE synchronous with the external clock *ATE clock*. When the Huff-decoder detects a codeword, the *code* line is high and the binary code is output on the *data* lines. The *special* input to the CGU is used to differentiate between the two types of patterns, runs of 0s only and runs of 0s ending in 1. After loading the code, the CGU generates the pattern and the internal *scan clock* for the CUT. If the decoding unit generates a new code while the CGU is busy processing the current one, the *ATE sync* line is low notifying the ATE to stop sending data and the *sync FSM clk* is disabled forcing the Huff-decoder to maintain its current state. It should be noted that if the ATE responds to the stop signal with a given latency (i.e. it requires a number of clock cycles before the data stream is stopped), the device interface board (DIB) between the ATE and the system will have to account for the latency with a first in first out (FIFO)—like structure. Dividing the VIHC decoder in Huff-decoder and CGU, allows the Huff-decoder to continue loading the next codeword while the CGU generates the current pattern. Thus, the Huff-decoder is interrupted *only if necessary*, which is in contrast to the Golomb [3] and the FDR [2] serial decoders. This leads to reduction in TAT when compared to the Golomb and the FDR, as it will be shown in Section 4.

The FSM for the Huffman decoder from the example in Fig. 1 is illustrated in Fig. 5. Starting from state S_1, depending on the value of *data_in* the Huff-decoder change its state. It is important to note the following:

- After the Huff-decoder detects a code word it goes back to state S1; for example, if the *data in* stream is 001 (last bit first) the Huff-decoder first changes its state to S2 then to S4 after which to S1, and setting *data* to the corresponding binary code and the *vector* line high

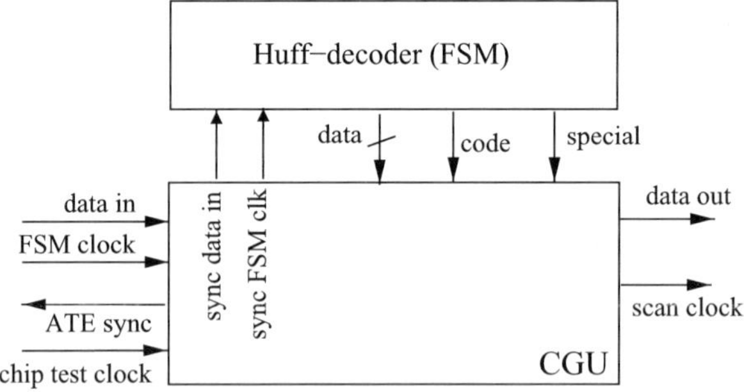

Fig. 4 VIHC decoder

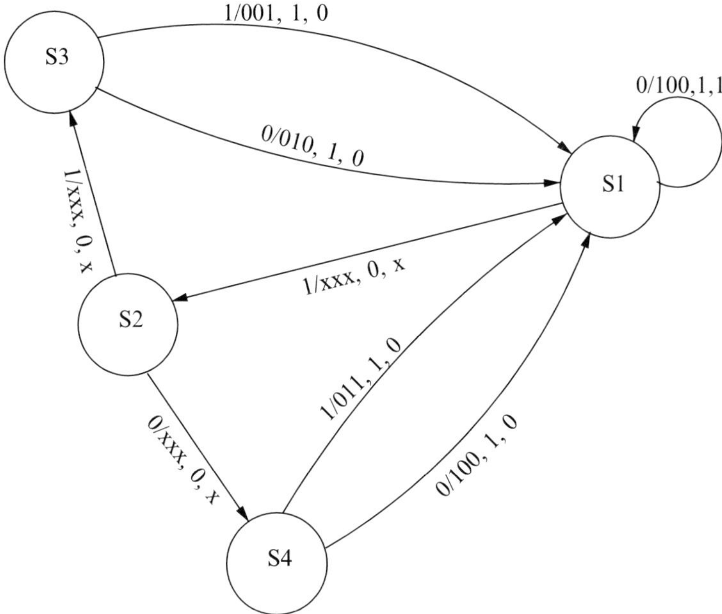

Fig. 5 FSM for example in Fig. 1

- The number of ATE clock cycles needed to detect a code is equal to the length of the code; for example, if the *data in* stream is 001 (last bit first), the Huff-decoder identifies the code in three ATE clock cycles
- The Huff-decoder has a maximum of m_h states for a group size of m_h; the number of states in a Huff-decoder is given by the number of leafs in the Huffman tree minus one; since there are a maximum of $m_h + 1$ leafs for a group size of m_h, the number of states is m_h

When compared to the SC decoder [10], which has an internal buffer and a shift register of size b and a Huffman FSM of at least b states, the proposed decoder has only a $\lceil log_2(m_h + 1) \rceil$ counter, two extra latches and a Huffman FSM of at most m_h states. Thus, for the same group size ($m_h = b$) significant reduction in area overhead is obtained, as demonstrated in Section 4.

A detailed view of the CGU is given in Fig. 6. The CGU is formed from a $\lceil log_2(m_h + 1) \rceil$ bits counter and additional logic. When the *code* line is high, the values at the *data* inputs are loaded into the counter. With the data loaded, the counter decrements to 1 setting the *data out* to 0. When the value of the counter is 1, *data out* is set to 0 if the *special* line is high, or to 1 if the *special* line is low. The *FSM clock* signal assures that the FSM will reach a stable state after one internal clock cycle, therefore it has to be generated as a one internal clock cycle for each external clock-cycle period. This can be easily achieved by using the same technique as in [7] for the *serial scan enable* signal.

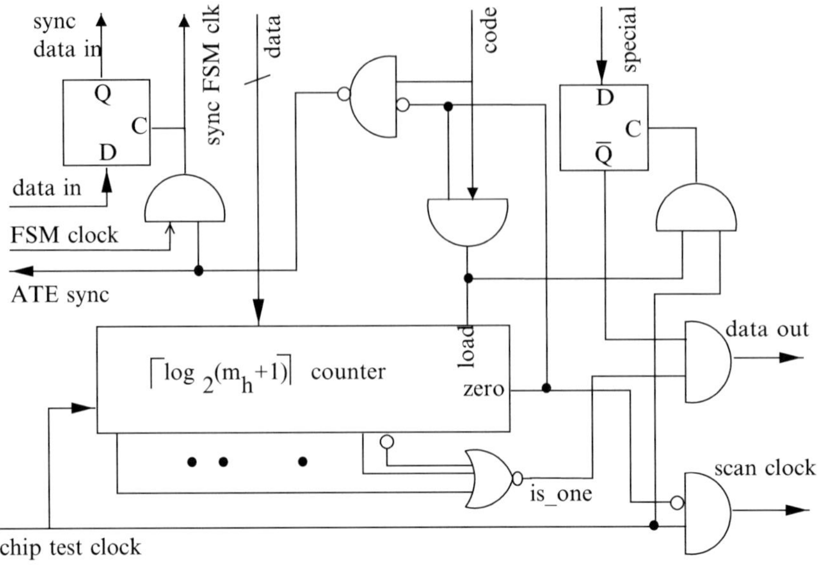

Fig. 6 CGU for VIHC decoder

3.2.2 Test Application Time Analysis

In this section a TAT analysis, for the proposed method, with respect to the ratio between the on chip test frequency and the ATE operating frequency is given. It is considered that the data is fed into the FSM with the ATE operating frequency and the FSM reaches a stable state after one on-chip test clock cycle. Analyzing the FSM for the Huffman decoder it can be observed that the number of ATE clocks needed to identify a codeword c_i is equal to the size of the codeword w_i (see Section 3). On the other side, the number of internal clock cycles needed to generate a pattern is equal to the size of the pattern.

Consider the ATE clock cycle of 50 ns, the on-chip test clock cycle of 10 ns, the next codeword of length 1, and the current pattern size of length 4. The Huff-decoder will identify the next codeword in 1 ATE clock cycle (50 ns) however, the CGU will generate the current pattern in 40 ns. Since the Huff-decoder reaches a stable state after one internal clock cycle, from the current ATE clock cycle only 10 ns where used, the rest of 40 ns can be used to generate the pattern. Thus, the CGU has enough time to generate the current pattern without having to set *ATE sync* low and stop the ATE from sending data.

Formally, if $d = \dfrac{f_{chip}}{f_{ate}}$ is the ratio between the on-chip test frequency (f_{chip}) and the ATE operating frequency (f_{ate}), then after a Huffman code is identified, $\alpha - 1$ internal clock cycles from the current ATE cycle can be used to generate the pattern. Thus, in order for the Huff-decoder to run without being stopped by the CGU, the CGU has to be able to generate the pattern L_i in the number of internal clock cycles remained from the ATE clock in which it was started plus the number of internal clock cycles needed for the Huff-decoder to identify the next codeword. Or, $|L_i| <$ $(\alpha - 1) + 1 + \alpha * (w_j - 1)$, where w_j is the length of the next code word in bits. The smallest TAT is given by $n_h + \delta$, where n_h is the number of bits in the compressed test set in ATE clock cycles and δ is the number of extra ATE clocks needed to decode the last code word. With $max(|L_i|) = m_h$ and $min\{w_i\} = w_{min}$, the frequency ratio to obtain the smallest TAT is given by $\alpha_{max} = \dfrac{f_{chip}}{f_{ate}} = \dfrac{m_n}{w_{min}}$ (*optimum frequency ratio*). When the frequency ratio is greater than the *optimum frequency ratio*, an increase in the compression ratio will lead to reduction in test application time. Thus, for better compression of VIHC when compared SC [10], an increase in frequency ratio will also lead to lower test application time as shown in Section 4.

4 Experimental Results

To validate the efficiency of the proposed method, experiments were performed on the full-scan version of the largest ISCAS 89 benchmark circuits. Test sets used in this experiments were obtained using MinTest [6] dynamic compaction (also used in [3]). The experiments were performed on a Pentium III 500 MHz workstation with 128 MB DRAM. Golomb [3], SC [10], FDR [2] and the proposed VIHC method were implemented in C++. To provide an uniform basis for the comparison between Golomb, SC and VIHC, as the three methods use a group size, we use the group size for which [3] reports the best results. The VIHC and SC compression ratios where computed for those group sizes. The results are provided in Table 1. It should be noted that [10] did not report compression ratios for the test sets used in this paper [6], however the results were computed by implementing SC method and applying it to MinTest.

Table 1 provides a compression comparison for the T_{diff} test sets. The compression ratios are relative to the size of the dynamic compacted test set from MinTest [6]. With the exception of one particular case when FDR has similar compression ratio as VIHC (s38417), the table clearly shows that the proposed method leads to better compression ratios than previous approaches. When compared to Golomb, the proposed method obtains up up to 66% better compression ratios (s35932). When compared to FDR, the method obtains 56% reduction for s35932 and up to 8% reduction for the other circuits. The reason for a much better compression ratio

Table 1 Compression comparison for T_{diff}

Circuit	Group size	SC [10]	Golomb [3]	FDR [2]	VIHC
s5378	4	34.79	40.70	48.19	**51.52**
s9234	4	35.52	43.34	44.88	**54.84**
s13207	16	77.73	74.78	78.67	**83.21**
s15850	4	40.16	47.11	52.87	**60.68**
s35932	16	65.72	N/A	10.19	**66.47**
s38417	4	37.11	44.12	**54.53**	54.51
s38584	4	37.72	47.71	52.85	**56.97**

Table 2 Area overhead comparison

Compression method	Area overhead in tu*		
	Group size		
	4	8	16
SC [10]	349	587	900
Golomb [3]	**125**	227	307
FDR [2]		320	
VIHC	136	**201**	**296**

*technology units for the lsi10k library (Synopsys Design Compiler)

in the case of circuit s35932 is that the test set for this circuit has a large number of bits '1', which is a problem for both Golomb and FDR methods. When compared to SC, the proposed method obtains up to 20% in compression ratio (s15850). It should be noted that a major contribution to this considerable savings is not only coding algorithm but also the pre-processing step using the *mapping and reordering algorithm* described in Section 3.1.

Table 2 illustrates the area overhead computed with the Synopsys Design Compiler for the four on-chip decoders. For all methods, the entire decoder including buffers, shift registers and counters was synthesized. For SC, VIHC, and Golomb, the area overhead was computed for a group size of 4, 8, and 16. Without loss of generality the decoders for s5378 were synthesized. As shown in Table 2, the area overhead for SC is *up to 3 times greater* than the area overhead for VIHC. The area overhead for Golomb is almost equal to the one of VIHC, and the area overhead for FDR is constantly greater than the area overhead for VIHC.

For TAT comparison, a simulator was implemented based on the TAT analysis for SC [10], Golomb [3], FDR [2] and the VIHC decoder (Section 3.2.1). For all decoders it was assumed that the data is fed into the decoder at ATE operating frequency and the internal FSM reaches a stable state after one internal clock cycle. In order to provide an accurate comparison, we used the same group size as in Table 1, and compressed and simulated all the test sets. The results are reported in Table 3. The table shows, the circuit, the compression method and the TAT obtained for four frequency ratios: $\alpha = 2, 4, 6$, and 8. Analyzing columns 3 to 6 from Table 3 it can be seen that for smaller frequency ratios SC has slightly better TAT than the proposed method (e.g. s35932 or s38417 and s38584 and $\alpha = 2$). However, it can be observed that when frequency ratio increases the proposed method leads up to 34% (s15850 and $\alpha = 6,8$) savings in TAT when compared to the SC method [10]. Comparing the proposed method with Golomb and FDR, TAT reduction up to 55% (when compared to Golomb) and up to 45% (when compared to FDR) is obtained for s35932. The reason for significantly better TAT for s35932 is the very high compression ratio obtained with the proposed method (see row s35932 in Table 1). For the rest of the circuits, the TAT ranges from similar values (when frequency ratio increases) to reduction of up to 35% when compared to Golomb (s15850), and reduction of up to 31% when compared to FDR (s9234).

Table 3 TAT comparison for T_{diff} compressed test sets

Circuit	Comp. method	TAT (ATE clock cycles)			
		$\alpha = 2$	$\alpha = 4$	$\alpha = 6$	$\alpha = 8$
s5378	SC [10]	**15491**	15491	15491	15491
	Golomb [3]	21432	16662	11892	11892
	FDR [2]	22263	14678	12968	12968
	VIHC	15763	**11516**	**11516**	**11516**
s9234	SC [10]	25324	25324	25324	25324
	Golomb [3]	33651	25693	17735	17735
	FDR [2]	36135	24128	21381	21381
	VIHC	**24743**	**17735**	**17735**	**17735**
s13207	SC [10]	**89368**	53490	45137	36784
	Golomb [3]	104440	66381	47783	47481
	FDR [2]	107059	63011	49858	41989
	VIHC	89865	**52769**	**44180**	**36065**
s15850	SC [10]	46067	46067	46067	46067
	Golomb [3]	63989	47164	30339	30339
	FDR [2]	62419	39628	33767	33767
	VIHC	**45765**	**30264**	**30264**	**30264**
s35932	SC [10]	**16870**	**11749**	**10710**	9671
	Golomb [3]	37869	32886	30509	30509
	FDR [2]	32509	20605	18438	17045
	VIHC	17584	12076	10857	**9645**
s38417	SC [10]	**103604**	103604	103604	103604
	Golomb [3]	143844	109801	75758	75758
	FDR [2]	144811	93450	81578	81578
	VIHC	106411	**74943**	**74943**	**74943**
s38584	SC [10]	124011	124011	124011	124011
	Golomb [3]	169356	127512	85668	85668
	FDR [2]	170143	110982	96677	96677
	VIHC	**123693**	**85668**	**85668**	**85668**

Table 4 Previous approaches compared to VIHC

	SC[10]	Golomb[3]	FDR[2]	VIHC
Compression	X	X	√	√
Area overhead	X	√	X	√
Test application time	√	√	X	√

Finally a comparison of the previous approaches [2, 3, 10] and the proposed VIHC is given in Table 4. While FDR [2] gives compression ratios comparable with VIHC, it leads to both lower test application time and greater area overhead. On the other hand, Golomb [3] has similar area overhead when compared to VIHC at the expense of lower compression ratio and greater TAT for small frequency ratios. For high frequency ratios, Golomb's TAT approaches VIHC's TAT, which is unlike SC [10] where TAT is comparable to VIHC only for small frequency ratios. However, this is achieved at a very high penalty in area overhead which is the main shortcoming of the parallel decoder based on *fixed-length* Huffman coding.

5   Conclusions

This paper has presented a new compression method called Variable-length Input Huffman Coding (VIHC). Unlike previous approaches [2,3,10] which reduce some test parameters at the expense of the others, the proposed compression method is capable of minimizing test parameters simultaneously. This is achieved by accounting for multiple interrelated factors that influence the results, such as pre-processing the test set, the size and the type of the input patterns to the coding algorithm, and the type of the the decoder. The results in Section 4 show that the proposed method obtains constantly better compression ratios than [2, 3, 10]. Furthermore, the parallel decoder leads to savings in TAT when compared serial decoders [2, 3]. Moreover, by the exploiting the variable-length input approach, great savings in area overhead are achieved (up to threefold reduction when compared to fixed-length approach [10]). Therefore, it was shown that the proposed method decreases the ATE memory and channel capacity requirements by obtaining good *compression ratios*, and reduces TAT through its parallel on-chip decoder with small *area overhead*. Thus, it is an effective solution for test data compression/decompression for SoCs.

Acknowledgments The authors wish to thank Anshuman Chandra and Dr. Krishnendu Chakrabarty from Duke University for providing the test sets used in their papers.

References

[1] The International Technology Roadmap for Semiconductors, 1999 Edition, ITRS.

[2] A. Chandra and K. Chakrabarty. Frequency-Directed Run-Length (FDR) Codes with Application to System-on-a-Chip Test Data Compression. In *Proceedings IEEE VLSI Test Symposium*, 114–121, Apr. 2001.

[3] A. Chandra and K. Chakrabarty. System-on-a-Chip Test Data Compression and Decompression Architectures Based on Golomb Codes. *IEEE Transactions on Computer-Aided Design*, 20:113–120, Mar. 2001.

[4] T. M. Cover and J. A. Thomas. *Elements of Information Theory*. 1991.

[5] A. El-Maleh, S. al Zahir, and E. Khan. A Geometric-Primitives-Based Compression Scheme for Testing Systems-on-Chip. In *Proceedings IEEE VLSI Test Symposium*, 114–121, Apr. 2001.

[6] I. Hamzaoglu and J. H. Patel. Test set compaction algorithms for combinational circuits. In *Proceedings International Conference on Computer-Aided Design*, 283–289, Nov. 1998.

[7] D. Heidel, S. Dhong, P. Hofstee, M. Immediato, K. Nowka, J. Silberman, and K. Stawiasz. High-speed Serialiazing/Deserializing Design-for-Test Methods for Evaluating a 1ghz microprocessor. In *Proceedings IEEE VLSI Test Symposium*, 234–238, Apr. 1998.

[8] M. Ishida, D. S. Ha, and T. Yamaguchi. Compact: A Hybrid Method for Compressing Test Data. In *Proceedings IEEE VLSI Test Symposium*, 62–69, Apr. 1998.

[9] V. Iyengar, K. Chakrabarty, and B. Murray. Deterministic Built-in Pattern Generation for Sequential Circuits. *Journal of Electronic Testing: Theory and Applications*, 15:97–114, Aug./Oct. 1999.

[10] A. Jas, J. Ghosh-Dastidar, and N. A. Touba. Scan Vector Compression/Decompression Using Statistical Coding. In *Proceedings IEEE VLSI Test Symposium*, 114–121, Apr. 1999.

[11] A. Jas and N. Touba. Test Vector Decompression Via Cyclical Scan Chains and Its Application to Testing Core-Based Designs. In *Proceedings IEEE International Test Conference*, 458–464, Oct. 1998.

[12] A. Jas and N. Touba. Using an Embedded Processor for Efficient Deterministic Testing of Systems-on-a-Chip. In *Proceedings International Conference on Computer Design*, 418–423, Oct. 1999.

[13] B. Koenemann, C. Barnhart, B. Keller, T. Snethen, O. Farnsworth, and D. Wheater. A SmartBIST Variat with Guaranteed Encoding. In *Proceedings of the Asian Test Symposium*, 325–330, Nov. 2001.

[14] E. J. Marinissen, Y. Zorian, R. Kapur, T. Taylor, and L. Whetsel. Towards a Standard for Embedded Core Test: An Example. In *Proceedings IEEE International Test Conference*, 616–627, Sept. 1999.

[15] I. Pomeranz, L. Reddy, and S. Reddy. Compactest: A Method to Generate Compact Test Set for Combinational Circuits. *IEEE Transactions on Computer-Aided Design*, 12:1040–1049, July 1993.

[16] Y. Zorian, S. Dey, and M. J. Rodgers. Test of Future System-on-Chips. In *Proceedings International Conference on Computer-Aided Design*, 392–400, Nov. 2000.

An Effective Technique for Minimizing the Cost of Processor Software-Based Diagnosis in SoCs

P. Bernardi, E. Sánchez, M. Schillaci, G. Squillero, and M. Sonza Reorda

Abstract The ever-increasing usage of microprocessor devices is sustained by a high-volume production that in turn requires a high-production yield, backed by a controlled process. Fault diagnosis is an integral part of the industrial effort towards these goals. This paper presents a novel cost-effective approach to the construction of diagnostic software-based test sets for microprocessors. The methodology exploits an existing post-production test set, designed for software-based self-test, and an already developed infrastructure IP to perform the diagnosis. An initial diagnostic test set is built, and then iteratively refined resorting to an evolutionary method. Experimental results are reported in the paper showing the feasibility and effectiveness of the approach for an Intel i8051 processor core.

1 Introduction

Microprocessor and microcontroller technology nowadays is virtually ubiquitous. The ever-increasing usage of such devices is sustained by a high-volume production. This in turn demands for a high production yield, backed by a controlled process. Fault diagnosis is an integral part of the industrial effort towards these goals.

Correct identification of the most common defective sections in a die helps to characterize the technological process, and localization of a fault allows to effectively direct physical investigation of the underlying defects. Moreover, most high-volume devices undergo several revisions. During these updates designers may decide to change some characteristics of a particular chip section in order to lower the impact of physical defects on the part (e.g. by increasing the minimum separation between active areas, or by changing the metal routing to obtain a flatter surface). The two activities of fault localization and design update are key in enhancing the final yield.

The high-production volumes of many relatively simple devices call for an economically sound diagnostic methodology, since this has to be applied to a great number of faulty devices. It is therefore important to be able to devise a relatively low-cost diagnostic process.

Politecnico di Torino, Dipartimento di Automatica e Informatica, Torino, Italy {paolo.bernardi, edgar.sanchez, massimiliano.schillaci, giovanni.squillero, matteo.sonzareorda}@polito.it

R. Lauwereins and J. Madsen (eds.), *Design, Automation, and Test in Europe–The Most Influential Papers of 10 Years DATE*, 497–509
© Springer 2008

Even more than test set construction, diagnostic set construction is a time-consuming activity. Most of the effort in diagnosis was directed towards combinational circuits, while sequential circuits received less attention, due to the widespread usage of scan-chain methodologies for test. Hard-to-test faults require a high-computational effort for their coverage, but once detected they are usually easy to diagnose; easy-to-test ones, on the other hand, may be difficult to discriminate from each other and require a special effort for diagnosis. This difficulty can lead to long diagnostic tests, with correspondingly long application times and high costs.

In this paper we concentrate on a software-based diagnosis (SBD) methodology particularly suitable for microprocessor cores embedded in SoCs.

The main novelty of the proposed method is the automatic generation of a diagnostic test set using an existing post-production test set. Starting from it, we build an initial diagnostic test set which we progressively improve using an evolutionary method.

We propose to exploit an existing Infrastructure IP (IIP) [1] whose original purpose was software-based self test (SBST) [11], which is also able to provide the processor with the diagnostic test code, and to gather and store the compressed results. This solution has a very low hardware overhead, since the IIP circuitry has already been included to perform the post-production test. The advantages are numerous: the construction of the diagnostic test set exploits an existing test set, leveraging the economic benefit of reusing already performed work; the process is automated, effectively saving resources; being software-based, the diagnosis can be performed at-speed, both increasing the diagnostic capability and reducing the test application time; thanks to the reuse of the existing IIP the methodology does not require the use of high-performance (and high cost) test equipment.

As a case study we tackled SBD on a well-known microprocessor, widely used as a microcontroller core. The performed experiments allow to uniquely diagnose 61.83% of single suck-at faults, and to classify 84.30% in equivalence classes containing less than 10 faults.

The paper is organized as follows: Section 2 provides a concise background in fault diagnosis for digital circuits; Section 3 details the proposed approach, with a discussion of its key characteristics; Section 4 presents some experimental results; finally, Section 5 concludes the paper.

2   Diagnosis Background

Fault diagnosis of VLSI circuits is one of the more investigated arguments in the test discipline. In the 1960s, [2, 3] pioneered the field by introducing the first classification structures and test generation algorithms. Recently, some formal definitions aimed at unifying the fault diagnosis notation have been given in [4, 5] (i.e. the concepts of *Diagnostic Resolution, Diagnostic Power,* and *Equivalence Class*). These measures and concepts are currently used to characterize diagnostic test generation tools [6, 7], together with the usage of diagnostic trees [16].

In practical terms, an equivalence class (EC) is a set of faults exactly causing the same faulty behavior for each applied pattern; diagnostic test generation aims at determining a pattern set able to partition the circuit faults in a set corresponding to ECs as small as possible. Up to now, the methods to determine ECs and to generate suitable patterns can be classified in *structural* and *functional* [8] techniques, both analyzing the circuit structure. These inspections permit to identify the possible logic value allowing the separation of two potentially equivalent faults by propagating different response values on the observation points.

With respect to diagnostic techniques based on pattern generation, software-based diagnosis of processors presents additional problems:

- There is usually a common part of logic excited each time one instruction is executed, since test programs rely on instructions rather than on test patterns.
- A test program suited to excite a specific module of the processor could also cover a wide number of faults not belonging to the pinpointed part, and therefore offer a very reduced classification ability.
- Many of the internal processor circuit elements cannot be accessed directly using a specific instruction, thus resulting hardly diagnosable.

Chen and Dey [9] tackled software-based diagnosis for the 2k-gate processor called PARWAN. They proposed the following:

- A great number of short-test programs are generated in order to partition the fault universe in as many subspaces as possible.
- Each program presents a reduced set of instructions to isolate faults related to different processor functional parts.
- Multiple copies of the same program are created, each propagating errors on different observable points in order to distinguish the faults affecting the processor outputs.
- At the end of the test set creation a binary tree is built for use in the actual diagnosis process.

This technique is based on the processor functional characteristics instead of a pure structural analysis. Anyway, the effort required to generate a test set following these guidelines is not trivial, and grows with the complexity of the considered processor.

Differently from [9], the purpose of our paper is to define a workflow for the automatic generation of a diagnostic test set starting from an existing post-production test set. Also, our approach uses an *n*-ary tree for fault classification, increasing the diagnostic capability of the set.

3   Proposed Approach

The software-based diagnosis (SBD) methodology discussed herein is an automated method able to generate a suitable diagnostic set of programs starting from an initial test set built for post-production testing. Additionally, it exploits an evolutionary approach to improve the final results. Figure 1 graphically describes the workflow of the method.

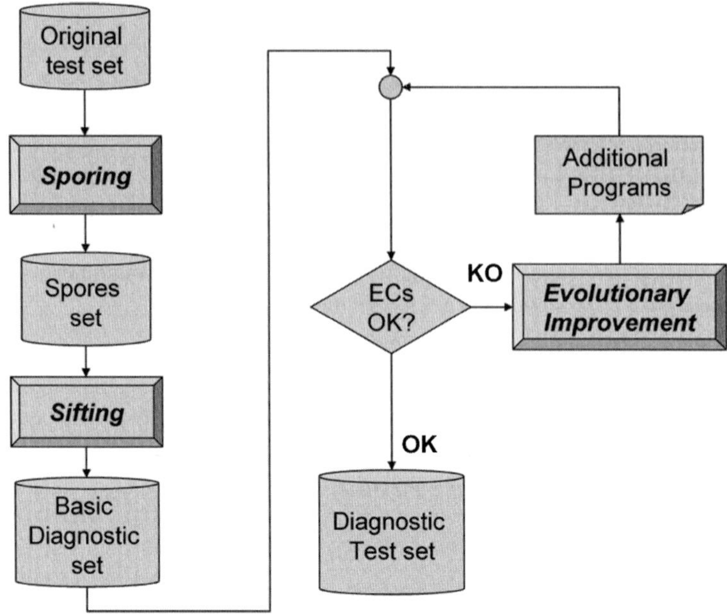

Fig. 1 Method workflow

In synthesis, the workflow is divided in the following steps:

- *Sporing*: the initial test set of programs is split up, generating a vast set of small programs.
- *Sifting*: following a heuristic analysis, only the most promising programs are kept in the test set.
- *Evolutionary improvement*: resorting to an automatic tool, the diagnostic ability of the test set is improved.

The following sections will better detail all the above steps. Moreover, although it cannot be considered part of the algorithm, the evaluation of the diagnostic capability of the generated test set (*diagnostic assessment*) is fundamental and will be described as well.

3.1 Sporing

A test set of programs for post-production testing is usually devised to cumulatively cover the highest possible number of faults. Each program is written with a specific target, for example to cover the faults belonging to a functional module of the processor. A conventional set of programs could be generated by using different approaches: by hand following some deterministic method (as in [11]); exploiting the test engineer expertise by writing test programs to cover corner cases; using automatic approaches (as in [13] or [14]); or even exploiting a random generation and compaction. In any case, the diagnostic set construction method presented here is independent on the origin of the initial test set of programs.

As mentioned before, the original test set is supposed to guarantee a high FC% of the processor but is likely unsuitable for diagnostic purposes, because its only goal is to cover faults, not distinguish them. A typical post-production test program is normally written with a definite set of goals, mainly compactness, short application time and high fault coverage. For pattern-based diagnosis it has been demonstrated that it is better to use many test sets each of which covers few faults [15]; likewise, for software-based diagnosis it is useful to have many very small programs that cover the smallest possible fault set.

Usually the high fault coverage is obtained by the stimulation of the functional modules of the processor with a great amount of input data, generated in a looping section of the test program. Many faults, especially in arithmetic units, need specific bit patterns on the execution unit inputs to be covered. If we could write a program that delivers only those data needed to detect a specific fault, and not others, that program would exhibit a very low fault covering ability, but a fairly good diagnostic capacity.

The basic idea is therefore, to obtain a number of very small programs, each one covering a small number of faults (ideally the smallest possible number, indeed) not covered by other programs in the same set.

The *spores* are very small programs that put the processor in a specific state in order to control some specific part of the processor, execute a target instruction and propagate the results to the primary outputs. Each spore represents a completely independent program, able to excite some processor function, observe the results, and possibly signal fault occurrence. It is worth noting that the spores are not written from scratch, but generated by automatically breaking an existing test set in very small fragments, thus fully exploiting its covering ability.

The *sporing* process is based on an ad hoc instruction set simulator able to trace the execution data flow of each instruction of the test program. Its goal is the generation of independent and small programs able to exactly replicate the behavior of the processor while executing a target instruction. Clearly, each program belonging to the initial test set is fragmented in a huge number of spores. More details can be found in [10].

A first fault simulation is required to analyze the generated spores. This preliminary fault simulation is aimed at determining the fault coverage figure for each generated *spore*. Figure 2a refers to this process: the result of this step is a *coverage matrix* storing for each *spore* the information about covered faults.

3.2 *Sifting*

At this point we are left with an inordinate number of spores, of the order of tens of thousands. It would be impractical and wasteful to simply apply all these programs to a faulty processor to diagnose it, so an effort is appropriate to reduce this diagnostic set. This goal is obtained by a *sifting* process, detailed below.

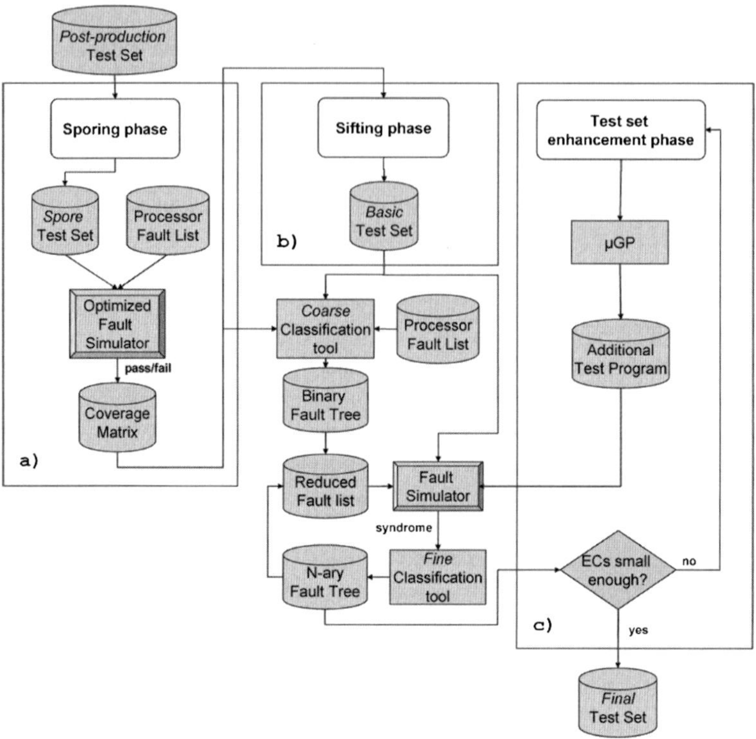

Fig. 2 Proposed workflow

First of all, the fault covering ability of each spore is considered in the context of the entire diagnostic set; the important thing for a spore is not covering a large number of faults, but covering faults which are not detected by other spores: all the spores able to do this have to be retained in the final diagnostic set.

Every fault is detected by a certain number of spores, depending on whether it is easy-to-detect or a hard-to-detect. This leads to the concept of *density* of the fault, that is, the number of spores able to detect it.

The diagnostic capability of a spore is then evaluated with respect to the whole set, using the density concept. Every spore is assigned a preliminary fitness value $f_s(d_F, NF_s)$:

$$f_s = \left(\sum_F \frac{1}{d_F} \right) \cdot \frac{1}{NF_s}$$

where F is the fault index over the covered faults, d_F is the corresponding fault density and NF_s is the number of faults covered by the spore. The value of f_s ranges from 0 to 1 and the higher its value the higher the diagnostic capability of the spore. The spores are then sorted by fitness value in decreasing order. In this way the spores with higher diagnostic capability are preferred for inclusion in the final diagnostic set. It is possible to note that the highest fitness value equals 1, and it is assigned to one spore covering only one fault of density 1. Finally, starting from the top of this list, only the spores that cover faults not detected by the previous ones are kept, while the others are discarded as redundant. These test programs compose the *initial* test set.

3.3 Diagnostic Assessment

Immediately after the sifting phase, the diagnostic ability of the selected *initial* test set is evaluated. This computation has been divided in two steps, graphically shown in Fig. 2b. Similarly to the process formalized in [9], a first *coarse* classification is based on the construction of a compact diagnostic tree obtained by processing only the pass/fail information related to each test program included in the initial test set; this information is simply extracted from the coverage matrix. The *binary* tree structure is shown in Fig. 3a.

A *fine* classification is then performed for all the equivalence classes (ECs) isolated by the coarse classification and still composed of more than one fault. This

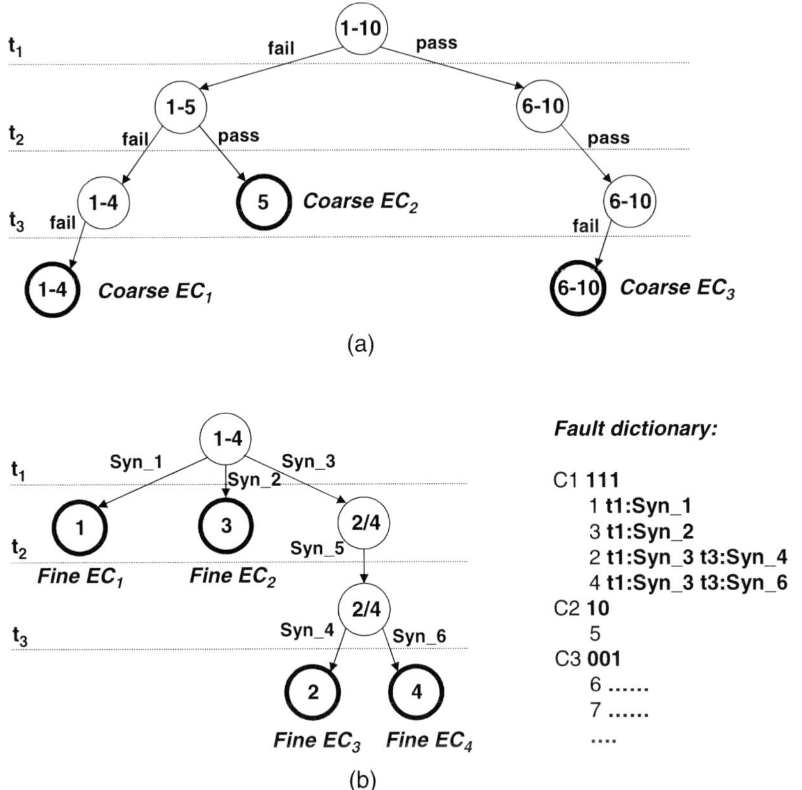

Fig. 3 Two steps, *coarse* and *fine*, of the diagnostic tree construction and the resulting fault dictionary

second classification is done by using the faulty circuit responses on primary outputs (also called *syndromes*) to build in parallel an n-ary tree for each EC to be further divided. The n-ary tree structure is in Fig. 3b. To reduce the impact of this second fault simulation process aimed at retrieving the faulty syndromes, an incremental approach has been used whereas:

- The n-ary tree structure is updated at the end of the fault-simulation of each test program in the initial test set and faults included in ECs of size 1 definitively dropped out from the fault list
- Faults not yet classified and not covered by the next test program to be fault simulated are temporarily dropped out from the fault list as their syndrome is not useful for classification

This process allows the generation of a compact fault dictionary composed of

- Pass/fail sequence leading to a *Coarse* EC
- The set of discriminating syndromes for each *Fine* EC

3.4 Evolutionary Improvement

At this phase of the workflow, there is still a set of unsatisfactorily large ECs. Thus, an effort is appropriate to partition these large classes.

To improve the diagnostic ability of the test set we resort to an evolutionary tool called μGP [12]. It is an evolutionary approach to generic optimization problems with a focus on the generation of test programs for microprocessors. It is based on an evolutionary core, an instruction library that allows targeting a specific microprocessor and an external evaluator to provide the core with the necessary feedback.

The evolutionary approach receives the information about the large ECs and generates new assembly programs able to split them. The fitness function provided as feedback to the evolutionary core is the size of the largest EC found using the fine classification. During this phase, the ability of a generated program in dividing the large EC is directly obtained by analyzing the faulty syndrome.

The fine classification tree is directly updated after each program generation. This third part of the diagnostic evaluation process is in Fig. 2c.

This process, using an n-ary diagnostic tree instead of a binary one, is totally new with respect to the work reported in [9]. The time it takes can be traded with the quality of the obtained results.

4 Experimental Evaluations

The feasibility and the effectiveness of the proposed approach for the software-based diagnosis of processor cores have been proved considering as a case of study an Intel i8051 microcontroller, supposed to be embedded into a SoC.

One of the major problems in testing a SoC is the reduced accessibility of each single embedded core: to improve the observability of the i8051 to be diagnosed, we resorted to the solution proposed in [1] to support the software-based self-test procedure of processor cores. The i8051 core is complemented with a 24-bit multiple input signature register (MISR) connected to its parallel output ports. The faulty signature, or syndrome, stored by the MISR can be read at the end of each test program execution. The synthesized microcontroller, obtained using a generic home-developed library, contains 37,417 equivalent gates, and the collapsed fault list counts 12,642 faults.

Let $D(n)$ be the percentage of faults that are classified into a class of cardinality less than or equal to n by the diagnostic test set. $D(1)$ is the percentage of faults that are uniquely classified; $D(10)$ is the percentage of faults that can be considered *correctly* classified, because the exact analysis of equivalence between faults cannot be performed for medium or large sequential circuits.

We started from a post-production test set composed of 8 test programs written by a skilled test engineer, reaching a fault coverage of about 92% on the collapsed list. 7 out of 8 programs aim at covering the ALU faults and are generated by following the deterministic approach described in [11]; the remaining program has been written resorting to the same technique, but it aims at the coverage of those faults in the decoding and control units still not covered. The *sporing* process generates about 60k test programs and the average number of instruction for these programs is 7: each program includes an initial and a final part, required to obtain the syndrome saving. More details about the program structure and *sporing* can be found in [10].

The fault simulation process required to proceed to the *sifting* phase has been done exploiting an in-house developed tool and required about 75 h on 3 SUN Blade processor-based workstations.

The sifting phase, including the fault simulation, required about 100 h on the same hardware. And the test set obtained processing coverage matrix is composed of 7,231 test programs (about 12% of the original set).

Table 1 shows the main characteristics of the three test sets: post-production; initial (sifted); final (enhanced). The rows reports: the number of test programs in the test set; the test set size; the maximum length of a test program in clock cycles; the result of the diagnostic assessment D(1) and D(10). Table 2, on the other hand, details the size of the equivalence classes for the three test sets.

It is interesting to note that the initial test set demonstrates a $D(10)$ equal to 58% of the processor faults. The biggest class, composed of 3,755 faults, is the one including those faults detected by all the test programs of the test set. Apart from this class, the greatest class has size 84.

The diagnostic enhancement of the test set eventually consisted in the generation of 35 new test programs. Each generation process is started by evolving an initial population of 20 test programs. The biggest class obtained by this enhancement process has size 1,092.

If the diagnostic test set is considered as a monolithic block to be entirely uploaded and executed on the Automatic Test Equipment (ATE), it is composed of

Table 1 The equivalence class summary for the analyzed processor core

	Post-production test set	*Initial* test set	*Final* test set
Programs [#]	8	7,231	7,266
Test set size [KB]	4	165	177
Max clock cycles	1.00M	1.93M	2.02M
$D(1)$ [%]	11.56	35.70	61.93
$D(10)$ [%]	32.90	58.02	84.30

Table 2 Equivalence class summary for the analyzed processor core

EC size	*Post-production* test set [# fault]	*Sifted* test set [# fault]	*Enhanced* test set [# fault]
1	1,334	4,120	7,148
2	802	872	1,106
3	543	522	546
4	360	264	336
5	255	150	155
6	138	150	180
7	112	154	91
8	80	224	72
9	63	81	36
10	110	160	60
11 – 100	2,335	1,090	720
>100	5,410	3,755	1,092

all 7,266 programs and requires 2,020,656 clock cycles. However, if an intelligent/ interactive ATE enabling the diagnostic process to be stopped as soon as the fault (or the faults) responsible for the wrong behavior has been individuated, the diagnostic process can be reduced to the execution of an average of 3,762 programs (corresponding to 900,858 clock cycles).

The Figs. 4–6 graphically show fault classification obtained at the end of each phase.

5 Conclusions and Future Work

A novel automated approach for the generation of software-based diagnostic sets for microprocessors has been presented. It exploits an existing post-production software-based test set and uses the existing Infrastructure IP designed for applying it, thus it is cost-effective and can be seamlessly fit into an existing design flow.

Faults by EC size

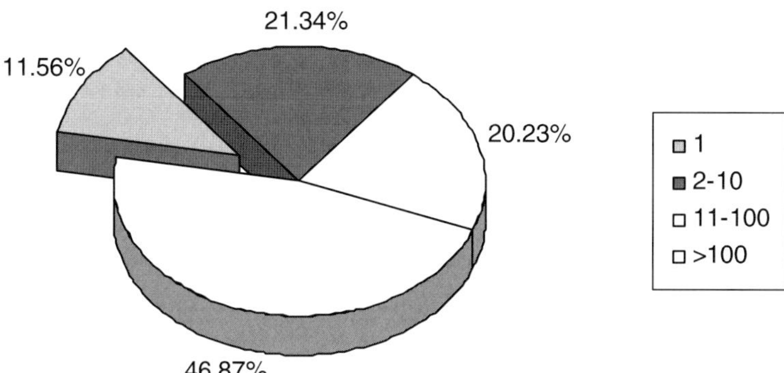

Fig. 4 *Post-production* test set diagnostic properties

Faults by EC size

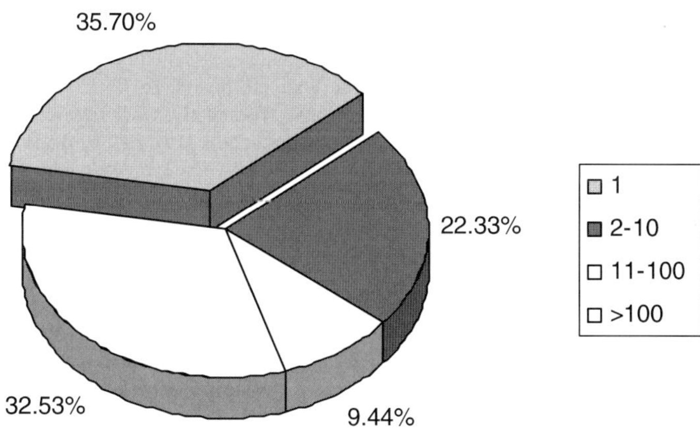

Fig. 5 *Initial* test set diagnostic properties

The reported results clearly show the effectiveness of the method on a widely used microprocessor core, thus highlighting the industrial relevance of the approach.

Work is currently under way to improve the workflow, with particular effort upon the reduction of the computational effort required.

Faults by EC size

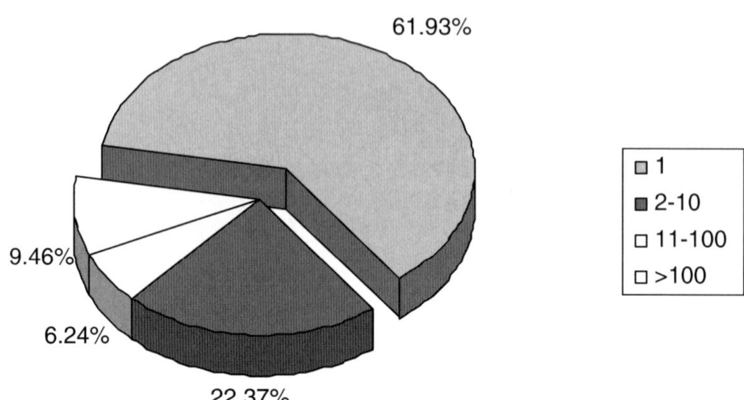

Fig. 6 *Final* test set diagnostic properties

References

[1] P. Bernardi, M. Rebaudengo, M. Sonza Reorda, Using Infrastructure IPs to support SW-based Self-Test of Processor Cores, IEEE International Workshop on Microprocessor Test and Verification, 2004, pp. 22–27

[2] H. Y. Chang, A Distiguishability criterion to selecting efficient diagnostic tests, Proc. Spring Joint Computer Conference, 1968, pp. 529–534,

[3] A. D. Friedman, Fault Detection in Redundant Circuits, IEEE Trans. Computers, Vol. EC-16, February 1967, 99–100

[4] P. Camurati, D. Medina. P. Prinetto, M. Sonza Reorda, A diagnostic test pattern generation algorithm, IEEE International Test Conference, 1990, pp. 52–58

[5] T. Griining, U. Mahlstedt, H. Koopmeiners DI- ATEST: A Fast Diagnostic Test Pattern Generator for Combinational Circuits, ICCAD-91: IEEE International Conference on Computer Aided Design, Santa Clara, CA (USA), November 1991, pp. 194–197

[6] A. Veneris, R. Chang, M.S. Abadir, M. Amiri, Fault Equivalence and Diagnostic Test Generation Using ATPG IEEE ISCAS 2004, Vol. 5, V-221–V-224

[7] S. Chakravarty, A Sampling Technique for Diagnostic Fault Simulation, IEEE International VLSI Test Symposium, 1996, pp. 192–197

[8] N. Jha and S. Gupta, Testing of Digital Systems, Cambridge University Press, 2003

[9] Li Chen, S. Dey, Software-based Diagnosis for Processors, Design Automation Conference, 2002, pp. 259–262

[10] E. Sánchez, M. Sonza Reorda, G. Squillero, On the transformation of Manufacturing Test Sets into On-Line Test Sets for Microprocessors, IEEE International Symposium on Defect and Fault Tolerance in VLSI Systems, 2005, pp. 494–504

[11] N. Kranitis, G. Xenoulis, D. Gizopoulos, A. Paschalis, Y. Zorian, Low-cost Software-Based Self-Testing of RISC Processor Cores IEE Proceedings of Computers and Digital Techniques, 2003, Vol. 150, Issue 5, 355–360

[12] G. Squillero, MicroGP – An Evolutionary Assembly Program Generator Genetic Programming and Evolvable Machines. Springer Science + Business Media B.V. ISSN: 1389–2576

[13] F. Corno, E. Sánchez, M. Sonza Reorda, G. Squillero Automatic Test Program Generation - A Case Study IEEE Design & Test, Special Issue on Benchmarking for Design and Test, 2004, Vol. 21, Issue 2, 102–109

[14] P. Parvathala, K. Maneparambil, W. Lindsay, FRITS - A Microprocessor Functional BIST Method, IEEE International Test Conference, 1990, pp. 52–58

[15] I. Pomeranz, S. M. Reddy, A Diagnostic Test Generation Procedure Based on Test Elimination by Vector Omission for Synchronous Sequential Circuits, IEEE Transactions on Computer-Aided Design of Integrated Circuits and Systems, May 2000, Vol. 19, Issue 5, 589–600

[16] V. Boppana, I. Hartanto, W.K. Fuchs, Full Fault Dictionary Storage Based on Labeled Tree Encoding, IEEE VLSI Test Symposium, 1996, pp. 174–179.

Appendix
Shortlist of Most Influential Papers

This list presents all 65 papers that were selected by the respective program chairs as the most influential. A subset of 30 papers from this list was selected to appear in the book.

Date 1998

- Address Bus Encoding Techniques for System-Level Power Optimization L. Benini, G. De Micheli, E. Maccii, D. Sciuto, and C. Silvano
- Scheduling of Conditional Process Graphs for the Synthesis of Embedded Systems P. Eles, K. Kuchcinski, Z. Peng, A. Doboli, and P. Pop
- Fast Sequential Circuit Test Generation Using High-Level and Gate-Level Techniques E.M. Rudnick, R. Vietti, A. Ellis, F. Corno, P. Prinetto, and M. Sonza Reorda
- Interconnect Tuning Strategies for High-Performance ICs A.B. Kahng, S. Muddu, E. Sarto, and R. Sharma
- FRIDGE: A Fixed-Point Design and Simulation Environment H. Kending, M. Willems, M. Coors, and H. Meyr

Date 1999

- EXPRESSION: A Language for Architecture Exploration Through Compiler/ Simulator Retargetability A. Halambi, P. Grun, V. Ganesh, A. Khare, N. Dutt, and A. Nicolau
- Dynamic Power Management for Nonstationary Service Requests E.Y. Chung, L. Benini, A. Bogiolo, and G. De Micheli
- MOCSYN: Multiobjective Core-Based Single-Chip System Synthesis R.P. Dick and N.K. Jha
- Combinational Equivalence Checking Using Satisfiability and Recursive Learning J. Marques-Silva and T. Glass

- Kernel Scheduling in Reconfigurable Computing R. Maestre, F.J. Kurdahi, N. Bagherzadeh, H. Singh, R. Hermida, and M. Fernandez
- On Programmable Memory Built-in Self-Test Architectures K. Zarrineh and S.J. Upadhyay
- Algorithms for Solving Boolean Satisfiability in Combinational Circuits L. Guerra e Silva, L.M. Silveira, and J. Marques-Silva
- Battery-Powered Digital CMOS Design M. Pedram and Q. Wu

Date 2000

- A Generic Architecture for On-Chip Packet-Switched Interconnections P. Guerrier and A. Greiner
- Cost Reduction and Evaluation of a Temporary Faults-Detecting Technique L. Anghel and M. Nicolaidis
- Quantitative Comparison of Power Management Algorithms Y. Lu, E. Chung, T. Simunic, L. Benini, and G. De Micheli
- A Bist Scheme for On-Chip ADC and DAC Testing J. Huang, C. Ong, and K. Cheng
- Fast Evaluation of Sequence Pair in Block Placement by Longest Common Subsequence Computation X. Tang, R. Tian, and D. Wong
- Standards for System-Level Design: Practical Reality or Solution in Search of a Question? C. Lennard, P. Schaumont, G. de Jong, A. Haverinen, and P. Hardee
- A Discrete-Time Battery Model for High-Level Power Estimation L. Benini, G. Castelli, A. Macii, E. Macii, M. Poncino, and R. Scarsi
- A Power Reduction Technique with Object Code Merging for Application Specific Embedded Processors T. Ishihara and H. Yasuura

Date 2001

- Efficient Inductance Extraction via Windowing M.W. Beattie and L.T. Pileggi
- An Efficient Architecture Model for Systematic Design of Application-Specific Multiprocessor SoC A. Baghdadi, D. Lyonnard, N.E. Zergainoh, and A.A. Jerraya
- Efficient Spectral Techniques for Sequential ATPG A. Giani, S. Sheng, and M.S. Hsiao, and V.D. Agrawal
- Slicing Tree is a Complete Floorplan Representation M. Lai and M.D.F. Wong
- Trace-Driven Application Modeling for System-Level Performance Analysis R. Marculescu and A. Nandi
- SystemC-SV – An Extension of SystemC for Mixed Multi-Level Communication Modelling and Interface-Based System Design R. Siegmund and D. Mueller

Date 2002

- BerkMin: A Fast and Robust Sat-Solver E. Goldberg and Y. Novikov
- Minimum Energy Fixed-Priority Scheduling for Variable Voltage Processor G. Quan and X.S. Hu
- Assigning Program and Data Objects to Scratchpad for Energy Reduction S. Steinke, L. Wehmeyer, B. Lee, and P. Marwedel
- Networks on Silicon: Combining Best-Effort and Guaranteed Services K. Goossens, P. Wielage, A. Peeters, and J. van Meerbergen
- Improving Compression Ratio, Area Overhead, and Test Application Time for System-on-a-Chip Test Data Compression/Decompression P. Gonciari, B. Al-Hashimi, and N. Nicolici
- Low Power Error Resilient Encoding for On-Chip Data Buses D. Bertozzi, L. Benini, and G. De Micheli

Date 2003

- Trade Offs in the Design of a Router with Both Guaranteed and Best-Effort Services for Networks On Chip E. Rijpkema, K.G.W. Goossens, A. Radulescu, J. Dielissen, J. van Meerbergen, P. Wielage, and E. Waterlander
- Exploiting the Routing Flexibility for Energy-Performance Aware Mapping of Regular NoC Architectures J. Hu and R. Marculescu
- RTOS Modeling for System Level Design A. Gerstlauer, H. Yu, and D.D. Gajski
- Validating Sat-Solvers Using an Independent Resolution-Based Checker-Practical Implementations and Other Applications L. Zhang and S. Malik
- Statistical Timing Analysis Using Bounds A. Agarwal, D. Blaauw, V. Zolotov, and S. Vrudhula

Date 2004

- XpipesCompiler: A Tool for Instantiating Application-Specific Networks-on-Chip A. Jalabert, S. Murali, L. Benini, and G. De Micheli
- An Efficient On-Chip Network Interface Offering Guaranteed Services, Shared-Memory Abstraction, and Flexible Network Configuration A. Radulescu, J. Dielissen, K. Goossens, E. Rijpkema, and P. Wielage
- Synthesis and Optimization of Threshold Logic Networks with Application to Nanotechnologies R. Zhang, P. Gupta, L. Zhong, and N.K. Jha
- Wrapper Design for Testing IP Cores with Multiple Clock Domains Q. Xu and N. Nicolici

- Poor Man's TBR: A Simple Model Reduction Scheme J. Phillips and L.M. Silveira
- Digital Ground Bounce Reduction by Phase Modulation of the Clock M. Badaroglu, G. Gielen, H. De Man, P. Wambacq, G. Van der Plas, S. Donnay

Date 2005

- Soft-Error Tolerance Analysis and Optimization of Nanometer Circuits Y.S. Dhillon, A.U. Diril, and A. Chatterjee
- Network Traffic Generator Model for Fast Network-on-Chip Simulation S. Mahadevan, F. Angiolini, M. Storoaard, R.G. Olsen, J. Sparsoe, and J. Madsen
- Energy Efficiency of the IEEE 802.15. 4 Standard in Dense Wireless Microsensor Networks: Modeling and Improvement Perspectives B. Bougard, F. Catthoor, D.C. Daly, A. Chandrakasan, and W. Dehaene
- GalsC: A Language for Event-Driven Embedded Systems E. Cheong and J. Liu
- A Study of the Speedups and Competitiveness of FPGA Soft Processor Cores Using Dynamic Hardware/..., R. Lysecky and F. Vahid
- Accurate Reliability Evaluation and Enhancement via Probabilistic Transfer Matrices S. Krishnaswamy, G. F. Viamontes, I.L. Markov, and J.P. Hayes
- Automatic Timing Model Generation by CFG Partitioning and Model Checking I. Wenzel, B. Rieder, R. Kirner, and P. Puschner
- An Application-Specific Design Methodology for STbus Crossbar Generation S. Murali and G. De Micheli
- Design Optimization of Time- and Cost-Constrained Fault-Tolerant Embedded Systems V. Izosimov, P. Pop, P. Eles, and Z. Peng

Date 2006

- A Reconfigurable HW/SW Platform for Computation Intensive High-Resolution Real-Time Digital Film Applications A. do Carmo Lucas, S. Heithecker, R. Ernst, P. Rueffer, H. Rueckert, W. Huther, G.Wischermann, K. Gebel, R. Fach, S. Eichner, and G. Scheller
- An Effective Technique for Minimizing the Cost of Processor Software-Based Diagnosis in SoCs P. Bernardi, E. Sanchez, M. Schillaci, G. Squillero, and M. Sonza Reorda
- A Single Photon Avalanche Diode Array Fabricated in Deep-Submicron CMOS Technology C. Niclass, M. Sergio, and E. Charbon
- Equivalence Verification of Arithmetic Datapaths with Multiple Word-Length Operands N. Shekhar, P. Kalla, and F. Enescu

- Two Phase Resonant Clocking for Ultra-Low-Power Hearing Aid Applications F. Carbognani, F. Buergin, N. Felbar, H. Kaeslin, and W. Fichtner
- A Tool for Automated Generation of NoC Architectures using Multi-Port Routers for FPGAs B. Sethuraman and R. Vemuri

Date 2007

- Statistical Blockade: A Novel Method for Very Fast Monte Carlo Simulation of Rare Circuit Events, and Its Application A. Singhee and R.A. Rutenbar
- Compositional Specification of Behavioral Semantics K. Chen, J. Sztipanovits, and S. Neem
- Synthesis of Task and Message Activation Models in Real-Time Distributed Automotive Systems W. Zheng, M. Di Natale, C. Pinello, P. Giusto, and A. Sangiovanni Vincentelli
- Soft Error Rate Analysis for Sequential Circuits N. Miskov-Zivanov and D. Marculescu
- On-Chip Router Modeling for MPSoC Communication Analysis U.Y. Ogras and R. Marculescu
- What if You Could Design Tomorrow's System Today? N. Wingen